INSTRUMENTATION FOR ENGINEERING MEASUREMENTS

INSTRUMENTATION FOR ENGINEERING MEASUREMENTS

Second Edition

JAMES W. DALLY
University of Maryland

WILLIAM F. RILEY
KENNETH G. McCONNELL
Iowa State University

JOHN WILEY & SONS, INC.

Acquisitions Editor	Cliff Robichaud
Marketing Manager	Susan Elbe
Production Manager	Lucille Buonocore
Production Supervisors	Savoula Amanatidis and Richard Blander
Copyediting Supervisor	Deborah Herbert
Designer	Pedro A. Noa
Illustration Coordinator	Sigmund Malinowski
Manufacturing Manager	Andrea Price

This book was set in Times Roman by Publication Services, Inc. and printed and bound by Hamilton Printing Company.

Recognizing the importance of preserving what has been written, it is a policy of John Wiley & Sons, Inc. to have books of enduring value published in the United States printed on acid-free paper, and we exert our best efforts to that end.

Library of Congress Cataloging in Publication Data:

Dally, James W.
Instrumentation for engineering measurements/ James W. Dally.
William F. Riley, Kenneth G. McConnell. — 2nd edition
p. cm

Includes bibliographical references and index.
ISBN 0-471-55192-9
1. Engineering instruments. I. Riley, William F. (William
Franklin), 1925-   . II. McConnell, Kenneth G. III. Title.

TA165.D34   1993
681'.2–dc20   92-32673
                                                                  CIP
Printed in the United States of America

20 19 18 17 16 15 14 13

PREFACE

Since publication of the first edition of this textbook in 1984, many changes have occurred that affect instruction of instrumentation and methods for performing engineering measurements. The changes that we consider important and that are addressed in this revision are as follows:

1. Rapid advancements in digital instruments, ranging from simple voltmeters to complex spectrum analyzers.
2. Widespread availability of personal computers and workstations that allow data acquisitions and data processing to be combined.
3. Development of improved methods for vibration analysis utilizing spectrum analyzers for mapping between the time and frequency domains.
4. Recognition of the need to better integrate a course in measurements with other courses and other educational objectives in engineering curricula.

We have added new material in almost every chapter of the first edition of the book and have added three new chapters. In Chapter 1, the treatment of process control is expanded and new information is included on heaters, motors, and actuators used to control different processes. Chapter 2, "Analysis of Circuits," is new and includes a brief review of electrical and electronic principles important to understanding the operation of instrument systems. Also new is Chapter 4, which covers digital recording systems and contains detailed descriptions of the analog-to-digital and digital-to-analog conversion processes. The chapters on analog voltage recording instruments; sensors for transducers; signal conditioning circuits; electric resistance strain gages; force, torque, and pressure measurements; fluid measurements; and statistical methods have been updated to reflect the more modest changes in these areas over the past decade. The coverage of displacement, velocity, and acceleration measurements in Chapter 9 has been extensively revised, in the hope of enhancing the reader's understanding of a difficult topic. A new Chapter 10, "Analysis of Vibrating Systems," covers methods used in processing displacement, velocity, and acceleration data. Of particular importance is characterization of signals and processing of vibration data with a digital frequency analyzer. Significant changes have also been made in Chapter 11, "Temperature Measurements," including a description of the new international standard for temperature, added detail on using thermocouples, and a discussion of sources of error in temperature measurement. Discussion of several measurement techniques that are now obsolete has been removed.

The role of the computer in design and in process control was important in 1984. Today the computer is even more important, as its use has expanded significantly with improvements in hardware performance and software for CAD, CAE, and CAM. Carefully performed experiments that test reliability and endurance and evaluate performance are an intergral part of the computer-assisted product development process. The computer is rapidly becoming an essential part of the experiment. Analog-to-digital converters are available on plug-in cards that utilize the computer's power supply and counters to power the transducers and measure the output voltage. Output signals from the sensor are stored in the computer's memory and are processed either in real time or on completion of the experiment. Continued improvements in computer hardware (particularly increasing the clock speeds) will expand the use of the computer in measuring dynamic signals at modest frequencies.

Material included in this revision significantly extends the treatment of principles and instrumentation for engineering measurements. There are too many topics to be covered well in a single semester course, particularly if the theory is supplemented with laboratory practice. The first course, usually taught in the third year, after students have completed an introductory circuits course, might consist of a lecture/laboratory sequence covering Chapters 1 through 7, 11, and 13. The second course would then consist of a lecture/project sequence, using material from Chapters 7 through 12 to cover the instrumentation principles needed to execute the projects. In spite of the costs of modern laboratory instruments and faculty time, laboratory exposure with hands-on experience is essential for a thorough understanding of the topic.

At the graduate level, because the students are more experienced and have a better background in mathematics, most of the material included in this revision can be covered in a typical 15-week semester. The emphasis on applications (Chapters 7 through 12) will depend on the interests of the instructor and of the class.

The exercises included for each chapter have been revised: some original problems have been recast and new problems have been added. We encourage the use of spreadsheets and Mathcad software in performing many of these exercises. Use of these codes eliminates tedium and permits the student to explore a broad range of solutions rather than grind out a single answer. We hope that students and instructors alike will find the presentation clear and informative and that many of the mysteries of the laboratory and of instrument systems will be clarified.

Finally, we thank the reviewers who provided valuable suggestions and comments that were important in the final draft of the revision.

James W. Dally
William F. Riley
Kenneth G. McConnell

CONTENTS

LIST OF SYMBOLS

a	Acceleration
a_x, a_y, a_z	Cartesian components of acceleration
A	Area
$\mathscr{B}$	Magnetic flux density
c	Velocity of light in a vacuum
c_p	Specific heat at constant pressure
c_v	Specific heat at constant volume
C	Capacitance
C	Calibration constant
C	Count
C	Discharge coefficient
C	Viscous damping constant
C_C	Contraction coefficient
C_D	Drag coefficient
C_e	Equivalent capacitance
C_f	Feedback capacitance
C_G	Galvanometer constant
C_L	Lead-wire capacitance
C_p	Pitot tube coefficient
C_t	Transducer capacitance
C_v	Coefficient of variation
C_V	Coefficient of velocity
d	Damping ratio
d	Deviation
d	Displacement
d^*	Full-scale displacement
D	Damping coefficient
D^*	Specific detectivity
e	Electron charge
e	Junction potential per unit temperature
E	Modulus of elasticity
E_m	Back electromotive force
E_λ	Radiation power
$\mathscr{E}$	Electric field strength
$\mathscr{E}$	Error
$\mathscr{E}_a$	Accumulated error for a system
$\mathscr{E}_A$	Amplifier error
$\mathscr{E}_R$	Recorder error
$\mathscr{E}_R$	Resolution error
$\mathscr{E}_{sc}$	Signal-conditioner circuit error
$\mathscr{E}_T$	Transducer error
f	Focal length
f	Cyclic frequency
f_{bw}	Bandwidth

f_c	Cutoff frequency
f_D	Doppler-shift frequency
f_n	Natural frequency
f_r	Resonant frequency
f_s	Sampling frequency
f^*	Nyquist frequency
f_{sh}	Self-heating factor
g	Gravitational constant
G	Conductance
G	Gain
G	Shear modulus of elasticity
G_c	Gain for a common-mode voltage
G_d	Gain for a difference voltage
h	Convective heat-transfer coefficient
h	Planck's constant
H	Irradiance
i	Current
i_f	Feedback current
i_g	Gage current
i_G	Galvanometer current
i_i	Input current
i_i^*	Full-scale input current
i_m	Meter current
i_m^*	Full-scale meter current
i_o	Output current
i_p	Photoelectric current
i_s	Source current
i_s	Steady-state current
i'	Current density
I	Intensity of light
I	Moment of inertia
J	Polar moment of inertia
k	Adiabatic exponent
k	Boltzmann's constant
k	Spring constant or stiffness
K	Dielectric constant
K	Torsional spring constant
K_t	Transverse sensitivity factor for a strain gage
L	Inductance
$\mathcal{L}$	Loss factor
m	Mass
M	Mach number
M	Moment
M_x, M_y, M_z	Cartesian components of a moment
N	Number of charge carriers
N_{dB}	Number of decibels
p	Power
p	Pressure
p_d	Dynamic pressure
p_d'	Measured dynamic pressure
p_D	Power density
p_g	Power dissipated by a strain gage
P_s	Stagnation pressure
p_T	Power dissipated by a transducer

q	Charge
q	Resistance ratio
Q	Volume flow rate
Q_i	Input quantity
Q_o	Output quantity
r	Frequency ratio
r	Resistance ratio
R	Range
R	Resistance
R	Resolution
R	Universal gas constant
R_A	Amplifier resistance
R_b	Ballast resistance
R_B	Equivalent bridge resistance
R_e	Equivalent resistance
R_f	Feedback resistance
R_g	Gage resistance
R_G	Galvanometer resistance
R_L	Lead-wire resistance
R_L	Load resistance
R_M	Recording instrument resistance
R_p	Parallel resistance
R_s	Series resistance
R_s	Source resistance
R_{sh}	Shunt resistance
R_T	Transducer resistance
R_x	External resistance
Re	Reynolds number
s	Span
S	Calibration constant
S	Stagnation recovery factor
S	Sensitivity
s_a	Axial strain sensitivity of a strain gage
S_A	Alloy sensitivity
S_c	Circuit sensitivity
S_{cc}	Constant-current circuit sensitivity
S_{cv}	Constant-voltage circuit sensitivity
S_f	Fatigue strength of a material
S_g	Gage factor, strain sensitivity of a gage
S_g^*	Corrected gage factor for a strain gage
S_G	Galvanometer sensitivity
S_i	Current sensitivity
S_N	Strouhal number
S_q	Charge sensitivity of a piezoelectric crystal
S_q^*	Charge sensitivity of a piezoelectric transducer
S_R	Reciprocal sensitivity
S_R	Recorder sensitivity
S_s	Shear strain sensitivity of a strain gage
S_t	Transverse strain sensitivity of a strain gage
S_v	Voltage sensitivity
S_x	Standard deviation
$S_{\bar{x}}$	Standard error
S_τ	Torsional yield strength of a material
t	Time

t_r	Rise time
T	Period
T	Temperature
T	Torque
T_n	Natural period
T_o	Reference temperature
T_s	Stagnation temperature
T_s^*	Nyquist interval
v	Specific volume
v	Velocity
v_x, v_y, v_z	Cartesian components of velocity
v	Voltage
v_b	Bias voltage
v_d	Voltage drop
v_i	Input voltage
v_o	Output voltage
v_r	Reference voltage
v_s	Source voltage
v_z	Breakdown voltage
v^*	Full-scale voltage
v'	Potential gradient
V	Fluid velocity
V_o	Centerline velocity
w	Specific weight of a solid
W	Weight
$\bar{x}$	Sample mean
Z	Impedance
Z_i	Input impedance
Z_o	Output impedance
Z_o	Zero offset
α	Angle of incidence
$\alpha, \beta, \gamma, \theta, \phi$	Angles
α	Angular acceleration
α, β	Coefficient of thermal expansion
β	Time constant for a sensor
γ	Shear strain
γ	Specific weight of a fluid
γ	Temperature coefficient of resistance
γ	Temperature coefficient of resistivity
$\gamma_{xy}, \gamma_{yz}, \gamma_{zx}$	Shear strain components in Cartesian coordinates
δ	Deflection
δ	Displacement
ε	Emissivity
ε	Normal strain
ε'	Apparent normal strain
ε_a	Axial strain
ε_c	Calibration strain
ε_t	Transverse strain
$\varepsilon_{xx}, \varepsilon_{yy}, \varepsilon_{zz}$	Cartesian components of normal strain
$\varepsilon_1, \varepsilon_2, \varepsilon_3$	Principal normal strains
η	Nonlinear term
θ	Absolute temperature
θ	Angle of rotation
θ_s	Steady-state rotation of a galvanometer

λ	Wavelength
μ	Arithmetic mean
μ	Mobility of charge carriers
ν	Poisson's ratio
π	Piezoresistive proportionality constant
ρ	Mass density
ρ	Resistivity coefficient
ρ	Specific resistance
σ	Normal stress
σ	Stefan–Boltzmann constant
σ	True standard deviation
σ_n	Normal stress component
$\sigma_{xx}, \sigma_{yy}, \sigma_{zz}$	Cartesian components of normal stress
$\sigma_1, \sigma_2, \sigma_3$	Principal normal stresses
τ	Shear stress
τ	Time constant
τ_e	Effective time constant
$\tau_{xy}, \tau_{yz}, \tau_{zx}$	Cartesian components of shear stress
ϕ	Magnetic flux
ϕ	Phase angle
ω	Circular frequency
ω_n	Undamped natural frequency

Chapter 1

Applications of Electronic Instrument Systems

1.1 INTRODUCTION

The primary objective of this book is to introduce electronic instrumentation systems in a manner sufficiently complete that the reader will acquire an ability to make accurate and meaningful measurements of mechanical and thermal quantities. The mechanical quantities include strain, force, pressure, moment, torque, displacement, velocity, acceleration, flow velocity, mass flow rate, volume flow rate, frequency, and time. The thermal quantities include temperature, heat flux, specific heat, and thermal conductivity.

Most readers have a conceptual understanding of these quantities through exposure in previous mechanics or physics courses, such as statics, dynamics, strength of materials, or thermodynamics. The student's experience in actually measuring these quantities by conducting experiments, however, is usually quite limited. An objective of this book is to introduce methods that are commonly employed to make such measurements. Through this exposure to the experimental aspects of a problem, the student will improve his or her understanding of many of the concepts that were introduced in the analytically oriented courses. Another objective of this book is to introduce all the elements of an electronic instrumentation system to enable the student to improve his or her ability to design effective experiments and to use measurement methods that can provide solutions to many practical engineering problems.

Emphasis in the text is on electronic instrumentation systems rather than mechanical measurement systems. In most cases, electronic systems provide data that more accurately and more completely characterize the design or process being experimentally evaluated. Also, the electronic system provides an electrical output signal that can be either used directly for analog control of processes or digitized for automatic data reduction. These advantages of the electronic measurement system over the mechanical measurement system are so significant that mechanical methods of measurements are now rarely used.

1.2   THE ELECTRONIC INSTRUMENT SYSTEM

A complete electronic instrument system usually contains at least six of the eight subsystems or elements indicated in Fig. 1.1.

The *transducer* is an analog device that converts a change in the mechanical or thermal quantity being measured into a change of an electrical quantity. For example, a strain sensor (gage) bonded to a specimen converts a change in strain $\Delta\varepsilon$ in the specimen to a change in electrical resistance ΔR in the gage. The change in resistance ΔR can then be converted to a change in voltage Δv, which can be measured accurately with relative ease. Since the voltage is proportional to the strain, the strain sensed by the transducer (the sensor and specimen) can be determined when the instrument system is properly calibrated.

The *power supply* provides the energy to drive the transducer. For instance, a differential transformer, which is a transducer used to measure displacement, requires an ac voltage supply to create a fluctuating magnetic field that excites two sensing coils. Power supplies, such as constant dc voltage sources, constant dc current sources, and ac voltage sources, are selected to satisfy the requirements of the transducer being employed.

Signal conditioners are electronic circuits that convert, compensate, or manipulate the output from the transducer into a more usable electrical quantity. The Wheatstone bridge used with a strain sensor converts the change in gage resistance ΔR to a change in voltage Δv. Filters, compensators, modulators, demodulators, integrators, and differentiators are other examples of signal conditioning circuits commonly used in electronic instrument systems.

Amplifiers are required in the system when the voltage output from the transducer–signal conditioner combination is small (signals of a millivolt or less are common). Amplifiers with gains of 10 to 1000 are used to increase these signals to levels (1 to 10 V) that are compatible with the voltage-measuring devices used in the system.

Recorders are voltage-measuring devices used to display the measurement in a form that can be read and interpreted. Recorders may be analog or digital. The voltage from the amplifier is an analog signal that is the input to the recorder. Analog recorders, such as oscilloscopes, oscillographs, and magnetic tape recorders, display or store the analog signals. Digital recorders accept an analog input and convert this signal to a digital code that is then displayed in a numerical array or stored on magnetic media.

Data processors are used with instrument systems that incorporate analog-to-digital converters (A/D) and provide the output signal representing the measurement in a digital code. The data processors are usually microcomputers that accept the digital input and then perform computations in accordance with programmed instructions.

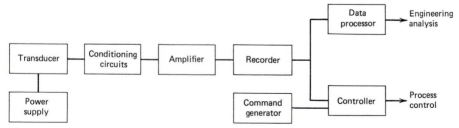

Figure 1.1 Block diagram representing an electronic instrumentation system for either engineering analysis or process control.

The output from the processor is displayed in graphs and tables that illustrate the salient results from the experimental study.

The *command generator* is a device that provides a control voltage that represents the variation (usually with respect to time) of an important parameter in a given process. As an example, the time–temperature profile for an oven must be controlled in curing plastics. The command generator provides a voltage signal that varies with time in exact proportion to the time–temperature profile required of the curing oven.

Process controllers are used to monitor and adjust any quantity that must be maintained at a specified value to produce a material or product in a controlled process. The signal from the instrumentation system is compared with a command signal that reflects the required value of the quantity in the process. The process controller accepts both the command signal and the measured signal and forms the difference to give an error signal. The error signal is then used to automatically adjust the process.

Electronic instrument systems are used in three different areas of application, which include the following:

1. Engineering analysis of machine components, structures, and vehicles to ensure efficient and reliable performance.
2. Monitoring processes to provide on-line operating data that allow an operator to make adjustments and thereby control the process.
3. Automatic process control to provide on-line operating data that are used as feedback signals in closed-loop control systems to continuously control the process.

Each of these applications is described in the sections that follow.

1.3   ENGINEERING ANALYSIS

An engineering analysis is conducted to evaluate new or modified designs of a machine component, structure, electronic system, or vehicle to ensure efficient and reliable performance when the prototype is placed in operation. Two approaches can be followed in performing the engineering analysis: theoretical modeling or experimental investigation.

In the *theoretical approach,* an analytical model of the component is formulated and assumptions are made pertaining to the operating conditions, the loads imposed on the component, the properties of the material, and the mode of failure. Equations describing the behavior of the analytical model are written and then solved using either exact mathematical methods or, more frequently, numerical computation. The results of the theoretical analysis provide the designer with an indication of the adequacy of the design and an estimate of the probable performance of the component or structure in service.

Uncertainties often exist pertaining to the validity of results from either the analytical model or the numerical procedures. Does the model accurately reflect all aspects of the prototype design? Do the assumed operating conditions cover the full range of loadings imposed on the component? Are the boundary conditions properly represented in the model? Have significant errors been introduced in the analysis through use of the numerical procedures?

In the *experimental approach,* a prototype or a scale model of the component is fabricated and a test program is conducted to evaluate the performance of the component in service by making direct measurements of the important quantities that control the adequacy of the design. This approach eliminates two serious uncertainties of the theoretical approach: an analytical model is not required, and the assumptions regarding operating conditions and material properties are not necessary. However, the experimental approach also has serious shortcomings. In comparison with the theoretical approach, it is extremely expensive. Also, uncertainties arise owing to inevitable experimental error in the measurements. Finally, there is always a question as to whether the transducers were placed at the correct locations to record the quantities that actually affect the adequacy of the design.

The preferred approach is a combination of the theoretical and the experimental methods. Theoretical analysis should be conducted to ensure a thorough understanding of the problem. The significance of the results of the theoretical analysis should be completely evaluated, and any shortcomings of the analysis should be clearly identified. An experimental program should then be designed to verify the analytical model, to check the validity of assumptions pertaining to operating conditions and material properties, and to ensure the accuracy of the numerical procedures.

The results of the theoretical analysis are extremely important in the design of the experimental program, as they enable the locations and orientations of the transducers to be specified more accurately and the number of measurements to be reduced appreciably. The number of tests required to cover the full spectrum of operating conditions may also be reduced when results from a verified theoretical model are available.

The results from the experimental program are intended to verify the analytical model and to check the validity of the assumptions and numerical procedures. If significant differences exist, the analytical model must be modified and new results developed for comparison with the experimental findings. After the theoretical approach is verified and confidence in the analysis is established, it is possible to optimize the weight, strength, or cost of the component.

The combined theoretical–experimental approach to engineering analysis provides the most cost-effective method to ensure efficient and reliable performances of new or modified designs of mechanical, structural, or electronic systems.

1.4   PROCESS CONTROL

Electronic instrumentation systems are used in two types of process control: open-loop, or monitoring, control and closed-loop, or automatic, control.

Open-loop control, involving a process that is being monitored with several transducers, is illustrated in Fig. 1.2. Data from the transducers are displayed continuously on an instrument panel containing charts, meters, and digital displays. An operator observes the quantities displayed and, if necessary, makes adjustments to the process input parameters to maintain control of the process. The operator serves to close the loop in this type of process control. The accuracy and reliability of the data displayed on the instrument panel are extremely important, as they provide the basis for the operator's decisions in adjusting the process. Most ships are operated with open-loop control. An operator in the engine room monitors measurements of ship speed, engine speed, engine temperature, oil pressure, fuel consumption, and the like and manually makes the adjustments to the engine controls needed to maintain the required speed.

A second type of process control, known as *closed-loop,* or *automatic, control,* is illustrated in Fig. 1.3. In the closed-loop control system, the operator has been eliminated. Instead, the signals from the electronic instrumentation system are compared with command signals that represent voltage–time relationships for the important quan-

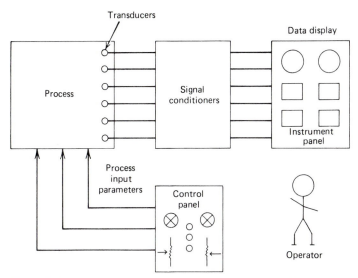

Figure 1.2 Schematic diagram of open-loop process control that requires the operator to monitor and adjust the process parameters.

tities associated with the process. The first controller measures the difference between the command signal and the transducer signal and develops an error, or feedback, signal. The feedback signal is then transmitted to the second controller, where it is amplified and used to drive devices that correct the process.

As an example of closed-loop control, consider a screw-actuated positioning mechanism that moves an engine block, during machining, through a battery of drilling and tapping machines. The desired position of the engine block along a track and the time required at each position are used by the command generator to establish

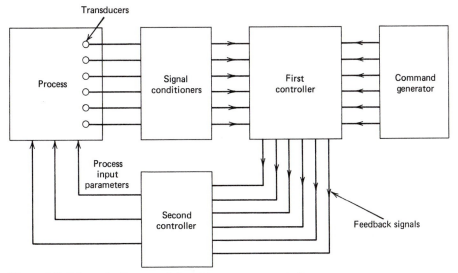

Figure 1.3 Schematic diagram of closed-loop process control.

a voltage–time trace that represents the required position of the block at any time. The actual position of the engine block is measured with a displacement transducer. The difference between the command signal and the measured displacement signal is used by the first controller to generate a feedback signal that is proportional to the adjustment needed to correct the position. The feedback signal is amplified and used to drive a current amplifier in the second controller. The current from this amplifier drives a servo-controlled motor, which turns the positioning screw. The screw drive moves the engine block and resets the feedback signal to zero so that the block is correctly positioned for the subsequent machining operation.

1.4.1   Process Control Devices

Control of a process requires frequent adjustment of the quantities involved in the process. Devices used in closed-loop control are similar in many respects to those used in open-loop control. Fluid flow is controlled by a valve that is opened or closed manually by an operator in open-loop control or automatically with a servomotor in closed-loop control. To indicate the type of hardware used in adjusting the parameters involved in a process, description of several control devices follows.

A. DC Motors

Motors are, with few exceptions, rotating machines that employ axial conductors that move in a magnetic field in a cylindrical gap between two iron cores. The magnetic field is produced by stationary coils that are wound around pole pieces incorporated into the stator, as shown in Fig. 1.4. In practice, an even number of poles (alternating N and S around the circumference of the stator) are excited by direct current flowing in the field windings. The rotor, or armature, is a cylindrical iron core that carries axial conductors embedded in slots and connected to the segments of a commutator. The current to the armature coils is provided through fixed brushes that slide on the rotating commutator.

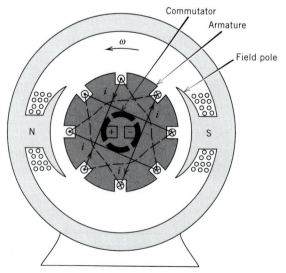

Figure 1.4 Configuration of an electrical rotating machine.

The speed–torque relations for all dc motors are

$$v = E + iR_A = K\phi\omega + iR_A \tag{1.1}$$

$$T = K\phi i \tag{1.2}$$

where

  v is the terminal voltage
  E is the emf produced by the rotating armature
  ϕ is the magnetic flux per pole (in webers)
  i is the armature current
  R_A is the resistance of the armature winding and other resistances
  K is a constant for a specific motor

If both mechanical and electrical losses are ignored, the electrical power input to the motor equals the mechanical power output. Thus,

$$Ei = T\omega \tag{1.3}$$

DC motors are classified as shunt, series, or compound according to the field-coil connections. Wiring diagrams and speed–torque characteristics of shunt and series motors are shown in Fig. 1.5. Compound motors (shunt and series fields connected so that their emf's add) have operating characteristics that are intermediate between those of series and shunt motors, as indicated in Fig. 1.6.

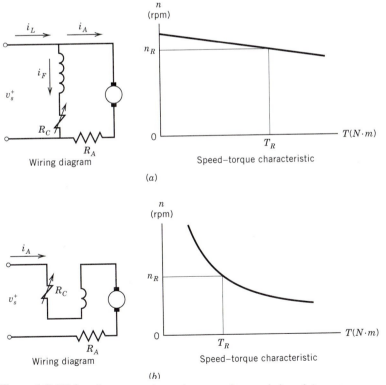

Figure 1.5 Wiring diagrams and speed–torque characteristics of dc motors:
(a) shunt motor, and (b) series motor.

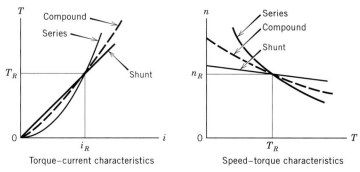

Figure 1.6 Operating characteristics of dc motors.

In control systems, dc motors are used to operate valves for controlling flow, to rotate a screw drive, to adjust dampers, and so forth. In these applications, control of the speed and torque delivered by the motor is important. DC motors can be controlled in many different ways. From Fig. 1.6 it is evident that the torque of these three types of dc motors can be adjusted by varying the current. With a shunt motor, a variable resistor is used in series with the field coils to adjust the speed. The variable resistance is placed in series with the armature coils of a series motor to adjust the speed. In all three types of motors, the speed and torque can be controlled by varying the terminal voltage.

B. Stepping Motors

Stepping motors differ in several ways from dc motors. As shown in Fig. 1.7, the stator incorporates a large number of small field poles deployed around the circumference. Moreover, each pole piece is fabricated with several (five in the example) teeth. The armature is constructed with one or two rows of teeth uniformly spaced around the periphery as shown in Fig. 1.7c. Fabrication of the armature core from permanent-magnet materials eliminates the need for the armature coils, the commutator, and the brushes.

The stepping motor is driven by a train of pulses delivered to the windings of the field coils. The pulse characteristics (rise time, duration, fall time, and amplitude) are matched to the inertia of the motor so that the armature rotates one step (related to the angle between two teeth on the armature) for each pulse. Angular position control can be accomplished with stepping motors simply by counting the number of input pulses applied to each set of field coils. The speed of a stepping motor is controlled by adjusting the pulse rate.

A typical permanent-magnet stepping motor uses a 50-tooth pitch with a four-pole single-phase stator to produce increments of 200 steps per revolution. This construction provides for a rotation of 1.8 deg per step or per pulse. When a second row of teeth is used on the armature (see Fig. 1.7c) and this second row is displaced by $\frac{1}{2}$ pitch with respect to the first row, the resolution of the stepping motor can be improved to 0.9 deg per step.

The precision of a stepping motor is excellent if its torque capacity is not exceeded. Regardless of the number n of steps, the position of the armature will be 1.8 n deg with an error of less than 3 percent of 1 step of 3.24 min. The speed and torque capabilities depend on the size and construction of the motors. Large motors with stacked rotors produce torques up to 10N · m. Small single-rotor motors can operate at very low torques at speeds up to 15,000 pulses per second (4500 rpm).

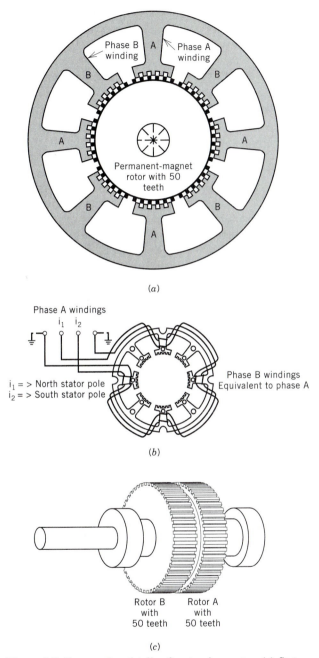

Figure 1.7 Construction details of a stepping motor. (a) Stator, (b) field-coil windings, and (c) permanent-magnet armature.

C. Solenoids

A solenoid is a coil consisting of many turns of magnet wire wound on a cylindrical core. When a current passes through the coil, a magnetic field develops with the N and S poles aligned with the axis of the coil. An iron cylinder inserted into the core of the solenoid responds to the magnetic field, and a magnetic force is developed. The force on the solenoid plunger depends on the current, the number of turns of wire, and the geometry of the coil. Significant forces can be developed even though the stroke length of the plunger is limited.

Solenoids are often used to open and close valves by coupling the plunger to the valve stem and a return spring. These valves control the flow of liquids in many different processes. This open or shut state of the valve is often termed bang-bang control, since one either has full flow or no flow depending on the position of the solenoid plunger.

D. Motorized Valves

In many instances the degree of control of a flow by a solenoid-operated valve is not adequate. In these cases, motorized valves are employed in which a dc motor acts through a screw mechanism to position the valve stem. With this arrangement, the valve opening is adjustable and the flow rate can be accurately controlled. Closed-loop control is achieved when the flow rate is measured and compared with the specified flow rate. The difference in these two rates produces an error signal with polarity. The error signal is then used with the appropriate controller to adjust the dc motor to modify the flow rate until it is within acceptable limits. The acceptable limits are indicated by control bands above and below the specified quantity.

Motorized valves are effective in controlling flow in quasi-static processes where the rate of change in the flow rate is small. In these cases the relatively slow valve adjustments are acceptable. In many applications, however, flow control by motorized valves is not sufficient to accommodate the rapidly changing requirements.

E. Servo Control Valves

Servo control valves are electromagnetic devices used to control flow. They differ from motorized valves in that they can respond quickly to a change in the command signals. Frequency response for smaller valves is usually in the range of 20 to 40 Hz.

Construction details of a typical servo valve are shown in Fig. 1.8. The valve contains two interacting subsystems. The first is the electromagnetic unit located at the top of Fig. 1.8. Control signals from a current amplifier energize the coils to produce a force that translates the coil insert. The coil insert moves the drive arm, with the motion constrained by the flexure tube, between the control nozzles in proportion to the coil current. This motion unbalances the four-arm hydraulic bridge circuit formed by the two fixed control orifices and the two control nozzles in the hydraulic subsystem. The hydraulic unbalance results in flow of fluid to one end of the valve spool and the subsequent movement of the spool to the position where one of the output ports is uncovered. The hydraulic amplification incorporated into the subsystem permits a large valve spool to be moved with very small electromagnetic forces applied to the drive arm. This reduced force permits the servo valve to be driven by small control currents that usually range from 10 to 50 mA.

F. Positioning Devices

Rotational position is easily accomplished by using either a dc motor with feedback control or a stepping motor with or without feedback control. Linear positioning, on the other hand, requires additional mechanical subsystems to produce linear move-

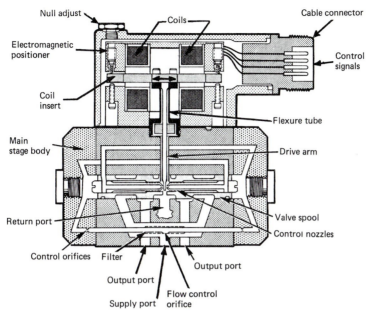

Null adjust

Coils

Cable connector

Electromagnetic positioner

Control signals

Coil insert

Flexure tube

Main stage body

Drive arm

Return port

Valve spool

Control nozzles

Control orifices   Filter

Output port

Output port

Supply port   Flow control orifice

Figure 1.8 Cutaway view illustrating the operating principles of a servo valve. (Courtesy of Schenck Pegasus Corporation.)

ment. The simplest method used for precision control of linear motion is to attach a servomotor to a drive screw as illustrated in Fig. 1.9*a*.

The precision in the position x that can be achieved with a drive screw is excellent. For example, if a stepping motor can be positioned with limits of ± 3.24 min, the drive nut on a screw with a 1-mm pitch can be positioned to within 0.15 μm. Moreover, the range of motion can be large because it is not difficult to produce long lead screws.

There are two disadvantages in employing lead screws for positioning. First, they are slow; velocities $\dot{x}$ of 100 mm/s are near the upper limit. Second, when the direction of motion is changed, the clearance between the nut and the screw produces backlash and the precision of positioning is lost.

For high-speed positioning, a hydraulic cylinder can be used, as indicated in Fig. 1.9*b*. In this approach, the position x on the end of the piston rod is controlled by the servo valve, which adjusts the flow to one side of the piston or the other. The velocity $\dot{x}$ of the rod depends on the flow rate of the servo valve and the area A of the piston. For small-bore cylinders equipped with high-flow-rate valves, rod velocities of 10 m/s can be achieved. The forces developed by hydraulic actuators are a function of the hydraulic pressure (usually 3000 to 5000 psi) and the area A of the piston. The forces are large even with the relatively large pressure drop that occurs across the servo valve when it is operating at high frequencies.

G. Resistance Heaters

When it is necessary to increase the temperature of a body or a mass of fluid in controlling a process, resistance heaters are usually employed. The power p dissipated by these resistive elements is

$$p = i^2 R = \frac{v^2}{R} \tag{1.4}$$

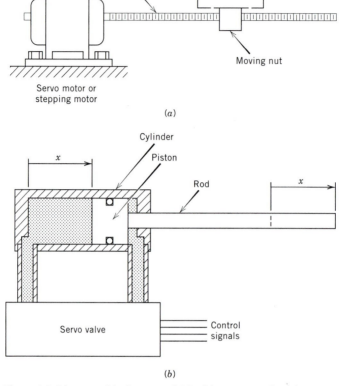

Figure 1.9 Linear positioning control (*a*) with a motor and a drive screw, and (*b*) with a servo-controlled hydraulic actuator.

where

> i is the current
> R is the resistance
> v is the applied voltage

For liquids, the resistances are deployed in the flow as immersion heaters, which incorporate the resistive element in a protective sheathing. For solid objects in an oven, the resistances may be glow bars (rods of SiC) or quartz lamps. Both of these resistive elements provide a very high temperature (in excess of $1000°$ C). Heat is transferred to the body by infrared radiation. In all cases, control is achieved by using a feedback signal with a current amplifier to adjust the power level until the temperature of the body is within the error bands of the command temperature.

1.5   EXPERIMENTAL ERROR

Error is the difference between the true value and the measured value of a quantity, such as displacement, pressure, temperature, and the like. Well-designed electronic instrumentation systems limit the error, which is inevitable in any measurement, to a value that is acceptable in terms of the accuracies required in an engineering analysis or the control of a process. Errors result from the following causes:

1. Accumulation of accepted error in each element of the system.
2. Improper functioning of any element in the system.
3. Effect of the transducer on the process.
4. Dual sensitivity of the transducer.
5. Other less obvious sources.

Each of these causes is described in terms of the general characteristics of the elements in the instrumentation system.

1.5.1   Accumulation of Accepted Error

All elements of an instrumentation system have accuracy limits specified by the manufacturer. For instance, a recorder may have an accuracy specified as ± 2 percent of full-scale values. The recorder can be expected to operate within these limits if it is used with care, properly maintained, and periodically calibrated. Because of limits on accuracy, the recorder will introduce error in a measurement when it is placed in an instrumentation system. However, this error is known, provided the recorder is operating within specifications.

The specified accuracy limits should be clearly understood, since an instrument accurate to within ± 2 percent can introduce larger errors than these limits seem to imply. Consider the input–output curve shown in Fig. 1.10 that characterizes the recorder. The *deviation d* is defined as the product of the accuracy and the full-scale value of the response of the recorder. Lines drawn parallel to the true response of the recorder, but displaced by $\pm d$, form the upper and lower accuracy bounds defining the actual response of the instrument. The shaded area between these two bounds gives the region where the recorder (or any other element) is operating within the manufacturer's specifications. If an instrument is operated at one-half scale, the deviation d

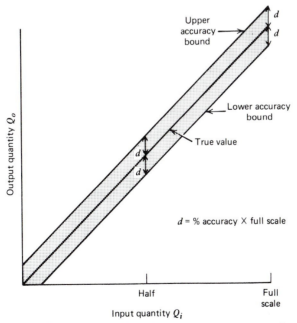

Figure 1.10 Accuracy bounds for an instrument operating within specifications.

remains constant: however, the true value is reduced by a factor of 2. Thus, the *error,* which is defined as the deviation divided by the true value, is doubled. This example indicates that errors of ± 4 percent would be within specifications if the recorder is operated at half scale. Operation of instruments at less than full scale is sometimes convenient; however, any reduction in scale should be carefully considered, since the error increases rapidly as the percent of scale used is reduced. Instruments should not normally be used at less than one third to one half of full scale without confirming that the magnification of the errors involved in the measurement does not invalidate the purpose of the experiment.

Since an instrumentation system normally contains several elements, with each element introducing error even when it operates within specifications, error accumulates. It is possible to estimate the accumulated error $\mathscr{E}_a$ for the instrument system as

$$\mathscr{E}_a = \sqrt{\mathscr{E}_T^2 + \mathscr{E}_{SC}^2 + \mathscr{E}_A^2 + \mathscr{E}_R^2} \qquad (1.5)$$

where the element errors are random and

$\mathscr{E}_T$ is the transducer error
$\mathscr{E}_{SC}$ is the signal conditioner error
$\mathscr{E}_A$ is the amplifier error
$\mathscr{E}_R$ is the recorder error

It is evident from Eq. 1.5 that small but acceptable errors for each element can accumulate and become unacceptably large for critical measurements that require high accuracy.

1.5.2 Improper Functioning of Instruments

If any element in the instrumentation system is not properly maintained or adjusted prior to use, calibration, zero-offset, or range errors can occur. Before discussing these errors, let us consider the response curve for a typical instrument, shown in Fig. 1.11. Here the output quantity Q_o is measured as the input quantity Q_i is varied. A significant portion of the response curve can be represented by a straight line that is fitted to the data by using a least-squares method (i.e., the instrument response is linear). The slope of the straight line is the calibration constant or sensitivity S of the instrument. Thus,

$$S = \frac{\Delta Q_o}{\Delta Q_i} \qquad (1.6)$$

For a strip-chart recorder, the sensitivity S is given in units of pen displacement per volt. For a piezoelectric pressure gage, the sensitivity S is given as the voltage or charge output per unit of pressure.

If the response line does not pass through the origin, the deviation d measured at the intercept with the ordinate is called the *zero offset* Z_0. It is evident from Fig. 1.11 that

$$Q_o = SQ_i + Z_0 \qquad (1.7)$$

Most instruments have a capability for adjusting the zero offset so that Z_0 can be set equal to zero. The relationship for the output quantity Q_o then reduces to

$$Q_o = SQ_i \qquad (1.8)$$

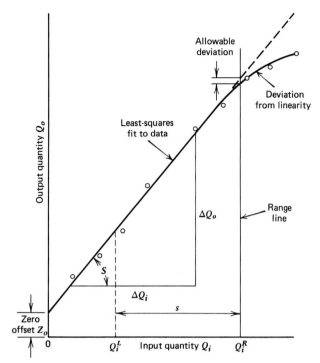

Figure 1.11 Input–output response curve for a typical instrument element.

For large values of the input quantity, the typical response curve frequently deviates from a straight line (linear relationship), as shown in the upper right portion of Fig. 1.11. When this deviation becomes excessive, say, 1 or 2 percent, Eqs. 1.7 and 1.8 are no longer valid because the range of the instrument has been exceeded. If an allowable deviation is specified, a range line can be drawn on the response graph and the range of the instrument Q_i^R can be established (see Fig. 1.11). The value Q_i^R defines the upper limit of operation of the instrument. The lower limit of operation Q_i^L is determined by excessive scale error (operation of the instrument at less than full scale). The difference between the upper limit of operation and the lower limit of operation defines the *span s* of the instrument. Thus,

$$s = Q_i^R - Q_i^L \qquad (1.9)$$

Errors in the measurement of Q_o will occur if the instrument is not properly calibrated and zeroed. Errors will also occur if the input Q_i is greater than the range of the instrument Q_i^R or less than the scale limit Q_i^L. Illustrations of calibration error, zero-offset error, and range error are shown in Fig. 1.12.

1.5.3   Effect of the Transducer on the Process

The transducer must be selected and placed in the process in such a manner that it does not affect or change the process. If the installation of the transducer does affect the process, serious errors can result and the measurements may be meaningless or misleading. For most measurements, the size and weight of the transducer should be

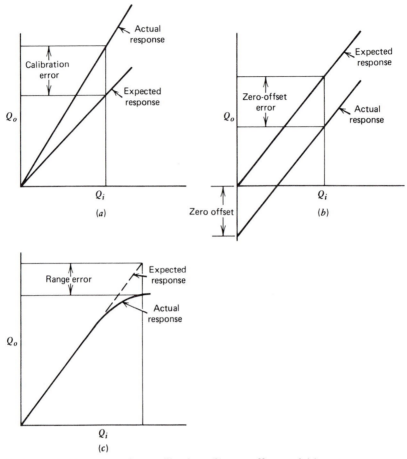

Figure 1.12 Illustration of (*a*) calibration, (*b*) zero-offset, and (*c*) range errors.

small relative to the size and weight of the component or process. Also, the transducer should require small forces and draw little energy from the process for its operation.

To illustrate the errors that can occur as a result of the presence of the transducer, consider an experiment designed to measure the frequency associated with the fundamental mode of vibration of a circular plate of uniform thickness h with clamped edges. The equation[1] governing the frequency of the first mode of this plate with an additional concentrated mass at the center of the plate is

$$\omega = \frac{\lambda^2}{a^2\sqrt{\rho/D}} \tag{1.10}$$

where

    ω is the circular frequency
    λ is a constant that depends on the ratio of the concentrated mass to the plate mass
    a is the radius of the plate

---

[1]A. W. Leissa, "Vibration of Plates," NASA SP–160, 1969, p. 19.

ρ is the mass density per unit area of the plate

D is the flexural rigidity of the plate. $D = Eh^3/12(1 - \nu^2)$

E is the modulus of elasticity of the plate material

ν is Poisson's ratio of the plate material

For this experiment, the value of the constant λ^2 depends on the ratio of the concentrated mass of the accelerometer m_a to the mass of the plate m_p. For m_a/m_p equal to 0, 0.05, and 0.10, the constant λ^2 equals 10.214, 9.012, and 8.111, respectively. Thus, the error in this measurement of the first natural frequency, owing to the mass of the accelerometer, is

$$\left(\frac{10.214}{9.012} - 1\right)100 = 13.3\% \qquad \text{for} \quad \frac{m_a}{m_p} = 0.05$$

and

$$\left(\frac{10.214}{8.111} - 1\right)100 = 25.9\% \qquad \text{for} \quad \frac{m_a}{m_p} = 0.10$$

It is clear from this example that the mass of the transducer has a profound effect on the vibratory process and that significant errors may occur as a result of the presence of the transducer. To avoid excessive errors, the mass of the transducer in this instance should not exceed 1 percent of the mass of the plate.

1.5.4 Dual Sensitivity Errors

Transducers are usually designed to measure a single quantity, such as pressure; however, they usually exhibit some sensitivity to one or more other quantities, such as temperature or acceleration. If a transducer is employed to measure some quantity, say, pressure, in a process, and if the temperature also changes as the measurement is made, error will result from the dual sensitivity of the transducer. The effect of the dual sensitivity is illustrated in the input–output response graph of Fig. 1.13. As shown in this figure, two errors arise owing to dual sensitivity when both quantities that affect the transducer or instrument are changing simultaneously during the time period of the measurement. First, a zero shift occurs because of the change in the secondary quantity. Second, a change in the sensitivity of the transducer occurs. These errors can be significant in poorly designed transducers.

In some experiments, the secondary quantity changes as a function of time. In these cases, the zero offset and the sensitivity also vary as a function of time. The changing zero offset is referred to as *zero drift,* and the varying sensitivity is termed *sensitivity drift.* It is extremely difficult to make accurate measurements under these conditions, since the continuous changes in the zero base and calibration constant of the instrument system preclude any possibility of making a single correction for the effect of the secondary quantity. A better approach is to carefully select a transducer with a negligible secondary sensitivity and to house the remaining elements of the instrumentation system in a temperature-controlled environment if possible.

Although this discussion has centered on the influence of dual sensitivity of the transducer, it should be recognized that all elements in the instrumentation system exhibit dual sensitivity. This dual sensitivity of the other elements becomes particularly important if the study is of long duration (several days or weeks). Time then becomes

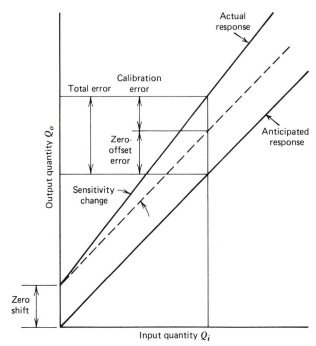

Figure 1.13 Change in response of an instrument as a result of dual sensitivity.

the secondary parameter, and the stability characteristics of the signal conditioner, power supply, amplifiers, and recorder affect the accuracy of the measurements. Since zero drift will occur in most instruments, particularly amplifiers, provision must be made in any experiment of long duration to periodically check and reestablish the zero base (rezero) or correct for the zero shift.

1.5.5   Other Sources of Error

Other important sources of error include lead-wire effects, electronic noise, and the human operator.

The effects of lead wires, which are used to connect the transducer to the instrumentation system, can be significant if the transducer contains resistive sensing elements. Lead wires, which are long and have a small diameter, exhibit a resistance that is significant relative to the transducer resistance. The added resistance of the lead wires changes the sensitivity or calibration constant of the transducer. The lead wires also produce erroneous signals because of temperature-induced resistance changes in the wires. When long lead wires are placed in the arms of a Wheatstone bridge that is being used for strain measurements, the accuracy of the measurements can be easily compromised.

Electronic noise usually results from spurious signals that are picked up by the lead wires. When lead wires are positioned in close proximity to electrical devices, such as motors or lights, the fluctuating magnetic fields in the vicinity of the devices generate small voltages in the adjoining lead wires that superimpose on the measurement signal. Since the measurement signal is usually small, the error produced by lead-wire noise can be significant. The noise picked up by the lead wires can be minimized with proper shielding, which isolates the lead wires from the effects of the fluctuating magnetic fields. In certain measurements with a very small measurement signal, noise from a properly shielded lead-wire installation may still be objectionable. In these cases, notched filters that block passage of a narrow band of frequencies can be employed to eliminate most of the noise, since it usually exhibits the 60-Hz power-line frequency.

Another source of error is the operator. The operator must properly record the sensitivity S of each element in the instrumentation system, and he or she must accurately zero each element. Finally, the operator must read the output that is displayed on the recorder. Reading errors of 1 or 2 percent resulting from parallax and trace width are common. Fortunately, digital displays that provide numerical readout are eliminating parallax errors. The count error on a digital display is analogous to the trace-width error on a strip chart.

1.6   MINIMIZING EXPERIMENTAL ERROR

In the preceding section, many sources of experimental error were identified so that the reader would become aware of some of the difficulties commonly encountered in conducting relatively simple experiments. Measurement systems designed to yield accuracies of 0.1 or even 1 percent are usually unrealistic when the cost of the system and the time required to make the measurements are considered. Accuracies of 2 to 5 percent can usually be achieved at reasonable cost; however, procedures must be followed to minimize error at each step of the experiment. A single mistake can easily degrade the system beyond acceptable limits of error. In the worst case, the mistake will degrade the system to the point at which the data are meaningless or even misleading. Accepted procedures for minimizing error in a measurement system are the following:

1. Carefully select the transducer. Pay particular attention to its size, weight, and energy requirements to ensure that it does not affect the process.
2. Check the accuracy of each element in the instrumentation system, and determine the accumulated accepted error.
3. Calibrate each instrument in the system to verify that it is operating within specifications.
4. Examine the process and the environment in which the instrumentation system must operate. Pay particular attention to temperature variations and the time required for the measurement. Estimate the errors that will be produced owing to dual sensitivity of each element in the instrumentation system.
5. Connect the system with properly shielded and terminated lead wires, using wiring procedures that minimize lead-wire errors. Estimate errors that may be introduced by the lead wires.
6. Check the system for electronic noise. If necessary, reroute the lead wires, add shielding where required, and insert suitable filters to minimize the noise.
7. Perform a system calibration by measuring the variable in a known process. This procedure, illustrated in Fig. 1.14, gives a single calibration constant for the entire instrument system.
8. Estimate the total error in the system from all known sources.

This systematic method of minimizing error does not ensure a perfect measurement, because some error is inherent in any experimental determination of unknown quantities. However, it does provide an organized approach to reducing error and to estimating the error involved at each step in the process.

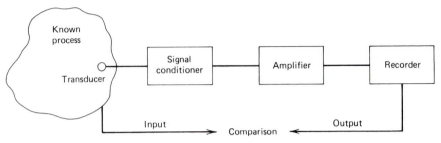

Figure 1.14 System calibration.

1.7 SUMMARY

An electronic instrumentation system usually contains a transducer, a power supply, a signal conditioner, an amplifier, an analog-to-digital converter, and a recorder. Such a system is used to experimentally determine unknown quantities, such as force, pressure, displacement, and temperature. The data or output signals from an instrumentation system are used for engineering analyses of machine components or structures, for process monitoring, for open-loop process control, or for closed-loop process control.

Engineering analyses are conducted to evaluate new or modified designs of machine components, structures, electronic systems, vehicles, and so on to ensure efficient and reliable performance of the prototype. A combined theoretical–experimental approach to engineering analysis provides the most thorough and most cost-effective method.

Instrument systems are widely used to monitor or control processes. Open-loop control requires an operator to adjust valves, rheostats, and other control devices to maintain the process within the control limits. Closed-loop control eliminates the operator. The output signal from an instrument system is compared with a command signal and the difference (the error signal) is used to adjust the process. The error signal is amplified and used to drive process-control devices (motors, solenoids, valves, and resistance heaters). Adjustments of the process are made to reduce the error signal and to maintain the process within the control limits.

Error will always occur in an experimental determination of an unknown quantity. The error, which accumulates, may have many causes, such as summation of accepted error, improper functioning of an instrument, transducer interaction with the process, or dual sensitivity of the transducer. Measurement systems that yield accuracies of 0.1 to 1 percent are often too costly for most applications. Accuracies of 2 to 5 percent are more realistic and can be achieved if careful procedures are employed in the design and installation of the instrumentation system. It is imperative that meticulous attention to detail be observed so that errors can be kept within acceptable bounds.

REFERENCES

1. Ambrosius, E. E., R. D. Fellows, and A. D. Brickman: *Mechanical Measurement and Instrumentation,* Ronald Press, New York, 1966.
2. Bartholomew, D.: *Electrical Measurements and Instrumentation,* Allyn & Bacon, Boston, 1963.
3. Beckwith, T. G., and R. D. Marangoni: *Mechanical Measurements,* 4th ed., Addison–Wesley, Reading, Mass., 1990.
4. Bendat, J. S., and A. G., Piersol: *Random Data: Analysis and Measurement Procedures.* 2nd ed., Wiley, New York, 1986.

5. Benedict, R. P.: *Fundamentals of Temperature, Pressure, and Flow Measurements,* 3rd ed., Wiley, New York, 1984.

6. Bibbero, R. J.: *Microprocessors in Industrial Control,* Instrument Society of America, Research Triangle Park, N.C., 1982.

7. Brophy, J. J.: *Basic Electronics for Scientists,* 5th ed., McGraw–Hill, New York, 1990.

8. Cerni, R. H., and L. E. Foster: *Instrumentation for Engineering Measurement,* Wiley, New York, 1962.

9. Considine, D. M. (Ed.): *Process Instruments and Controls Handbook,* 3rd ed., McGraw–Hill, New York, 1985.

10. Cook, N. H., and E. Rabinowicz: *Physical Measurement and Analysis,* Addison–Wesley, Reading, Mass., 1963.

11. Dally, J. W., and W. F. Riley: *Experimental Stress Analysis,* 3rd ed., McGraw–Hill, New York, 1991.

12. Doebelin, E. O.: *Measurement Systems,* 4th ed., McGraw–Hill, New York, 1990.

13. Dove, R. C., and P. H. Adams: *Experimental Stress Analysis and Motion Measurement,* Merrill, Columbus, Ohio, 1964.

14. Holman, J. P.: *Experimental Methods for Engineers,* 4th ed., McGraw–Hill, New York, 1984.

15. Hughes, T. A.: *Measurement and Control Basics,* Instrument Society of America, Research Triangle Park, N.C., 1988.

16. Kallen, H. P. (Ed.): *Handbook of Instrumentation and Controls,* McGraw–Hill, New York, 1961.

17. Keast, D. N.: *Measurements in Mechanical Dynamics,* McGraw–Hill, New York, 1967.

18. Lion, K. S.: *Instrumentation in Scientific Research,* McGraw–Hill, New York, 1959.

19. Malmstadt, H. V., C. G. Enke, and S. R. Crouch: *Electronics and Instrumentation for Scientists,* Benjamin/Cummings, Menlo Park, Calif., 1981.

20. Neubert, H. K. P.: *Instrument Transducers: An Introduction to their Performance and Design,* 2nd ed., Oxford University Press (Clarendon), London/New York, 1975.

21. Schenck, H., and R. J. Hawks: *Theories of Engineering Experimentation,* 3rd ed., Hemisphere, Washington, D.C., 1987.

22. Stein, P. K.: *Measurement Engineering,* Stein Engineering Services, Phoenix, Ariz., 1964.

23. Sweeney, R. J.: *Measurement Techniques in Mechanical Engineering,* Wiley, New York, 1953.

24. Tuve, G. L., and L. C. Domholdt: *Engineering Experimentation,* McGraw–Hill, New York, 1966.

25. Wilson, E. B., Jr.: *An Introduction to Scientific Research,* McGraw–Hill, New York, 1966.

EXERCISES

1.1 Why have electronic measurement systems largely replaced mechanical measurement systems?

1.2 (a) List the subsystems usually contained in a complete electronic instrumentation system. (b) Describe the function of each of these subsystems.

1.3 Discuss three different types of applications for electronic instrumentation systems.

1.4 Write a one-page description of "engineering analysis" that can be read and understood by an accountant.

1.5 Describe the preferred approach to performing an engineering analysis.

1.6 List the two common types of process control and describe the characteristic features of each type.

1.7 Prepare a block diagram illustrating the generic elements in a typical closed-loop control system.

1.8   Describe the differences in performance characteristics of series- and shunt-type dc motors.

1.9   Write a short engineering brief describing the mechanical features incorporated in a stepping motor.

1.10 Compare the operating characteristics of stepping motors and dc motors used in control applications.

1.11 Describe the flow control devices used in automatic control of processes. List the advantages and disadvantages of each device.

1.12 A drive screw in a linear positioning device has a pitch of 1 thread per mm. If the screw is driven with a stepping motor with a 50-tooth pitch and a four-pole single-phase stator, determine the movement per pulse of the positioning mechanism.

1.13 Discuss the advantages and disadvantages of lead-screw positioning devices.

1.14 Determine the maximum velocity of the piston rod in a 4-in.-diameter hydraulic cylinder if the oil flow is through a servo valve with a maximum capacity of 10 gal/min.

1.15 For the cylinder described in Exercise 1.14, determine the maximum force exerted by the piston rod if the hydraulic pump provides oil at 3000 psi and the pressure drop across the servo valve is 400 psi.

1.16 Describe three types of resistance heaters and discuss their relative merits for different applications.

1.17 List several sources of error that must be considered in the design of an instrumentation system.

1.18 A recorder is specified accurate to ± 1 percent of full scale and full scale is set at 200 mV. Determine the deviation that can be anticipated. Compute the probable percent error when the instrument is used at 3/4, 1/2, 1/4, and 1/8 scale. State the conclusion that can be drawn from the results of your computation.

1.19 An instrumentation system that is composed of a transducer, power supply, signal conditioner, amplifier, and recorder will exhibit what accumulated error $\mathscr{E}$ if the accuracies of the individual elements are:

	Case 1	Case 2	Case 3	Case 4
Transducer	0.05	0.01	0.01	0.02
Power supply	0.01	0.01	0.01	0.02
Signal conditioner	0.01	0.02	0.01	0.05
Amplifier	0.01	0.02	0.01	0.02
Recorder	0.01	0.03	0.01	0.02

1.20 Define range and span of an instrument.

1.21 Determine the error produced by a zero offset Z_0 if it is not taken into account in determining the output quantity Q_o.

1.22 Determine the error produced if an instrument sensitivity is S_1 instead of the anticipated sensitivity S.

1.23 Determine the error produced if an instrument sensitivity is S_1 instead of the anticipated sensitivity S and if a zero offset Z_0 is not taken into account in determining the output quantity Q_o.

1.24 Give an example of a transducer that produces error because of its influence on the quantity being measured.

1.25 Give an example of an instrument with dual sensitivity and explain how it may produce unanticipated error in a measurement.

1.26 (a) Explain why it is often difficult to accurately measure quantities over long periods of time. (b) What procedures are employed to improve the accuracy if the measurement must be made over a long period of time (weeks or months)?

1.27 An amplifier in an instrumentation system exhibits a zero drift of $\frac{1}{2}$ percent of full scale per hour. Determine the error if the measurement of Q_o is taken six hours after the initial zero was established and if the amplifier is operated at one half of full scale.

1.28 A pressure transducer exhibits a temperature sensitivity of 0.05 units per degree Celsius and a pressure sensitivity of 5 units per MPa. If the temperature changes 30°C during a measurement of a pressure of 200 MPa, determine the error resulting from the dual sensitivity of the transducer.

1.29 The sensitivity of an electrical resistance strain gage is defined as

$$S = \frac{\Delta R/R}{\epsilon}$$

where ΔR is the resistance change of the gage resulting from an applied strain ϵ and R is the resistance of the gage. If the sensitivity $S = 2.0$ for a gage with a resistance of 350 Ω, compute the sensitivity if the gage is connected to the instrument system with lead wires having a total resistance of 4 Ω.

1.30 Determine the apparent strain indicated by the strain gage lead-wire system described in Exercise 1.29 if the lead wires are subjected to a temperature change of 20°C after the initial zero is established for the system. Note that the lead wires change resistance with temperature according to:

$$\Delta R = R\gamma \, \Delta T$$

where γ is the temperature coefficient of resistance (0.0039/°C for copper).

1.31 Describe other common sources of error in electronic instrumentation systems.

1.32 Place a weight limit on a transducer used to determine the natural frequency of a clamped circular plate fabricated from aluminum and having a diameter of 200 mm and a thickness of 0.7 mm.

1.33 Describe calibration procedures for:
(a)   A power supply   (d)   An amplifier
(b)   A pressure transducer   (e)   A voltmeter
(c)   A Wheatstone bridge

1.34 Describe a calibration procedure to check the entire instrumentation system if the quantity being measured is:
(a)   Strain   (d)   Displacement
(b)   Pressure   (e)   Acceleration
(c)   Temperature

1.35 How is it possible to reduce noise in an electronic measurement system?

Chapter 2

Analysis of Circuits

2.1 INTRODUCTION AND DEFINITIONS

In the design and application of an instrument system a number of analog and digital circuits are used. Although it is not necessary to design complex electrical circuits in order to use instrument systems, it is important to understand the basic laws that govern the behavior of both ac and dc circuits. It is also important to analyze signals in order to describe the effect of the instrument system and its response on the quantity being measured. This chapter contains a review of the basic electronic concepts and laws that are useful in using and understanding modern instrumentation systems.

The analysis of circuits begins with defining the SI system of units, where the meter is the unit of length, the kilogram the unit of mass, and the second the unit of time. Other units of importance are absolute temperature in degrees Kelvin, relative temperature in degrees Celsius, and electric current in amperes. Quantities that will be used throughout this book, together with their standard symbol, units, and abbreviations, are defined in Table 2.1.

Brief definitions of each quantity follow:

Force. A force of 1 N causes a mass of 1 kg to accelerate at 1 m/s^2.

Energy. An object weighing 1 N receives 1 J of potential energy when it is elevated 1 m. Alternatively, a mass of 2 kg moving with a velocity of 1 m/s possesses 1 J of kinetic energy.

Table 2.1 Electrical Quantities Important in Instrumentation Systems

Quantity	Symbol	Unit	Abbreviation	(Alternate)
Force	f	newton	N	(kg · m/s^2)
Energy	w	joule	J	(N · m)
Power	p	watt	W	(J/s)
Charge	q	coulomb	C	(A · s)
Current	i	ampere	A	(C/s)
Voltage	v	volt	V	(W/A)
Electric field strength	$\mathcal{E}$	volt/meter	V/m	(N/C)
Magnetic flux density	$\mathcal{B}$	tesla	T	(Wb/m^2)
Magnetic flux	ϕ	weber	Wb	(T · m^2)

Power. Power represents the time rate at which energy is transformed. The transformation of 1 J of energy in 1 s represents an average power of 1 W. Instantaneous power is

$$p = \frac{dw}{dt} \tag{2.1}$$

Charge. An electric charge is the integral of the current with respect to time.

$$q = \int_0^t i \, dt \tag{2.2}$$

A charge of 1 C is transferred in 1 s by a current of 1 A.

Current. A current is the net rate of flow of positive charges.

$$i = \frac{dq}{dt} \tag{2.3}$$

A current of 1 A involves the transfer of a charge at the rate of 1 C/s.

Voltage. A charge of 1 C receives (or delivers) an energy of 1 J in moving through a voltage of 1 V. In general

$$v = \frac{dw}{dq} \tag{2.4}$$

Electric Field Strength. The electric field strength $\overline{\mathscr{E}}$ is defined by the magnitude and direction of the force $\overline{f}$ on a unit positive charge in the electric field.

$$\overline{f} = q\overline{\mathscr{E}} \tag{2.5}$$

It is easy to show that electric field strength is equal and opposite to the voltage gradient.

$$\mathscr{E} = -\frac{dv}{d\ell} \tag{2.6}$$

Magnetic Flux Density. A magnetic field develops in the region around a moving charge carrier or a current. The intensity of the magnetic effect is determined by

$$\overline{f} = q\overline{u} \times \overline{\mathscr{B}} \tag{2.7}$$

where

$\overline{u}$ is the velocity of the charge q
$\overline{\mathscr{B}}$ is the magnetic flux density
$\mathbf{x}$ is the symbol representing the vector cross product

A force of 1 N is developed by a charge of 1 C moving with a velocity of 1 m/s normal to a magnetic field with a flux density of 1 T.

Magnetic Flux. Magnetic flux is obtained by integrating the magnetic flux density over an area $\overline{A}$.

$$\phi = \int \overline{\mathscr{B}} \cdot d\overline{A} \tag{2.8}$$

where the symbol · represents the vector dot product.

The power p and the energy w transmitted in a circuit in terms of current i and voltage v are obtained as follows. From Eqs. 2.1, 2.3, and 2.4 we note that

$$p = \frac{dw}{dt} = \frac{dw}{dq}\frac{dq}{dt}$$

$$v = \frac{dw}{dq} \quad \text{and} \quad i = \frac{dq}{dt}$$

Therefore,

$$p = vi \tag{2.9}$$

$$w = \int p\, dt = \int vi\, dt \tag{2.10}$$

2.2   BASIC ELECTRICAL COMPONENTS

Analog and digital circuits are developed using several different components that affect the behavior of the current flow and the voltage at different locations in the circuit. This section introduces three of these basic components and gives the laws that govern the effect of the component on the circuit.

A. *Resistance*

The symbol for resistance is illustrated in Fig. 2.1a where the resistor R is shown inserted in a circuit. Ohm's law (in honor of Georg Ohm)

$$v = Ri \tag{2.11}$$

defines the relation between the voltage drop across the resistor and the current flow. When v is expressed in volts and i in amperes, R is given in ohms Ω.

The conductance G of a component is the reciprocal of the resistance. Thus

$$G = \frac{1}{R} \tag{2.12}$$

where G is expressed in terms of units known as siemens S.

B. *Capacitance*

The symbol for capacitance is illustrated in Fig. 2.1b where the capacitor C is shown inserted in a simple circuit. Physically, the capacitor consists of two electrodes separated by a dielectric that serves as an insulator. When a voltage is applied, a positive

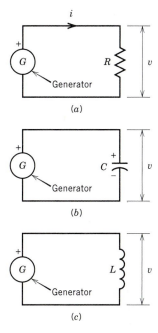

(a)

(b)

(c)

Figure 2.1 (a) Resistor inserted in a circuit. (b) Capacitor inserted in a circuit. (c) Inductor inserted in a circuit.

voltage develops on the upper plate and a negative voltage develops on the lower plate. The capacitor stores the charge q according to

$$q = Cv \tag{2.13}$$

and the energy stored w is determined from Eqs. 2.13 and 2.4 as

$$w = \int_0^V Cv \, dv = \frac{1}{2}CV^2 \tag{2.14}$$

When the voltage v is constant, the charge on the capacitor is maintained and no current flows. However, if the voltage changes with time, current flows through the dielectric according to

$$i = C\frac{dv}{dt} \tag{2.15}$$

In Eqs. 2.13 through 2.15, the capacitance is expressed in farads F (named for Michael Faraday).

C. Inductance

The symbol for inductance is illustrated in Fig. 2.1c where the inductor L is shown inserted in a simple circuit. Physically, the inductor is a multiturn coil of small-diameter wire. The coil has a very small resistance to steady current flow; however, when the current varies with time, a significant voltage drop develops across the coil as indicated by

$$v = L\frac{di}{dt} \tag{2.16}$$

where the inductance L is expressed in henrys H (named for Joseph Henry).

The energy stored in the inductor is determined from Eqs. 2.10 and 2.16 as

$$w = \int_0^I Li\ di = \frac{1}{2}LI^2 \qquad \textbf{(2.17)}$$

2.3 KIRCHHOFF'S CIRCUIT LAWS

More than two centuries ago Gustav Kirchhoff developed two circuit laws that provide the foundations for network theory. The first is the *current law,* illustrated in Fig. 2.2*a,* which states that the algebraic sum of the currents flowing into a junction point at any instant is zero.

$$\sum i = 0 \qquad \textbf{(2.18)}$$

The arrows representing the currents in Fig. 2.2 specify both magnitude and sign. Current flow into the junction is considered positive; flow away from the junction is considered negative.

The second of Kirchhoff's circuit laws is the *voltage law,* which states that the algebraic sum of the voltages around a loop at any instant of time is zero.

$$\sum v = 0 \qquad \textbf{(2.19)}$$

To show the application of Eq. 2.19, refer to the circuit loop presented in Fig. 2.2*b* and write the voltage change across each of the four legs of the loop as

$$\sum v = v_{ba} + v_{cb} + v_{dc} + v_{ad} = 0$$

where v_{ba} indicates the voltage at point b measured with respect to point a. If v_{ba} is positive, point b is at a higher potential than point a. Now start at point a, go clockwise around the loop, and note

$$\sum v = v_s - v_R - v_C + 0 = 0$$

or

$$v_s = v_R + v_C$$

at any instant of time. In this case, v_s represents the voltage source and v_R and v_C represent voltage drops in the direction of current flow.

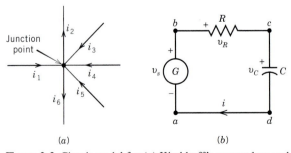

(a)   (b)

Figure 2.2 Circuit model for (*a*) Kirchhoff's current law, and (*b*) Kirchhoff's voltage law.

2.4 DIODES, TRANSISTORS, AND GATES

The basic components of circuits, resistors, capacitors, and inductors, were reviewed in Section 2.2. Diodes, transistors, and gates are more advanced components, usually fabricated from semiconductors, that are employed with the basic components in most analog and digital circuits.

2.4.1 Diodes

A *diode* is a two-terminal component, shown symbolically in Fig. 2.3*a*. The ideal diode presents no resistance to current flow when a positive voltage (bias) is applied, as indicated in Fig. 2.3*a*. However, when the voltage is reversed (negative bias), the diode offers infinite resistance. Essentially, the diode acts as a selective switch, which is closed for reverse bias voltages and open for forward bias voltage. The voltage current characteristic of the diode, presented in Fig. 2.3*b*, is identical to that obtained with a selective switch.

In practice, most diodes are fabricated with a P/N junction in silicon and require a forward bias exceeding a threshold voltage of 0.6 *V* before conduction occurs.

2.4.2 Transistors

Transistors are semiconductor devices used either as amplifiers or as high-speed electronic switches. The most widely used amplifier is based on the bipolar junction transistor, illustrated in Fig. 2.4. The devices are planar; therefore, they can be fabricated by using lithographic methods in P- and N-doped silicon. They are extremely small, with areas of $10^{-9} m^2$.

The transistors are three-terminal devices, with the base represented by B, the collector by C, and the emitter by E. The theory of operation of the bipolar transistor is beyond the scope of this book. Basically, the transistor will act as a current amplifier because relatively small base currents i_B produce large collector currents i_C. For example, when an NPN transistor is connected in a common emitter configuration with voltage sources and a resistive load R_L, as shown in Fig. 2.5, the transistor acts to amplify the input signal i_i. The signal current i_i causes a variation in the base current i_B, which in turn produces a variation in the collector current i_C along the load line as shown in Fig. 2.5*b*. The time-varying part of the collector current represents the amplified output current that is drawn from the source V_{CC} and the flow through the load resistance R_L. The gain G is

$$G = \frac{i_o}{i_i} \tag{2.20}$$

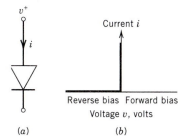

Figure 2.3 (*a*) Symbol for a diode showing current flow with a positive bias. (*b*) Voltage–current characteristics of an ideal diode.

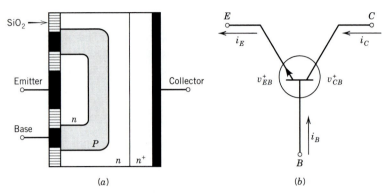

Figure 2.4 Representation of an NPN bipolar transistor. (*a*) Planar structure in silicon, and (*b*) circuit symbol.

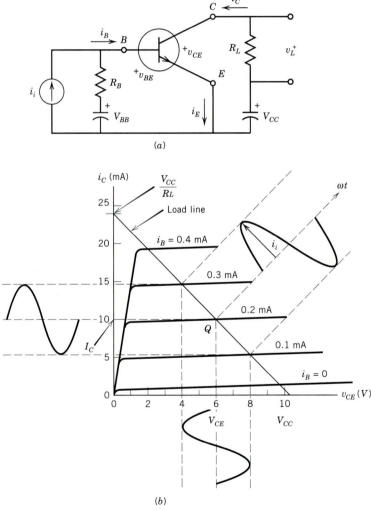

Figure 2.5 A basic NPN-transistor current amplifier. (*a*) Circuit diagram, and (*b*) operating characteristics.

where i_o is the sinusoidal component of i_C. The gain G depends on the base and collector characteristics of the transistor and V_{CC} and R_L. Signal gains for a single transistor are in the range from 10 to 100.

Transistors are also employed as very high-speed electronic switches, which are open or closed depending on the voltage applied to the base. When operated as a switch, the transistor is connected into the simple circuit as shown in Fig. 2.6a. Since both P/N junctions are reverse biased, practically no collector current flows and the transistor is operated in the cutoff region of Fig. 2.6b at point 1 when the input voltage (current) to the base is zero. At point 1 the collector current is small (5 μA), with an applied voltage $V_{CC} = 5$ V. This condition corresponds to a cutoff resistance of 1 MΩ, and the switch, whose contacts are the collector and emitter terminals, is open.

When a positive voltage is applied at the input, the base current increases (say, to 0.3 mA) and the operation of the transistor moves along the load line of Fig. 2.6b to point 2. At this point the transistor is operating in a saturated condition and the voltage drop v_{CE} across the transistor is very small. The collector current is about 30 mA at a saturation voltage of 0.3 V, which yields a switch resistance of about 10 Ω. In this state, the transistor is considered a closed switch.

2.4.3   Gates

In processing digital signals, the information is expressed as a digital code and is transmitted through logic operations, which transform and manipulate this information.

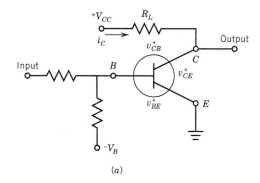

(a)

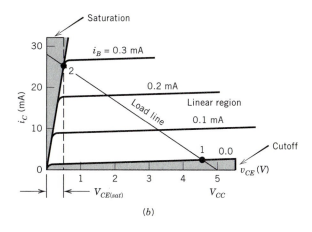

(b)

Figure 2.6 The transistor in a switching application. (a) Switching circuit, and (b) operating regions.

The logic *gate* is the device that controls the flow of information in a logic network. Although there are many different gates that perform specialized logic operations, all of these more complex gates are made from three elementary gates, namely AND, OR, and NOT gates.

The AND gate may be represented by the circuit shown in Fig. 2.7a, where two switches A and B are placed in the line from the source to the load. The voltage v_s is applied to the load only if switch A and switch B are both closed. The possibilities for the AND gate are listed in a truth table shown in Table 2.2. Note that 0 is used to represent a false statement and 1 to represent a true statement. With regard to the voltage applied to the load, 1 indicates that it is true that v_s is applied to the load.

The OR gate is represented by the circuit shown in Fig. 2.7b, where two switches A and B are placed parallel to one another in the line between the voltage source and the load. When A is closed or when B is closed, the voltage is applied to the load (T = 1). The truth table for a two-switch OR gate is presented in Table 2.3.

The NOT gate, which is illustrated in Fig. 2.7c, is an inverter. In this case, the mechanical switch has been replaced by a transistor that is turned on (closed) by a positive input voltage. If the input signal to the transistor is, say, 0, the transistor acts as an open switch, no current flows, and the output voltage is v_s or 1. When the input signal goes to 1, the transistor conducts, acting like a closed switch, and the output is grounded, giving the low state, or 0. It is clear from this description that when the input is high (A), the output is low ($\overline{A}$), and changing the input to low ($\overline{A}$) results in an output that is high (A).

These basic gates are arranged in circuits to perform digital functions. A digital system is composed of many of these digital functions and may contain a million or more of the simple basic gates. The number of chips used to build the logic circuits

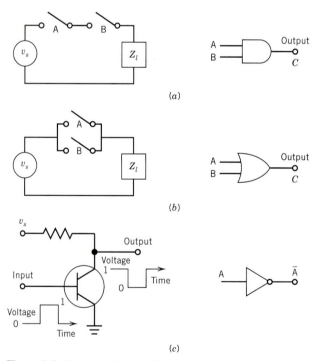

(a)

(b)

(c)

Figure 2.7 Circuits and symbols for the three basic logic gates.
(a) AND gate. (b) OR gate. (c) NOT gate.

Table 2.2 Truth Table for the AND Gate A·B = T

Switches or Inputs		Output
A	B	T
0	0	0
0	1	0
1	0	0
1	1	1

Table 2.3 Truth Table for the OR Gate A + B = T

Switches or Inputs		Output
A	B	T
0	0	0
0	1	1
1	0	1
1	1	1

depends on the scale of integration used to fabricate the circuits. With VLSI (very-large-scale integration) it is possible to place on the order of 10^4 gates on a single chip of silicon, thus permitting the development of extremely large digital systems with only 100 to 1000 chips.

2.5 DC CIRCUITS

In dc circuits, the current flow is constant with respect to time. This fact simplifies the circuit analysis because the voltage drop across an inductor is zero ($di/dt = 0$) and the current flow through a capacitor is zero ($dv/dt = 0$). The resistor is the only component that produces a voltage drop in accordance with Ohm's law.

Consider resistors R_1 and R_2 arranged in series in a dc circuit, as shown in Fig. 2.8a. Kirchhoff's voltage law, given by Eq. 2.19, and Ohm's law, given by Eq. 2.11, yield

$$v_s = v_{d1} + v_{d2} = iR_1 + iR_2 = iR_e \qquad (2.21)$$

where $R_e = R_1 + R_2$ is the equivalent closed-loop resistance as illustrated in Fig. 2.8a.

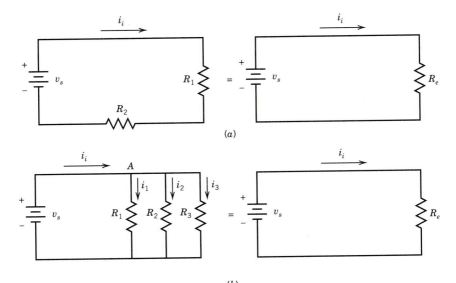

Figure 2.8 Resistances in a circuit loop. (a) Series resistances, and (b) parallel resistances.

Next, consider the parallel circuit illustrated in Fig. 2.8b and apply Kirchhoff's current law, Eq. 2.18, to point A to obtain

$$i_i = i_1 + i_2 + i_3 \tag{a}$$

Substituting Eq. 2.11 into Eq. a gives

$$\frac{v_s}{R_e} = \frac{v_d}{R_1} + \frac{v_d}{R_2} + \frac{v_d}{R_3} \tag{b}$$

Since $v_s = v_d$, the equivalent resistance for a group of three parallel resistors is

$$\frac{1}{R_e} = \frac{1}{R_1} + \frac{1}{R_2} + \frac{1}{R_3} \tag{2.22}$$

2.6 PERIODIC FUNCTIONS

When the current or voltage varies with time in a circuit, the signal has some type of waveform. The many different types of waveforms are all considered either periodic or transient. Periodic signals are repetitive and can be represented by sinusoidal functions or by a series of sinusoidal components by means of Fourier analysis. Transient signals are one-time events; they do not repeat. They are more difficult to analyze and will be discussed in Chapter 10.

A special type of periodic function is the sinusoid ($\sin n\omega t$ or $\cos n\omega t$). Sinusoidal functions are important in describing the dynamic response of instrument systems where the ratio of the input voltage to the output voltage is a function of frequency. The sinusoid is also used extensively in Fourier analysis of other periodic signals that have a more complex waveform.

To illustrate two sinusoidal functions, consider a point rotating about a circle centered at point O in the xy plane as shown in Fig. 2.9. Since the line OR rotates with a constant magnitude A_0 in a counterclockwise direction with an angular velocity ω, the projection of OR onto the x axis gives the position of point P as

$$x = A_0 \cos \omega t \tag{2.23}$$

and the projection of OR onto the y axis gives the position of point Q as

$$y = A_0 \sin \omega t \tag{2.24}$$

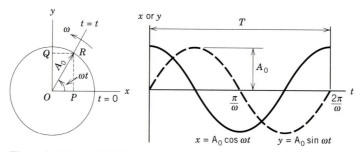

Figure 2.9 Sinusoidal functions $x = A_0 \cos \omega t$ and $y = A_0 \sin \omega t$.

The sinusoidal function is repetitive, as indicated in Fig. 2.9, with the values of x and y repeated every period (T seconds). The angular frequency ω, the period T (s), and the cyclic frequency f (Hz) are related by the expression

$$\omega = \frac{2\pi}{T} = 2\pi f \tag{2.25}$$

and from Eq. 2.25 it is evident that

$$f = \frac{1}{T} \tag{2.26}$$

The velocities[1] of points P and Q, illustrated in Fig. 2.9, are given by

$$v_P = \dot{x} = \frac{d}{dt}(A_0 \cos \omega t) = -A_0\omega \sin \omega t = A_0\omega \cos\left(\omega t + \frac{\pi}{2}\right)$$
$$\tag{2.27}$$
$$v_Q = \dot{y} = \frac{d}{dt}(A_0 \sin \omega t) = A_0\omega \cos \omega t = A_0\omega \sin\left(\omega t + \frac{\pi}{2}\right)$$

and the accelerations of points P and Q are

$$\begin{aligned} a_P = \ddot{x} &= -A_0\omega^2 \cos \omega t \\ &= A_0\omega^2 \cos(\omega t + \pi) \end{aligned}$$
$$\tag{2.28}$$
$$\begin{aligned} a_Q = \ddot{y} &= -A_0\omega^2 \sin \omega t \\ &= A_0\omega^2 \sin(\omega t + \pi) \end{aligned}$$

Equations 2.27 and 2.28 show that the magnitudes of the velocities and accelerations of points P and Q can be obtained by multiplying the positions x and y by ω and ω^2, respectively. Note that there is a phase difference and the velocities and accelerations lead the displacements (positions) by $\pi/2$ and π, respectively.

From Eqs. 2.23, 2.24, 2.27, and 2.28, the maximum values of the velocities and accelerations can be written as

$$\begin{aligned} v_P &= -\omega y & v_Q &= \omega x \\ a_P &= -\omega^2 x & a_Q &= -\omega^2 y \end{aligned} \tag{2.29}$$

Since this motion is proportional to the displacement from a fixed point and the velocity $\bar{v}$ and acceleration $\bar{a}$ are directed toward that fixed point, the motion is classified as simple harmonic motion.

More complex waveforms that are periodic can be represented by a Fourier series of sinusoids. Thus,

---

[1]The dot notation above a variable is used to indicate differentiation with respect to time. One dot indicates the first derivative, and two dots represent the second derivative.

$$x = \frac{A_0}{2} + \sum_{n=1}^{\infty} A_n \cos n\omega t + \sum_{n=1}^{\infty} B_n \sin n\omega t \qquad (2.30)$$

where A_0, A_n, and B_n are the harmonic amplitudes and ω is the fundamental frequency. If a sufficient number of terms are employed in the Fourier series representation, the periodic motions x, $\dot{x}$, or $\ddot{x}$ can be accurately described with a sum of simple harmonic motions of frequencies that are multiples of the fundamental frequency (i.e., 2ω, 3ω, $\cdots$, $m\omega$).

A second method of analysis of signals uses phasors in the complex plane, where the phasor, projected on the real and imaginary axes, exhibits real and imaginary parts of the signal. Consider a phasor A, shown in Fig. 2.10, and expressed in exponential form as

$$A = A_0 e^{j\omega t} \qquad (2.31)$$

where $j = \sqrt{-1}$. Recall the identity that

$$e^{j\omega t} = \cos \omega t + j \sin \omega t \qquad (2.32)$$

Then Eq. 2.31 can be written as

$$A = A_0 \cos \omega t + j A_0 \sin \omega t \qquad (2.33)$$

Note that the first term in Eq. 2.33 is real and the second term is imaginary. A graph of the complex plane, shown in Fig. 2.10, illustrates the phasor, the real and imaginary terms, and the angle (ωt) that is taken as positive counterclockwise. A comparison of Figs. 2.9 and 2.10 indicates that the phasor represents the rotating line OR, which provides an example of simple harmonic motion (or signal).

Differentiation of the phasor A to obtain dA/dt gives

$$\dot{A} = j\omega A_0 e^{j\omega t} \qquad (2.34)$$

Note from Eq. 2.32 that $j = e^{j\pi/2}$ and $j^2 = -1 = e^{j\pi}$. Substituting the first of these equalities in Eq. 2.34 yields

$$\dot{A} = \omega A_0 e^{j(\omega t + \pi/2)}$$
$$= \omega A_0 \cos\left(\omega t + \frac{\pi}{2}\right) + j\omega A_0 \sin\left(\omega t + \frac{\pi}{2}\right) \qquad (2.35)$$

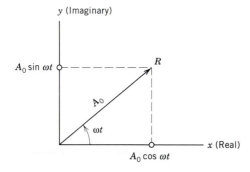

Figure 2.10 Representation of the phasor $A = A_0 e^{j\omega t}$ in a complex plane.

A comparison of Eq. 2.35 with Eq. 2.27 shows that the real part of $\dot{A}$ corresponds to v_P and the imaginary part represents v_Q. Differentiating Eq. 2.34 to obtain d^2A/dt^2 gives

$$\ddot{A} = -\omega^2 A_0 e^{j\omega t} \tag{2.36}$$

By using Eq. 2.32 with Eq. 2.36, it is clear that

$$\begin{aligned}
\ddot{A} &= \omega^2 A_0 e^{j(\omega t + \pi)} \\
&= \omega^2 A_0 \cos(\omega t + \pi) + j\omega^2 A_0 \sin(\omega t + \pi)
\end{aligned} \tag{2.37}$$

A comparison of Eq. 2.37 with Eq. 2.28 indicates that the real part of A corresponds to a_P and the imaginary part represents a_Q. The phase angle for $\dot{A}$ is $\phi = \pi/2$ and for $\ddot{A}$ is $\phi = \pi$ relative to the reference phasor. These phase angles are leading, as shown in the complex plane in Fig. 2.11. Note that leading phase angles are positive (counterclockwise) and lagging phase angles are negative (clockwise) with respect to the reference line.

The amplitude A_0 of the phasor is

$$A_0 = \sqrt{(Re)^2 + (Im)^2} \tag{2.38}$$

where

Re is the amplitude of the real part
Im is the amplitude of the imaginary part

The phase angle ϕ is

$$\phi = \tan^{-1} \frac{(Im)}{(Re)} \tag{2.39}$$

The important advantages of using a phasor representation of sinusoidal motion include the ease of differentiating and integrating the exponential function and the presence of both magnitude and phase information. Differentiation is accomplished by multiplying by $j\omega$, and integration is accomplished by dividing by $j\omega$. Amplitude and phase angles for A, $\dot{A}$, and $\ddot{A}$ can be determined easily by using Eqs 2.38 and 2.39. Because of these advantages, the exponential notation will be used throughout this book.

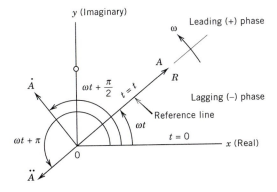

Figure 2.11 Phase angles of $\dot{A}$ and $\ddot{A}$ relative to A, showing velocity leading displacement by $\pi/2$ and acceleration leading displacement by π.

2.7   AC CIRCUITS

The three fundamental electrical components used to describe the behavior of ac circuits are inductance L, resistance R, and capacitance C. These three components are illustrated individually in Fig. 2.12, where they are connected to a sinusoidal input voltage v_s. The voltage drop v_d across each of these components was covered in Section 2.2. To determine the effect of each component in a circuit powered with an ac signal let

$$i = i_0 e^{j\omega t} \qquad\qquad (2.40)$$

which can be substituted into Eqs. 2.11, 2.13, and 2.16 to give

$$v_d = jL\omega i_0 e^{j\omega t} = jL\omega i = Z_L i \qquad \text{for the inductor}$$

$$v_d = Ri_0 e^{j\omega t} = Ri = Z_R i \qquad \text{for the resistor} \qquad (2.41)$$

$$v_d = \left(\frac{i_0}{j\omega C}\right) e^{j\omega t} = \frac{i}{j\omega C} = Z_C i \qquad \text{for the capacitor}$$

where Z_L, Z_R, and Z_C are the impedances for the basic components. Thus,

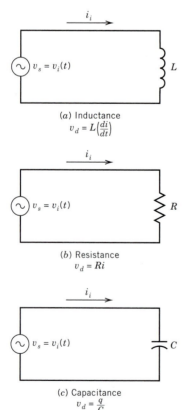

(a) Inductance
$v_d = L\left(\frac{di}{dt}\right)$

(b) Resistance
$v_d = Ri$

(c) Capacitance
$v_d = \frac{q}{C}$

Figure 2.12 Elementary circuits illustrating voltage drop across the three basic components L, R, and C.

$$Z_L = j\omega L$$
$$Z_R = R \tag{2.42}$$
$$Z_C = \frac{1}{j\omega C} = -\frac{j}{\omega C}$$

From Eq. 2.42 it is evident that the inductance voltage leads the resistance voltage and current with a phase angle of $\pi/2$.

To illustrate the use of the impedance relations and to show elementary methods of ac circuit analysis, consider the circuit shown in Fig. 2.13. Applying Kirchhoff's law, Eq. 2.19, and with Eqs. 2.11 and 2.13 gives

$$v_s = v_i(t) = iR + \frac{q}{C} \tag{a}$$

Since the output voltage $v_o = q/C$, Eq. a can be reduced to

$$v_i(t) = iR + v_o(t) \tag{b}$$

From Eq. 2.15 it is evident that

$$i = C\frac{dv_o(t)}{dt} = C\dot{v}_o(t) \tag{c}$$

Substituting Eq. c into Eq. b gives a first-order differential equation

$$RC\dot{v}_o(t) + v_o(t) = v_i(t) = v_i e^{j\omega t} \tag{2.43}$$

Let $v_o(t) = v_o e^{j\omega t}$ and substitute this relation into Eq. 2.43 to obtain

$$v_o(t) = \frac{1}{1 + j\omega RC} v_i e^{j\omega t} \tag{2.44}$$

Eliminating j from the denominator of Eq. 2.44 gives

$$v_o(t) = \frac{1 - j\omega RC}{1 + (\omega RC)^2} v_i e^{j\omega t} \tag{d}$$

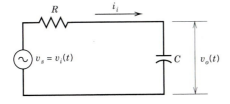

Figure 2.13 An RC circuit with an output voltage $v_o(t)$ representing the voltage across a capacitor.

By using Eqs. 2.38 and 2.39 with Eq. d, the output voltage across the capacitor in Fig. 2.13 can be represented by

$$v_o(t) = \frac{v_i e^{j(\omega t - \phi)}}{\sqrt{1 + (\omega RC)^2}} = v_o e^{j(\omega t - \phi)} \tag{2.45}$$

where

$$v_o = \frac{v_i}{\sqrt{1 + (\omega RC)^2}}$$

and the phase angle ϕ is

$$\phi = \tan^{-1} \omega RC$$

Inspection of Eq. 2.45 shows that the amplitude v_o and the phase ϕ of the output voltage are a function of a single combined term ωRC.

A second method of analysis for the circuit shown in Fig. 2.13 utilizes the impedances defined in Eq. 2.42. With this approach, the input and output voltages are written as sinusoids with the $e^{j\omega t}$ notation. The voltage drop across a given element is taken as Zi. For example, the output voltage $v_o(t)$ across the capacitor in Fig. 2.13 is given by Eq. 2.41 as

$$v_o(t) = Z_C i \tag{e}$$

But

$$i = \frac{v_i e^{j\omega t}}{Z_R + Z_C} \tag{f}$$

Substituting Eq. f into Eq. e yields

$$v_o(t) = \frac{Z_C}{Z_R + Z_C} v_i e^{j\omega t} \tag{g}$$

Next use Eq. 2.42 with Eq. g to obtain

$$v_o(t) = \frac{1}{1 + j\omega RC} v_i e^{j\omega t} \tag{2.46}$$

Comparison of Eq. 2.46 with Eq. 2.44 shows that the results are identical and that both methods can be used to determine the dynamic performance of ac circuits; however, the approach using impedances is easier and requires less time.

2.7.1 Impedance

In ac circuits, the impedance Z is related to a complex function that depends on the frequency of the signal. To show the impedance in the most general way, consider the circuit shown in Fig. 2.14. The circuit is driven with a sinusoidal voltage $v_s(t)$.

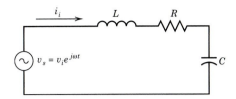

Figure 2.14 An ac circuit containing the three basic components L, R, and C.

If the voltage drops across the components are summed and equated to the supply voltage, then

$$v_i = \left[R + j \left(\omega L - \frac{1}{\omega C} \right) \right] i_i \qquad (2.47)$$

Examination of Eq. 2.47 shows that a complex function with real and imaginary parts is involved in the relation between v_i and i_i. If this complex function is divided into real and imaginary parts, as illustrated in Fig. 2.15, a more useful expression for v_i in terms of i_i is obtained.

$$v_i = Z i_i \qquad (2.48)$$

where

$$Z = \left[R^2 + \left(\omega L - \frac{1}{\omega C} \right)^2 \right]^{1/2} \qquad (2.49)$$

is the total impedance of the circuit.

Eqs. 2.48 and 2.49 define the amplitude of the voltage and the current, but the phase of the voltage relative to the current remains to be determined. Reference to Fig. 2.15 and Eqs. 2.39 and 2.47 shows the phase angle ϕ as

$$\phi = \tan^{-1} \left(\frac{\omega L - 1/\omega C}{R} \right) \qquad (2.50)$$

When the phase angle $\phi > 0$, as shown in Fig. 2.15, the voltage leads the current and the complete expression for $v_s(t)$ is

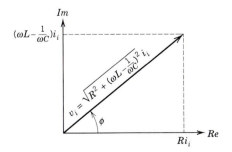

Figure 2.15 Representing the impedance components in the complex plane.

$$v_s(t) = \left[R^2 + \left(\omega L - \frac{1}{\omega C} \right)^2 \right]^{1/2} i_i\, e^{j(\omega t + \phi)} \tag{2.51}$$

Of particular importance is the effect of the frequency ω on both the impedance Z and the phase angle ϕ.

2.8   FREQUENCY RESPONSE FUNCTION

The *frequency response function*,[2] often termed the FRF, for a circuit or instrument is defined as a ratio of output to input over a frequency range. Thus,

$$H(\omega) = \frac{v_o(\omega)}{v_i(\omega)} \tag{2.52}$$

where $v_o(\omega)$ and $v_i(\omega)$ are the frequency spectra of the output and the input signals. The frequency response function for the circuit of Fig. 2.13 can be written from Eq. 2.44 as

$$H(\omega) = \frac{1}{1 + j\omega RC} = \frac{e^{-j\phi}}{\sqrt{1 + (\omega RC)^2}} \tag{2.53}$$

where Eqs. 2.38 and 2.39 were used in the manipulation.

From Eq. 2.53 it is clear that the magnitude of the FRF is

$$|H(\omega)| = \frac{1}{\sqrt{1 + (\omega RC)^2}} \tag{2.54}$$

and the phase ϕ is

$$\phi = -\tan^{-1} \omega RC \tag{2.55}$$

It is clear from Eq. 2.54 that the frequency response function $H(\omega)$ gives the ratio for the amplitude of the output voltage to the input voltage, and Eq. 2.55 gives the phase shift ϕ of the output voltage relative to the input voltage. The negative phase angle indicates that the output signal lags behind the input signal.

The frequency response function and other parameters are often expressed as a relative number in terms of decibels N_{dB}. The number of decibels is defined as

$$N_{dB} = 10 \log \left(\frac{p}{p_r} \right) \tag{2.56}$$

where

p is the measured power
p_r is a reference power

---

[2]The transfer function and the frequency response function are often confused with one another. The transfer function is dependent on the LaPlace operator S, whereas the frequency response function is dependent on ω and is a more restricted relationship. In fact, the frequency response function is obtained from the transfer function when S is replaced by $j\omega$.

The decibel can also be expressed in terms of a voltage ratio by substituting $p = v^2/R$ into Eq. 2.56 to obtain

$$N_{dB} = 20 \log\left(\frac{v}{v_r}\right) \tag{2.57}$$

When describing the dynamic behavior of a measurement system in N_{dB} it is essential to specify the reference quantity p_r or v_r.

The magnitude and phase of $H(\omega)$ are represented graphically on a Bode diagram, where $|H(\omega)|$ and ϕ are shown individually as functions of ωRC. The magnitude $|H(\omega)|$ is represented in terms of decibels on a Bode diagram and the ωRC parameter is shown on a $\log_{10}$ scale. To illustrate the construction of a Bode diagram, rewrite Eq. 2.54 in terms of decibels by using Eq. 2.57 to obtain

$$N_{dB} = 20 \log\left[\left(1 + (\omega RC)^2\right)^{-1/2}\right]$$
$$= -10 \log\left[1 + (\omega RC)^2\right] \tag{2.58}$$

The phase ϕ is given directly by Eq. 2.55.

The Bode diagrams corresponding to Eqs. 2.58 and 2.55 are shown in Fig. 2.16. The magnitude of $H(\omega)$ is down 3 dB when $\omega RC = 1$, and when $\omega RC \gg 1$, the magnitude decays linearly at 20 dB per decade. This example illustrates that the Bode diagrams provide a visual representation of a wide dynamic range of circuit or instrument characteristics. They are thus useful in determining the behavior of instruments in application to dynamic measurements.

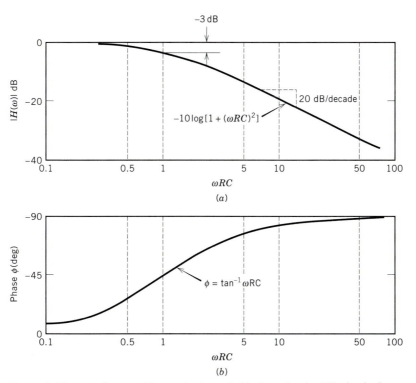

Figure 2.16 Bode diagram (*a*) magnitude, and (*b*) phase for the *RC* circuit shown in Figure 2.13.

2.9   SUMMARY

In the design and application of an instrumentation system, a number of analog and digital circuits are used. An understanding of the basic laws that govern the behavior of these circuits is required to make effective use of the systems. In this chapter, brief definitions are provided for all of the electrical quantities encountered in the remaining chapters of the book together with a listing of their standard symbols, units, and abbreviations. Fundamental relationships among these electrical quantities are summarized in Table 2.4.

Three basic components (resistors, capacitors, and inductors) affect the behavior of current flow and voltage in all electrical circuits. Ohm's law defines the relation between voltage drop across a resistor and current flow. A capacitor stores electric charge. When a constant voltage is applied to a capacitor, the charge is maintained and no current flows. When the voltage changes with time, current flows. An inductor exhibits only a very small resistance to a steady flow of current. When the current varies with time, however, a significant voltage drop develops across an inductor.

Diodes, transistors, and gates are advanced components that are employed with the basic components in most analog and digital circuits. A diode acts as a selective switch. It offers no resistance to current flow if a positive voltage is applied, and infinite resistance to current flow if the voltage is reversed. Transistors are used as amplifiers or as high-speed switches. Signal gains for a single transistor can range from 10 to 100. Gates are used in logic networks to control the flow of information. Circuits designed to perform specialized operations may contain large numbers of AND, OR, and NOT basic gates on a single silicon chip.

Kirchhoff's two laws, the voltage law and the current law, provide the foundations for circuit analysis. Analysis procedures are presented for both dc and ac circuits.

Table 2.4 Summary of Basic Relations

Element	Unit	Symbol	Characteristics
Resistance (Conductance)	ohm (siemens)		$v = Ri$ ($i = Gv$)
Inductance	henry		$v = L\dfrac{di}{dt}$ $i = \dfrac{1}{L}\displaystyle\int_0^t v\,dt + I_0$
Capacitance	farad		$i = C\dfrac{dv}{dt}$ $v = \dfrac{1}{C}\displaystyle\int_0^t i\,dt + V_0$
Short circuit			$v = 0$ for any i
Open circuit			$i = 0$ for any v
Voltage source	volt		$v = v_s$ for any i
Current source	ampere		$i = i_s$ for any v

In dc circuits, where the current flow is constant, the resistor is the only component producing a voltage drop. In ac circuits, the current varies with time; therefore, voltage drops develop across resistors, capacitors, and inductors. Impedance in an ac circuit was defined and the magnitude and phase of the voltage drops with respect to the current was determined.

Signal analysis is an important part of any experimental investigation. Periodic signals are repetitive; transient signals are one-shot events that do not repeat. Sinusoidal functions, which are a special type of periodic function, are very important in describing the dynamic response of instrument systems where the ratio of the input voltage to the output voltage is a function of frequency. Also, the sinusoid is used extensively in the Fourier analysis of other periodic signals with a more complex wave form.

A second method of analysis of signals uses phasors in the complex plane. The advantages of using a phasor representation of sinusoidal signals include ease of differentiating and integrating and the presence of both magnitude and phase information. Differentiation is accomplished simply by multiplying by $j\omega$; integration is accomplished by dividing by $j\omega$.

The frequency response function for a circuit or for an instrument gives the amplitude ratio of output voltage to input voltage and the phase shift ϕ of the output voltage relative to the input voltage. The frequency response function and other parameters are often expressed as a relative number in terms of decibels N_{dB}. Bode diagrams provide a visual representation of a wide dynamic range of circuit or instrument characteristics. Thus, they are useful in determining the behavior of instruments in application to dynamic measurements.

REFERENCES

1. Burns, S. G., and P. R. Bond: *Principles of Electronic Circuits,* West Publishing Co., St. Paul, Minn., 1987.
2. Greenbaum, J.: *Analysis and Design of Electronic Circuits using PCs,* Van Nostrand–Reinhold, New York, 1988.
3. Grob, B.: *Direct and Alternating Current Circuits,* McGraw–Hill, New York, 1985.
4. Johnson, D. E., J. L. Hilburn, and J. R. Johnson: *Basic Electric Circuit Analysis,* 4th ed., Prentice–Hall, Englewood Cliffs, N.J., 1989.
5. Leach, D. P.: *Basic Electric Circuits,* 3rd ed., Wiley, New York, 1984.
6. Nilsson, J. W.: *Electric Circuits,* 3rd ed., Addison–Wesley, Reading, Mass., 1990.

EXERCISES

2.1 List the symbol, units, and abbreviations for

(a) Force, charge, and electric field strength
(b) Energy, current, and magnetic flux density
(c) Power, voltage, and magnetic flux

2.2 For a resistance $R = R_o$ placed across a voltage supply v, determine the power dissipated as v is increased from 0 to 1000 V. Prepare a graph showing these results if

(a) $R_o = 50,000 \ \Omega$
(b) $R_o = 100,000 \ \Omega$
(c) $R_o = 250,000 \ \Omega$
(d) $R_o = 500,000 \ \Omega$

2.3   A voltage v_o is switched across a capacitor C at time $t = 0$. Derive an equation that describes the current flow with time during charging of the capacitor if the voltage source has an internal resistance R.

2.4   Using the results of Exercise 2.3, prepare a graph showing $i(t)$ if

(a) $v_o = 10$ V   and $C = 10\ \mu F$
(b) $v_o = 5$ V   and $C = 200$ pF
(c) $v_o = 100$ V   and $C = 1$ F
(d) $v_o = 3$ V   and $C = 1500$ pF

and the source resistance is $10\ \Omega$.

2.5   Determine the charge and energy stored by the capacitor in each case listed in Exercise 2.4.

2.6   An ac voltage described by the expression $v = a \sin 2\pi f t$ is switched across an inductor L at $t = 0$. Derive an expression for the current flow with time through the inductor.

2.7   Using the results of Exercise 2.6,

(a) prepare a graph showing $v(t)$ on the abscissa and $i(t)$ on the ordinate. Let

$$(1)\ f = 60\ \text{Hz}, L = 10\ \text{mH}, a = 10\ \text{V}$$
$$(2)\ f = 1\ \text{MHz}, L = 10\ \mu\text{H}, a = 5\ \text{V}$$

(b) What is the shape of the curve you have plotted?
(c) Is the shape stationary with respect to time?

2.8   Determine the energy stored in the inductor for the two cases given in Exercise 2.7.

2.9   In your own words describe Kirchhoff's first law. Indicate why it is an important principle in circuit analysis.

2.10   Repeat Exercise 2.9 for Kirchhoff's second law.

2.11   Sketch the symbol for a diode. Explain what is implied by positive bias voltage and negative bias voltage.

2.12   A voltage $v_1 = 10 \sin(120\pi t)$ is applied to a diode as shown in Fig. E2.12. Prepare a graph showing $v_2(t)$.

2.13   What are the two primary applications for transistors?

2.14   Using the characteristic curves for a transducer given in Fig. 2.5b, determine the maximum and minimum values of the collector current i_c if $i_i = (0.2 - 0.1 \sin \omega t)(10^{-3})$. Note: $V_{cc} = 10$ V and $R_L = 400\ \Omega$.

2.15   Sketch the circuit for an electronic switch that employs a bipolar transistor. Use your own words to describe its operation.

2.16   For a transistor employed as a switch, as illustrated in Fig. 2.6, determine the current flow through the switch when it is open and when it is closed. Compare these values to those obtained using a mechanical switch. Are these differences important? If so, when?

2.17   Sketch the symbol and write a truth table for the AND gate and the OR gate.

2.18   Sketch the circuit using transistors as switches for the following basic gates:

(a) AND
(b) OR
(c) NOT

2.19   With reference to Fig. 2.8, verify the relations for R_e given in Eqs. 2.21 and 2.22.

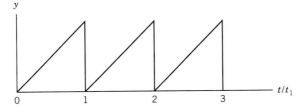

Figure E2.12

2.20 For a displacement given by $x = A_0 \cos \omega t$, show that the velocity $\dot{x}$ and the acceleration $\ddot{x}$ exhibit a phase difference with respect to the displacement of $\pi/2$ and π, respectively.

2.21 For the sawtooth function shown in Fig. E2.21, write $y(t)$ in terms of a Fourier series expansion.

2.22 A quantity $Q = A_0 e^{j(\omega t - \phi)}$. (a) Find $\dot{Q}$ and $\ddot{Q}$. (b) What is the effect of the phase angle ϕ on these unknowns.

2.23 A phasor has a real part $Re = R$ and an imaginary part $Im = \omega L - 1/\omega C$. Find the amplitude of the phasor and its phase angle.

2.24 For the circuit shown in Fig. 2.13, (a) verify Eq. 2.45. (b) Prepare a graph showing the amplitude of v_o/v_i as a function of ωRC. (c) What happens to the impedance Z_c as ω becomes very large?

2.25 For the circuit shown in Fig. E2.25, derive an expression for $v_o(t)$. Define the amplitude and the phase angle in this expression.

2.26 For the circuit shown in Fig. E2.26 determine the impedance and the phase angle between the current and the voltage.

2.27 For the circuit shown in Fig. 2.14, prepare a graph of $Z(\omega)$ if L, C, and R have the following values

	$L(\mu H)$	$C(\mu F)$	$R(k\Omega)$
(a)	0.01	0.50	10,000
(b)	0.05	0.20	1000
(c)	0.10	0.10	500
(d)	0.20	0.20	200
(e)	0.50	0.50	100

2.28 Determine the magnitude $|H(\omega)|$ and the phase angle ϕ of the frequency response function for the circuit shown in Fig. 2.13 with the values of R and C listed in Exercise 2.27.

2.29 Determine the decibel equivalents for the following ratios of p/p_r and v/v_r.

	p/p_r	v/v_r
(a)	1000	15
(b)	2.0	0.001
(c)	0.003	3000

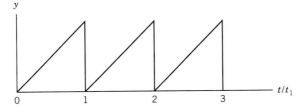

Figure E2.21

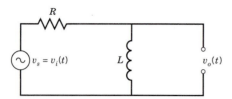

Figure E2.25

2.30   Construct a Bode diagram, in terms of decibels, for Eq. 2.54. Describe the results shown in this diagram.

2.31   Construct a phase diagram for the *RC* circuit shown in Fig. 2.13.

2.32   Construct a Bode diagram, in terms of decibels, for the *RL* circuit of Exercise 2.25.

2.33   Construct a phase diagram for the *RL* circuit of Exercise 2.25.

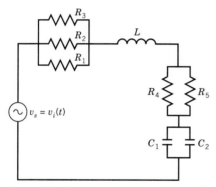

Figure E2.26

Chapter **3**

Analog Recording Instruments

3.1 INTRODUCTION

Recording instruments are used in an electronic measurement system to display, in a form that may be read easily by an operator, an output that is proportional to the quantity Q_i being measured. If the quantity Q_i is constant with respect to the time (static), voltmeters or ammeters are usually employed. Self-balancing potentiometers or digital voltmeters may be used if the voltage varies slowly with time (quasi-static). Quantities that vary rapidly with time (transient signals) must be displayed with oscillographs, oscilloscopes, or magnetic tape or disk recorders that can respond to the dynamic input.

In the design of an instrumentation system, it is important that the correct recorder be selected so that the output displayed is accurate, easily interpreted, and rapidly processed. The general characteristics of recorders must be understood before the best recording instrument can be specified for a specific instrumentation system. The characteristics of analog recording instruments will be treated in this chapter; those of digital recording instruments will be covered in Chapter 4.

3.2 GENERAL CHARACTERISTICS OF RECORDING INSTRUMENTS

The general characteristics that describe the behavior of a recording instrument are input impedance, sensitivity, range, zero drift, and frequency response. Each of these characteristics is described in the following subsections.

A. *Input Impedance*

Input impedance Z controls the energy removed from the system by the recording instrument in order to display the input voltage. Consider a simple dc voltmeter used to measure the voltage v of a source. The power loss p through the meter is given by

$$p = \frac{v^2}{Z_m} \tag{3.1}$$

where Z_m is the input impedance of the meter. In most instances, the input impedance can be modeled by a resistance and a capacitance in parallel so that

$$Z_m = \frac{Z_C Z_R}{Z_C + Z_R} = \frac{R_m}{1 + j\omega R_m C} \tag{3.2a}$$

From Eq. 3.2a and Eq. 2.38 the magnitude of Z_m is

$$|Z_m| = \frac{R_m}{\sqrt{1 + (\omega R_m C)^2}} \tag{3.2b}$$

For dc and quasi-static measurements of voltage $\omega \to 0$ and the input impedance $Z_m \to R_m$ where R_m is the resistance of the meter. It is evident from Eq. 3.1 that an ideal voltmeter should have an input impedance $Z_m = R_m \to \infty$ to reduce the power loss and resulting measurement error to zero. In most meters $R_m = 10^6$ to $10^8\ \Omega$; therefore, power losses are small and measurement errors are usually negligible. In a few meters, the input impedance is low, with $R_m < 10^4\ \Omega$, and the presence of the meter significantly affects the accuracy of the measurements.

To determine the error produced by a finite input impedance, it is necessary to consider the interaction between two adjacent elements in the instrumentation system. Consider, for example, the Wheatstone bridge–voltmeter combination shown in Fig. 3.1a. The Wheatstone bridge converts the resistance change ΔR_1 to a voltage v_i with a source resistance R_s. By applying Thevenin's theorem to the bridge and the voltmeter, the equivalent circuit shown in Fig. 3.1b is obtained. The Wheatstone bridge is replaced by a voltage generator with a potential v_i and a source resistance $R_s = R_1$. A current i flows in the loop and the voltmeter resistance R_m acts as a load on the source in series with R_s. The voltage displayed by the recorder v_m is the $i R_m$ drop across the resistor R_m. Thus, from Eq. 2.11

$$v_m = i R_m \tag{a}$$

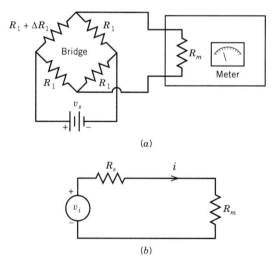

(a)

(b)

Figure 3.1 (a) Combination of a Wheatstone bridge and voltage recorder. (b) Equivalent circuit by Thevenin's theorem.

From Eq. 2.21, it is evident that

$$i = \frac{v_i}{R_s + R_m}$$ (b)

Substituting Eq. b into Eq. a gives

$$v_m = \frac{v_i}{1 + (R_s/R_m)}$$ (3.3)

Inspection of Eq. 3.3 shows that the meter indication v_m will be less than the source potential v_i. The error $\mathscr{E}$ is

$$\mathscr{E} = \frac{v_i - v_m}{v_i} = \frac{R_s/R_m}{1 + (R_s/R_m)}$$ (3.4)

The load error as a function of the ratio of source impedance to meter impedance is shown in Fig. 3.2. The results indicate that $R_s/R_m < 0.01$ gives a load error of less than 1 percent. The rule that the input impedance should be 100 times the source impedance is based on Eq. 3.4 and limits load error to less than 1 percent.

B. Sensitivity

The sensitivity S of a voltage recording instrument is given by Eq. 1.6 as

$$S = \frac{d}{v_i}$$ (3.5)

where

    d is the displacement of the pointer or pen for an analog instrument
    v_i is the voltage being measured

Sensitivity of the recorder is important when measurements of small voltages are to be made. High sensitivity is required to give a sufficiently large pen displacement d for accurate readout.

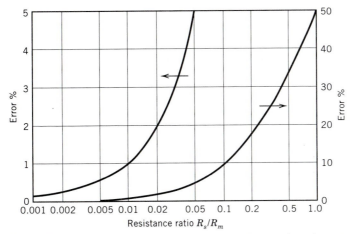

Figure 3.2 Load error as a function of the ratio of source impedance to meter impedance.

From Eq. 3.5 it is clear that the voltage v_i is determined by measuring d and dividing by S. Since division is more difficult than multiplication, most manufacturers of recording instruments define a voltage sensitivity S_R as

$$v_i = d \, S_R \tag{3.6}$$

where $S_R = 1/S$ is expressed in terms of volts per division of displacement.

High sensitivity S or low reciprocal sensitivity S_R is usually achieved with amplifiers that are incorporated into the recorder. The gain of the amplifier is varied to provide a recorder with several different sensitivities to accommodate a wide range of input voltages.

C. Range

The range, which represents the maximum voltage that can be recorded, is determined from Eq. 3.5 as

$$v^* = \frac{d^*}{S} = d^* S_R \tag{3.7}$$

where

v^* is the maximum voltage or range
d^* is the width of the chart (fixed for a given instrument)

Eq. 3.7 illustrates the trade-off that must be made between range and sensitivity. When the sensitivity S is high, the range v^* will be low; conversely, if the range is high, the sensitivity will be low. A voltage amplifier with a variable gain extends the applicability of a recorder by matching appropriate sensitivity with the input voltage, thereby extending the useful range of the instrument.

D. Zero Drift

Most recorders have provisions for adjusting the zero offset so that the pen (pointer) displacement is zero when the input voltage is zero. However, the position of the zero on the chart may change with time, owing to instabilities in one or more of the circuits in the recorder. Zero drift usually results from circuit instabilities in the amplifier that occur with temperature fluctuations, variations in line voltage, and time.

Zero drift is specified for most recording instruments and can be minimized by using a regulated line voltage, by turning instruments on for a suitable time period before recording (warm-up), and by controlling the temperature of the room in which the instrument is housed. If measurements are to be made over a long period of time, provisions should be made to periodically check and adjust the zero position, to account for the drift.

E. Frequency Response

If the voltage being recorded is dynamic, the recorder should reproduce the transient input without amplitude or phase distortion. The ability of a recorder to respond to transient signals is determined by its frequency response, which is based on the recorder's steady-state response to a sinusoidal input

$$v_i = A_i e^{j\omega t} \tag{3.8}$$

The output v_o of the recorder is of the form

$$v_o = A_o e^{j(\omega t + \phi)} \tag{3.9}$$

The frequency response function $H(\omega)$ for the recorder is obtained from Eqs. 2.54 and 2.55, which show that the amplitude ratio A_o/A_i and the phase angle ϕ both change as a function of the frequency ω of the input signal. Curves such as those shown in Fig. 3.3 for A_o/A_i and ϕ as a function of ω define the frequency response of a recording system. A more complete discussion of the frequency response of voltage recorders, which are modeled as second-order systems, is given in Section 3.5.

Specifications for recorders frequently give the amplitude ratio A_o/A_i in terms of decibels. It should be noted from Eq. 2.57 that

$$N_{dB} = 20 \log_{10}(A_o/A_i) \tag{3.10}$$

where

N_{dB} is the number of decibels
A_o and A_i are voltages

Results from Eq. 3.10, shown in Table 3.1, indicate that significant errors result in recording dynamic signals even for relatively small values of N_{dB}. For instance, a recorder specification that indicates that the frequency response is within ± 3 dB from 0 to 100 Hz implies an error of $+41$ percent (1.413) for $+3$ dB and -29 percent

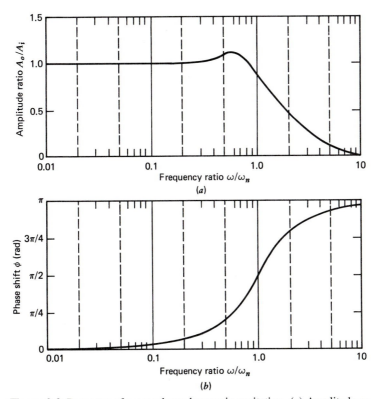

Figure 3.3 Response of a recorder to harmonic excitation. (*a*) Amplitude as a function of frequency. (*b*) Phase shift as a function of frequency.

Table 3.1 Conversion of Voltage Ratio
A_o/A_i to N_{dB}

A_o/A_i	N_{dB}	A_o/A_i	N_{dB}
1	0	1	0
1.01	0.086	0.99	−0.087
1.02	0.172	0.98	−0.175
1.05	0.424	0.95	−0.446
1.10	0.827	0.90	−0.915
1.20	1.583	0.80	−1.938
1.50	3.522	0.707	−3.012
2.00	6.020	0.500	−6.021

(0.708) for −3 dB over the range of frequencies specified. Limits of ±0.4 dB should be maintained to reduce recorder error to less than ±5 percent.

3.3 VOLTMETERS FOR STEADY-STATE MEASUREMENTS

There are three types of analog voltmeters in general use for the measurement of static or dc voltages: the analog voltmeter, the amplified analog voltmeter, and the potentiometer. The first two of these meters use the D'Arsonval galvanometer to indicate voltage.

3.3.1 D'Arsonval Galvanometer

The *D'Arsonval galvanometer,* illustrated in Fig. 3.4, is the basic device used in detecting and measuring dc current. The galvanometer design incorporates a coil of wire that is supported in a magnetic field with either jeweled bearings or torsion springs. When a current i flows in the coil, it rotates in the magnetic field until restrained by springs that are a part of the suspension system.

The torque T_1 developed by the current flow in the coil is

$$T_1 = NB\ell Di \tag{3.11}$$

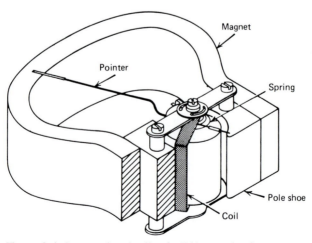

Figure 3.4 Construction details of a D'Arsonval galvanometer.

where

N is the number of turns in the coil
B is the flux density of the magnetic field
ℓ is the axial length of the field
D is the mean coil diameter

The torque T_2 developed by the restraining springs is

$$T_2 = K\theta \qquad \text{(3.12)}$$

where

θ is the angle of rotation of the coil
K is the spring constant

Equilibrium $(T_1 - T_2 = 0)$ leads to

$$\theta = \frac{NB\ell D}{K}i = Si \qquad \text{(3.13)}$$

where $S = NB\ell D/K$ is the sensitivity or calibration constant for the galvanometer.

The sensitivity of the galvanometer can be changed by varying the parameters N, B, ℓ, D, or K; however, for economic reasons, galvanometers are standardized and designed to measure small currents. A typical D'Arsonval galvanometer, with a coil resistance of 50 Ω, will exhibit a full-scale deflection at 1 mA. A wide variety of current and voltage measurements are produced by controlling the current i passing through the standardized galvanometer.

3.3.2 Ammeter

An *ammeter* consists of a D'Arsonval galvanometer with a shunt resistance, as shown in Fig. 3.5. The input current i_i divides, with i_m passing through the meter and i_{sh} passing through the shunt. It is clear from Eq. 2.11 and Eq. 2.18 that

$$i_m = \frac{i_i}{1 + (R_m/R_{sh})} \qquad \text{(3.14)}$$

A D'Arsonval galvanometer with a full-scale current capability of 1 mA and a resistance R_m of 50 Ω can be used to measure any current greater than 1 mA by properly selecting the shunt resistance from the expression

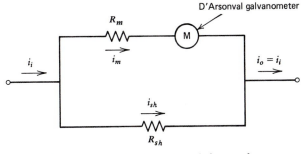

Figure 3.5 A D'Arsonval galvanometer being used as an ammeter.

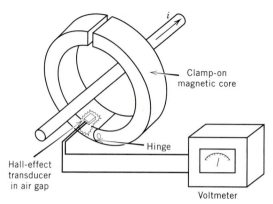

Figure 3.6 Schematic illustration of a clamp-on probe for an analog ammeter.

$$R_{sh} = \frac{i_m^*}{i_i^* - i_m^*} R_m \tag{3.15}$$

where i_i^* and i_m^* are full-scale input and meter currents, respectively. For example, the 1-mA–50-Ω galvanometer can be used to measure 5 A (full scale) if $R_{sh} = 0.01 \, \Omega$.

A significant disadvantage of using an ammeter of this type for measuring current i in existing circuits is the need to break the circuit to insert the ammeter in the path of current flow. In instances where the circuit exists, it is preferable to use an alternative method for measuring i that utilizes the magnetic field around the wire, which is proportional to the current. The clamp-on probe, shown clipped over the wire in Fig. 3.6, contains a magnetic core. A Hall-effect transducer mounted in an air gap in the magnetic core produces an output voltage proportional to the magnetic field and the current i.

3.3.3 DC Voltmeters

A D'Arsonval galvanometer is converted to a dc voltmeter by using a series resistor, as shown in Fig. 3.7. Application of Eq. 2.21 to the circuit shown yields

$$R_{sr} = \frac{v^*}{i_m^*} - R_m \tag{3.16}$$

where

v^* is the full-scale voltage
i_m^* is the full-scale current for the meter

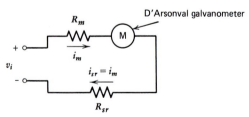

Figure 3.7 A D'Arsonval galvanometer being used as a voltmeter.

For example, a 20-μA–30-Ω galvanometer is converted to a 100-mV voltmeter with a series resistor $R_{sr} = 4970 \ \Omega$.

A dc analog multimeter incorporates several different series resistors that are switched to a single meter to give a multirange instrument. Typical *multimeters* have approximately nine ranges with full-scale readings from 100 mV to 1000 V. Analog multimeters are low-cost instruments with accuracies of ± 2 to 3 percent of full scale. When operated on the lower voltage ranges, the input impedance of the instrument is relatively low and loading errors can occur because of the voltmeter. The impedance is often given in terms of ohms per volt full scale with 20,000 Ω/V being common.

3.3.4 Voltmeter Loading Errors

Whenever a voltmeter draws current from the voltage source during a measurement, an error will result owing to voltmeter load. To illustrate voltmeter load, note that the total meter resistance is given by $(R_m + R_{sr})$. Substituting this resistance into Eq. 3.4 gives the error $\mathscr{E}$ resulting from voltmeter load as

$$\mathscr{E} = \frac{R_s/(R_m + R_{sr})}{1 + R_s/(R_m + R_{sr})} \tag{3.17}$$

where R_s is the output impedance of the voltage source. Reference to Fig. 3.2 shows that a ratio $(R_m + R_{sr})/R_s > 100$ is required to reduce voltmeter loading errors to less than 1 percent. Since R_{sr} decreases as the full-scale range is reduced, the input impedance is low when the multimeter is used on sensitive scales and the voltmeter readings must be corrected for loading errors.

3.3.5 Amplified Voltmeters

Difficulties in measuring very small voltages while maintaining a high input impedance can be resolved by using a high-gain amplifier in conjunction with the D'Arsonval galvanometer. A schematic of this circuit is shown in Fig. 3.8, where an amplifier is used between the voltage source and the meter. The voltage output from the amplifier is

$$v_o = G v_i \tag{3.18}$$

where G is the gain of the amplifier and v_i is the input signal to the amplifier.

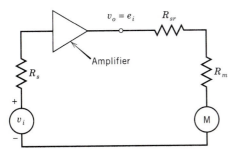

Figure 3.8 Equivalent circuit for a voltage source and an amplified voltmeter.

As the output voltage from the amplifier is imposed on the meter,

$$v_o = e_i = Gv_i$$

An analysis of the circuit in Fig. 3.8 gives

$$R_{sr} = \frac{Gv_i}{i_m} - R_m \tag{3.19}$$

Gains of $\cong 10^3$ are common, and it is possible to increase R_{sr} so that $R_{sr}/R_s \gg 100$ and loading error is essentially eliminated for amplified voltmeters.

The amplifier also permits the meter to be used to measure very small voltages, since

$$v_i = \frac{v_m^*}{G} \tag{3.20}$$

Thus, the meter sensitivity can be increased by a factor G while maintaining the input impedance of the meter.

3.3.6 Potentiometric Voltmeters

Potentiometers are null-balance instruments in which an unknown voltage v_x is compared with a precision reference voltage v_r. The basic potentiometer, shown in Fig. 3.9, contains a reference source that energizes a slide-wire resistor of length ℓ. As the wiper is moved along the slide-wire, an adjustable voltage v_{sw} is obtained, which is given by

$$v_{sw} = \frac{x}{\ell}v_r$$

where x is the wiper position along the slide-wire resistor. If $v_{sw} \neq v_x$, the galvanometer deflects to indicate a voltage difference. The slide-wire wiper is then adjusted until the galvanometer indicates the null position and

$$v_x = v_{sw} = \frac{x}{\ell}v_r \tag{3.21}$$

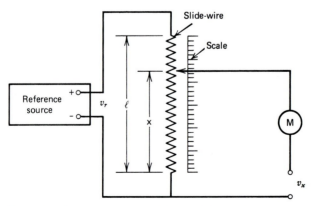

Figure 3.9 Potentiometer circuit for measuring voltage.

With the potentiometer balanced, no current flows; therefore, there is no voltage drop across the instrument. The slide-wires used in these instruments exhibit a uniform resistance (linear variation with wiper position); therefore, the scale reading x provides an accurate means of reading the voltage without a meter movement and an associated load error. Since the galvanometer is used only to indicate null or zero voltage, it can be very sensitive yet inexpensive. Extremely accurate readings of voltage can be made with potentiometric voltmeters, although this method of voltage measurement is very time consuming.

3.4   VOLTMETERS FOR SLOWLY VARYING SIGNALS

The basic approach for recording *quasi-static voltages,* which change relatively slowly with respect to time, utilizes the type of potentiometric circuit shown in Fig. 3.9. The potentiometric circuit is balanced automatically with servomotors that are driven with an amplified error signal. Although these servomotors are relatively slow, this method of voltage measurement remains important in two commercial instruments, strip-chart recorders and x–y recorders.

3.4.1   Strip-Chart Recorders

A *strip-chart recorder* utilizes a servomotor-driven null-balance potentiometric circuit similar to the one illustrated in Fig. 3.10. The input signal from the transducer is amplified and used as a command signal for a servo amplifier. The signal from the servo amplifier drives a servomotor that positions a wiper along a slide-wire resistor. Since the slide-wire resistor is across a reference voltage, the wiper picks up a feedback voltage from the slide-wire that is proportional to the position of the wiper along the wire. The servo amplifier receives the feedback voltage and compares it with the command signal. The output from the servo amplifier is proportional to the difference between the two signals. When the servomotor has adjusted the wiper so that the difference, or error signal, from the servo amplifier is zero, the system is in balance and the wiper position provides an indication of the input voltage.

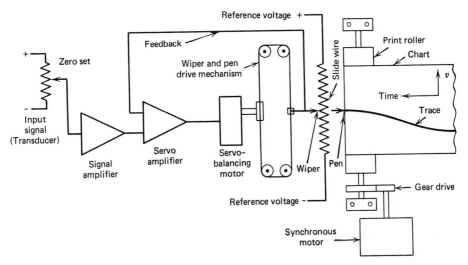

Figure 3.10 Components of a strip-chart recorder.

A permanent record of the input voltage is obtained by connecting a pen to the wiper. As the wiper is positioned along the slide-wire resistor, the voltage is recorded as an ink trace on a roll of chart paper. The chart paper is driven at a constant velocity in a direction perpendicular to the wiper motion by a print roller, which is driven by a clock motor and a suitable gear train. Distance along the length of the chart is then proportional to time.

Typical commercial strip-chart recorders employ charts with widths from 5 to 10 in. (120 to 250 mm). The *response time,* the time required for the servomotor to move the pen across the width of the chart, is typically 0.5 s. The sensitivity of strip-chart recorders can be varied by attenuating the output from the signal amplifier. Typical sensitivities range from 5 mV to 100 V for the full width of the chart paper. Chart speeds can be varied by changing the speed of the print roller. Speeds from 1 in./h to 8 in./min (25 mm/hr to 250 mm/min) are common. The input impedance will depend on the details of the design and may be potentiometric (i.e., no current flow at null balance) or about 1 MΩ, which corresponds to the input impedance of the signal amplifier. Accuracy of ± 0.2 percent is typical.

3.4.2 X–Y Recorders

The *x–y recorder* is a second type of instrument that utilizes servomotor-driven and null-balance potentiometric circuits. The operation of the *x–y* recorder is similar to that of the strip-chart recorder, except that the *x–y* recorder simultaneously records two voltages along orthogonal axes (usually referred to as the *x* and *y* axes). The *x–y* recorder uses sheets of graph paper (either $8\frac{1}{2}$ by 11 in. or 11 by 17 in.) instead of a strip chart for recording purposes.

The sensitivities of an *x–y* recorder depend on the design of the input amplifiers. Some models have plug-in modules that permit the characteristics of the recorder to be changed. A typical recorder has an attenuated input amplifier with about 10 different sensitivities, ranging from 0.5 mV/in. to 10 V/in. in conventional models or 0.025 mV/mm to 0.5 V/mm in models with metric calibration. The input impedance of both amplifiers is usually about 1 MΩ. The *deadband,* that small zone about the balance point where friction inhibits exact zeroing of the error signal, is about 0.1 percent of full scale. Accuracy, which includes deadband error, is typically 0.2 percent.

The dynamic recording characteristics of an *x–y* recorder depend on the amplitudes of the quantities being measured, the acceleration capabilities of the servo drives, and the slewing speed. The *slewing speed* is the maximum velocity of the pen when it is being driven by both servo drives. Slewing speeds of 20 in./s (500 mm/s), with peak accelerations of 1000 in./s^2 (25 m/s^2) of an individual servo drive, are common. Typical dynamic performance is represented in Fig. 3.11, where the linear region of operation is defined in terms of the amplitude and frequency of the input signals. Operation in the nonlinear region produces measurement error and should be avoided.

Many *x–y* recorders are equipped with a time base so that the recorder can also be used to record the variable *y* as a function of time *t* rather than of variable *x*. In this mode of operation, the *x–y* recorder (acting as a *y–t* recorder) is similar to a slow oscillograph. The time base provides the input signal for the *x*-axis servo system. Sweep speeds of 0.5 to 100 s/in. (0.025 to 5 s/mm) are available in eight calibrated ranges in a typical model.

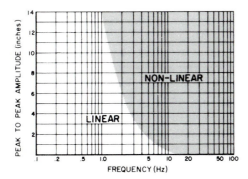

Figure 3.11 Dynamic frequency response of a typical x–y recorder.

3.5   VOLTMETERS FOR RAPIDLY VARYING SIGNALS

Measuring transient phenomena, where the signal from the transducer is a rapidly changing function of time, is the most difficult and most expensive measurement in experimental work. Frequency response is the dominant characteristic required of the recording instrument in dynamic measurements, and usually accuracy and economy are sacrificed in order to improve the response capabilities. Three markedly different instruments are used to record transient voltages: the oscillograph, the oscilloscope, and the magnetic tape recorder. The oscillograph utilizes a galvanometer, which incorporates a highly refined D'Arsonval movement, to drive a pen or a light beam over a moving strip of chart paper. The oscilloscope utilizes a focused beam of electrons to produce a voltage–time trace on a phosphor screen. The magnetic tape recorder stores the dynamic signal on magnetic tape for later playback and display on either an oscillograph or an oscilloscope.

3.5.1   Oscillograph Recorders

Oscillograph recorders employ galvanometers to convert the dynamic input signal to a displacement on a moving strip of chart paper. There are two types of oscillographs: the *direct-writing type,* where the galvanometer drives the pen or hot stylus used to write on a moving strip of chart paper, and the *light-writing type,* where the galvanometer drives a mirror that deflects the light beam so that it writes on a moving strip of photosensitive paper. The frequency responses of the two different types of oscillographs differ markedly. The relatively high inertia associated with the pen or hot stylus of the direct-writing type limits the frequency response to about 150 Hz. The inertia of the rotating mirror is much lower in a light-writing type, and a frequency response as high as 25 kHz is possible, with a writing speed of 1270 m/s.

Although both types of recorders can be used to record low-frequency signals, the direct-writing oscillograph is usually preferred because records made with pen or hot stylus on chart paper are less expensive, more permanent, and of higher quality than comparable recordings made on photosensitive paper.

The light-writing oscillograph is often used as the recording instrument for dynamic signals that contain frequency components between 0 and 25 kHz. A schematic diagram illustrating the operating principle of the light-writing oscillograph is shown in Fig. 3.12. A typical oscillograph utilizes several galvanometers, which are mounted in a row of holes in magnetic blocks that serve as the pole pieces. A mirror, mounted

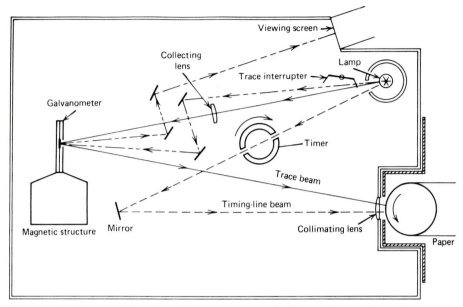

Figure 3.12 Schematic illustration of the components of a light-writing oscillograph.

on the filament suspension member of the galvanometer, reflects a focused beam of light onto a moving strip of photosensitive paper. Rotation of the mirror produces a deflection of the light beam, which is amplified optically to provide a trace on the photosensitive paper. The deflection of the trace from a null position is proportional to the dynamic input voltage. The speed of the strip of photosensitive paper is controlled by a motor and gear train. The paper speed can be adjusted to give a specified time scale on the abscissa of the record (i.e., along the length of the strip of paper).

The galvanometer for an oscillograph is a highly refined version of the D'Arsonval movement described in Section 3.3. The essential components of this galvanometer (see Fig. 3.13) include a filament suspension system, a rotating coil, a mirror, and a pair of stationary pole pieces. The design features of the components are varied to change the dynamic characteristics (sensitivity and frequency response) of different galvanometers. The characteristics of several commercially available galvanometers are presented in Table 3.2.

3.5.2 Transient Response of Galvanometers[1]

Since the galvanometer is a highly refined version of a D'Arsonval movement, its transient response can be studied by considering the equation of motion for the coil. Thus

$$T_1 - T_2 - T_3 = J\,\ddot{\theta} \qquad (3.22)$$

---

[1]The dynamic behavior of a galvanometer is modeled by a second-order mechanical system. Since the second-order system is an extremely important concept that applies to several topics covered in subsequent sections of this book, it should be studied in detail. The dynamic response of a galvanometer to a step-input voltage and to a sinusoidal signal will be covered here and will be used again in the discussions of dynamic response of transducers in Chapter 9.

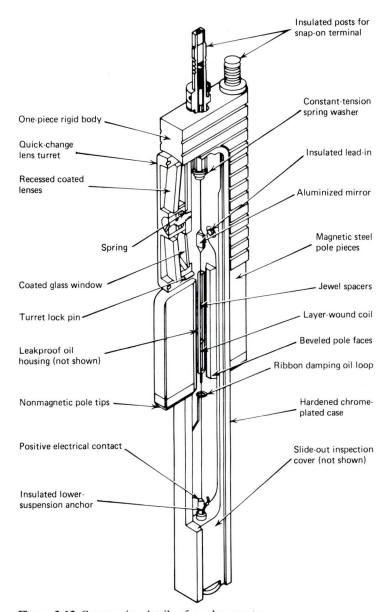

Figure 3.13 Construction details of a galvanometer.

where

T_1 is the applied torque (Eq. 3.11)
T_2 is the opposing torque resulting from the restraining spring (Eq. 3.12)
T_3 is the opposing torque caused by damping, given by

$$T_3 = D_1 \dot{\theta} \tag{3.23}$$

D_1 is the damping coefficient
J is the moment of inertia of the rotating mass

Table 3.2 Honeywell Series "M" Miniature Galvanometer Characteristics

Galvo Type No.	Nominal Undamped Natural Frequency	Flat (±5%) Frequency Response	Required External Damping Resistance	Nominal Coil Resistance	Sensitivity in 30-cm Optical Arm of all Honeywell Oscillographs						Max. Safe Current	Maximum Deflection		Balance	
					Current (±5%)		Voltage						AC Peak to Peak		
							Galvanometer		Circuit			DC		Standard	Precision
					i_g	$\frac{1}{i_g}$	V_g	$\frac{1}{V_g}$	V_C	$\frac{1}{V_C}$					
	Hz	Hz	Ω	Ω	µA/cm	cm/µA	mV/cm	cm/mV	mV/cm	cm/mV	mA	cm	cm	mm/g	mm/g
ELECTROMAGNETICALLY DAMPED TYPES															
*M40-120A	40	0– 24	120	30.0	3.15	0.317	0.094	10.6	0.472	2.12	10	10.0	20.0	0.89	0.46
*M40-350A	40	0– 24	350	66.0	1.61	0.621	0.160	9.43	0.673	1.49	10	10.0	20.0	0.89	0.46
*M100-120A	100	0– 60	120	52.2	3.94	0.254	0.206	4.85	0.677	1.48	10	10.0	20.0	0.41	0.21
*M100-350	100	0– 60	350	75.5	2.48	0.403	0.187	5.35	1.06	0.943	10	10.0	20.0	0.56	0.28
*M200-120	200	0– 120	120	62.0	10.0	0.100	0.622	1.61	1.82	0.546	10	10.0	20.0	0.41	0.21
*M200-350	200	0– 180	350	62.0	10.0	0.100	0.622	1.61	4.13	0.242	10	10.0	20.0	0.41	0.21
*M400-120	400	0– 240	120	125.0	30.3	0.033	3.79	0.264	7.44	0.134	10	10.0	20.0	0.41	0.21
*M400-350	400	0– 360	350	125.0	30.3	0.033	3.79	0.264	14.4	0.069	10	10.0	20.0	0.41	0.21
*M600-350	600	0– 540	350	320.0	51.2	0.020	16.4	0.061	34.3	0.029	15	10.0	20.0	0.41	0.21
	Hz	Hz	Ω	Ω	mA/cm	cm/mA	V/cm	cm/V			mA	cm	cm	mm/g	
FLUID DAMPED TYPES															
*M1000	1000	0– 600	150	39.0	1.04	0.961	0.041	24.4			70	10.0	20.0	0.21	
*M1650	1650	0– 1000	100	26.8	3.66	0.273	0.098	10.2			70	10.0	20.0	0.21	
*M3300	3300	0– 2000	100	32.0	7.87	0.127	0.252	3.97			70	5.0	14.5	0.21	
*M5000	5000	0– 3000	100	39.5	12.3	0.081	0.484	2.07			70	3.05	8.9	0.21	
*M8000	8000	0– 4800	100	35.0	15.7	0.064	0.551	1.81			70	2.0	5.8	0.21	
*M10000	10000	0– 6000±12%	100	35.0	15.7	0.064	0.551	1.81			70	2.0	5.8	0.21	
*M13000	13000	0–13000±8%	100	71.6	32.1	0.031	2.30	0.435			55	1.0	2.8	0.21	
M25K	22000	0–25000±15%	100	49.0	49.2 (±15%)	0.020	2.41	0.415			75	1.0	2.8	0.21	

Substituting Eqs. 3.11, 3.21, 3.13, and 3.23 into Eq. 3.22 yields:

$$J\ddot{\theta} + D_1\dot{\theta} + K\theta = SKi \tag{3.24}$$

where i is the instantaneous current in the coil.

The dynamic response of a galvanometer can be determined by considering the transient condition associated with a step-voltage input, which is applied as indicated in Fig. 3.14. For a step-voltage input, the instantaneous current i in the coil is given by

$$i = \frac{v - e_m}{R_s + R_G} \tag{3.25}$$

where

   v is the applied signal voltage, $v = v_s$ for $t \geq 0$
   R_s is the resistance of the source
   R_G is the resistance of the galvanometer coil
   e_m is the back electromotive force induced as the coil rotates in the magnetic field.

Thus

$$e_m = NB\ell D\,\dot{\theta} = SK\,\dot{\theta} \tag{3.26}$$

Substituting Eq. 3.26 into Eq. 3.25 and solving for the instantaneous current i gives

$$i = i_s - \frac{SK}{R_s + R_G}\,\dot{\theta} \tag{3.27}$$

where i_s is the steady-state current $i_s = v_s/(R_s + R_G)$. Substituting Eq. 3.27 into Eq. 3.24 and simplifying gives

$$J\ddot{\theta} + D_2\dot{\theta} + K\theta = SKi_s \tag{3.28}$$

where the damping coefficient D_2 is a combination of electromagnetic and fluid damping, which can be expressed as

$$D_2 = D_1 + \frac{(SK)^2}{R_s + R_G} \tag{3.29}$$

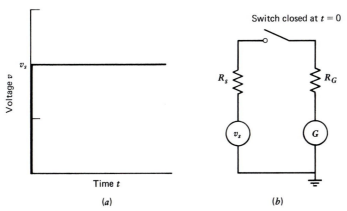

Figure 3.14 Circuit for applying a step pulse of voltage to a galvanometer. (*a*) Step-function input. (*b*) Galvanometer circuit.

It is important to note from Eq. 3.29 that the damping coefficient D_2 can be varied by changing the resistance of the source R_s; thus, the dynamic response of a galvanometer can be adjusted by changing D_2.

Equation 3.28 describes the angular movement of the pen arm or mirror of a galvanometer with respect to time. Since this is a second-order differential equation, the galvanometer is often referred to as a second-order instrument. Note that the undamped natural frequency ω_n is given by

$$\omega_n = \sqrt{\frac{K}{J}} \tag{3.30}$$

Also, the damping ratio d can be defined as

$$d = \frac{D_2}{2\sqrt{KJ}} \tag{3.31}$$

By using Eqs. 3.30 and 3.31, Eq. 3.28 can be rewritten as

$$\frac{1}{\omega_n^2}\ddot{\theta} + \frac{2d}{\omega_n}\dot{\theta} + \theta = S i_s = \theta_s \tag{3.32}$$

Substituting $\theta = e^{\lambda t}$ into Eq. 3.32 gives the auxiliary equation associated with the complementary solution of Eq. 3.32 as

$$\frac{1}{\omega_n^2}\lambda^2 + \frac{2d}{\omega_n}\lambda + 1 = 0 \tag{3.33}$$

which exhibits roots of

$$\lambda = \omega_n(-d \pm \sqrt{d^2 - 1}) \tag{3.34}$$

Inspection of Eq. 3.34 shows that three different solutions of Eq. 3.32 must be considered. These solutions are as follows:

Case 1. Overdamped when $d > 1$ (λ real roots)
Case 2. Critically damped when $d = 1$ (λ double roots)
Case 3. Underdamped when $d < 1$ (λ complex conjugate roots)

The solutions of Eq. 3.32 for these three cases with the step input and initial conditions

$$\theta(0) = \dot{\theta}(0) = 0$$

are as follows.

Case 1. Overdamped ($d > 1$).

$$\frac{\theta}{\theta_s} = 1 + \frac{d - \sqrt{d^2 - 1}}{2\sqrt{d^2 - 1}}e^{-[d+(d^2-1)^{1/2}]\omega_n t}$$

$$- \frac{d + \sqrt{d^2 - 1}}{2\sqrt{d^2 - 1}}e^{-[d-(d^2-1)^{1/2}]\omega_n t} \tag{3.35}$$

In the overdamped case, the response of the galvanometer is sluggish, as indicated by the response curve of Fig. 3.15. The time required for the galvanometer to reach the steady-state position ($\theta = \theta_s$) is so long that the overdamped condition is avoided in most applications.

Case 2. Critically damped ($d = 1$).

$$\frac{\theta}{\theta_s} = 1 - (1 + \omega_n t)e^{-\omega_n t} \tag{3.36}$$

When the galvanometer is critically damped, the response is improved with respect to the overdamped condition. The rotation θ approaches the steady-state rotation θ_s, but never exceeds this value. The response curve for a critically damped galvanometer is also shown in Fig. 3.15.

Case 3. Underdamped ($d < 1$).

$$\frac{\theta}{\theta_s} = 1 - \frac{e^{-d\omega_{n_t}}}{\sqrt{1 - d^2}} \cos(\omega_d t - \phi) \tag{3.37}$$

where

$$\phi = \tan^{-1}\left(\frac{d}{\sqrt{1 - d^2}}\right)$$

Underdamped galvanometers respond quickly to a transient signal, initially overshoot the steady-state rotation, and then oscillate with decaying amplitude about the steady-state rotation θ_s for some time. The response curve for an underdamped galvanometer is also shown in Fig. 3.15.

The amount of overshoot in the underdamped case depends on the damping ratio used with the galvanometer. The damping ratio that reduces the response time of the galvanometer to a minimum depends on the overshoot that is permitted. For example, if the accuracy required in a transient measurement is ± 5 percent, a damping ratio is selected such that the overshoot gives $(\theta/\theta_s)_{\max} = 1.05$. This example, illus-

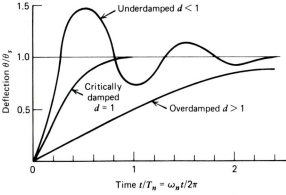

Figure 3.15 Response curves for underdamped, critically damped, and overdamped galvanometers for a step-pulse input.

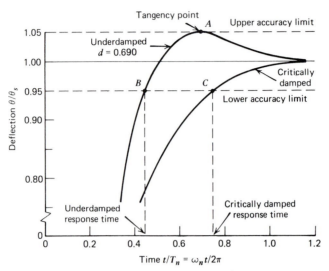

Figure 3.16 Accuracy limits superimposed on the response curves of a galvanometer.

trated in Fig. 3.16, shows that the response curve for the underdamped galvanometer is tangent to the upper accuracy limit at point A. The response time is defined as the time when the response curve first intersects the lower accuracy limit (see point B of Fig. 3.16). For this example, the response time t equals $0.454T_n$, where T_n is the natural period of oscillation. It should be noted that the response time for the critically damped galvanometer (point C in Fig. 3.16) is considerably longer ($t = 0.754T_n$) than that for the underdamped case.

In most applications, galvanometers are employed in an underdamped condition such that the first overshoot and the subsequent oscillations about θ_s do not exceed the limits imposed by the error bounds. The damping ratio required to bound the oscillation amplitude is given by the expression

$$d = \left[\frac{\ln^2 \mathscr{E}}{\pi^2 + \ln^2 \mathscr{E}} \right]^{1/2} \qquad (3.38)$$

where $\mathscr{E}$ is the error associated with the accuracy limits. Use of Eq. 3.38 ensures that the overshoot at the peak value of θ / θ_s will be tangent to the upper accuracy limit and that the response time is minimum for this level of accuracy.

3.5.3 Periodic Signal Response of Galvanometers

When a galvanometer is used to monitor a periodic signal, such as a sinusoidal voltage, the initial transient response discussed in the previous subsection is not an important consideration. Instead, the particular solution is of interest because it shows whether the galvanometer is precisely following the input signal. In some cases, the galvanometer will distort the amplitude of the sinusoidal signal and will introduce a nonlinear phase shift.

The response of a galvanometer to a periodic input i_i of the form

$$i_i = i_a e^{j\omega t} \qquad (3.39)$$

can be studied by considering the equation of motion for the galvanometer (Eq. 3.32), which has been modified to account for the periodic input. Thus

$$\frac{1}{\omega_n^2}\,\ddot{\theta} + \frac{2d}{\omega_n}\,\dot{\theta} + \theta = \theta_a e^{j\omega t} \tag{3.40}$$

where $\theta_a = S I_a$ is the amplitude of the sinusoidal oscillation.

The particular solution of Eq. 3.40 is required, since it describes the steady-state response of the galvanometer to a periodic input. The particular solution is obtained by substituting

$$\theta = A e^{j\omega t} \tag{3.41}$$

into Eq. 3.40 to obtain

$$\frac{\theta}{\theta_s} = \left[\frac{1 - (\omega/\omega_n)^2}{B} - j\frac{2d\,(\omega/\omega_n)}{B} \right] e^{j\omega t} \tag{3.42}$$

where

$$B = [1 - (\omega/\omega_n)^2]^2 + 4d^2(\omega/\omega_n)^2$$

The amplitude of the phasor in Eq. 3.42 is obtained from Eq. 2.38 as

$$A_0 = |H(\omega)| = \frac{1}{\sqrt{[1 - (\omega/\omega_n)^2]^2 + 4d^2(\omega/\omega_n)^2}} \tag{3.43}$$

Note that the amplitude A_0 is the same as the magnitude of the frequency response function. The phase angle is obtained from Eq. 2.39 as

$$\phi = \tan^{-1}\frac{2d\,(\omega/\omega_n)}{1 - (\omega/\omega_n)^2} \tag{3.44}$$

Finally, the frequency response function of the galvanometer is

$$\frac{\theta}{\theta_s} = H(\omega) = \frac{e^{-j\phi}}{\sqrt{[1 - (\omega/\omega_n)^2]^2 + 4d^2(\omega/\omega_n)^2}} \tag{3.45}$$

The amplitude is a function of the frequency ratio ω/ω_n and the damping ratio d. The results of Eq. 3.45 are used in Fig. 3.17 to plot the frequency response function $|H(\omega)|$ as a function of ω/ω_n for different damping ratios. The frequency response of the galvanometer depends on the damping ratios and the accuracy required. In Fig. 3.18, an error band of ± 5 percent has been superimposed on a response curve ($d = 0.59$) for a large range of frequencies. Inspection of the response curve relative to the error bands shows that $d = 0.59$ optimizes the frequency response of the galvanometer for these particular accuracy limits. As illustrated in Fig. 3.18, the response curve stays within the error band for the range of frequencies $0 < \omega/\omega_n < 0.87$ and is tangent to the upper accuracy limit at $\omega/\omega_n = 0.60$. As the allowable error band is decreased, the range of the frequency response of the galvanometer also decreases, as indicated in Table 3.3.

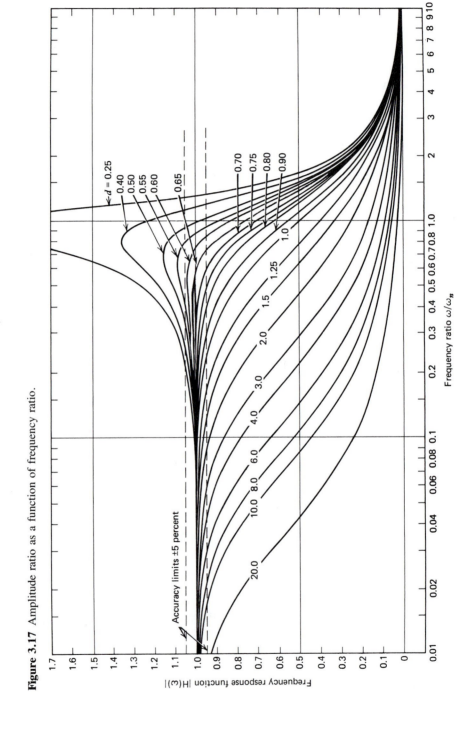

Figure 3.17 Amplitude ratio as a function of frequency ratio.

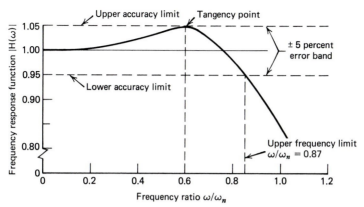

Figure 3.18 Response curve for $d = 0.59$.

Fidelity of recording output signals that exhibit a phase shift ϕ relative to the input signal depends on the type of input signal and the number of output signals being recorded. If the input is harmonic (a single frequency), it repeats with time and the phase shift is not important for a single signal. In this instance, the phase shift produces a shift of the trace along the time axis of the oscillograph record, which can be ignored.

However, if the system is used to record a transient signal or several signals, the phase angle becomes very important because it can produce signal distortion and relative time shift between signals. To illustrate the effect of phase angle, consider an input signal composed of many frequencies and given by

$$\theta_i = \sum_{k=1}^{N} A_k \sin k\omega t \tag{3.46}$$

The output signal is given by Eq. 3.45 as

$$\theta_o = \sum_{k=1}^{N} a_k \sin(k\omega t - \phi_k) \tag{3.47}$$

where

$$a_k / A_k = |H(\omega)|$$

Table 3.3 Frequency Response, Accuracy, and Optimum Damping for Galvanometers Responding to a Periodic Input

Allowable Error (%)	Optimum Damping d	Frequency Response $(\omega/\omega_n)_{\max}$
±10	0.540	1.028
± 5	0.589	0.870
± 2	0.634	0.692
± 1	0.655	0.585

It is evident that $|H(\omega)|$ must be equal to a constant for each of the $k\omega$ frequencies in Eq. 3.46 in order to avoid transient signal distortion caused by amplitude changes with frequency. To examine the influence of phase angle ϕ rewrite Eq. 3.47 as

$$\theta_o = a_1 \sin(\omega t - \phi_1) + a_2 \sin(2\omega t - \phi_2)$$
$$+ \cdots + a_N \sin(N\omega t - \phi_N) \tag{3.48}$$

If ϕ_1, ϕ_2, $\cdots$, ϕ_N vary with frequency in a nonlinear manner, phase distortion will occur because each term will be shifted in time by a different amount.

Examination of the relationship for the phase angle ϕ, Eq. 3.44 and Fig. 3.19, shows that ϕ is not usually linear with respect to ω/ω_n. Only when $d = 0.64$ is ϕ linear with respect to ω/ω_n over the range $0 < (\omega/\omega_n) < 1$. In this case, the phase angles ϕ_1, ϕ_2, and ϕ_N may be written as

$$\phi_1 = c\omega, \qquad \phi_2 = 2c\omega, \qquad \phi_N = Nc\omega \tag{a}$$

where c is the constant of proportionality associated with the linear phase shift. Substituting Eq. a into Eq. 3.48 gives

$$\theta_o = a_1 \sin\omega(t - c) + a_2 \sin 2\omega(t - c) + \cdots + a_N \sin N\omega(t - c) \tag{b}$$

It is clear from Eq. b that the time shift c in the recording of each harmonic is the same. The composite record of all of the harmonics will be shifted in time by the amount $c = \phi_1/\omega$. No signal distortion results from phase angle since the time shift for each term is identical.

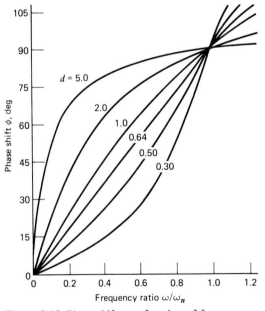

Figure 3.19 Phase shift as a function of frequency ratio.

The damping ratios normally associated with extended frequency response (see Table 3.3) also provide a linear or near-linear phase angle with respect to frequency; thus the damping ratios for extended frequency recording and for distortion-free recording are approximately the same.

In more modern oscillographs, the galvanometers and associated optical components have been replaced with a fiber-optic cathode-ray tube. The faceplate of this special-purpose tube consists of approximately 10×10^6 glass fibers that have been fused together to form a 0.2-by-8-in. (5-by-200-mm) rectangular area. The fibers transmit light from the phosphor coating on the inside surface of the cathode-ray tube to the photosensitive direct-print paper, which is driven past the faceplate of the tube. This arrangement provides excellent trace resolution (sharp traces) at fast writing speeds.

Each input channel in this type of oscillograph contains a voltage-to-time converter that produces a pulse with a duration proportional to the input voltage. These pulses activate the beam of the cathode-ray tube, which continuously sweeps at a fixed frequency of 50 kHz. A dotted-trace representation is avoided at high frequencies by using a memory circuit to produce a continuous trace.

This major advance in oscillographic recording, which replaces the galvanometers and associated optical components with a special-purpose fiber-optic cathode-ray tube, eliminates the need for impedance matching and current amplifiers. Also, most of the concerns related to sensitivity and frequency response are eliminated. The new systems can be used to record up to 18 channels of data on an 8-in.-wide (200-mm) strip of photosensitive paper at speeds up to 120 in./s (3000 mm/s). Frequency response is flat from dc to 5 kHz, and with a high-gain differential amplifier, sensitivities as low as 1 mV per division can be attained.

3.5.4 Oscilloscopes

The *cathode-ray-tube oscilloscope* is a voltage-measuring instrument capable of recording extremely high-frequency signals (exceeding 1 GHz). The cathode-ray tube (CRT), which is the most important component in an oscilloscope, is illustrated in Fig. 3.20. The CRT is an evacuated tube in which electrons are produced, controlled, and used to provide a voltage–time record of a transient signal. The electrons produced by heating a cathode are collected, accelerated, and focused onto the face of the tube. The impinging stream of electrons forms a bright point of light on a fluorescent screen at the inside face of the tube. Voltages are applied to horizontal and vertical deflection plates in the CRT (see Fig. 3.20) to deflect the stream of electrons and thus move the point of light over the face of the tube. It is this ability to deflect the stream of

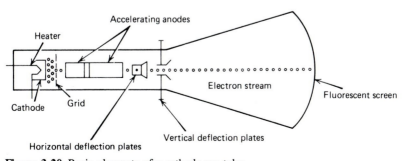

Figure 3.20 Basic elements of a cathode-ray tube.

electrons that enables the CRT to act as a dynamic voltmeter recorder system with essentially zero inertia.

An oscilloscope can be used to record a signal y as a function of time or it can be used to simultaneously record two unknown signals x and y. A block diagram of an oscilloscope, presented in Fig. 3.21, shows the inputs and the connections to the deflection plates in the CRT. The y and the x inputs are connected to the vertical and horizontal deflection plates through amplifiers. Since the sensitivity of the CRT is relatively low (approximately 100 V are required on the deflection plates to deflect the beam of electrons 1 in. [25 mm] on the face of the tube), high-gain amplifiers are used to increase the voltage of the input signal.

When the oscilloscope is used as a y–t recorder, the input to the horizontal amplifier is switched to a sawtooth generator. The sawtooth generator produces a voltage–time output in the form of a ramp function, where the voltage increases uniformly with time from zero to a maximum and then almost instantaneously returns to zero so that the process can be repeated. When this ramp function is imposed on the horizontal deflection plates, it causes the electron beam to sweep from left to right across the face of the tube. When the voltage from the sawtooth generator goes to zero, the electron beam is returned almost instantaneously to its starting point. The frequency of the sawtooth generator can be varied to give different sweep times. The sweep rates depend on the bandwidth of the oscilloscope. For high-frequency oscilloscopes with a bandwidth of 1 GHz, the sweep rate can be varied from 200 ps per division to 0.2 s per division in calibrated steps in a 1–2–5–10 sequence. For lower frequency oscilloscopes with a bandwidth of, say, 2 MHz, the sweep rate can be varied from 100 ns per division to 5 s per division. Since the horizontal dimension of the face of the CRT is divided into 10 divisions (usually a division equals 10 mm), the observation period is 10 times the sweep rate.

Since the observation time can be relatively short, the horizontal sweep must be synchronized with the event to ensure that a recording of the signal from the transducer is made at the correct time. Three different triggering modes are used to synchronize the oscilloscope with the event: the external trigger, the line trigger, and the internal trigger. Trigger signals from any one of these three sources activate the sawtooth generator and initiate the horizontal sweep. The *external trigger* requires an independent triggering pulse from an external source usually associated with the dynamic event

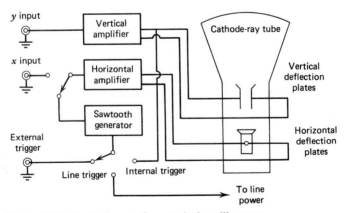

Figure 3.21 Block diagram for a typical oscilloscope.

being measured. A sharp front pulse of 2 to 5 V is recommended for the input to the external trigger.

The *line trigger* utilizes the signal from the power line to activate the sawtooth generator. Since the line-trigger signal is repetitive at 60 Hz, the horizontal sweep can trigger up to 60 times each second; therefore, the trace on the CRT appears continuous. The line trigger is quite useful when the oscilloscope is used to measure periodic waveforms that exhibit a fundamental frequency of 60 Hz.

The *internal trigger* makes use of the y input signal to activate the sweep. The level of the trigger signal required to initiate the sawtooth generator can be adjusted to very low levels; therefore, only a small region of the record is lost in measuring a transient pulse. If the input signal is repetitive, the frequency of the sawtooth generator can be adjusted to some multiple of the frequency of the input signal. The sawtooth generator is then synchronized with the input signal, and the trace appears stationary on the CRT screen.

The trace on the screen of the CRT is produced when the electron beam impinges on a phosphor coating on the inside face of the tube. Of the total energy of the beam, 90 percent is converted to heat and 10 percent to light. The light produced by fluorescence of the phosphor has a degree of persistence that enhances the visual observation of the trace. For instance, the phosphor identified as P31 produces a yellowish green trace that requires 38 μs to decay to 10 percent of its original intensity. Permanent records of the traces can be made with special-purpose oscilloscope cameras that attach directly to the main frame of the oscilloscope.

Recording high-speed traces from the CRT depends on the type of phosphor used in fabricating the CRT, the camera lens and magnification ratio, and the speed (ASA rating) of the film. Writing rates of 10 to 20 mm/ns can be achieved with P11 blue phosphor CRTs, using a camera with a magnification ratio of 0.5, an f-1.2 lens, and a 20,000-ASA film. Of course, recording at lower writing rates is much less demanding and the phosphor, film, and camera requirements can be relaxed.

Many cathode-ray tubes are of the storage type and continue to display the trace after the input signal ceases. The period of retention varies from a few seconds to several hours, depending on the type of phosphor used in fabricating the screen of the CRT. Most storage tubes utilize a bistable phosphor that permits the tube to be used in both the storage mode and the conventional (nonstorage) mode. The advantages of trace storage are numerous. Storage permits easy and accurate evaluation of slowly changing events that would appear as slowly moving dots on the conventional CRT. Storage is useful also in observing rapidly changing nonrepetitive signals that would flash across the screen too quickly to be evaluated on a conventional CRT. Operation in the storage mode also permits the careful selection of the data to be recorded photographically. Unwanted displays can be erased, and the expense of photographing is avoided. Finally, many storage oscilloscopes have split-screen viewing, which allows each half of the screen (top and bottom) to be used independently for stored trace displays. With the split screen, a reference trace can be stored on one half of the screen and the other half can be used to display the unknown trace. In this manner, comparisons can be made quickly and accurately.

The writing rate that can be achieved varies considerably with the technology used in producing the stored image. Relatively expensive high-frequency oscilloscopes can store at writing rates of 10^4 mm/μs. Lower-cost oscilloscopes can store at writing rates of 2.5 mm/μs.

The amplifier used in an oscilloscope is quite important since it controls the sensitivity, bandwidth, rise time, input impedance, and the number of channels that can be recorded with the instrument. Because of the importance of the amplifier to the operational characteristics of the oscilloscope, many instruments are designed to accept plug-in amplifiers. This arrangement permits the amplifier to be changed quickly and easily to alter the characteristics of the oscilloscope.

Bandwidth is defined as the frequency range over which signals are recorded with less than 3-dB loss compared with midband performance. Since modern amplifiers perform very well at low frequencies (down to dc), bandwidth refers to the highest frequency that can be recorded with an error less than 3 dB (29 percent). Bandwidth f_{bw} and rise time t_r are related such that

$$f_{bw}t_r = 0.35 \qquad \textbf{(a)}$$

Since good practice dictates utilization of a vertical amplifier capable of responding five times as fast as the rise time of the applied signal, Eq. a is modified in practice to

$$f_{bw}t_r = 1.70 \qquad \textbf{(3.49)}$$

It is apparent from Eq. 3.49 that an amplifier–oscilloscope combination with a bandwidth of 10 MHz is capable of recording signals with a rise time of approximately $0.17 \ \mu$s.

A significant factor to consider in bandwidth specification for an oscilloscope is cost. Costs increase modestly with bandwidth over the range from 0.5 to 100 MHz; however, costs increase dramatically as the bandwidth is increased from 100 MHz to 1 GHz. Increasing bandwidth also increases noise pickup and decreases sensitivity. For these reasons a high-bandwidth oscilloscope should not be specified unless very short rise times are anticipated. For general-purpose mechanical measurements, rise times of less than 0.2 μs are rare and a bandwidth of 10 MHz permits measurements with $t_r \geq 0.17\mu$s accurate to 2 percent. Very-high-bandwidth oscilloscopes are required for high-speed electronic or optic measurements where frequencies of the dynamic events are extremely high.

Reciprocal sensitivity S_R refers to the voltage of the input signal required to produce a specified vertical deflection on the CRT. The sensitivity of a typical amplifier used to control the y deflection ranges from 1 mV per division to 5 V per division in calibrated steps arranged in a 1–2–5 sequence. Higher sensitivity amplifiers are available with S_R equal to 10 μV per division to 50 mV per division; however, the bandwidth is reduced as sensitivity is increased. A sensitivity-bandwidth trade-off occurs in the selection of an amplifier, and for lower-frequency mechanical measurements it is usually advisable to specify the minimum bandwidth required and to work with higher-sensitivity lower-noise amplifiers.

The input amplifier also controls the number of traces displayed on the oscilloscope screen. In some models, an electronic switch is housed in the amplifier that alternatively connects two input signals to the vertical deflection system in the CRT. The principal advantages of using this feature to produce a dual-trace oscilloscope are lower cost and better comparison capabilities. These advantages exist because only one horizontal amplifier and one set of deflection plates is used in making both traces. However, high-speed transient events are difficult to record in this manner, because a significant variation might occur on one channel while the beam is tracing on the other channel. Since the electronic switch operates at a frequency of approximately 250 kHz, dynamic events with frequencies between 25 and 50 kHz ($\frac{1}{10}$ to $\frac{1}{5}$ of the switching frequency) can be recorded. Whenever two nonrecurrent signals of very short duration must be recorded together, dual-beam oscilloscopes are employed. A dual-beam oscilloscope has independent deflection plates within the CRT for each beam and employs independent horizontal and vertical amplifiers for each beam. The dual-beam system is superior to the dual-trace system, since it can display two signals separately and simultaneously; however, it is more costly.

The modular oscilloscope with plug-in amplifiers and time bases provides a versatile instrument that can be adapted to a wide range of applications in a large multipurpose laboratory. For more specific applications the portable oscilloscopes without plug-in amplifiers may be more suitable, since they are lower in cost and simpler to operate. Also, many portable oscilloscopes are battery powered, small, and lightweight, which permits their application in remote-site field measurements. A typical portable oscilloscope weighs 3.5 lb, occupies a volume of 151 in.[3], and is designed to accommodate rugged use in the field.

3.5.5   Magnetic Tape Recorders

Magnetic tape recorders are used to store dynamic signals when the frequency components in the signals range from dc to up to a maximum of about 400 kHz. The recording is accomplished by applying a magnetizing field to a magnetic film coating on a Mylar tape. The recording process is illustrated in Fig. 3.22. The magnetic flux in the record head fluctuates as a result of variations in the input signal, and a magnetic record of these variations is permanently imposed on the coating. The data are retrieved by moving the tape under the reproduce head, where the variations in the magnetic field stored on the tape induce a voltage in the windings of the reproduce head, which produces the output signal.

In a magnetic tape recorder, the tape is driven at a constant speed by a servo-type dc capstan motor over either the record or reproduce heads, as shown in Fig. 3.23. The speed of the capstan motor is monitored with a photocell and tone wheel. The frequency of the signal from the photocell is compared with the frequency from a crystal oscillator to produce a feedback signal that is used in a closed-loop servo

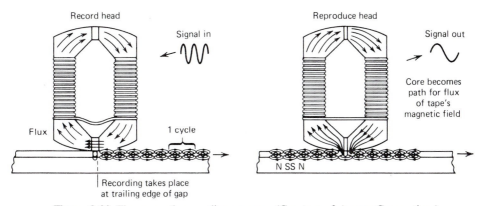

Figure 3.22 The magnetic recording process. (Courtesy of Ampex Corporation.)

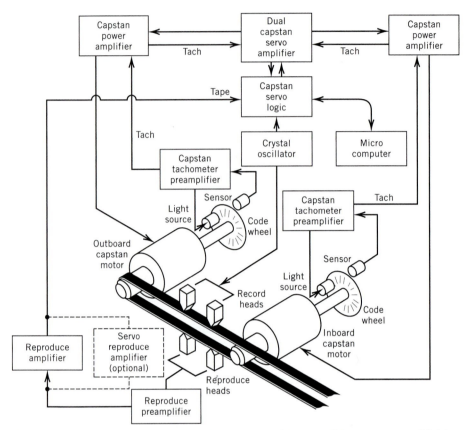

Figure 3.23 DC capstan magnetic tape drive. (Courtesy of Honeywell Test Instruments Division, Denver, Colorado.)

system to maintain a constant tape speed. Tape speeds are variable and can be selected from the standardized values shown in Table 3.4. Table 3.4 illustrates that the bandwidth of the tape recorder depends on tape speed and that higher tape speeds extend the bandwidth capabilities.

Tape is available in several sizes, depending on the particular recorder. Common sizes are $\frac{1}{2}$-in.–7 tracks and 1-in.–14, –28, and –32 tracks. Recently, L-type beta-

Table 3.4 Standardized Tape Speeds and Bandwidths

Tape Speed in./s	Bandwidth[a] kHz within 1 dB	Bandwidth[b] kHz +1, −2 dB
120	0–40	400
60	0–20	200
30	0–10	100
15	0– 5	50
7.5	0– 2.5	25
3.75	0– 1.25	12.5
1.87	0– 0.625	6.25
0.937	0– 0.313	

[a]System electronics designed for ±40% FM IRIG, intermediate bandwidth.
[b]System electronics designed for ±30% FM IRIG, wideband group II.

format videocassette tape has been used to record 21 tracks of data and one voice channel. Tape thickness varies from 0.7 to 1.5 mils and can be obtained in reels up to 15 in. in diameter. The maximum recording time depends on reel size, tape thickness, and tape speed. For example, a 0.7-mil tape on a 10.5-in.-diameter reel provides 7200 ft of tape, which gives a recording time of 25.6 h at a tape speed of 0.937 in./s.

Multichannel recorders employ four stacked-head assemblies that are precisely positioned on a single base plate to ensure alignment of the tape. Two of the heads are for recording and the other two are for reproducing, as indicated in Fig. 3.23. As the tape passes the first head assembly, odd-channel data are recorded; as it passes the second head assembly, even-channel data are recorded. This recording procedure minimizes interchannel cross talk by maximizing the spacing between individual heads in the stacked-head assembly. Use of two recording-head stacks permits up to 32 channels on 1-in. (25.4-mm) tape.

Three different types of recording are commonly used: direct or AM, FM, and digital. The characteristics of each type are shown in Fig. 3.24. The direct, or AM (amplitude modulation), recording method is the most frequently used, since it is simple, low cost, and suitable for most audio (speech and music) recordings. The signal to be recorded is amplified, mixed with a high-frequency bias, and then used to drive the record head. In playback or reproduction, the tape is driven under the reproduce head at the same speed that was used in recording. The output from the

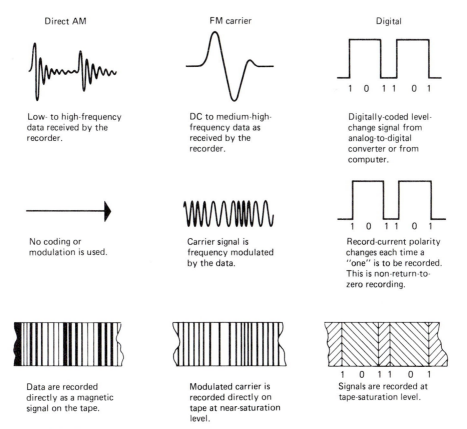

Figure 3.24 Features of direct or AM, FM carrier, and digital methods for magnetic recording.

reproduce head is proportional to the frequency of the recorded signal. This output is then fed into a reproduce amplifier with a frequency response that is the inverse of the frequency response of the reproduce head. This procedure compensates for the frequency distortion generated in the reproduce head and provides a flat frequency response for the system.

There are two serious disadvantages to AM-type recording. First, the lowest frequency that can be recorded is about 50 Hz; therefore, dc or slowly varying signals cannot be stored. Second, imperfections in the coating on the tape can produce significant reductions in signal levels, which can cause serious errors in the recording of transient signals. This type of error can be tolerated in speech or music recording, but not in data recording, where precision is critical. Because of these limitations, AM recording is used only on one track of a multitrack recorder for voice commentary relative to the event being recorded (identification or experimental description).

The method used most frequently to record data is the FM (frequency modulation) method, which overcomes both of the limitations of AM recording. With FM recording, the input signal is used to drive a voltage-controlled oscillator (VCO). The center frequency output from the VCO corresponds to a zero input voltage. A positive dc input signal produces an increase in the frequency of the carrier signal issued from the VCO, and an ac input signal produces carrier frequencies on both sides of the center frequency. Thus, with frequency modulation (FM) the voltage–time data are recorded on the tape in the frequency domain. Low-frequency or dc input signals can be recorded, and amplitude instabilities resulting from imperfections in the coating on the tape do not markedly affect the output.

The signal is retrieved from the magnetic tape in a playback process that includes filtering and demodulation. The signal from the reproduce head is filtered to remove the carrier frequency and then demodulated to give the output signal. This output signal is displayed on an oscillograph or an oscilloscope. The magnetic tape system is used only to record and store data. However, the display of the data can be enhanced, since the recording can be reproduced at a different tape speed than that used for recording. This procedure alters the time base, and voltage–time traces can be stretched or compressed. Data stored on magnetic tape can also be processed automatically without visual display. For FM records, the output signal can be fed into an analog-to-digital converter. The digitized data can then be processed on a computer according to programmed instructions.

Digital recording involves storing two-level data (0 and 1) and is accomplished by magnetizing the tape to saturation in either one of the two possible directions. In one type of digital recording (return to zero), the positive state of saturation represents the binary digit 1 and the negative state of saturation represents the digit 0. Data are recorded as a series of pulses that represent the decimal number (expressed in binary code) of the voltage input averaged over a sampling interval. Bits that express the number are recorded simultaneously in parallel across the width of the tape, with each bit on a separate track.

Although digital recording is sensitive to tape dropouts, and thus requires high-quality tape and tape transports that ensure excellent head-to-tape contact, other aspects of the process make digital recording easier than FM recording. The output is not strongly dependent on tape speed, and the record and reproduce amplifiers are simple and therefore inexpensive. Also, since the output is in digital form, it can be processed directly on a computer.

The primary disadvantage of digital recording has been the need to digitize the input data prior to recording. This disadvantage is being overcome as significant improvements are being made in high-speed analog-to-digital converters. It is possible that in the near future developments in high-speed analog-to-digital converters will permit digital recorders to replace FM recorders for dynamic recording of long-term events.

3.6 SUMMARY

A voltage recording instrument, the final component in a measuring system, is used to convert a voltage representing the unknown quantity Q_i into a display for visual readout or a digital code suitable for automatic data processing. The instruments in use today (1992) range from simple analog voltmeters to complex digital oscilloscopes with auxiliary magnetic storage as discussed in Section 4.8. Although all five of the general characteristics of a recording instrument (input impedance, sensitivity, range, zero drift, and frequency response) are important, the single characteristic that dominates selection of a recorder is frequency response.

Static measurements, where frequency response is not important and the unknown is represented by a single number independent of time, can be made quickly and accurately with relatively inexpensive voltmeters. When the phenomena being studied begin to vary with time, the recording instrument becomes more complex, less accurate, and more expensive. The major difficulty experienced in measuring unknown parameters associated with time-dependent phenomena is the need to record and display the data with respect to time without distortion of the signal.

Display of the data for quasi-static measurements, where the unknown is varying with a frequency of only a few hertz, can be accomplished with a servo-driven potentiometer recorder. The servo-driven potentiometer provides an accurate and inexpensive means of measuring and displaying a voltage; the time display of the voltage is presented on a strip of chart paper that is driven by a simple clock motor. For very slowly varying quantities (one cycle every few minutes), switching can be used with a single recorder to handle multiple inputs. For many studies, however, an independent channel is needed for each unknown quantity being measured. The multichannel requirement greatly increases the complexity and cost of the measurements.

As the frequency of the unknown quantity increases to about 25 kHz, recording instruments with adequate frequency response must be used in the measurement system. Oscillographs with galvanometer-driven pens or hot styli are used for frequencies between 0 and 150 Hz. For frequencies between 0 and 25 kHz, light-writing oscillographs with galvanometer-driven mirrors or fiber-optic cathode-ray tubes are used. Magnetic tape recorders are useful for frequencies up to about 400 kHz. Signals with frequencies above 100 kHz must be recorded with either digital or conventional oscilloscopes. As the frequency increases, it is evident that the recording instrument becomes more sophisticated, more difficult to operate, less accurate, and more expensive. The data obtained are usually in the form of voltage–time traces that require considerable time for analysis.

Recent advances in digital electronics are resulting in new instruments that offer significant advantages in making measurements at both ends of the frequency spectrum. Digital voltmeters and data-acquisition systems are easy to use and offer a relatively low-cost method for acquiring and processing large amounts of low-frequency data. The digital oscilloscope or waveform recorder has the capacity to store high-speed transient signals that can be easily displayed, compared, and processed externally on a computer.

The digital methods of recording and displaying both static and dynamic signals are covered in Chapter 4.

REFERENCES

1. Ahmed, H., and P. J. Spreadbury: *Analogue and Digital Electronics for Engineers,* 2nd ed., Cambridge Univ. Press, London/New York, 1984.
2. Collins, T. H.: *Analog Electronics Handbook,* Prentice–Hall, Englewood Cliffs, N.J. 1989.
3. Hickman, I.: *Analog Electronics,* Heinemann Newnes, London, 1990.
4. Malmstadt, H. V., C. G. Enke, and S. R. Crouch: *Electronic Analog Measurements and Transducers,* W. A. Benjamin, Menlo Park, Calif., 1973.
5. Northrop, R. B.: *Analog Electronic Circuits: Analysis and Applications,* Addison–Wesley, Reading, Mass., 1990.

EXERCISES

3.1 List the general characteristics of a recording instrument.

3.2 Prepare a graph showing power loss p through a voltmeter as a function of the voltage v. Let v range from 0 to 100 V. Let the input impedance Z_m equal

(a) 10^3 Ω (b) 10^4 Ω (c) 10^5 Ω
(d) 10^6 Ω (e) 10^7 Ω

3.3 The input impedance of a meter is modeled by a resistance $R_m = 10^6$ Ω in parallel with a capacitance $C = 100$ pF. Prepare a graph showing the magnitude of the impedance Z_m as a function of ωRC.

3.4 Determine the error involved in using the meter described in Exercise 3.3 to measure an ac voltage of 5 V if the output impedance of the source is $R_s = 10$ Ω and if ω is

(a) 10^2 (b) 10^4 (c) 10^6

3.5 What limit must be placed on the resistance ratio R_s/R_m if the acceptable voltmeter load error is

(a) 0.5%   (c) 2%
(b) 1%   (d) 5%

3.6 Determine the reciprocal sensitivity S_R and the input voltage v_i if the deflection exhibited by an oscilloscope and the sensitivity of the oscilloscope are

	d (div)	S (div/V)	S_R (V/div)	v_i (V)
(a)	2.5	0.2	?	?
(b)	6.1	0.5	?	?
(c)	3.7	2.0	?	?
(d)	7.6	0.1	?	?

3.7 The oscilloscope described in Exercise 3.6 has eight divisions in the vertical direction on the face of the tube. For the sensitivities given in Exercise 3.6, determine the ranges. Can both sensitivity and range be increased simultaneously?

3.8   Describe in your own words the relationship between sensitivity and range of an instrument.

3.9   An amplifier that is being used in an instrumentation system to measure a voltage of 12 mV over a period of 15 days exhibits a drift of 0.05 mV/h. Determine the error that may result from zero drift.

3.10   Specifications for a recorder indicate that it is down 1.5 dB at 1000 Hz. Determine the error if the recorder is used to measure a signal with a frequency of 1000 Hz.

3.11   Tests with a recorder at frequencies of 10, 20, 40, 60, 80, and 100 Hz provided the following output-to-input ratios C_o/C_i: 1.01, 1.03, 1.05, 1.00, 0.93, and 0.80. Determine the amplitude ratio in decibels at each frequency.

3.12   Find the amplitude ratio A_o/A_i for a recorder if N_{dB} is

    (a) −5.0   (c) −0.25

    (b) +2.5   (d) +0.70

3.13   What limits on N_{dB} must be maintained to reduce recorder error to less than

    (a) 5%   (c) 1%

    (b) 2%   (d) 0.5%

3.14   The sensitivity of a galvanometer is listed by the manufacturer as 25-μA full scale. The full-scale rotation of the pointer is 45 deg. Determine the sensitivity S of the galvanometer by using the definition of sensitivity given in Eq. 3.13.

3.15   A galvanometer with a 40-Ω coil is rated at 10-mA full scale. Determine the required shunt resistance if it is to be used to measure a current of

    (a) 100 mA   (c) 2 A

    (b) 0.5 A   (d) 20 A

3.16   Prepare a graph showing the series resistance needed to convert a 50-μA full-scale galvanometer with a coil resistance of 40 Ω to a multimeter with full-scale voltages that range from 10 mV to 100 V.

3.17   Prepare a graph showing the loading error for the multimeter of Exercise 3.16 if it is used to measure voltage from a source with $R_s = 10\ \Omega$. Why does this error differ, and for what scale is it the largest?

3.18   Outline the advantages of the amplified voltmeter when compared with the conventional voltmeter.

3.19   What is the most significant advantage of a null-balance instrument? What is the most significant disadvantage?

3.20   Determine the input voltages when a potentiometer with a slide-wire resistor 10 in. long and a reference voltage of 2 mV is balanced with the wiper at the following positions

    (a) 4.5 in.   (c) 2.1 in.

    (b) 7.2 in.   (d) 8.6 in.

3.21   Describe how a single-channel strip-chart recorder can be converted so that it can record data from several different sources.

3.22   Define the terms *deadband* and *slewing speed* as they apply to an x–y recorder.

3.23   Describe how an x–y recorder can be used to record the stress–strain diagram from a tension test in order to determine the modulus of elasticity, yield strength, and ultimate tensile strength of a material.

3.24 Write an engineering brief describing both pen- and light-writing oscillographs. Compare the features of these two instruments and cite the advantages and disadvantages of each.

3.25 Select a galvanometer from Table 3.2 that can be employed to measure a transient signal containing frequency components as high as

(a) 80 Hz   (c) 800 Hz

(b) 300 Hz   (d) 9000 Hz

3.26 For the galvanometer selections in Exercise 3.25 determine the required external damping resistance and the sensitivity S.

3.27 Verify Eq. 3.35.

3.28 Verify Eq. 3.36.

3.29 Verify Eq. 3.37.

3.30 For an error band of ± 2 percent, show that the response time of a critically-damped galvanometer to a step-function input is $t/T_n \approx 0.93$.

3.31 Derive Eq. 3.38.

3.32 Verify Eq. 3.45.

3.33 Verify the results presented in Table 3.3.

3.34 A current represented by the expression

$$i = 10 \sin 400t + 2 \sin 800t + \sin 1200t$$

is recorded with an oscillograph. Compare the input and output signals and determine amplitude and phase distortions of the recorded pulse for a galvanometer with

	d	ω_n
(a)	0.55	1100
(b)	0.60	1200
(c)	0.65	1300

3.35 Describe measurements that would be made with each of the following trigger modes:

(a) Internal   (b) Line   (c) External

3.36 Write an engineering brief describing the cathode-ray tube (CRT) used in an instrument-type oscilloscope.

3.37 Describe the differences between an oscilloscope used in the laboratory for general-purpose measurements and one used in a hospital to monitor transducers attached to patients.

3.38 Explain why the oscilloscopes of Exercise 3.37 are so different.

3.39 Define bandwidth and describe its importance in measuring transient signals.

3.40 Prepare a graph showing rise time as a function of bandwidth for use with oscilloscopes.

3.41 Compare the differences between a bench-top oscilloscope and a portable oscilloscope.

3.42 Discuss the advantages and disadvantages of the direct or AM method of tape recording.

3.43 Discuss the advantages and disadvantages of the FM method of tape recording.

3.44 Discuss the advantages and disadvantages of the digital method of tape recording.

3.45 Prepare a graph showing the relation between tape speed and bandwidth for
   (a) the ±40% FM IRIG, intermediate bandwidth
   (b) the ±30% FM IRIG, wideband group II

3.46 Prepare a two-page (400-word) summary of this chapter.

3.47 Prepare a 3 × 5 index card with what you consider to be the five most important equations in this chapter.

3.48 Cite the reasons for your selections in Exercise 3.47.

Chapter 4

Digital Recording Systems

4.1 INTRODUCTION

During the past two decades enormous progress has been made in developing digital instrumentation. The combination of analog instrumentation with digital processing, accomplished through analog-to-digital conversion, has added new dimensions to both engineering analysis and process control because of the many advantages associated with digital processing of analog data.

Measurements of physical quantities as described in Chapter 1 are analog measurements. An analog instrument system, illustrated in Fig. 4.1, is used to obtain an output signal v_o that is proportional to the quantity Q being monitored. The analog system interfaces with the digital system through an analog-to-digital (A/D) interface. The key element, which represents the interface between the analog acquisition system and the digital processing system, is an A/D converter. This A/D converter takes the voltage v_o from the acquisition system as an input and converts it to an equivalent digital code. Once digitized, the signal (i.e., a digital code) can be displayed, processed, stored, or transmitted. The arrangement of a combined analog and digital system, shown in Fig. 4.1, depends on the function of the system. A simple system, involving direct display of a single quantity, incorporates a transducer, a power supply, a number of integrated circuits (chips), and a numerical display. A more involved digital system would include a computer for real-time processing, disk drives for storage of raw and processed data, cathode-ray-tube (CRT) or panel displays, printers for readout and hard copy, and data-transmission lines to off-site facilities.

4.2 DIGITAL CODES

Digital systems contain many logic gates (AND, OR, and NOT gates), which act like switches that can be turned on or off. Since the logic gates have only two states (on or off), digital words consist of binary elements, called bits, which are either 0, for off, or 1 for on. Consider a digital word consisting of a 4-bit array, for example, 1011. The 1 at the extreme left is the most significant bit (MSB) and the 1 at the extreme right is the least significant bit (LSB). With a binary code, the least significant bit has a weight of 2^0, the next bit has a weight of 2^1, the next 2^2, and

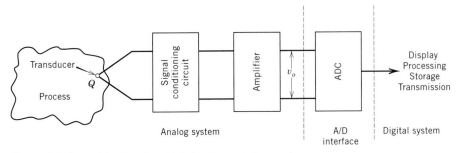

Figure 4.1 A combined analog-digital instrumentation system.

the most significant bit m has a weight 2^{m-1}. When all four bits are 0 (i.e., 0000), the equivalent count is 0. When all four bits are 1 (i.e., 1111), the equivalent count is $2^3 + 2^2 + 2^1 + 2^0 = 8 + 4 + 2 + 1 = 15$. A complete listing of a 4-bit binary code is presented in Table 4.1.

It is evident from Table 4.1 that a 4-bit binary word permits a count of 2^4, or 16, arranged from 0 to 15. The maximum count C is determined from

$$C = (2^n - 1) \tag{4.1}$$

where n is the number of bits in the digital word. It is clear that $C = 255$ for an 8-bit word and $C = 1023$ for a 10-bit word.

The least significant bit represents the resolution R of a digital count containing n bits, which can be written as

$$R = \frac{2^0}{2^n - 1} = \frac{1}{2^n - 1} = \frac{1}{C} \tag{4.2a}$$

Table 4.1 Equivalent Count for a 4-Bit Binary Code

MSB[a]	Bit 2	Bit 3	LSB[a]	MSB		Bit 2		Bit 3		LSB		Count
0	0	0	0	0	+	0	+	0	+	0	=	0
0	0	0	1	0	+	0	+	0	+	2^0	=	1
0	0	1	0	0	+	0	+	2^1	+	0	=	2
0	0	1	1	0	+	0	+	2^1	+	2^0	=	3
0	1	0	0	0	+	2^2	+	0	+	0	=	4
0	1	0	1	0	+	2^2	+	0	+	2^0	=	5
0	1	1	0	0	+	2^2	+	2^1	+	0	=	6
0	1	1	1	0	+	2^2	+	2^1	+	2^0	=	7
1	0	0	0	2^3	+	0	+	0	+	0	=	8
1	0	0	1	2^3	+	0	+	0	+	2^0	=	9
1	0	1	0	2^3	+	0	+	2^1	+	0	=	10
1	0	1	1	2^3	+	0	+	2^1	+	2^0	=	11
1	1	0	0	2^3	+	2^2	+	0	+	0	=	12
1	1	0	1	2^3	+	2^2	+	0	+	2^0	=	13
1	1	1	0	2^3	+	2^2	+	2^1	+	0	=	14
1	1	1	1	2^3	+	2^2	+	2^1	+	2^0	=	15

[a]Where MSB and LSB are the most and least significant bits, respectively.

Table 4.2 Count C, Resolution R, and Error $\mathscr{E}_R$ as a Function of the Number of Binary Bits n.

n	C	$R(\text{ppm})^a$	$\mathscr{E}_R(\%)$
4	15	66666	6.6666
5	31	32258	3.2258
6	63	15873	1.5873
7	127	7874	0.7874
8	255	3922	0.3922
9	511	1957	0.1957
10	1023	978	0.0978
11	2047	489	0.0489
12	4095	244	0.0244
13	8191	122	0.0122
14	16383	61	0.0061
15	32767	31	0.0031
16	65535	15	0.0015

^{a}Parts per million.

This result indicates that the resolution that can be achieved with logic gates arranged to yield an 8-bit digital word (a byte) is (1/255), or 0.392 percent of full scale.

Resolution is an important concept in digital instrumentation, because it defines the number of bits required for a specified error in a measurement or for the conversion of an analog voltage into a digital count representing that voltage. Values for C, R, and resolution error $\mathscr{E}$ as a function of n are presented in Table 4.2. An 8-bit digital word that provides resolution to within ± 1 count out of a total or full-scale count of 256 will limit the resolution error to 0.39 percent.

4.3   CONVERSION PROCESSES

The analog-to-digital interface, illustrated in Fig. 4.1, shows an analog-to-digital converter (ADC), which transforms the analog voltage into a digital count. The ADC is a one-way device converting an analog signal to a digital code. To convert a digital code to an analog, voltage requires a digital-to-analog converter (DAC) which is also a one-way device. Since the DAC and ADC are the key functional elements in analog–digital instrument systems, they will be described in detail.

First, consider a 4-bit DAC where the input is a digital code ranging from 0000 to 1111, as listed in Table 4.1. The digital input (the independent variable) is plotted along the abscissa of Fig. 4.2. The analog voltage output, ranging from 0 to $\frac{15}{16}$ of full scale, is shown along the ordinate. Although full scale, 16, is not available as a digital input, it represents the reference quantity to which the analog voltage is normalized. For example, if 10 V is the full-scale voltage, then the digital code 1000 will give, under ideal conditions, an analog voltage of $\left(\frac{8}{16}\right)(10) = 5$ V.

Next, consider an ADC with the analog voltage as the independent variable, as shown in Fig. 4.3. Since all analog voltages between zero and full scale can exist, they must be quantized by dividing the range of voltage into subranges. If FSV is the full-scale analog voltage input, the quantization increment is equal to $\text{FSV} \times \text{LSB}$ where $\text{LSB} = 2^{-n}$. For a 4-bit ADC, $\text{FSV} \times \text{LSB} = \frac{1}{16}$ FSV or 0.0625 FSV. All analog voltages within a given quantization increment are represented by the same

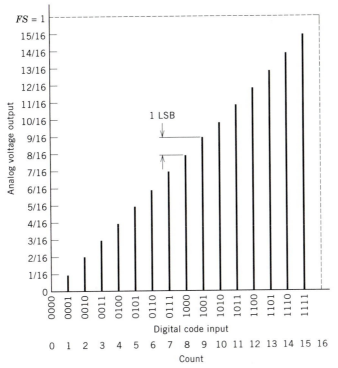

Figure 4.2 Relation between digital count and analog voltage for a digital-to-analog converter (DAC)

digital code. The illustration in Fig. 4.3 shows that the digital code corresponds to the midpoint in each increment. The quantizing process, which replaces a linear analog function with a staircase digital representation, results in a quantization uncertainty of $\pm \left(\frac{1}{2}\right)$LSB and a quantization error as shown at the top of Fig. 4.3. The average value of the quantization error is zero; however, if it is assumed that it is equally probable that v/FSV takes any value within the quantization increment, it can be shown that the standard deviation of a number of measurements is

$$\sigma = \frac{1}{2\sqrt{3}}\text{LSB} = \frac{2}{\sqrt{3}}^{-(n+1)} \tag{4.2b}$$

This statistic clearly indicates that the effective resolution of the ADC is much better $(1/\sqrt{12})$ than the usually specified resolution given by Eq. 4.2a.

Converters of either type (DAC or ADC) are not ideal, and errors can occur because of offset, gain, and scale factor. These three types of conversion errors are illustrated graphically in Fig. 4.4 for both the DAC and the ADC.

4.4 DIGITAL-TO-ANALOG CONVERTERS

Many different circuits are employed in the design of digital-to-analog converters. Although the more sophisticated circuits are beyond the scope of this book, a simple circuit, which shows the essential features involved in the digital-to-analog conversion process, is described. The circuit for a 4-bit DAC is illustrated in Fig. 4.5. A voltage

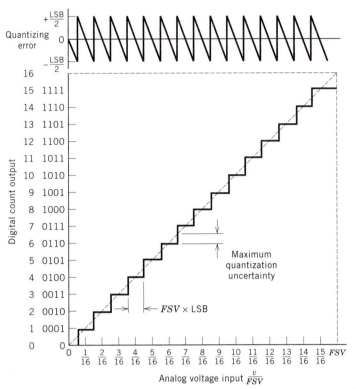

Figure 4.3 Relation between analog voltage input and digital count–code output for a 4-bit analog-to-digital converter (ADC).

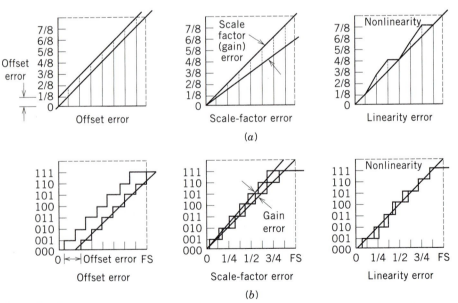

Figure 4.4 Typical sources of error in (*a*) digital-to-analog converters and (*b*) analog-to-digital converters.

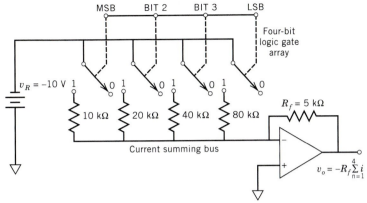

Figure 4.5 Schematic diagram of a simple 4-bit digital-to-analog converter.

reference v_R is connected to a set of precision resistors by a set of switches. The switches are gates in a digital logic circuit, with 0 representing an open switch and 1 representing a closed switch. The resistors are binary weighted, which means that resistance is doubled for each higher switch or bit so that

$$R_n = 2^n R_f \tag{4.3}$$

where

R_n is the resistance of the n_{th} bit
R_f is the feedback resistance on the operational amplifier

When the switches are closed, a binary-weighted current i_n flows to the summing bus. This current is

$$i_n = \frac{v_R}{R_n} = \frac{v_R}{2^n R_f} \tag{4.4}$$

The operational amplifier converts the currents to voltages and furnishes a low-impedance output. The analog output voltage v_o is given by

$$v_o = -R_f \sum_{n=1}^{k} i_n \tag{4.5}$$

where i_n is summed only if the switch n is closed.

Consider as an example the digital code 1011 (equivalent to 11) for the circuit shown in Fig. 4.5, with $R_f = 5$ kΩ and $v_s = -10$ V. From Eq. 4.4 it is clear that $i_1 = -1$ mA, $i_2 = 0$, $i_3 = -\frac{1}{4}$ mA, and $i_4 = -\frac{1}{8}$ mA. Summing these currents and multiplying by R_f gives $v_o = 6.875$ V, which is $\frac{11}{16}$ of the full-scale (reference) voltage.

Commercial DACs are more complex than the schematic shown in Fig. 4.5 because they contain more bits (8, 12, and 16 are common), and have several regulated voltages, integrating circuits for switching, and on-chip registers. The large number of bits are serviced with a system of parallel conductors called an input bus. The voltage on each conductor in the bus is either high or low and activates each of the switches (gates) to provide the digital code as input to the device. The analog voltage

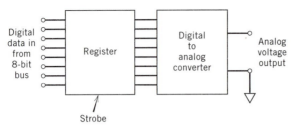

Figure 4.6 A digital-to-analog converter with a register to control data flow during conversion.

output v_o is constant with respect to time as long as the digital code is held at the same value on the input.

In many digital systems several functions occur together, and proper sequencing of these different functions is mandatory if a common bus is used in the data-distribution system. In these applications, the DAC is preceded by a register, as indicated in Fig. 4.6. The register is a memory device where the input data may be stored and held. With a common data bus, which serves several digital devices, only select signals are to be converted by the DAC; the other signals are to be ignored as they are intended for different devices. Sorting the signals from the bus is accomplished with the strobe, which is activated when the DAC has been addressed and given a signal to write. The strobe signal enables the register to read the data on the bus during the period of this signal. The new data take the form of a digital code, which replaces the old data in an update. The register is then latched, and the updated data held in memory. The new digital code is then converted to an analog voltage in the DAC, and the analog voltage is held constant until the next update.

4.5 ANALOG-TO-DIGITAL CONVERTERS

Conversion of analog signals to digital code is extremely important in any instrument system that involves digital processing of the analog output signals from the signal conditioners. Of the many circuits available for analog-to-digital conversion, three of the most common will be described here: the successive-approximation method, the integration method, and the flash-conversion method.

4.5.1 Successive-Approximation Method

The method of successive approximation is illustrated in Fig. 4.7, which shows a bias voltage v_b as a close approximation to an unknown analog voltage v_u, given by

$$v_b = v_u + \frac{1}{2^{n+1}}\text{FSV} \qquad (4.6)$$

The term $(\frac{1}{2^{n+1}})$FSV in Eq. 4.6 is added to the unknown voltage to place the DAC output in the center of a quantization increment, as illustrated in Fig. 4.3. In the case illustrated in Fig. 4.7, a 4-bit DAC is employed to convert a fixed analog input voltage $v_u = \frac{10}{16}$FSV. The bias voltage is given by Eq. 4.6 as

$$v_b = \left(\frac{10}{16} + \frac{1}{2^5}\right)\text{FSV} = \frac{21}{32}\text{FSV}$$

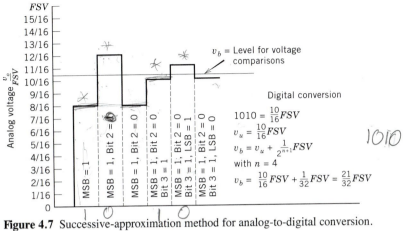

Figure 4.7 Successive-approximation method for analog-to-digital conversion.

The bias voltage v_b is compared with a sequence of precise voltages generated by an input-controlled DAC. The digital input to the DAC is successively adjusted until the known output v_o from the DAC compares closely with v_b. The accuracy of this method of successive approximation will be defined after the conversion process is described.

At the start of the conversion process, the input to the DAC is set at 1000 (i.e., the most significant bit is on, and all other bits are off); the analog voltage output $v_o = \frac{8}{16}FSV$, and a voltage comparison shows $v_o < v_b$. Since this first approximation underestimates v_b, the MSB is locked at 1 and bit 2 is turned on to give a digital input of 1100 to the DAC. The output from this second approximation is $v_o = \frac{12}{16}FSV$, as shown in Fig. 4.7. The voltage comparison shows $v_o > v_b$ (an overestimate), which causes bit 2 to be turned off and locked at 0. The third approximation involves turning bit 3 on to give a digital input of 1010 and $v_o = \frac{10}{16}FSV$. Since the voltage comparison shows $v_o < v_b$, bit 3 is locked on. The final approximation turns bit 4 on to give the digital input of 1011 and $v_o = \frac{11}{16}FSV$. Since $v_o > v_u$, bit 4 is turned off to give a final result of 1010 as the input to the DAC after four approximations. It should be noted that this method of successive approximations is analogous to weighing an unknown mass on a balance by using a set of n binary weights.

The input of the DAC, in this case 1010, is transferred to the output register of the ADC. In this simple example the conversion process was exact, since the unknown voltage v_u was selected at a 4-bit binary value $\frac{10}{16}FSV$. In general, an uncertainty in the conversion will occur when the unknown voltage differs from a binary value. To determine the uncertainty with an ADC involving n approximations consider

$$v_o = v_b - \left(\frac{1}{2^n}\right)FSV \tag{4.7}$$

Substituting Eq. 4.6 into Eq. 4.7 gives

$$v_o = v_u - \left(\frac{1}{2^{n+1}}\right)FSV \tag{4.8}$$

The relative difference between v_u and v_o in reference to the full-scale voltage is

$$\frac{v_u - v_o}{FSV} = \frac{1}{2^{n+1}} \tag{4.9}$$

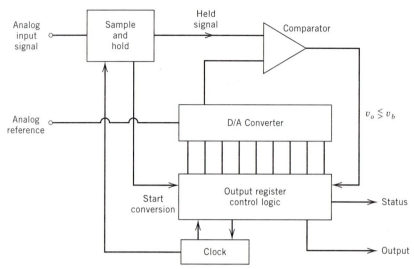

Figure 4.8 Successive-approximation converter with a sample-and-hold device and control logic.

where n is the number of bits used in the approximation. Equation 4.9 shows that analog-to-digital conversion results in a maximum uncertainty equivalent to $\frac{1}{2}$LSB. The quantization error, shown at the top of Fig. 4.3, varies between $\pm\frac{1}{2}$LSB and is bounded by the maximum uncertainty.

In this example, the analog input voltage did not change during the conversion process. Accurate conversion cannot be accomplished when the input voltage changes with time. To avoid problems associated with fluctuating voltages during the conversion period, the ADC utilizes an input device that samples and holds the signal constant for the time required for conversion. The complexity of the process increases because it is necessary to sample, hold, convert, and release from hold when the conversion is complete. The block diagram shown in Fig. 4.8 illustrates the key elements in the successive-approximation converter. The clock is used to time each portion of the conversion process. For instance, a fast ADC will track or sample for 50 ns and hold the signal constant for 5 μs. Conversion is completed during the 5 μs hold period. The status indication is high during the conversion to indicate that the DAC is busy.

When the conversion is complete, the 8-, 12-, or 16-bit digital code is transferred to a register and the conversion process is repeated. Conversion rates depend on the number of bits, design of the circuit, and speed of the transistors used in switching. In 1992, a low-cost 12-bit ADC typically requires 15 to 20 μs to convert an analog signal. For a single-channel application, this conversion time gives a data-acquisition rate of 50,000 to 66,667 samples per second.

4.5.2   Integration Method

Analog-to-digital conversion by integration is based on counting clock pulses. A typical circuit for a dual-slope ADC is shown in Fig. 4.9a. At the start of a conversion, the unknown input voltage v_u is applied together with a reference voltage v_R to a summing amplifier, which gives an output voltage

$$v_a = -\frac{1}{2}(v_u + v_R) \tag{a}$$

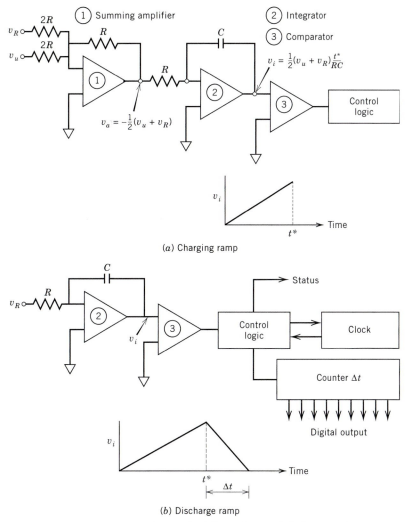

Figure 4.9 Dual-slope integration method for analog-to-digital conversion.

This voltage is imposed on an integrator, which integrates v_a with respect to time to obtain

$$v_i = \frac{(v_u + v_R)t^*}{2RC} \tag{4.10}$$

where t^* is a fixed time of integration, as shown in Fig. 4.9a. On completion of the integration at time t^*, a switch on the input of the integrator is activated to disconnect the summing amplifier and connect the reference voltage v_R to the integrator, as shown in Fig. 4.9b. The output voltage of the integrator then decreases with a slope of

$$\frac{\Delta v_i}{\Delta t} = -\frac{v_R}{RC} \tag{4.11}$$

The comparator monitors the output voltage v_i and issues a signal to the control logic when v_i goes to zero. This zero crossing occurs when

$$\frac{(v_u + v_R)t^*}{2RC} = \frac{v_R \, \Delta t}{RC} \tag{b}$$

Reducing Eq. b yields

$$\frac{\Delta t}{t^*} = \frac{1}{2}\left(\frac{v_u}{v_R} + 1\right) \tag{4.12}$$

It is clear from Eq. 4.12 that $\Delta t/t^*$, a proportional count of clock pulses, is related to v_u/v_R. If a counter is started by a switch on the integrator, the counter will give a binary number representing the unknown voltage v_u.

The integration method for analog-to-digital conversion has several advantages. Its output is independent of R, C, and the clock frequency, because these quantities affect both the up and down ramps in the same way, and so it is extremely accurate. The influence of noise on the unknown signal is markedly attenuated because the noise signal, which occurs at high frequency, is averaged toward zero during the integration period t^*. The primary disadvantage of the integration method is the speed of conversion, which is less than $\frac{1}{2}t^*$ conversions per second. In order to attenuate the effect of 60 Hz noise, $t^* \geq 16.67$ ms; therefore, the speed of conversion must be less than 30 samples per second. This conversion rate is too slow for large, multipurpose, high-performance data-acquisition systems, but it is satisfactory for digital voltmeters and smaller, lower-cost data-logging systems.

The conversion speed of the ADC determines the frequency of the unknown analog signal that can be measured. To determine this frequency, let the input signal to an ADC be a sinusoid with a frequency f given by

$$v_u = \frac{v_u^*}{2} \sin 2\pi f t \tag{4.13}$$

where $v_u^*/2$ is the amplitude $=$ FSV/2. The maximum rate of change of this input is obtained by differentiating Eq. 4.13 and letting $\cos 2\pi f t = 1$. Then

$$\left(\frac{dv_u}{dt}\right)_{\max} = \pi f v_u^* \tag{4.14}$$

The term $(dv_u/dt)_{\max}$ is called the slew rate of the signal. If the ADC is to convert the signal into a digital code of n bits to within 1 LSB, then the change in input voltage Δv must be limited to $\Delta v < (\text{LSB} \times \text{FSV})$ during the conversion time T. It is clear from Eq. 4.14 that

$$\Delta v_{\max} = \pi f v_u^* T < (\text{LSB} \times \text{FSV}) \tag{4.15}$$

Solving Eq. 4.15 for the frequency limit f gives

$$f < \frac{\text{LSB} \times \text{FSV}}{v^*_u}\left(\frac{1}{\pi T}\right)$$

which reduces to

$$f < \frac{2^{-n}}{\pi T} \tag{4.16}$$

If the signal is unipolar, n is the number of bits; however, if the signal is bipolar, an additional bit is necessary to give the sign. For this reason, a unipolar signal can be sampled with twice the frequency of a bipolar signal. For example, a 10-bit ADC converting at 20 readings per second can monitor a unipolar signal with a frequency $f < 0.0031$ Hz. This signal with $v_u^* = 10$ V corresponds to a maximum slew rate of 0.098 V/s in an ADC with FSV $= 10$ V.

4.5.3   Parallel or Flash Method

Parallel, or flash, analog-to-digital conversion is the fastest but most expensive method for designing ADCs. The flash converter, illustrated in Fig. 4.10, employs $(2^n - 1)$ voltage comparators arranged in parallel. Each comparator is connected to the unknown voltage v_u. The reference voltage is applied to a binary resistance ladder so that the reference voltage applied to a given comparator is 1 LSB higher than the reference voltage applied to the lower adjacent comparator.

When the analog signal is presented to the comparator bank, all the comparators with $v_R^* < v_u$ will go high and those with $v_R^* > v_u$ will stay low. Since they are latching-type comparators, they hold high or low until they are downloaded to a system of digital logic gates that convert the parallel-comparator word into a binary-coded word.

The illustration shown in Fig. 4.10 is deceptively simple, since an 8-bit parallel ADC contains $2^8 - 1 = 255$ latching comparators and resistances and about 1000 logic gates to convert the output to binary code. Also, the accuracy is improved by placing a sample-and-hold amplifier before the ADC so that the input voltage does not change over the period required to operate the comparators.

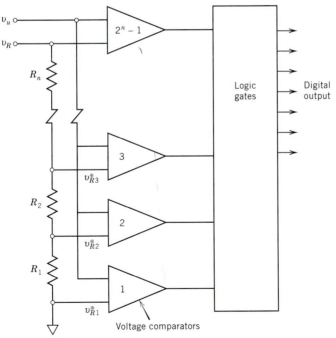

Figure 4.10 Schematic diagram for a parallel or flash analog-to-digital converter.

Parallel comparators have improved in performance at reduced cost in recent years with the availability of very large scale integrated circuits that accommodate all of the components on a single chip of silicon.

Flash ADCs are employed on very high speed digital acquisition systems, such as waveform recorders or digital oscilloscopes. Conversion in these applications is made at 50 to 200 million samples per second, which gives sampling times of 20 to 5 ns.

4.6 DATA DISTRIBUTION

When conversion of an analog voltage to a digital code is completed, the digital code is stored in a register until the digital word is needed in the instrument system. The transmission of the digital word from the ADC register to another component in the digital system is performed with a data bus. For the simple digital system illustrated in Fig. 4.11, the data bus is a series of parallel wires, with one wire for each bit, which connects the register to the display driver. This is a dedicated data bus, because it services only one ADC and one display. In this case, the control logic is designed to operate the ADC, transmit the digital data, and operate the display. The control logic is driven by a clock, and the process begins with a conversion signal from the logic device to the ADC. The ADC then indicates a status (busy), samples and holds the analog input, completes the conversion, and downloads to the register. The control logic activates the strobe, and the digital word is transmitted to the display driver over the parallel data bus. The display driver interprets the digital code and activates the correct light segments to display the output signal for observation by the experimentalist. The process is repeated at a frequency controlled by the conversion speed of the DAC.

As the digital instrumentation system increases in complexity, data transmission becomes more involved and several additional devices must be introduced to control the conversion process and to direct the flow of data in the system. The multichannel data-acquisition system, illustrated in Fig. 4.12, is used to introduce several devices employed in the more complex digital systems. The input to this system is m channels of analog data, which are to be converted to digital code and stored in memory or transmitted on a data bus. Note that the system contains an address bus as well as a data bus to accommodate control and flow of the data. As the processing of the data may change from one application to another, the system is equipped with a microprocessor and an interface device. The microprocessor is programmed to provide instructions that control the data-conversion process and direct the flow of data. The interface device serves to address the appropriate analog channel and to receive data from the ADC.

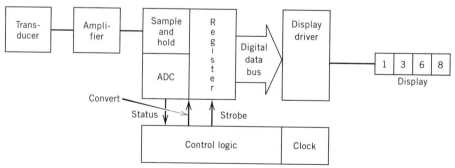

Figure 4.11 A simple dedicated single-channel system for acquiring, converting, and displaying data.

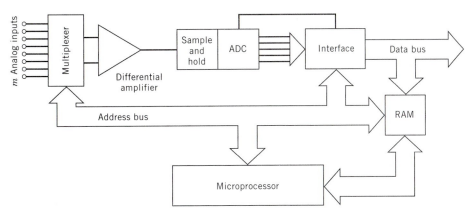

Figure 4.12 Data flow and control through a multichannel analog-to-digital conversion system.

The m analog channels interface with the digital system through a multiplexer (MUX). The multiplexer is an m-channel switch that connects one of the analog signals to the ADC. An address signal containing n bits for 2^n channels determines which data channel is being sampled. The microprocessor or interface unit addresses the MUX and causes it to switch to the appropriate channel.

The analog signal is amplified before conversion is initiated so that the maximum voltage corresponds to the range of the ADC. A conversion command imposed on the strobe input of the ADC initiates the analog-to-digital conversion with a sample-and-hold operation. After the conversion is complete, the microprocessor issues a read pulse with the channel address. The digital word is then transferred with its n-bit address to a register in the interface unit. The addressed digital word is then transferred to memory (RAM), the microprocessor, or the data bus, depending on the application.

4.6.1 Bus Structures

In the preceding discussion, the bus structure employed was parallel with one conductor for each bit of digital code and one conductor for each address bit. Parallel bus structure with either a multiconductor cable or a data-acquisition board is the proper approach if the analog instrument is close to the processor. However, in some applications, the analog sources and the processor are too far apart for a parallel bus structure to be effective, and one- or two-wire transmission is required. In this case the data are transmitted in a serial manner, where a series of pulses transmitted over a single conductor carries a coded analog or digital signal. The serial and parallel bus structure have been standardized so that a wide variety of digital instruments can be easily connected to a measurement/control system.

RS-232

The RS-232 is a standard bus structure that employs a serial protocol where a single transmitter sends one bit (0 or 1) of information at a time to a single receiver. This method of transmission is used when data-transfer rates are low or when data must be transferred over long distances. The serial protocol can be converted by a modem, placed on a standard telephone line, and converted back to serial data by a second modem at the receiving end of the line.

A typical example of connecting a computer to a digital instrument is illustrated in Fig. 4.13. Transmission of data with a serial protocol is also illustrated in Fig. 4.13 by

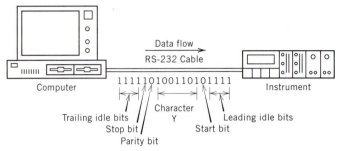

Figure 4.13 Connecting a computer and a digital instrument with an RS-232 cable using a serial protocol.

a train of bits that include the leading idle bits, the start bit, the character bits, the parity bit, the stop bit, and the trailing idle bits. The start bit is the first bit in the transmission sequence; the parity bit is an optional bit that ensures that the correct number of character bits have been sent and received; the stop bit signifies the end of transmission. In addition to serial sequencing of the data flow, the RS-232 standard defines the baud rate, hardware and software handshakes, and the communication path.

The RS-232 bus structure is popular because most personal computers used in the laboratory have at least one RS-232 port and many digital instruments are available with an RS-232 bus structure.

The major disadvantage of the RS-232 is its relatively low baud rate. The maximum rate of data transmission is only 19K bits per second. Another limitation is that a serial bus can only communicate with a single receiver. If communication must be effected with several receivers, either a circuit board with multiple serial ports or a serial-port multiplexing device must be added to the computer.

IEEE-488

In 1965, Hewlett–Packard designed an instrumentation interface bus, designated as HP-IB, that gained widespread acceptance. This bus was later standardized as the IEEE-488 bus, sometimes called the General Purpose Interface Bus (GPIB).

The IEEE-488 bus is a parallel protocol that transmits eight bits of information simultaneously over a standard cable. The bus contains eight data lines, three handshake lines, five interface-management lines, and eight ground and shield wires, as illustrated in Fig. 4.14. The bus carries information between a system controller and one or more digital instruments. Unlike the RS-232 bus, which is limited to a single instrument, the IEEE-488 bus can connect up to 15 compatible instruments simultaneously.

The advantages of the IEEE-488 bus include ease of use, because the user does not have to set communication parameters and speed. The IEEE-488 bus has an extremely

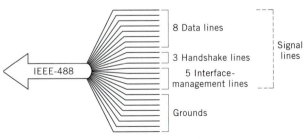

Figure 4.14 IEEE-488 parallel bus structure.

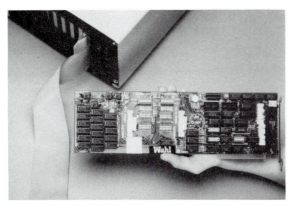

Figure 4.15 Plug-in data-acquisition board for an IBM-compatible personal computer. (Courtesy of Wahl Instruments, Inc.)

high baud rate (up to 8×10^6 bits per second). Also, the three-wire interlocked handshake guarantees that message bytes on data lines are sent and received without transmission error.

The disadvantages of this bus include a limitation of 2 m per device up to a maximum of 20 m on cable length. Also, most personal computers are not equipped with a 488 interface, although 488 interfaces are available as plug-in cards that can be installed at relatively low cost.

Plug-in Boards

Cables to transmit the digital code can be eliminated completely if the data-acquisition system is available on a plug-in circuit board. Boards, such as the one shown in Fig. 4.15, are inserted directly into the computer's bus and transfer data directly to the computer's memory. Plug-in boards are usually less expensive and more adaptable than stand-alone instruments, because they utilize computer hardware (power supply, back panel, fan, case, and so on). Plug-in boards can be designed to perform several functions. For example, the features of the board illustrated in Fig. 4.15 include 16-bit resolution, 16 channels of differential analog input, 16 channels of digital input/output, and a 16-bit counter/timer. Analog input signals from ± 25 mV to ± 5 V can be accommodated. The counter/timer can be used to accumulate up to 64,000 input counts from an external frequency source at the rate of 2 MHz. The counter can also be used as a timer to generate precise control signals after a predetermined elapsed time.

Plug-in circuit cards effectively eliminate stand-alone digital instruments by collecting the functions of these individual instruments on a single circuit board. This process eliminates the need for connecting cables and bus structures. The bus structure still exists, but it is contained in the personal computer. The only requirements are that the plug-in board must be designed for the particular bus structure used in the computer and must be unaffected by the computer's internal electrical noise.

4.7   INTERFACES

Digital instrument systems typically interface with other digital devices, such as controllers, terminals, printers, and microcomputers. Communication across these inter-

Table 4.3 American Standard Code for Information Interchange

Character	Value Decimal	Hex	Character	Value Decimal	Hex	Character	Value Decimal	Hex
NULL	0	00	*	42	2A	T	84	54
SOH,CTRL A	1	01	+	43	2B	U	85	55
STX,CTRL B	2	02	,	44	2C	V	86	56
ETX,CTRL C	3	03	—	45	2D	W	87	57
EOT,CTRL D	4	04	.	46	2E	X	88	58
ENQ,CTRL E	5	05	/	47	2F	Y	89	59
ACK,CTRL F	6	06	0	48	30	Z	90	5A
BELL,CTRL G	7	07	1	49	31	[	91	5B
BS	8	08	2	50	32	\	92	5C
TAB	9	09	3	51	33	]	93	5D
LF	10	0A	4	52	34	^	94	5E
VT	11	0B	5	53	35	←	95	5F
FF, CTRL L	12	0C	6	54	36	SPACE	96	60
CR	13	0D	7	55	37	a	97	61
SO, CTRL N	14	0E	8	56	38	b	98	62
SI, CTRL O	15	0F	9	57	39	c	99	63
DLE	16	10	:	58	3A	d	100	64
DC1	17	11	;	59	3B	e	101	65
DC2	18	12	<	60	3C	f	102	66

Table 4.3 (*Continued*)

Character	Decimal	Hex
DC3	19	13
DC4	20	14
NAK,CTRL U	21	15
SYN,CTRL V	22	16
ETB,CTRL W	23	17
CAN,CTRL X	24	18
EM,CTRL Y	25	19
SUB,CTRL Z	26	1A
ESC	27	1B
FS	28	1C
GS	29	1D
RS	30	1E
US	31	1F
sp	32	20
!	33	21
"	34	22
#	35	23
$	36	24
%	37	25
&	38	26
'	39	27
(	40	28
)	41	29

Character	Decimal	Hex
=	61	3D
>	62	3E
?	63	3F
@	64	40
A	65	41
B	66	42
C	67	43
D	68	44
E	69	45
F	70	46
G	71	47
H	72	48
I	73	49
J	74	4A
K	75	4B
L	76	4C
M	77	4D
N	78	4E
O	79	4F
P	80	50
Q	81	51
R	82	52
S	83	53

Character	Decimal	Hex
g	103	67
h	104	68
i	105	69
j	106	6A
k	107	6B
l	108	6C
m	109	6D
n	110	6E
o	111	6F
p	112	70
q	113	71
r	114	72
s	115	73
t	116	74
u	117	75
v	118	76
w	119	77
x	120	78
y	121	79
z	122	7A
{	123	7B
|	124	7C
}	125	7D
~	126	7E
del	127	7F

faces between digital devices is essential to process and display information in an efficient manner.

The ASCII code (American Standard Code for Information Interchange) is the most widely used code for transmitting alphanumerics, special characters, and control characters. The ASCII code listing in Table 4.3 contains 128 characters with their decimal and hex equivalents. The ASCII characters are transmitted in serial or in parallel as a 7-bit digital word. In the serial format, the bits that define the ASCII character are sent as a synchronous train of binary (on/off) levels with the LSB transmitted first. A total of 11 pulses are used to transmit a 7-bit character, as indicated in Fig. 4.16. The first pulse is the start bit, which goes low to indicate that transmission of a new character is about to occur. The next seven pulses define the 7-bit binary coding for one of the 128 characters. The data bits are followed by a parity bit and two stop bits to indicate that transmission is complete. The signal remains high until the next character is transmitted. In the example shown in Fig. 4.16, the binary number 1001010, which is 74, was transmitted to indicate the ASCII character J.

The speed of transmission (baud rate) depends on the digital devices, the capabilities of the transmission line, and the distance transmitted. The rate of information transmission is expressed in terms of baud, which is the unit for bits per second. Common information rates are 300 baud for a maximum distance of 10,000 ft and 1200 baud for 4,000 ft with a 20-mA driver operating into a twisted-wire pair.

When the IEEE-488 parallel bus is employed, only seven wire conductors carry the 7-bit ASCII code over relatively short transmission distances while three handshake lines asynchronously control the transfer of information between devices. The NRFD (not-ready-for-data) line indicates that a device is ready or not to receive a message byte. The NDAC (No-Data-accepted) line indicates that a device has or has not accepted the message byte. The DAV (data-valid) line indicates when the data on the lines are valid and acceptable for receiving. With a relatively short transmission distance and a parallel bus, the information-transmission rate is controlled by the slowest device involved in the data transfer.

ASCII characters can also be transmitted using the RS-232C standard. This standard is widely employed in many digital systems and can be used with properly shielded cable for distances up to 50 ft. The RS-232C standard specifies the method for encoding information at the sending end (see Fig. 4.13), and the method for decoding at the receiving end. The interfacing of two digital devices, each equipped with an RS-232C port, is not always a simple matter of connecting the appropriate cable to each device. The signals used in both devices must correspond with respect to voltage levels, timing, control sequence, and handshake options, and the wiring of data and control lines with up to 25 assignable wires must be consistent between devices.

4.8   DIGITAL VOLTMETERS

Digital voltmeters (DVM) offer many advantages over analog meters, such as speed in reading, increased accuracy, better resolution, and the capability of automatic operation. Digital voltmeters display the measurement with lighted numerals, as shown in Fig. 4.17, rather than as a pointer deflection on a continuous scale as with analog meters. Digital multimeters are available to read current, resistance, and ac and dc voltages. The DVM may also be used with a multiplexer and a digital printer to provide a simple but reliable automatic data-logging system.

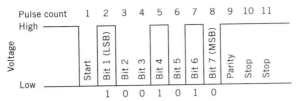

Figure 4.16 Serial transmission of the binary code 1001010 = 74 for ASCII character *J*.

The range of a DVM is determined by the number of full digits in the display. For example, a 4-digit DVM can record a count to 9999. If the full scale of the DVM is set at 1 V, the count of 9999 provided by the four digits would register a reading of 0.9999 V. Some DVMs are equipped with partial digits to extend the range. The partial digit can only display a limited range of numbers. Although zero and one are common for the $\frac{1}{2}$ or most significant digit, some newer models are capable of displaying partial digits of 2, 3, or 4. The partial digit extends the resolution of the DVM. For example, consider use of a 4-digit DVM for measuring 10.123 V. Since only four digits are available, the meter set on the 10-V scale would read 10.12 V. The last digit (3) would be truncated and lost. If a $4\frac{1}{2}$-digit DVM is used for the same measurement, the extra partial digit permits 100 percent overranging and a maximum count of 19999. The display of the $4\frac{1}{2}$-digit meter would show five numbers (10.123), provided the partial digit (1) is not exceeded.

Overranging may be expressed as a percentage of full scale. For instance, a 4-digit DVM with 100 percent overrange displays a maximum reading of 19999. Similarly, with a 200 percent overrange, the maximum display is 29999. In some instances, the overrange capability of the DVM is expressed in terms of the specified range. The 4-digit DVM with 100 percent overrange, which has a maximum display of 19999, could be specified with full-scale ranges of 2 V, 20 V, 200 V, and so on, and with no overrange specification.

Resolution of a DVM is determined by the maximum count displayed. For example, a 4-digit DVM with a maximum count of 10,000 has a resolution of 1 part in 10,000 or 100 ppm (see Eq. 4.2a).

The sensitivity of a DVM is the smallest increment of voltage that can be detected and is determined by multiplying the lowest full-scale range by the resolution. Therefore, a 4-digit DVM with a 100-mV lowest full-scale range has a sensitivity of 0.0001×100 mV $= 0.01$ mV.

Figure 4.17 A $5\frac{1}{2}$-digit digital multimeter/scanner. (Courtesy of Keithley Instruments, Inc.)

Accuracy of a DVM is usually expressed as $\pm x$ percent of the reading $\pm N$ digits. A typical value for a $5\frac{1}{2}$-digit instrument operating on a 2-V range is $x = 0.0015$ percent and $N = 2$. Accuracy depends heavily on calibration and instrument stability. A modern DVM utilizes electronic calibration, where calibration constants are stored in nonvolatile memory and calibration is accomplished without internal adjustments. Improved stability, which reduces fluctuations in the last digit, is the key to enhanced accuracy. Hermetically sealed resistance networks in the circuits have greatly improved stability in more modern designs.

The simplified block diagram for a digital multimeter, shown in Fig. 4.18, illustrates the overall features of integrating digital meters. The input to the multimeter may be an ac voltage, a dc voltage, a current, or a resistance; however, in all cases, the input is immediately converted to a dc voltage. The signal is then amplified with two variable-gain amplifiers, identified as the input and post amplifiers in Fig. 4.18. To avoid overload, the gain of these amplifiers is automatically adjusted by control logic so that the voltage applied to the analog-to-digital converter is within specifications.

The A/D converter changes the dc voltage input to a proportional clock count by using the dual-slope integration technique illustrated in Fig. 4.19. There are three different operations in the dual-slope integration technique for A/D conversion. First, during autozero, the potential at the integrator output is zeroed for a fixed time, such as 100 ms. Second, the dc input is integrated with respect to time for a fixed period, again, such as 100 ms. The output of the integrator is a linear ramp with respect to time as shown in Fig. 4.19. At the end of the run up, the dc input voltage is disconnected from the integrator and the third operation, run down, is initiated. Run-down time may vary from 0 to about 200 ms and will depend on the charge developed on the integrating capacitor during run up. Since the discharge rate is fixed during run down, the larger the charge on the integrating capacitor, the longer the discharge time. Since both run up and run

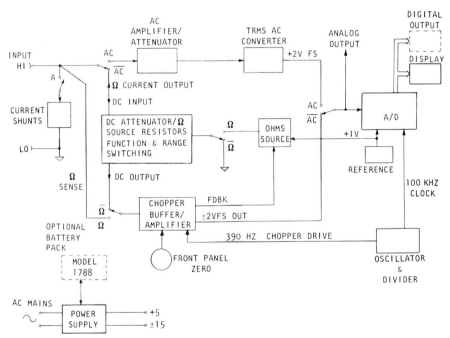

Figure 4.18 Simplified signal-flow diagram for a digital multimeter. (Courtesy of Keithley Instruments, Inc.)

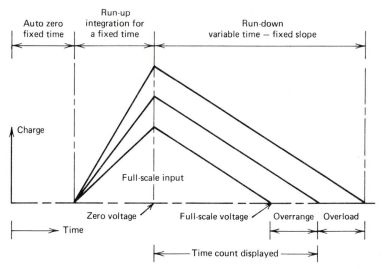

Figure 4.19 Dual-slope integration technique for analog-to-digital conversion.

down produce slopes on the voltage–time trace, this conversion method from voltage to time is called dual-slope integration (See Section 4.5.2).

A counter is started at the beginning of a run down and operates until the output voltage from the integrator crosses zero. The accumulated time is proportional to the dc voltage applied to the integrator. This time count is then displayed as the voltage. Polarity, range, and function information provided by the controller are also displayed.

The characteristics of a digital voltmeter may be altered by changing the number of digits, the time interval for integrator run up, and the frequency of the clock. A professional hand-held multimeter with $3\frac{1}{2}$ digits and a maximum count of 1999 is designed with five different ranges: ±199.9 mV, ±1.999 V, ±19.99 V, ±199.9 V, and ±1199 V. The highest sensitivity is 100 μV on the 200-mV range. Accuracy is ±0.1 percent of the reading plus one digit. The clock frequency is 200 kHz, and the integration time is 100 ms. The reading rate varies from 2.4 to 4.7 readings per second, depending on the input.

System DVMs are more complex than bench-type DVMs, since the former are provided with a microprocessor and local memory to facilitate interfacing with other components of an automated data-processing system. A typical data-processing system incorporating a system-type DVM includes a scanner or multiplexer for switching input voltages into the DVM for analog-to-digital conversion and a bus that is compatible with a personal computer (PC). The memory in the PC (RAM or disk) is used to store the acquired data. Reduction, manipulation, and analysis of the data is performed according to programmed instructions, and results are often presented in easily understood graphic form from a digital printer.

System multimeters are higher performance devices than are bench multimeters. The number of digits is usually increased to $6\frac{1}{2}$ or $7\frac{1}{2}$, which increases the count and improves the resolution to 1 or 0.1 ppm, respectively. Clock frequencies are increased and advanced conversion techniques are employed to give a reading rate of 500 readings per second. Microprocessors are added to control the different measurements and to control the interface. System multimeters are available with either IEEE-488 or RS-232 bus structures. The internal microprocessor has modest computing capability that permits the user to add, subtract, multiply, and divide as well as store and compare numerical information.

Table 4.4 Characteristics of a System-type DVM Hewlett–Packard model 3457A

Number of Digits[a]	Maximum Readings per Second	Resolution (ppm)
$3\frac{1}{2}$	1350	333.3
$4\frac{1}{2}$	1250	33.3
$5\frac{1}{2}$	360	3.3
$6\frac{1}{2}$	53	0.3

[a]The number of digits can be selected, ranging from $3\frac{1}{2}$ to $6\frac{1}{2}$, with the model 3457A. The $\frac{1}{2}$ digit on this model may read 0, 1, or 2.

4.9   DATA-LOGGING SYSTEMS

A basic data-logging system consists of a scanner or multiplexer, a digital voltmeter, and a recorder. Such a system can be employed to record the output from a large number of transducers (1000 or more) at a sampling rate that depends on the capability of the DVM and the resolution required. Conversion rates for a modern DVM that utilizes a multislope analog-to-digital converter are shown in Table 4.4.

Since the DVMs are relatively fast, a system controller is needed to direct the scanner to each new channel, to control the integration time for the DVM, and to transfer the output from the DVM to the recorder. A block diagram of a typical multichannel data-logging system is shown in Fig. 4.20.

The system controller is a microprocessor that uses two separate buses: one for data transfer and the other for memory addressing. The software, which directs the operation of the controller, is stored in read-only memories (ROMs) and a random-

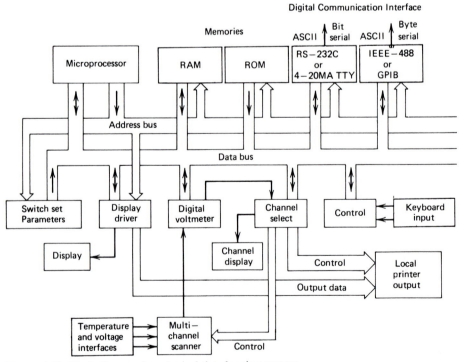

Figure 4.20 Block diagram for a typical data-logging system.

access memory (RAM). The system operating programs are permanently stored in the ROMs, which are programmed during manufacture of the instrument. Operator input, individual channel parameters, and other program routines are entered through a keyboard and stored in RAM.

The scanner contains a bank of switches (usually three-pole) that serve to switch two leads and the shield from the input cable to the integrating digital voltmeter. In most cases, high-speed (1000 channels per second) solid-state switching devices (J field-effect transistors) are employed. The scanner operation is directed by the system controller. The modes of operation available include single-channel recording, single scan of all channels, continuous scan, and periodic scan. In the single-channel mode, a preselected channel is continuously monitored at the reading rate of the system. In the single-scan mode, the scanner makes a single sequential sweep through a preselected group of channels. The continuous-scan mode is identical to the single-scan mode, except that the system automatically resets and recycles on completion of a scan. The periodic scan is simply a single scan that is initiated at preselected time intervals, such as 1, 5, 15, 30, or 60 min. The scanner also provides a signal for the visual display of the channel number and a code signal to the controller to identify the transducer being monitored.

The transducer signal is switched through the scanner to a high-quality integrating digital voltmeter that serves as the ADC. The speed of operation depends primarily on the capabilities of the DVM and the resolution required. As is evident in Table 4.4, high reading rates (in excess of 1000 per second) are possible even with a $4\frac{1}{2}$-digit ADC that provides a resolution of 33 ppm.

The output from a data logger is displayed with a digital panel meter that indicates the voltage output, its polarity, and the channel number. A permanent record is often made with a line printer, which records the output data and identification on a paper tape. However, the output of most data-logging systems can be recorded with several other devices, such as magnetic tapes, local disk memories, or remote disk memory associated with a central computing center. One of the principal advantages of a data-logging system is the capability of processing the data in what is essentially real time with an on-line computer.

4.10 DATA-ACQUISITION SYSTEMS

Data-acquisition systems are similar to data-logging systems in that they accept input from a large number of transducers and automatically process the data. There are, however, two principal differences between data-logging and data-acquisition systems. First, data-acquisition systems are much faster (sample rates from 20,000 to 250,000 per second). Second, the on-board microcomputer, memory capacity, and speed of a data-acquisition system are superior to those available with data-logging systems. Also, data-acquisition systems have additional software and graphics-processing capability for both data processing and report-quality graphic display of results. The higher sampling rates are achieved by replacing the integrating DVM with high-speed, successive-approximation, analog-to-digital converters. These converters utilize sample-and-hold amplifiers that sample and hold the input value fixed during the conversion period as described in Section 4.5.1. For extremely high speed operation (250 kHz) the flash-type A/D converters are used as described in Section 4.5.3.

Automatic data-acquisition systems usually involve many optional components and can be custom designed. All systems, however, contain six basic subsystems, the controller, the signal conditioner, the multiplexer/amplifier, the analog-to-digital converter

(ADC), the storage or memory unit, and the readout devices. A schematic illustration of the elements of a data-acquisition system is presented in Fig. 4.21.

To illustrate these six subsystems, consider the commercial data-acquisition system shown in Fig. 4.22. The controller is a microprocessor that serves as the interface between the operator and the data-acquisition system. The operator enters directions to the controller through the front-panel keypads. A liquid-crystal display (LCD) provides a readout of the system operating parameters and select readings of the quantities being measured. The controller is programmed with parameters that affect the data flow, such as the sampling rate, the sequence of channels to be monitored, signal levels to trigger and to stop recording, time limits for the same purpose, and activation limits for supervisory alarms.

The controller also directs the flow of data collected in the random-access memory. Depending on programming, the data stored in the RAM buffer can be held for one cycle, erased, or transferred to a permanent storage medium such as a hard disk or, on high-capacity systems, an optical disk.

The signal conditioner (for strain gages, for example) consists of the power supply, the Wheatstone bridges, and the terminals used to connect the output from a large number of bridges to the multiplexer. The bridges are usually contained on a plug-in circuit board, which can be modified by adding or deleting fixed resistors to provide for quarter-, half-, or full-bridge arrangements. A single power source is often used to power a number of individual bridges (from 10 to 100, depending on the design of the system). The power supply is a highly regulated, constant-current supply, which can be adjusted to provide about 4 mA to the bridge.

The multiplexer portion of the signal conditioner–scanner subassembly consists of two parts: (1) a bank of switches that connect the two output leads and the cable shield from each bridge to the differential amplifier. In modern systems, solid-state

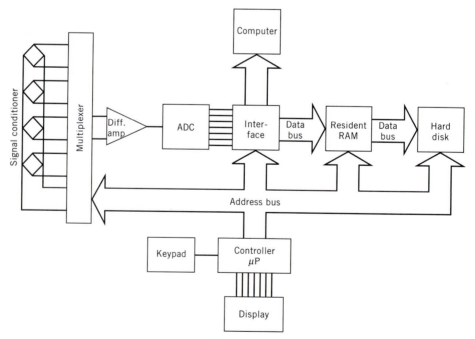

Figure 4.21 Schematic diagram showing the key elements in a digital data-acquisition system.

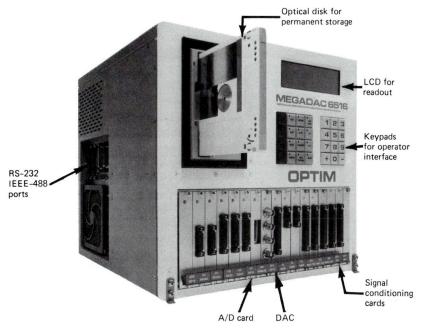

Figure 4.22 A 128-channel, 250,000-sample-per-second, portable data-acquisition system. (Courtesy of Optim Electronics Corporation.)

switching devices (J field-effect transistors) are employed. (2) Circuits that control the switching sequence as programmed in the controller. The low-voltage signal from the bridge is switched through the multiplexer to the differential amplifier. The signal is amplified to be consistent with the full-scale voltage of the ADC.

The amplified analog signal is converted into a digital signal using an analog-to-digital converter. Two different ADCs are commonly used: the dual-slope integrating type and the successive-approximation type. The dual-slope integrating ADC is relatively slow, usually less than 1K samples per second, and it is often used on the lower performance data-logging systems. The successive-approximation ADC is much faster and is used in more advanced acquisition systems, which operate at 20,000 to 250,000 samples per second.

The data from the ADC are output from the interface unit (see Fig. 4.21) on a parallel-wired data bus and are usually stored temporarily in RAM on a first-in–first-out basis. The data can be processed in real time on a host computer or, in less-critical experiments, downloaded from RAM to a permanent storage medium such as a hard disk. In very high capacity systems, data are stored on an optical disk. The address bus carries instructions and addresses from the controller and provides for the flow of the data and the organization of the memory devices.

The disks provide the input data, in digital format, to an off-line computer, which processes the data. Software is usually available from the suppliers of the data-acquisition system to assist in the organization of data files and the transfer of files to spreadsheets for subsequent processing. Data manipulation and graphics output often depend on the capabilities of commercially available spreadsheets or more powerful processing and graphing programs. Graphic output of processed data in a report-ready format is relatively easy to achieve and is a significant advantage of digital data-acquisition systems.

Rapid advances in developing higher performance and lower cost electronic components, such as ADCs, RAMs, multiplexers, and storage devices, have markedly improved the capability of digital data-acquisition systems. A decade ago these systems were very slow (10 to 20 samples per second) and considered suitable only for recording static phenomena. Today, with systems capable of 250K samples per second, a digital data-acquisition system can process several channels of dynamic signals with the capability of an oscillograph.

4.11 PC-BASED DATA-ACQUISITION SYSTEMS

In recent years relatively low cost circuit boards that contain many of the elements found in higher cost, more elaborate data-acquisition systems have become available. Some data-acquisition boards (similar to the circuit card shown in Fig. 4.15) are designed to interface with several different sensors[1], such as strain gages, resistance temperature detectors (RTDs), thermocouples, and thermistors. The sensor support includes sensor excitation, linearization, cold reference compensation, and conversion of the output to engineering units.

Multisensor plug-in boards typically contain four sections. The first section performs the signal conditioning for the sensors, multiplexes to the appropriate sensor, and amplifies the signal with a programmable gain. The second section performs A/D conversion. A typical card utilizes a 12-bit successive-approximation-type ADC with a conversion time of about 2 μs, or a sampling rate of 50,000 samples per second. The third section incorporates a microcomputer with on-board memory used in data processing to perform tasks such as linearization, reference junction compensation, and engineering unit conversion. The microprocessor also provides the logic to scan the sensors, adjust the amplifier gain, and transfer the data to the standard bus registers. The standard bus interface incorporates drivers and receivers to facilitate communication with a PC host computer. The input/output (I/O) ports vary from card to card, but a typical configuration that utilizes serial communication has three ports. One port is used to transfer data and instructions to the card, and the other two ports are used to transfer data and indicate status to the host computer.

The card is programmed from the PC host computer and the digitized data are transferred from the card to the memory of the PC using the standard bus in the host computer. External bus structure is not required for data transmission, because the entire data-acquisition system is contained within the PC. All further processing and preparation of graphics are performed on the host computer using commercially available software.

4.12 DIGITAL OSCILLOSCOPES

The digital oscilloscope is identical to the conventional oscilloscope except for (1) the manipulation of the input signal prior to display on the cathode-ray tube, (2) the permanent-storage capabilities of the instrument, and (3) the signal-processing capabilities. With a digital oscilloscope, the input signal is converted to digital form, stored in a buffer memory, and then transferred to the DAC, where it is reconverted

---

[1]These sensors are described in Chapter 6.

to an analog voltage for display on the CRT. A microprocessor controls the storage, transfer, and display of the data. A photograph of a modern digital oscilloscope is shown in Fig. 4.23.

Since input data are stored in addition to being displayed, operation of the digital oscilloscope differs from operation of the conventional analog oscilloscope. The display on the CRT of the digital oscilloscope is a series of points produced by the electron beam at locations controlled by the data in storage. The data collected during the sweep of the oscilloscope are in memory and can be recalled and analyzed either within the digital oscilloscope or by downloading to a host computer.

Several features establish the capabilities of a digital oscilloscope. First, the sample rate and the bandwidth are important in recording transient signals. The bandwidth requirements for analog oscilloscopes described in Chapter 3 also apply to digital oscilloscopes. An additional requirement is the sample rate, which determines the time interval between data points. For example, a digital oscilloscope with a 100 MHz sample rate can sample, hold, convert, and store a data point in 10 ns.

Second, the number of bits of resolution used in the A/D conversion determines the accuracy relative to full scale. Resolutions of 6, 7, 8, 10, and 12 bits are used in commercially available digital oscilloscopes, which produce resolution errors ranging from 1.59 to 0.024 percent (see Table 4.2). Higher sampling rates are available with lower resolution instruments, and a trade-off between sampling rate and resolution is necessary in selecting dynamic digital recorders.

Next, the size of the memory is important, because memory length controls the length of the signal that can be recorded. Memory width is the same as the number of bits provided by the ADC. For example, if a digital oscilloscope with 8-bit resolution contains an 8K-word memory and operates each of two channels at 100 MHz, the memory size is 12,768 words, with each word containing 8 bits, for a total memory of 262,144 bits. The memory is divided into two equal segments of 16,384 words with one segment for channel A and the second for channel B. At the maximum sampling rate of 100 samples/μs the recording period is $16,384/100 = 163.84$ μs when both channels are in use. Reducing the sampling rate to 20 samples/s extends the recording period to 1619.2 s. Clearly, the range of time of observation that can be covered with digital oscilloscopes with variable sampling rates is large (500,000/1 is common).

Figure 4.23 Photograph of a modern digital oscilloscope. (Courtesy of Nicolet Instrument Corporation.)

A final important feature is whether the A/D conversion method is designed for single-shot pulse measurements or measurements of repetitive periodic signals. Repetitive signals are easier to measure, as sampling can be repeated on the second and subsequent waveforms to give instruments with apparent sampling rates that are an order of magnitude higher than the real sampling rates. The delayed sequential sampling technique used to increase the number of data points (samples) that define a repetitive waveform is illustrated in Fig. 4.24. For pulse measurements, the signal occurs once and only once and repetitive measurements cannot be used to extend the sampling rate.

In operation, A/D conversion takes place continuously at a prescribed sampling rate with the words going to storage until the buffer is full. The buffer address for each data word is proportional to the time when the data were taken. After a sweep, when the buffer is full, the data are discarded, unless a trigger signal is received during the sweep and the conversion process continues. If a trigger signal is received during the sweep, the data in the buffer memory are transferred to mainframe memory and then processed and displayed as a voltage–time trace on the CRT.

The fact that the input signal has been stored offers many advantages for data display or data processing. The data are displayed on the CRT in a repetitive manner, so that even traces from transient events appear stationary. The trace can also be manipulated by expanding either the horizontal or vertical scales or both. This expansion feature permits a small region of the record to be enlarged and examined in detail, as illustrated in Fig. 4.25. Readout of the data from the trace is also much easier and more accurate with digital oscilloscopes. A pair of marker lines called cursors (one vertical and the other horizontal) can be positioned anywhere on the screen. The procedure is to position the vertical line at a time on the trace when a reading is needed. The horizontal marker (or cross hair) automatically positions itself on the trace. The coordinates of the cross hair, the intersection with the trace, are presented as a numerical display on the screen, as illustrated in Fig. 4.26.

Modern digital oscilloscopes are usually equipped with a microprocessor that provides several on-board signal-analysis features. These features include the following:

1. Pulse characterization—rise time, fall time, base-line and top-line width, overshoot, period, frequency, rms, standard deviation, and duty cycle.

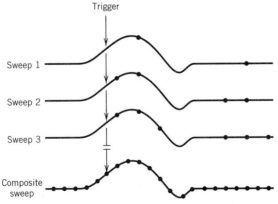

Figure 4.24 By varying the time between the trigger and the digitizer, a different data point can be recorded on each repetition (sweep).

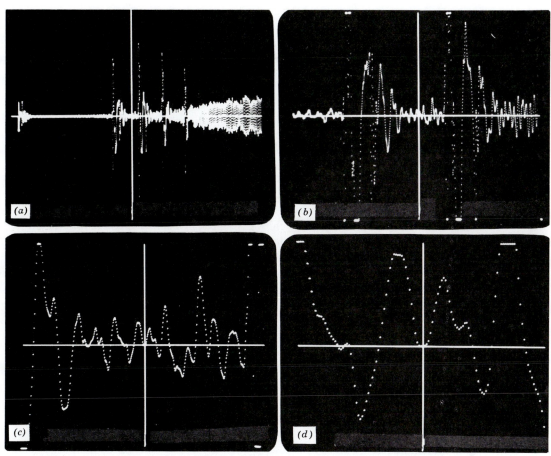

Figure 4.25 Expansion of the display on the screen of a digital oscilloscope. (*a*) Unmagnified. (*b*) Both axes (voltage and time) expanded by a factor of 4. (*c*) Both axes expanded by a factor of 16. (*d*) Both axes expanded by a factor of 64. (Courtesy of Nicolet Instrument Corp.)

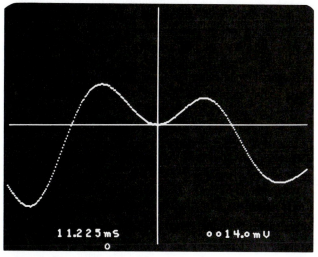

Figure 4.26 Numeric display of data on the screen of a digital oscilloscope. (Courtesy of Nicolet Instrument Corp.)

2. Frequency analysis—power, phase, and magnitude spectrum.
3. Spectrum analysis—100-to-50,000-point fast Fourier transforms.
4. Math package—add, subtract, multiply, integrate, and differentiate.
5. Smoothing—1, 3, 5, 7, or 9 point.
6. Counter—average frequency and event crossings.
7. Display control—x zoom, x position, y gain, y offset.
8. Plotting display.
9. Mass storage to floppy disk, hard disk, or nonvolatile bubble memory.

If additional processing is required, the data can be downloaded to a host computer for final analysis.

Digital oscilloscopes are relatively new; early models were introduced in 1972. Initially, performance of digital oscilloscopes was limited because of the low-bandwidth capability (10 kHz or less); however, improvements in A/D converters, microprocessors, and high-speed RAM memory chips have greatly enhanced speed of conversion, improved resolution, and expanded the amount of data that can be stored. Except for the recording of very high-speed transient signals with very short rise times, the digital oscilloscope is superior in every respect to the analog oscilloscope. Digital methods of recording are rapidly replacing the analog methods, because costs are decreasing while performance is improving.

4.13   WAVEFORM RECORDERS

Waveform recorders are similar to digital oscilloscopes in that they incorporate a high-speed ADC and store transient pulses or high-frequency waveforms in high-speed memory (RAM). They differ from digital oscilloscopes in that they require an auxiliary display device, an oscilloscope or the monitor of a computer, to display either the original or the processed output.

Waveform recorders usually have superior resolution to and longer memories than digital oscilloscopes with comparable sampling rates. They can be thought of as extremely high performance tape recorders with the tape replaced by silicon memory. A typical multichannel waveform recorder incorporates $2^{18} = 262,144$ words of memory with 12 bits per word. At a maximum sampling rate of 10^7 samples/s (corresponding to a minimum sampling time of 100 ns), with the memory divided equally between two channels, the recording time is $131,072/10^7 = 13.1$ ms. At the recorder's lowest sampling rate of 20 samples/s, the recording time is $131,072/20 = 6553$ s or 109 min.

The waveform recorder has three main components, as illustrated in Fig. 4.27. At the input side, the analog signal is digitized with an independent ADC for each

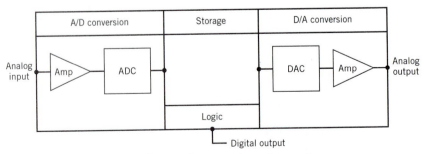

Figure 4.27 Block diagram of a waveform recorder and generator.

channel. The digitized output goes to the second component, a memory buffer (silicon RAM), where it is stored. The digital signal can be downloaded over a data bus to a host computer for processing and subsequent display. Alternatively, the digital signal can be reconverted to an analog signal and replayed repeatedly into a display device, such as an analog oscilloscope.

4.14 ALIASING

Digital data-acquisition systems contain an ADC that converts an analog signal to a digital signal at a specified sampling rate. This sampling rate is extremely important in dynamic measurements where high-frequency analog signals are being processed. For a well-defined representation of a dynamic waveform, the analog signal should be determined with a digital point (sample) 10 or more times during the period. This concept of using 10 samples to define a sine wave is illustrated in Fig. 4.28. As the number of digital data points decreases, the definition of the type of waveform and its characteristics degrade to the point where the digital representation can be misleading. When the sampling frequency f_s is

$$f_s \leq 2f \tag{4.16}$$

the waveform with frequency f takes on a false identity. Nyquist sampling theory is the basis for Eq. 4.16. The minimum sampling frequency is $f_s = 2f^*$ and the maximum conversion time is $T_s^* = \frac{1}{2}f^*$, where f^* is called the Nyquist frequency and T_s^* is the Nyquist interval.

If the frequency of the analog signal $f \geq f_s/2$, the sampling process is inadequate and the output from the ADC gives a false low-frequency waveform, called an *alias*, that differs from the true analog signal. To illustrate aliasing, consider a sinusoidal analog signal with a frequency f_1 given by

$$v_1(t) = \cos(2\pi f_1 t) \tag{4.17}$$

If the ADC has a sampling frequency f_s, the times at which the signal is sampled is

$$t = nT_s = \frac{n}{f_s} \qquad n = 0, 1, 2, \ldots \tag{4.18}$$

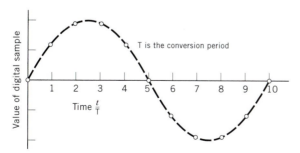

Figure 4.28 Digital representation of an analog waveform (sine) that illustrates the definition achieved with 10 samples.

Substituting Eq. 4.18 into Eq. 4.17 gives the sampled voltages as

$$v_1(nT_s) = \cos\left(\frac{2\pi n f_1}{f_s}\right) \tag{4.19}$$

Next, consider a second sinusoidal signal with a frequency f_2, which is greater than a cutoff frequency f_c, so that

$$f_2 = 2m f_c \pm f_1 \qquad m = 1, 2, \ldots \tag{4.20}$$

If the ADC used with both signals is the same, then $v_2(nT_s)$ is obtained by interchanging f_2 for f_1 in Eq. 4.19 to give

$$v_2(nT_s) = \cos\left[\frac{2\pi n(2m f_c \pm f1)}{f_s}\right] \tag{4.21}$$

Let the cutoff frequency be

$$f_c = \frac{f_s}{2} \tag{4.22}$$

as indicated by sampling theory. Substituting Eq. 4.22 into Eq. 4.21 gives

$$v_2(nT_s) = \cos\left(2\pi n m \pm \frac{2\pi n f_1}{f_s}\right)$$

$$= \cos\left(\frac{2\pi n f_1}{f_s}\right) \tag{4.23}$$

A comparison of Eqs. 4.23 and 4.19 shows that v_1 and v_2 are identical at each sampling time (nT_s) and that the two frequencies f_1 and f_2 cannot be distinguished. For example, if $f_c = 200$ Hz, then $f_s = 400$ Hz and an alias signal with a frequency $f_1 = 50$ Hz occurs whenever the input signal has frequencies f_2 of 350, 450, 750, 850, and so on. The relation showing the alias frequency is obtained by substituting Eq. 4.22 into Eq. 4.20 to give

$$\pm f_1 = f_2 - m f_s \qquad \text{if} \qquad f_2 \ge f_s/2 \qquad m = 1, 2, \ldots \tag{4.24}$$

where

f_1 is the frequency of the alias signal
f_2 is the frequency of the input signal

The results of Eq. 4.24 are illustrated in Fig. 4.29 for $f_s = (3/2)f_2$. Note that the recorded digital signal exhibits the alias frequency $f_1 = f_2/2$.

Aliasing can be avoided if the sampling frequency exceeds twice the maximum frequency in the analog signal.

$$f_s = 2f_c > 2f_2 \tag{4.25}$$

This relation is essentially the same as Eq. 4.22, which defines the cutoff frequency in terms of the sampling frequency. Clearly, if the maximum frequency in the analog

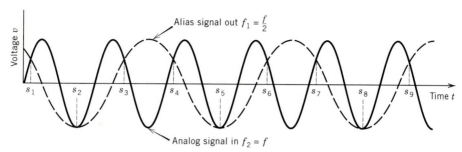

Figure 4.29 Effect of aliasing when a sinusoid is sampled at a frequency $f_s = 3/2 f_2$.

signal does not exceed the cutoff frequency, the identity of Eqs. 4.19 and 4.23 cannot be shown and aliasing will not occur.

4.14.1   Anti-aliasing Filters

Commercial instruments avoid the aliasing problem by using anti-aliasing analog filters to reduce the high-frequency components in the analog signals. These filters have a frequency response function that exhibits a relatively sharp drop beginning at $f = 0.4 f_s$, as shown in Fig. 4.30. The analog signal is attenuated 40 dB at the Nyquist frequency $f_s^* = f_s/2$. Since 40 dB is equivalent to a signal transmission of only 1 percent, the filter reduces the signal component that produces aliasing to an insignificant amount.

Unfortunately, anti-aliasing filters can severely distort transient signals with high-frequency components. The possibility of distortion on the one hand and the need to prevent aliasing of the signal on the other implies that care must be exercised when selecting the sampling rate of the ADC for a specific high-frequency measurement.

attenuate the signal below the threshold of the instrument

4.15   SUMMARY

Digital systems that employ analog-to-digital converters (ADCs) or digital-to-analog converters (DACs) are widely used in instrument systems for engineering analysis and for process control. The principal advantage of digital systems is the ability to store data and to process that data by using low-cost commercially available storage devices, computers, controllers, and a wide variety of application software.

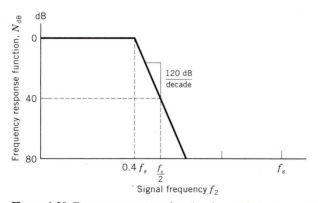

Figure 4.30 Frequency response function for an ideal analog anti-aliasing filter for an ADC with a sampling rate f_s.

Analog-to-digital conversion is the key element in systems utilized for data acquisition and for digital instruments such as digital voltmeters, digital oscilloscopes, and waveform recorders. Features such as sampling rate, resolution, accuracy, and cost are used to compare ADCs. For high resolution and low cost, ADCs usually employ the integration method for conversion; however, the sampling rate is relatively low. For very high sampling rates, the flash method of conversion is used with more limited resolution (eight bits) and significantly higher cost. The method of successive approximation is used in ADCs with moderate sampling rates, good resolution, and intermediate cost.

Digital-to-analog conversion is also employed in process control. For example, after digital processing of data according to programmed instructions, it may be concluded that a process must be modified by increasing the temperature of an oven. The digital signal for the temperature increase is converted to an analog signal by means of a DAC. This signal is then amplified and used to activate a temperature controller.

One of the most critical problems encountered with digital systems is interfacing. If a system is composed of several instruments produced by different companies, interfacing problems are likely to occur. Although standardized buses and connectors are common, differences still exist in wiring order, communication protocol, and the voltage levels in the logic circuits. It is hoped that in the future, improvements in standardization will occur and components produced by all manufacturers will be compatible.

Finally, developments in digital devices are continuing at a rapid rate as the technology used in manufacturing integrated circuits permits further reductions in device size. Continued improvements will certainly lead to higher sampling rates, more bits, higher resolution, and lower costs for digital hardware. Software written for specific measurements is currently expanding significantly, particularly with PC-based data-acquisition systems.

REFERENCES

1. Ahmed, H., and P. J. Spreadbury: *Analogue and Digital Electronics for Engineers*, 2nd ed., Cambridge Univ. Press, London/New York, 1984.
2. Artwick, B. A.: *Microcomputer Interfacing*, Prentice–Hall, Englewood Cliffs, N.J., 1980.
3. Bibbero, R. J., and D. M. Stern: *Microprocessor Systems: Interfacing and Applications*, Wiley, New York, 1982.
4. Floyd, Thomas L.: *Digital Fundamentals*, 4th ed., Merrill, Columbus, Ohio, 1990.
5. Hall, D. V.: *Digital Circuits and Systems*, McGraw–Hill, New York, 1989.
6. Hilburn, J. L., and P. M. Julich: *Microcomputers/Microprocessors*, Prentice–Hall, Englewood Cliffs, N.J., 1976.
7. Hoeschele, D. F.: *Analog-to-Digital/Digital-to-Analog Conversion Techniques*, Wiley, New York, 1968.
8. Johnson, D. E., J. L. Hilburn, and P. M. Julich: *Digital Circuits and Microcomputers*, Prentice–Hall, Englewood Cliffs, N.J., 1979.
9. Spencer, C. D.: *Digital Design for Computer Data Acquisition*, Cambridge Univ. Press, London/New York, 1990.

EXERCISES

4.1 Prepare a block diagram showing a combined analog-digital instrumentation system to measure pressure for an application involving

(a) engineering analysis   (b) process control

4.2   Prepare a table showing the maximum count C as a function of the number of bits n. Let n vary from 1 to 32.

4.3   Add a column showing the resolution R to the table in Exercise 4.2.

4.4   Add a column showing the resolution error $\mathcal{E}_R$ (%) to the table in Exercise 4.2.

4.5   What is the resolution error that can be expected from an instrument with 8-bit logic circuits?

4.6   Write an engineering brief describing an A/D converter (ADC).

4.7   Write an engineering brief describing a D/A converter (DAC).

4.8   Describe the difference between resolution error and quantizing error. Which is the most important?

4.9   Prepare a graph showing the standard deviation of a number of measurements of v/FSV as n increases from 6 to 16.

4.10   Prepare a graph showing C versus v/FSV that demonstrates offset error for

(a) a D/A converter   (b) an A/D converter

4.11   Prepare a graph showing C versus v/FSV that demonstrates scale-factor error for

(a) a D/A converter   (b) an A/D converter

4.12   Prepare a graph showing C versus v/FSV that demonstrates linearity error for

(a) a D/A converter   (b) an A/D converter

4.13   For the 4-bit DAC shown in Fig. 4.5, determine the output voltage v_o for the following digital codes:

(a) 1101   (c) 0110   (e) 1001
(b) 1010   (d) 0101   (f) 1110

4.14   In a brief paragraph describe a register and indicate some of its uses in a digital system.

4.15   What is a strobe signal, and how it is used in a D/A converter? Why is it necessary?

4.16   What are the three common systems used in designing A/D converters? List the advantages and disadvantages of each system.

4.17   Prepare an illustration, similar to Fig. 4.7, that demonstrates A/D conversion by the method of successive approximations if the fixed analog input voltage v_u is

(a) 1/8 FSV   (c) 3/4 FSV
(b) 7/16 FSV   (d) 13/16 FSV

4.18   Explain why the conversions of Exercise 4.17 were all exact.

4.19   Since A/D conversions require some time for switching and comparing, how are errors caused by voltage fluctuations during the conversion period avoided?

4.20   Using Fig. 4.8 as a guide, describe the operation of a successive-approximation A/D converter. Indicate the purpose of each block element and each input or output signal.

4.21   Beginning with Eq. 4.10 verify Eq. 4.12.

4.22   Using Fig. 4.9 as a guide, describe the operation of a dual-slope integrating A/D converter. Indicate the purpose of each block element and each input or output signal.

4.23   Verify Eq. 4.16 beginning with Eq. 4.13.

4.24   Determine the frequency limit of a 12-bit unipolar dual-slope integrating A/D converter capable of 20 readings per second.

4.25   Determine the frequency limit if the A/D converter of Exercise 4.24 is bipolar.

4.26   Determine the slew rate of the A/D converter in Exercise 4.24 if the FSV is

    (a) 1 V   (c) 5 V

    (b) 2 V   (d) 10 V

4.27   Using Fig. 4.10 as a guide, describe the operation of a flash-type A/D converter.

4.28   Describe instruments that use the flash-type A/D converter. What are the sampling rates employed in these instruments? How are the sampling rate, bandwidth, and sampling times related?

4.29   Digital instrumentation systems often use both an address bus and a data bus. Explain the purpose of these two buses.

4.30   Using Fig. 4.12 as a guide, describe the operation of a multichannel ADC system.

4.31   Explain why both parallel- and series-type bus structures are required.

4.32   Describe the RS-232 standard bus structure.

4.33   Describe the IEEE-488 standard bus structure.

4.34   Compare the RS-232 and IEEE-488 bus structures and cite the advantages and disadvantages of each.

4.35   How does the use of a plug-in data-acquisition board eliminate the need for external bus structures?

4.36   Why is a plug-in data-acquisition board so cost-effective?

4.37   Prepare an illustration, similar to Fig. 4.16, which demonstrates transmission of the ASCII characters

    (a) /   (c) S

    (b) 8   (d) z

4.38   What is baud rate and what factors affect it?

4.39   A 5-digit DVM is capable of what maximum count if it has

    (a) 0 overranging   (c) 200 percent overranging

    (b) 100 percent overranging

4.40   You are to measure a voltage v_i with a $5\frac{1}{2}$-digit DVM capable of 100% overranging. If the meter is specified with an accuracy of ±0.002 % and ±2 counts, determine the maximum and minimum readings anticipated if v_i is

    (a) 1.80000 V   (b) 2.50000 V   (c) 9.99996 V

4.41   Determine the error in each of the three cases of Exercise 4.40.

4.42   Describe the differences between system and bench-type multimeters.

4.43   The purchasing department of a state agency asks you to write a specification so that they can procure bids for a system multimeter. Prepare this specification.

4.44   What is the difference between a system multimeter and a data-logging system?

4.45   Explain the use of ROM and RAM memory incorporated into a data-logging system.

4.46   Describe the scanner employed in a data-logging system.

4.47   Explain the difference between data-logging systems and data-acquisition systems.

4.48 What are the six basic subsystems that are included in most data-acquisition systems?

4.49 Describe the function of a controller in a digital data-acquisition system.

4.50 What are the four sections usually incorporated into the design of a multisensor plug-in board?

4.51 Compare digital and analog oscilloscopes and cite the advantages and disadvantages of each.

4.52 What is the role of auxiliary storage in applying digital oscilloscopes to transient measurements?

4.53 Prepare a graph showing observation time in a digital oscilloscope as a function of sampling rate. Use memory size in words as a parameter and let the sampling rate vary from 10 samples/s to 10^7 samples/s. Let the memory size be 1000, 2000, 5000, 10,000, and 20,000 words.

4.54 In measuring periodic signals, it is possible to increase the apparent sampling rate of a digital oscilloscope. Explain how this is accomplished.

4.55 Explain how data preceding the trigger time can be recovered on a digital oscilloscope.

4.56 What are the common on-board signal-analysis features found on digital oscilloscopes equipped with a microprocessor?

4.57 Explain the differences between a waveform recorder and a digital oscilloscope.

4.58 Verify Eq. 4.24.

4.59 Prepare a graph, similar to the one in Fig. 4.29, showing the analog and the alias signals if

	f_c (Hz)	f_s (Hz)	f_2 (Hz)
(a)	200	400	850
(b)	300	600	1,500
(c)	1,000	2,000	5,000
(d)	5,000	10,000	30,000

4.60 What are the characteristics of an anti-aliasing filter?

4.61 List the advantages and disadvantages of using an anti-aliasing filter.

Chapter 5

Sensors for Transducers

5.1 INTRODUCTION

Transducers are electromechanical devices that convert a mechanical change, such as displacement or force, into a change in an electrical signal that can be monitored as a voltage after conditioning. A wide variety of transducers are available for use in measuring mechanical quantities. Transducer characteristics, which include range, linearity, sensitivity, and operating temperatures, are determined primarily by the sensor that is incorporated into the transducer to produce the electrical output. For example, a set of strain gages on a tension link provides a transducer that produces a resistance change $\Delta R/R$ in proportion to the load applied along the axis of the link. The strain gages serve as the sensor in this force transducer and play a significant role in establishing the characteristics of the transducer.

Sensors used in transducer design include potentiometers, differential transformers, strain gages, capacitors, piezoelectric and piezoresistive crystals, thermistors, and so on. The important features of these different sensors are described in this chapter.

5.2 POTENTIOMETERS

The simplest type of potentiometer, shown schematically in Fig. 5.1, is the slide-wire resistor. This sensor consists of a length ℓ of resistance wire attached across a voltage source v_i. The relationship between the output voltage v_o and the position x of a wiper, as it moves along the length of the wire, can be expressed as

$$v_o = \frac{x}{\ell}v_i \qquad \text{or} \qquad x = \frac{v_o}{v_i}\ell \tag{5.1}$$

Thus, the slide-wire potentiometer can be used to measure a displacement x.

Straight-wire resistors are not feasible for most applications, since the resistance of a short length of wire is low and low resistance imposes excessive power requirements on the voltage source. To alleviate this difficulty, high-resistance, wire-wound potentiometers are obtained by winding the resistance wire around an insulating core, as shown in Fig. 5.2. The potentiometer illustrated in Fig. 5.2*a* is used for linear-displacement measurements. Cylindrically shaped potentiometers, similar to the one

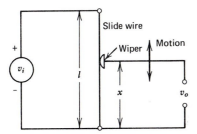

Figure 5.1 Slide-wire resistance potentiometer.

illustrated in Fig. 5.2*b,* are used for angular-displacement measurements. The resistance of a wire-wound potentiometer can range between 10 and 10^6 Ω, depending on the type and diameter of the wire used and the length of the coil.

The resistance of the wire-wound potentiometer increases in a stepwise manner as the wiper moves from one turn to the adjacent turn. This step change in resistance limits the resolution of the potentiometer to L/n, where n is the number of turns in the length L of the coil. Resolutions ranging from 0.05 to 1 percent are common, with the lower limit obtained by using many turns of very small diameter wire.

The range of the potentiometer is controlled by the active length L of the coil. Linear potentiometers are available in many lengths up to about 1 m. The range of the angular-displacement potentiometer can be extended by arranging the coil in the form of a helix. Helical potentiometers are commercially available with as many as 20 turns; therefore, angular displacements as large as 7200 deg can be measured quite easily.

In order to improve resolution, potentiometers have been introduced that utilize thin films with controlled resistivity instead of wire-wound coils. The film resistance on an insulating substrate exhibits high resolution, lower noise, and longer life. For example, a resistance of 50 to 100 Ω/mm can be obtained with the conductive plastic films that are used for commercially available potentiometers with a resolution of 0.001 mm.

The dynamic response of both the linear and the angular potentiometer is severely limited by the inertia of the shaft and wiper assembly. Since this inertia is large,

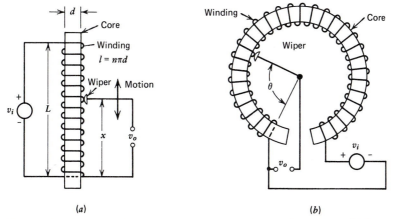

Figure 5.2 Wire-wrapped resistance potentiometer for (*a*) longitudinal displacements and (*b*) angular displacements.

the potentiometer is used only for static or quasi-static measurements where a high-frequency response is not required.

Electronic noise often occurs as the electrical contact on the wiper moves from one wire turn to the next. This noise can be minimized by ensuring that the coil is clean and free of oxide films and by lubricating the coil with a thin film of light oil. Under ideal conditions, the lives of wire-wound and conductive-plastic potentiometers exceed 1 and 10 million cycles, respectively.

Potentiometers are used primarily to measure large displacements, that is, 10 mm or more for linear motion and 15 deg or more for angular motion. Potentiometers are relatively inexpensive yet accurate; however, their main advantage is simplicity of operation, since only a voltage source and a DVM to measure voltage constitute the complete instrumentation system. Their primary disadvantage is limited frequency response, which precludes their use for dynamic measurements.

5.3   DIFFERENTIAL TRANSFORMERS

Differential transformers, based on a variable-inductance principle, are also used to measure displacement. The most popular variable-inductance sensor for linear-displacement measurements is the linear variable differential transformer (LVDT). An LVDT, illustrated in Fig. 5.3a, consists of three symmetrically spaced coils wound onto an insulated bobbin. A magnetic core, which moves through the bobbin without contact, provides a path for magnetic flux linkage between coils. The position of the magnetic core controls the mutual inductance between the center or primary coil and the two outer or secondary coils.

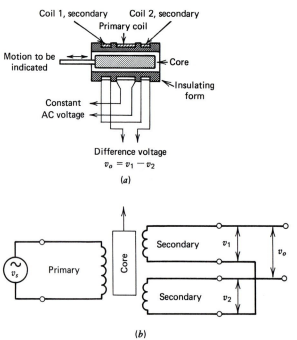

Figure 5.3 (*a*) Sectional view of a linear variable differential transformer (LVDT). (*b*) Schematic diagram of the LVDT circuit.

When an ac excitation is applied to the primary coil, voltages are induced in the two secondary coils. The secondary coils are wired in a series-opposing circuit, as shown in Fig. 5.3b. When the core is centered between the two secondary coils, the voltages induced in the secondary coils are equal but out of phase by 180 deg. Since the coils are in a series-opposing circuit, the voltages v_1 and v_2 in the two coils cancel and the output voltage is zero. When the core is moved from the center position, an imbalance in mutual inductance between the primary and secondary coils occurs and an output voltage, $v_o = v_2 - v_1$, develops. The output voltage is a linear function of core position, as shown in Fig. 5.4, as long as the motion of the core is within the operating range of the LVDT. The direction of motion can be determined from the phase of the output voltage relative to the input voltage.

The frequency of the voltage applied to the primary winding can range from 50 to 25,000 Hz. If the LVDT is used to measure transient or periodic displacements, the carrier frequency should be 10 times greater than the highest frequency component in the dynamic signal. Highest sensitivities are attained with excitation frequencies between 1 and 5 kHz. The input voltage can range from 5 to 15 V. The power required is usually less than 1 W. Sensitivities of different LVDTs vary from 0.02 to 0.2 V/mm of displacement per volt of excitation applied to the primary coil. At rated excitation voltages, sensitivities vary from 0.16 to 2.5 V/mm of displacement. The higher sensitivities are associated with short-stroke LVDTs, with an operating range of ±2 mm; the lower sensitivities are for long-stroke LVDTs, with an operating range of ±150 mm.

Since the LVDT is a passive sensor that requires ac excitation at a frequency different from common ac supplies, signal conditioning circuits are required for its operation. A typical signal conditioner (see Fig. 5.5) provides a power supply, a frequency generator to drive the LVDT, and a demodulator to convert the ac output signal from the LVDT to an analog dc output voltage. Finally, a dc amplifier is incorporated in the signal conditioner to increase the magnitude of the output voltage.

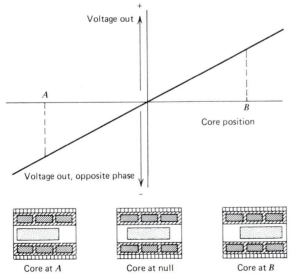

Figure 5.4 Phase-referenced output voltage as a function of LVDT core position.

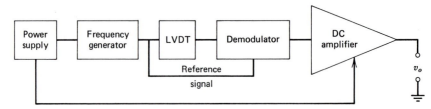

Figure 5.5 Block diagram of the signal conditioning circuit for an LVDT.

During the past decade, microelectronic circuits have been developed that permit miniaturization of the signal conditioners so that they can be packaged within the case of an LVDT. The result is a small self-contained sensor known as a direct-current differential transformer (DCDT). A DCDT operates from a battery or a regulated power supply and provides an amplified output signal that can be monitored on either a DVM or an oscilloscope. The output impedance of a DCDT is relatively low (about 100 Ω).

Both the LVDT and the DCDT are used to measure linear displacement. An analogous device known as a rotary variable differential transformer (RVDT) has been developed to measure angular displacements. As shown in Fig. 5.6, the RVDT consists of two primary coils and two secondary coils wound symmetrically on a large-diameter insulated bobbin. A cardioid-shaped rotor, fabricated from a magnetic material, is mounted on a shaft that extends through the bobbin and serves as the core. As the shaft rotates and turns the core, the mutual inductance between the primary and secondary windings varies and produces an output voltage versus rotation response that resembles a modified sinusoid.

Although the RVDT is capable of a complete rotation (360 deg), the range of linear operation is limited to ±40 deg. The linearity of a typical RVDT, having a range of ±40 deg, is about 0.5 percent of the range. Reducing the operating range improves the linearity, and an RVDT operating within a range of ±5 deg exhibits a linearity of about 0.1 percent of this range.

The LVDT, DCDT, and RVDT have many advantages as sensors for measuring displacement. There is no contact between the core and the coils; therefore, friction and hysteresis are eliminated. Since the output is continuously variable with

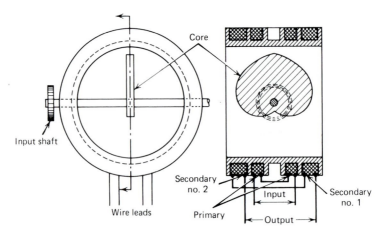

Figure 5.6 Simplified cross section of a rotary variable differential transformer (RVDT).

input, resolution is determined by the characteristics of the voltage recorder. Noncontact also ensures that life will be long, with no significant deterioration of performance over this period.[1] The small core mass and freedom from friction give the sensor a limited capability for dynamic measurements. Finally, the sensors are not damaged by overtravel; therefore, they can be employed as feedback transducers in servo-controlled systems where overtravel may occur owing to accidental deviations beyond the control band. Typical performance characteristics for LVDTs, DCDTs, and RVDTs are listed in Table 5.1.

5.4 RESISTANCE STRAIN GAGES

Electrical resistance strain gages are thin metal-foil grids (see Fig. 5.7) that can be adhesively bonded to the surface of a component or structure. When the component or structure is loaded, strains develop and are transmitted to the foil grid. The resistance of the foil grid changes in proportion to the load-induced strain. The strain sensitivity of metals, first observed in copper and iron by Lord Kelvin in 1856, is explained by the following simple analysis.

The resistance R of a uniform metallic conductor can be expressed as

$$R = \frac{\rho L}{A} \tag{5.2}$$

where

ρ is the specific resistance of the metal
L is the length of the conductor
A is the cross-sectional area of the conductor

Differentiating Eq. 5.2 and dividing by the resistance R gives

$$\frac{dR}{R} = \frac{d\rho}{\rho} + \frac{dL}{L} - \frac{dA}{A} \tag{a}$$

The term dA represents the change in cross-sectional area of the conductor resulting from the applied load. For the case of a uniaxial tensile stress state, recall that

$$\varepsilon_a = \frac{dL}{L} \quad \text{and} \quad \varepsilon_t = -\nu\varepsilon_a = -\nu\frac{dL}{L} \tag{b}$$

where

ε_a is the axial strain in the conductor
ε_t is the transverse strain in the conductor
ν is Poisson's ratio of the metal used for the conductor

If the diameter of the conductor is d_0 before application of the axial strain, the diameter of the conductor d_f after it is strained is

$$d_f = d_0\left(1 - \nu\frac{dL}{L}\right) \tag{c}$$

From Eq. c it is clear that

$$\frac{dA}{A} = -2\nu\frac{dL}{L} + \nu^2\left(\frac{dL}{L}\right)^2 \approx -2\nu\frac{dL}{L} \tag{d}$$

---

[1] Mean time between failures for a typical DCDT is 33,000 h.

Table 5.1 Performance Characteristics for LVDTs, DCDTs, RVDTs

A. Linear Variable Differential Transformers

Model Number	Nominal Linear Range (in.)	Linearity ± Percent — Percent of Full Range				Sensitivity [(mV/V)/0.001 in.]	Impedance (Ω)	
		50	100	125	150		Primary	Secondary
050 HR	± 0.050	0.10	0.25	0.25	0.50	6.3	430	4000
100 HR	± 0.100	0.10	0.25	0.25	0.50	4.5	1070	5000
200 HR	± 0.200	0.10	0.25	0.25	0.50	2.5	1150	4000
300 HR	± 0.300	0.10	0.25	0.35	0.50	1.4	1100	2700
400 HR	± 0.400	0.15	0.25	0.35	0.60	0.90	1700	3000
500 HR	± 0.500	0.15	0.25	0.35	0.75	0.73	460	375
1000 HR	± 1.000	0.25	0.25	1.00	1.30[a]	0.39	460	320
2000 HR	± 2.000	0.25	0.25	0.50[a]	1.00[a]	0.24	330	330
3000 HR	± 3.000	0.15	0.25	0.50[a]	1.00[a]	0.27	115	375
4000 HR	± 4.000	0.15	0.25	0.50[a]	1.00[a]	0.22	275	550
5000 HR	± 5.000	0.15	0.25	1.00[a]	—	0.15	310	400
10000 HR	±10.000	0.15	0.25	1.00[a]	—	0.08	550	750

[a] Requires reduced core length.

Table 5.1 Performance Characteristics for LVDTs, DCDTs, RVDTs (*continued*)

B. Direct-Current Differential Transformers

Model Number	Nominal Linear Range (in.)	Scale Factor (V/in.)	Response −3 dB (Hz)
050 DC-D	± 0.050	200	500
100 DC-D	± 0.100	100	500
200 DC-D	± 0.200	50	500
500 DC-D	± 0.500	20	500
1000 DC-D	± 1.000	10	200
2000 DC-D	± 2.000	5.0	200
3000 DC-D	± 3.000	3.3	200
5000 DC-D	± 5.000	2.0	200
10000 DC-D	±10.000	1.0	200

C. Rotary Variable Differential Transformers

Model Number	Linearity ± Percent			Sensitivity [(mV/V)/degree]	Impedance (Ω)	
	± 30°	± 40°	± 60°		Primary	Secondary
(@ 2.5 kHz)						
R30A	0.25	0.5	1.5	2.3	125	500
R36A	0.5	1.0	3.0	1.1	750	2000
(@ 10 kHz)						
R30A	2.5	0.5	1.5	2.9	370	1300
R36A	0.5	1.0	3.0	1.7	2500	5400

Source: Courtesy of Schaevitz Engineering

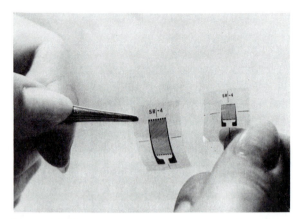

Figure 5.7 Electrical resistance strain gages. (Courtesy of BLH Electronics.)

Substituting Eq. d into Eq. a and simplifying yields

$$\frac{dR}{R} = \frac{d\rho}{\rho} + \frac{dL}{L}(1 + 2\nu) \tag{5.3}$$

which can be written as

$$S_A = \frac{dR/R}{\varepsilon_a} = \frac{d\rho/\rho}{\varepsilon_a} + (1 + 2\nu) \tag{5.4}$$

The quantity S_A is defined as the sensitivity of the metal or alloy used for the conductor.

It is evident from Eq. 5.4 that the strain sensitivity of a metal or alloy is a result of the changes in dimensions of the conductor, as expressed by the term $(1 + 2\nu)$, and the change in specific resistance, as represented by the term $(d\rho/\rho)/\varepsilon$. Experimental studies show that the sensitivity S_A ranges between 2 and 4 for most metallic alloys used in strain-gage fabrication. Because the quantity $(1 + 2\nu)$ is approximately 1.6 for most of these materials, the contribution owing to the change in specific resistance with strain varies from 0.4 to 2.4. This increase in specific resistance is the result of variations in the number of free electrons and their changing mobility with applied strain.

A list of the alloys commonly employed in commercial strain gages, together with their sensitivities, is presented in Table 5.2. The most commonly used strain gages are fabricated from the copper–nickel alloy known as Advance or constantan. The response curve for this alloy ($\Delta R/R$ as a function of strain) is shown in Fig. 5.8. This

Table 5.2 Strain Sensitivity S_A for Common Strain-Gage Alloys

Material	Composition (%)	S_A
Advance or Constantan	45 Ni, 55 Cu	2.1
Nichrome V	80 Ni, 20 Cr	2.1
Isoelastic	36 Ni, 8 Cr, 0.5 Mo, 55.5 Fe	3.6
Karma	74 Ni, 20 Cr, 3 Al, 3 Fe	2.0
Armour D	70 Fe, 20 Cr, 10 Al	2.0
Platinum–Tungsten	92 Pt, 8 W	4.0

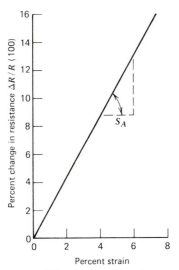

Figure 5.8 Change of resistance $\Delta R/R$ as a function of strain for Advance alloy.

alloy is widely used because its response is linear over a wide range of strain (beyond 8 percent), it has a high specific resistance, and it has excellent electrical stability with changes in temperature.

Most resistance strain gages are of the metal-foil type, where the grid configuration is formed by a photoetching process. Since the process is versatile, a wide variety of gage sizes and grid shapes can be produced. Typical examples are shown in Fig. 5.9. The shortest gage available is 0.20 mm; the longest is 102 mm. Standard gage resistances are 120 and 350 Ω; however, special-purpose gages with resistances of 500, 1000, and 5000 Ω are also available.

The etched metal–film grids are extremely fragile and easy to distort, wrinkle, or tear. For this reason, the metal grid is bonded to a thin plastic film, which serves as a backing or carrier, before photoetching. The carrier film, shown in Fig. 5.7, also provides electrical insulation between the gage and the component after the gage is bonded.

A strain gage exhibits a resistance change $\Delta R/R$ that is related to the strain ε in the direction of the grid by the expression

$$\frac{\Delta R}{R} = S_g \varepsilon \tag{5.5}$$

where S_g is the gage factor or calibration constant for the gage. The gage factor S_g is always less than the sensitivity of the metallic alloy S_A because the grid configuration of the gage with the transverse conductors is less responsive to axial strain than a straight uniform conductor.

The output $\Delta R/R$ of a strain gage is usually converted to a voltage signal with a Wheatstone bridge, as illustrated in Fig. 5.10. If a single gage is used in one arm of the Wheatstone bridge and equal but fixed resistors are used in the other three arms, the output voltage is

$$v_o = \frac{v_s}{4}\left(\frac{\Delta R_g}{R_g}\right) \tag{5.6}$$

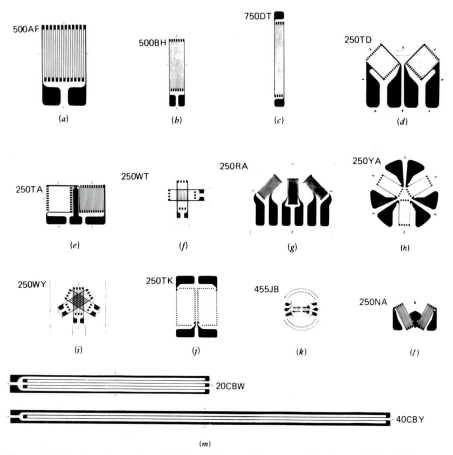

Figure 5.9 Configurations of metal-foil resistance strain gages. (Courtesy of Micro-Measurements Division, Measurements Group, Inc., USA). (*a*) Single-element gage. (*b*) Single-element gage. (*c*) Single-element gage. (*d*) Two-element rosette. (*e*) Two-element rosette. (*f*) Two-element stacked rosette. (*g*) Three-element rosette. (*h*) Three-element rosette. (*i*) Three-element stacked rosette. (*j*) Torque gage. (*k*) Diaphragm gage. (*l*) Stress gage. (*m*) Single-element gage for use on concrete.

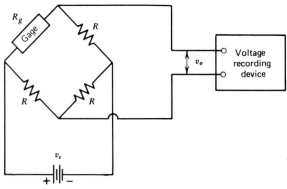

Figure 5.10 Wheatstone bridge circuit used to convert resistance change $\Delta R / R$ of a strain gage to an output voltage v_o.

Substituting Eq. 5.5 into Eq. 5.6 gives

$$v_o = \frac{1}{4} v_s S_g \varepsilon \tag{5.7}$$

The input voltage is controlled by the gage size (the power it can dissipate) and the initial resistance of the gage. As a result, the sensitivity, $S = v_o/\varepsilon = v_s S_g/4$, usually ranges from 1 to 10 $\mu V/(\mu m/m)$.

5.5   CAPACITANCE SENSORS

The *capacitance sensor,* illustrated in Fig. 5.11a, consists of a target plate and a second plate known as the sensor head. These two plates are separated by an air gap of thickness h and form the two terminals of a capacitor, which exhibits a capacitance C given by

$$C = \frac{kKA}{h} \tag{5.8}$$

where

    C is the capacitance in picofarads (pF)
    A is the area of the sensor head ($\pi D^2/4$)
    K is the relative dielectric constant for the medium in the gap ($K = 1$ for air)
    k is a proportionality constant; $k = 0.225$ for dimensions in inches and $k = 0.00885$ for dimensions in millimeters

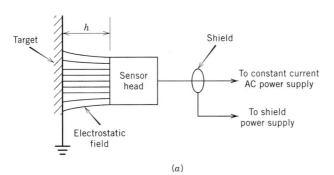

(a)

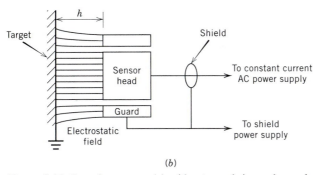

(b)

Figure 5.11 Capacitor sensor (a) without guard ring, where edge effects in the electrostatic field affect the range of linearity, and (b) with guard ring to extend the range of linearity.

If the separation between the head and the target is changed by an amount Δh, then the capacitance C becomes

$$C + \Delta C = \frac{kKA}{(h + \Delta h)} \tag{a}$$

which can be written as

$$\frac{\Delta C}{C} = -\frac{\Delta h/h}{1 + (\Delta h/h)} \tag{5.9}$$

This result indicates that $(\Delta C/C)$ is nonlinear, because of the presence of $\Delta h/h$ in the denominator of Eq. 5.9. To avoid the difficulty of employing a capacitance sensor with a nonlinear output, the change in the impedance owing to the capacitor is measured. Recall Eq. 2.42 where

$$Z_C = -\frac{j}{\omega C} \tag{2.42}$$

With a capacitance change ΔC,

$$Z_C + \Delta Z_C = -\frac{j}{\omega}\left[\frac{1}{C + \Delta C}\right] \tag{b}$$

Substituting Eq. 2.42 into Eq. b and solving for $\Delta Z_C/Z_C$ gives

$$\frac{\Delta Z_C}{Z_C} = -\frac{\Delta C/C}{1 + \Delta C/C} \tag{5.10}$$

Finally, substituting Eq. 5.9 into Eq. 5.10 yields

$$\frac{\Delta Z_C}{Z_C} = \frac{\Delta h}{h} \tag{5.11}$$

From Eq. 5.11 it is clear that the capacitive impedance Z_C is linear in h and that methods of measuring ΔZ_C will permit extremely simple plates (the target as ground and the sensor head as the positive terminal) to act as a sensor to measure the displacement Δh. Cylindrical sensor heads are linear and Eq. 5.11 is valid provided that $0 < h < D/4$ where D is the diameter of the sensor head. Fringing in the electric field produces nonlinearities if $(h + \Delta h)$ exceeds $D/4$. The linear range can be extended to $h \approx D/2$ if a guard ring surrounds the sensor, as shown in Fig. 5.11b. The guard ring essentially moves the distorted edges of the electric field to the outer edge of the guard, significantly improving the uniformity of the electric field over the sensor area and extending the linearity.

The sensitivity of the capacitance probe is given by Eqs. 2.42, 5.8, and 5.11 as

$$S = \frac{\Delta Z_C}{\Delta h} = \left|\frac{Z_C}{h}\right| = \left|\frac{1}{\omega C h}\right| = \left|\frac{1}{\omega k K A}\right| \tag{5.12}$$

Sensitivity can be improved by reducing the area A of the probe; however, as noted previously, the range of the probe is limited by linearity to about $D/2$. Clearly there is a range–sensitivity trade-off. Of particular importance is the circular frequency ω in Eq. 5.12. Low frequency improves sensitivity but limits frequency response of the instrument, another trade-off. It is also important to note that the frequency of the ac power supply must remain constant to maintain a stable calibration constant.

The capacitance sensor has several advantages. It is noncontacting and can be used with any target material, provided the material exhibits a resistivity less than 100 Ω/cm^2. The sensor is extremely rugged and can be subjected to high shock loads (5000 g) and intense vibratory environments. Its use as a sensor at high temperature is particularly impressive. Capacitance sensors can be constructed to withstand temperatures up to 2000°F, and they exhibit a constant sensitivity S over an extremely wide range of temperature (74°–1600°F). Examination of the relation for S in Eq. 5.12 shows that the dielectric constant K is the only parameter that can change with temperature. Since K is constant for air over a wide range of temperature, the capacitance sensor has excellent temperature characteristics.

The change in the capacitive impedance Z_C is usually measured with the circuit illustrated schematically in Fig. 5.12. The probe, its shield, and the guard ring are powered with a constant current ac supply. A digital oscillator is used to drive the ac supply and to maintain a constant frequency at 15.6 kHz. This oscillator also provides the reference frequency for the synchronous detector. The voltage drop across the probe is detected with a low-capacitance preamplifier. The signal from the preamplifier is then amplified again with a fixed-gain instrument amplifier. The high-voltage ac signal from the instrument amplifier is rectified and given a sign in the synchronous detector. The rectified signal is filtered to eliminate high-frequency ripple and give a dc output voltage related to Δh. A linearizing circuit is used to extend the range of the sensor by accommodating for the influence of the fringes in the electrostatic field. Finally, the signal is passed through a second instrument amplifier where the gain and zero offset can be varied to adjust the sensitivity and the zero reading of the DVM display. When the gain and zero offset are properly adjusted, the DVM reads Δh directly to the scale selected by the operator.

5.6 EDDY-CURRENT SENSORS

An eddy-current sensor measures distance between the sensor head and an electrically conducting surface, as illustrated in Fig. 5.13. Sensor operation is based on eddy currents that are induced at the conducting surface as magnetic flux lines from the sensor intersect with the surface of the conducting material. The magnetic flux lines

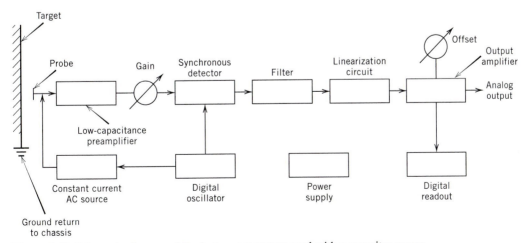

Figure 5.12 Schematic diagram of the instrument system used with a capacitor sensor.

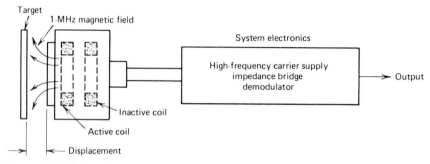

Figure 5.13 Schematic diagram for an eddy-current sensor.

are generated by the active coil in the sensor, which is driven at a very high frequency (1 MHz). The magnitude of the eddy current produced at the surface of the conducting material is a function of the distance between the active coil and the surface. The eddy currents increase as the distance decreases.

Changes in the eddy currents are sensed with an impedance (inductance) bridge. Two coils in the sensor are used for two arms of the bridge. The other two arms are housed in the associated electronic package, as shown in Fig. 5.13. The first coil in the sensor (active coil), which changes inductance with target movement, is wired into the active arm of the bridge. The second coil is wired into an opposing arm of the same bridge, where it serves as a compensating coil to balance and cancel the effects of temperature change. The output from the impedance bridge is demodulated and becomes the analog signal, which is linearly proportional to distance between the sensor and the target.

The sensitivity of the sensor is dependent on the target material, with higher sensitivity associated with higher conductivity materials. The output for a number of materials is shown as a function of specific resistivity in Fig. 5.14. For aluminum targets, the sensitivity is typically 100 mV/mil (4 V/mm). Thus, it is apparent that

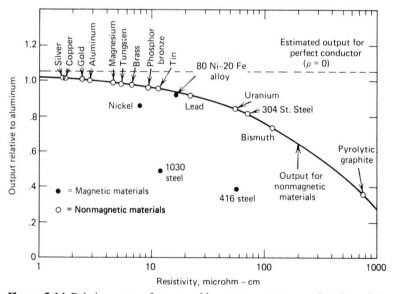

Figure 5.14 Relative output from an eddy-current sensor as a function of the resistivity of the target material.

eddy-current sensors are high-output devices if the specimen material is nonmagnetic. Figure 5.14 demonstrates that the sensitivity decreases significantly if the specimen material is magnetic.

For nonconducting, poorly conducting, or magnetic materials, it is possible to bond a thin film of aluminum foil to the surface of the target at the location of the sensor to improve the sensitivity. Since the penetration of the eddy currents into the material is minimal, the thickness of the foil can be as little as 0.7 mil (ordinary household aluminum foil).

The effect of temperature on the output of the eddy-current sensor is small. The sensing head with dual coils is temperature compensated; however, a small error can be produced by temperature changes in the target material, since the resistivity of the target material is a function of temperature. For instance, if the temperature of an aluminum target is increased by 500°F, its resistivity increases from 0.03 to 0.06 $\mu\Omega \cdot m$. From Fig. 5.14 it is evident that the bridge output is reduced by about 2 percent for this change in resistivity. For aluminum, the temperature sensitivity of the eddy-current sensor is 0.004 percent per °F.

The range of the sensor is controlled by the diameters of the coils, with the larger sensors exhibiting the larger ranges. The range-to-diameter ratio is usually about 0.25. Linearity is typically better than ± 0.5 percent and resolution is better than 0.05 percent of full scale. The frequency response is typically 20 kHz, although small-diameter coils can be used to increase this response to 50 kHz.

The fact that eddy-current sensors do not require contact for measuring displacement is quite important. As a result of this feature, they are often used in transducer systems for automatic control of dimensions in fabrication processes. They are also applied extensively to determine thickness of organic coatings that are nonconducting.

5.7 PIEZOELECTRIC SENSORS

A piezoelectric material, as its name implies, produces an electric charge when it is subjected to a force or pressure. Piezoelectric materials, such as single-crystal quartz or polycrystalline barium titanate, contain molecules with asymmetrical charge distributions. When pressure is applied, the crystal deforms and there is a relative displacement of the positive and negative charges within the crystal. This displacement of internal charges produces external charges of opposite sign on the external surfaces of the crystal. If these surfaces are coated with metallic electrodes, as illustrated in Fig. 5.15, the charge q that develops can be determined from the output voltage v_o, since

$$q = v_o C \tag{5.13}$$

where C is the capacitance of the piezoelectric crystal.

The surface charge q is related to the applied pressure p by

$$q = S_q A p \tag{5.14}$$

where

S_q is the charge sensitivity of the piezoelectric crystal
A is the area of the electrode

The charge sensitivity S_q is a function of the orientation of the sensor (usually a cylinder) relative to the axes of the piezoelectric crystal. Typical values of S_q for common piezoelectric materials are given in Table 5.3.

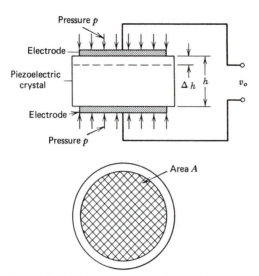

Figure 5.15 Piezoelectric crystal deforming under the action of an applied pressure.

The output voltage v_o developed by the piezoelectric sensor is obtained by substituting Eqs. 5.8 and 5.14 into Eq. 5.13. Thus,

$$v_o = \left(\frac{S_q}{kK}\right)h\,p \qquad \text{(a)}$$

The voltage sensitivity S_v of the sensor can be expressed as

$$S_v = \frac{S_q}{kK} \qquad \text{(5.15)}$$

The output voltage v_o of the sensor is then

$$v_o = S_v h\,p \qquad \text{(5.16)}$$

Again, voltage sensitivity S_v of the sensor is a function of the orientation of the axis of the cylinder relative to the crystallographic axes. Typical values of S_v are also presented in Table 5.3.

Most piezoelectric transducers are fabricated from single-crystal quartz because it is the most stable of the piezoelectric material and is nearly loss free both mechanically and electrically. Its properties are: modulus of elasticity, 86 GPa; resistivity, 10^{12} Ω·m;

Table 5.3 Typical Charge and Voltage Sensitivities, S_q and S_v, of Piezoelectric Materials

Material	Orientation	$S_q(pC/N)$	$S_v(V \cdot m/N)$
Quartz SiO$_2$	X-cut, length longitudinal	2.2	0.055
Single crystal	X-cut, thickness longitudinal	−2.0	−0.05
	Y-cut, thickness shear	4.4	0.11
Barium titanate BaTiO$_3$	Parallel to polarization	130	0.011
Ceramic, poled	Perpendicular to polarization	−56	−0.004

and dielectric constant, 40.6 pF/m. It exhibits excellent high-temperature properties and can be operated up to 550°C (1022°F). The charge sensitivity of quartz is low when compared with that of barium titanate; however, with high-gain charge amplifiers available for processing the output signal, the lower sensitivity is not a serious disadvantage.

Barium titanate is a polycrystalline material that can be polarized by applying a high voltage to the electrodes while the material is at a temperature above the Curie point (125°C) (257°F). The electric field aligns the ferroelectric domains in the barium titanate and it becomes piezoelectric. If the polarization voltage is maintained while the material is cooled well below the Curie point, the piezoelectric characteristics become permanent and stable after a short aging period.

The mechanical stability of barium titanate is excellent; it exhibits high mechanical strength and has a high modulus of elasticity (120 GPa). It is more economical than quartz and can be fabricated in a wide variety of sizes and shapes. Although its application in transducers is second to quartz, it is frequently used in ultrasonics as a driver. In this application, a voltage is applied to the electrodes and the barium titanate deforms and delivers energy to the work piece or test specimen.

Most transducers exhibit a relatively low output impedance (in the range of 100 to 1000 Ω). When piezoelectric crystals are used as the sensing elements in transducers, the output impedance is usually extremely high, but it is a variable. The output impedance of a small cylinder of quartz depends on the frequency ω associated with the applied pressure. Since the sensor acts like a capacitor, the output impedance is

$$Z_C = \frac{1}{j\omega C} = -\frac{j}{\omega C} \tag{2.42}$$

Clearly, the impedance ranges from infinity for static applications to about 10 kΩ for very high frequency applications (100 kHz). With this high output impedance, care must be exercised in monitoring the output voltage; otherwise, serious errors can occur.

A circuit diagram of a typical system used to measure a voltage produced by a piezoelectric sensor is shown in Fig. 5.16. The piezoelectric sensor acts as a charge generator. In addition to the charge generator, the sensor is represented by parallel components, including a capacitor C_p (about 10 pF) and a leakage resistor R_p (about

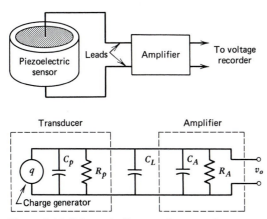

Figure 5.16 Schematic diagram of a measuring system with a piezoelectric sensor.

10^{14} Ω). The capacitance of the lead wires C_L must also be considered, because even relatively short lead wires have a capacitance larger than that of the sensor. The amplifier is either a cathode follower or a charge amplifier with sufficient input impedance to isolate the piezoelectric sensor in spite of its very large output impedance. If a pressure is applied to the sensor and maintained for a long period of time, the charge q developed by the piezoelectric material leaks off by way of a small current flow through R_p and the amplifier resistance R_A. The time available for readout of the signal depends on the effective time constant τ_e of the circuit, which is given by

$$\tau_e = R_e C_e = \frac{R_p R_A}{R_p + R_A}(C_p + C_L + C_A) \tag{5.17}$$

where

    R_e is the equivalent resistance of the circuit
    C_e is the equivalent capacitance of the circuit

Time constants ranging from 1000 to 100,000 s can be achieved with quartz sensors and commercially available charge amplifiers. These time constants are sufficient to permit measurement of quantities that vary slowly with time or measurement of static quantities for short periods of time. Problems associated with measuring the output voltage diminish as the frequency of the mechanical excitation increases and the output impedance Z_c of the sensor decreases.

The inherent dynamic response of the piezoelectric sensor is very high, since the resonant frequency of the small cylindrical piezoelectric element is so large. The resonant frequency of the transducer depends on the mechanical design of the transducer as well as the mass and stiffness of the sensor. For this reason, specification of frequency response will be deferred to Chapter 9. It should be noted here, however, that the most significant advantage of the piezoelectric sensor is its very high frequency response.

5.8 PIEZORESISTIVE SENSORS

Piezoresistive sensors, as the name implies, are fabricated from materials that exhibit a change in resistance when subjected to a pressure. The development of piezoresistive materials was an outgrowth of semiconductor research conducted by the Bell Telephone Laboratories in the early 1950s. This research eventually led to the transistor.

Piezoresistive sensors are made from semiconductive materials—usually silicon, with boron as the trace impurity for the P-type material and arsenic as the trace impurity for the N-type material. The resistivity ρ of a semiconducting material can be expressed as

$$\rho = \frac{1}{eN\mu} \tag{5.18}$$

where

    e is the electron charge, which depends on the type of impurity
    N is the number of charge carriers, which depends on the concentration of the
       impurity
    μ is the mobility of the charge carriers, which depends on strain and its direction
       relative to the crystal axes

Equation 5.18 shows that the resistivity of the semiconductor can be adjusted to any specified value by controlling the concentration of the trace impurity. The impurity concentrations commonly employed range from 10^{16} to 10^{20} atoms/cm^3, which permits a wide variation in the initial resistivity. For example, the resistivity for P-type silicon with a concentration of 10^{20} atoms/cm^3 is 500 $\mu\Omega\cdot$m, which is about 30,000 times higher than the resistivity of copper. This very high resistivity facilitates the design of miniaturized sensors.

Equation 5.18 also indicates that the resistivity changes when the piezoresistive sensor is subjected to either stress or strain because of the variations in the mobility μ. This change of resistivity is known as the piezoresistive effect and can be expressed by the equation

$$\rho_{ij} = \delta_{ij}\, p + \pi_{ijkl}\tau_{kl} \tag{5.19}$$

where

$i, j, k,$ and l range from 1 to 3

π_{ijkl} is a fourth rank piezoresistive tensor

τ_{kl} is the stress tensor

δ_{ij} is the Kronecker delta

Since silicon is a cubic crystal, the 36 piezoresistive coefficients reduce to 3 and Eq. 5.19 simplifies to

$$\rho_{11} = \rho[1 + \pi_{11}\sigma_{11} + \pi_{12}(\sigma_{22} + \sigma_{33})]$$

$$\rho_{22} = \rho[1 + \pi_{11}\sigma_{22} + \pi_{12}(\sigma_{33} + \sigma_{11})]$$

$$\rho_{33} = \rho[1 + \pi_{11}\sigma_{33} + \pi_{12}(\sigma_{11} + \sigma_{22})]$$

$$\rho_{12} = \rho\pi_{44}\tau_{12} \tag{5.20}$$

$$\rho_{23} = \rho\pi_{44}\tau_{23}$$

$$\rho_{31} = \rho\pi_{44}\tau_{31}$$

where the subscripts 1, 2, and 3 identify the axes of the single crystal.

These equations indicate that the piezoresistive crystal, when subjected to a general state of stress, becomes electrically anisotropic. The resistivity depends on the stresses τ_{kl}, which are referred to the axes of the semiconductor silicon. Because of this electrical anisotropy, Ohm's law is written as

$$v_i' = \rho_{ij} i_j' \tag{5.21}$$

where

v' is the potential gradient (V/m)

i' is the current density (A/m^2)

Substituting Eqs. 5.20 into Eq. 5.21 gives

$$\frac{v_1'}{\rho} = i_1'[1 + \pi_{11}\sigma_{11} + \pi_{12}(\sigma_{22} + \sigma_{33})] + \pi_{44}(i_2'\tau_{12} + i_3'\tau_{31})$$

$$\frac{v_2'}{\rho} = i_2'[1 + \pi_{11}\sigma_{22} + \pi_{12}(\sigma_{33} + \sigma_{11})] + \pi_{44}(i_3'\tau_{23} + i_1'\tau_{12}) \tag{5.22}$$

$$\frac{v_3'}{\rho} = i_3'[1 + \pi_{11}\sigma_{33} + \pi_{12}(\sigma_{11} + \sigma_{22})] + \pi_{44}(i_1'\tau_{31} + i_2'\tau_{23})$$

These results show that the voltage drop across a sensor depends on the current density i', the state of stress τ, and the three piezoresistive coefficients. The piezoresistive coefficients can be adjusted by controlling the concentration of the impurity and by optimizing the direction of the axis of the sensor with respect to the crystal axes. The sensitivity of a typical sensor is high (for example, a piezoresistive strain gage exhibits a gage factor of 100, whereas a metal-foil strain gage exhibits a gage factor of 2).

During the past decade, considerable progress has been made in applying fabrication techniques used in the microelectronics industry to the development of microminiature sensors. These solid-state sensors incorporate silicon as the mechanical element and piezoresistive sensors. In a piezoresistive pressure sensor, a diaphragm is etched from silicon to form the mechanical element of the transducer. Resistances that form the arms of a Wheatstone bridge are diffused or ion implanted directly into the silicon diaphragm.

Extremely small sensors may be produced using micromachining techniques that involve photolithography methods developed in the microelectronics industry. An accurate pattern of the mechanical element is transferred to a photoresist coating on a silicon wafer. Anisotropic etchants, which etch at different rates along different crystal axes, are used to produce three-dimensional shapes in exposed silicon areas.

An example of a solid-state sensor that utilizes micromachining and a number of integrated-circuit technologies is shown in Fig. 5.17. This neural microprobe was developed to record signals from single cells in the brain. Eight of the thirty-two recording sites on each probe can be selected for monitoring and amplification. The on-chip amplifiers, which require no off-chip components, each occupy only 0.06 mm^2 and provide a gain of 300. The overall probe is 4.7 mm long.

5.9 PHOTOELECTRIC SENSORS

In applications in which contact cannot be made with the specimen being examined, a photoelectric sensor can often be employed to monitor changes in the intensity of light that can be related to the quantity being measured. In these applications, the photoelectric device is the detector and the entire system, including the detector, amplifiers, signal conditioners, and readout devices, is often called a radiometer.

There are two basic types of detectors, thermal and photon, depending on the signal input to the detector. *Thermal detectors,* which utilize thermocouples and thermopiles (described in Section 5.12), employ heat transfer and temperature change to sense radiation. These detectors, with time constants on the order of 100 ms, respond very slowly when compared with photon detectors. Thermal detectors are used effectively to measure light intensity in continuous-wave laser beams, where the intensity of the radiation is extremely high and the intensity of the output is constant with time.

Photon detectors are used more widely as general-purpose sensors because they are more sensitive and respond more quickly with time. The three common types of photon detectors are (1) photoconductors, (2) photoemissive tubes, and (3) semiconductor photovoltaic diodes. Several characteristics of these photon detectors permit classification of their performance. The first is noise equivalent power (NEP), which is the radiant flux required to give an output signal equal to the detector noise. This definition gives a signal-to-noise ratio of one with

$$\text{NEP} = \frac{v_n}{R} \tag{5.23}$$

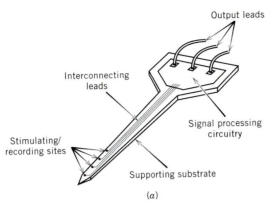

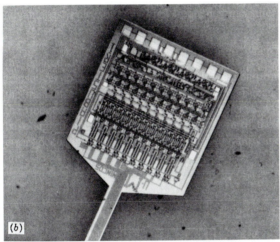

Figure 5.17 A multichannel neural microprobe produced using micromachining. The overall probe is 4.7 mm long. (Courtesy of Professor K. D. Wise, Center for Integrated Sensors and Circuits, The University of Michigan.)

where v_n is the noise signal in volts and R is the responsivity in V/W given by:

$$R = \frac{v_o}{HA_D} \qquad (5.24)$$

where

  v_o is the signal output
  H is the irradiance on the detector W/mm^2
  A_D is the sensitive area of the detector mm^2.

Combining Eqs. 5.23 and 5.24 gives

$$\text{NEP} = \frac{v_n}{v_o}HA_D \qquad \textbf{(a)}$$

The NEP value depends on wavelength, modulation frequency, detector area, temperature, and detector bandwidth. Detector bandwidth is usually 1 Hz, and NEP is sometimes expressed in units of $\text{W} \cdot \text{Hz}^{-1/2}$.

It is clear from Eq. a that NEP is proportional to detector area A_D, and it can also be shown that the output voltage is linear with respect to H. Thus,

$$v_o = RHA_D = v_n H \frac{A_D}{\text{NEP}} = k v_n H \qquad (5.25)$$

where $k = A_D/\text{NEP}$ is a constant.

The linearity of a photon detector is usually represented by the output current versus light intensity relationship illustrated in Fig. 5.18. The lower limit coincides with the NEP where the noise signal $v_n = v_o$. The linear range of the silicon photodiode shown in Fig. 5.18 extends from the NEP level to an upper limit, which depends on the load resistance in the external circuit. Typically, this linear range covers 8 to 10 orders of magnitude.

The normalized sensitivity, known as the specific detectivity D^*, is defined to allow comparisons of types of detectors independent of detector area A_D and bandwidth f_{bw}. Thus,

$$D^* = \frac{\sqrt{A_D f_{bw}}}{\text{NEP}} \qquad (5.26)$$

The bandwidth is incorporated into the sensitivity relation because noise power is proportional to f_{bw}. Noise signal v_n is then proportional to $\sqrt{f_{bw}}$. The sensitivity is a function of the wavelength λ of the radiation and the material used in the photon detector. The specific detectivity D^* for several detectors is presented in Fig. 5.19.

5.9.1 Vacuum-tube Detectors

The vacuum-tube detector, illustrated in Fig. 5.20, is based on the photoemissive effect. It contains a semitransparent photocathode and an anode mounted in a vacuum. When incident radiation (photons) impinges on the photocathode material, electrons are emitted from the surface and flow to the anode to produce a photoelectric current

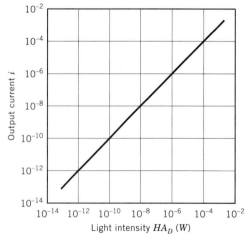

Figure 5.18 Output current as a function of light intensity for a silicon photodiode.

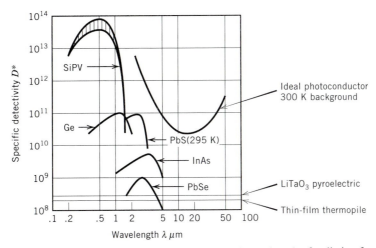

Figure 5.19 Specific detectivity as a function of wavelength of radiation for a number of photodetector materials.

i_p, which is proportional to the irradiance H. The voltage output v_o is given by Eq. 5.24 as

$$v_o = RHA_D = i_pR_L \qquad \textbf{(a)}$$

and

$$i_p = SH \qquad \textbf{(5.27)}$$

where the sensitivity S, the ratio of the photoelectric current in amperes (A) flowing in the photocathode to the incident light H in watts (W), is given by

$$S = R\frac{A_D}{R_L} \qquad \textbf{(5.28)}$$

where

R_L is a load resistor, as shown in Fig. 5.20b
R is the responsivity V/W

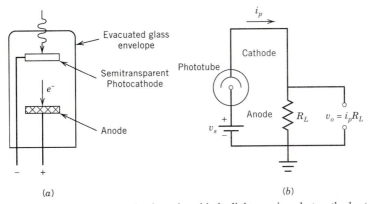

(a)   (b)

Figure 5.20 (a) Photoconducting tube with the light-sensing photocathode at the top end of the tube. (b) A typical circuit used to measure the photoelectric current i_p.

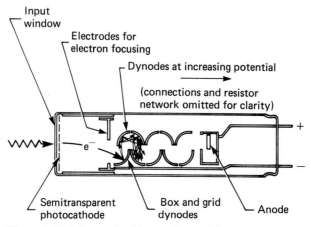

Figure 5.21 A box and grid-type photomultiplier tube.

The sensitivity S depends primarily on the photoemissive material deposited on the cathode surface. Photoemissive materials are usually compounds of the alkali metals, such as AgOCs, SbCs, NaKSbCs, and CsTe. The most common photoemissive material is SbCs, which is used in the range from ultraviolet to visible light. A typical sensitivity for SbCs is 50 mA/W with considerable variation in S as a function of the wavelength λ of the incident light.

Significant increases in sensitivity (up to 50×10^3 A/W) can be achieved by using photomultiplier tubes. These devices, illustrated in Fig. 5.21, contain a large-area photocathode followed by focusing electrodes, an electron multiplier, and an anode encased in a vacuum tube. When light (photons) impinges on the photo-cathodes, electrons are emitted into the vacuum. These electrons are collected and focused by circular anodes (much like those found in a CRT) to produce a beam of accelerated electrons. These electrons are driven into a box-and-grid electron multiplier. The multiplication occurs by the process of secondary emission, as indicated in Fig. 5.21. The gain achieved is related to the number of stages (dynodes) employed in constructing the tube, the interstage voltage in the dynode sequence, and the secondary electron emission ratio. Often the gain of the tube is expressed as

$$G = k_1 v_s^{k_2 n} \tag{5.29}$$

where

    k_1 and k_2 are constants for a given tube
    v_s is the applied voltage between the anode and cathode
    n is the number of stages

Because photomultiplier tubes usually incorporate 9 to 12 dynode stages, the gains are very large. For example, with $v_s = 1000$ V, $G = 2 \times 10^6$ is typical. It should be noted in Eq. 5.29 that the gain is strongly dependent on the supply voltage; therefore, the output signal will vary markedly with power-supply instabilities. Only the highest quality regulated high-voltage power supplies should be employed with photomultiplier tubes.

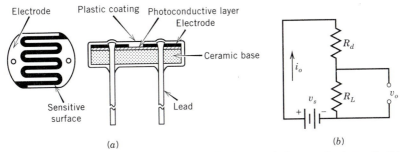

Electrode   Plastic coating   Photoconductive layer
Electrode

Ceramic base

Sensitive surface   Lead

R_d

i_o

v_s R_L v_o

(a)   (b)

Figure 5.22 (a) Construction details for a plastic-coated photoconduction cell. (b) Equivalent electrical circuit with R_d replacing the photoconduction cell.

5.9.2 Photoconductive Cells

Photoconductive cells, illustrated in Fig. 5.22, are fabricated from semiconductor materials, such as cadmium sulfide (CdS) or cadmium selenide (CdSe), which exhibit a strong photoconductive response. When a photon with sufficient energy strikes a molecule of, for example, CdS, an electron is driven from the valence band to the conduction band and a hole or vacancy remains in the valence band. This hole and the electron both serve as charge carriers, and with continuous exposure to light, the concentration of charge carriers increases and the resistivity decreases. A circuit used to detect the resistance change ΔR of the photoconductor is shown in Fig. 5.22b. The resistance of the detector changes over about three orders of magnitude as the incident radiation varies from dark to very bright.

When a photoconductor is placed in a dark environment, its resistance is high and only a small dark current is produced. If the sensor is exposed to light, the resistance decreases significantly (the ratio of maximum to minimum resistance for R_d in Fig. 5.22b ranges from 100 to 10,000 in common commercial sensors); therefore, the output current i_o can be quite large. The sensitivity depends on cell area, type of cell (CdS or CdSe), and power dissipation capacity of the cell. If the maximum voltage v_s is used to supply the cell at its maximum dissipation capacity, then sensitivities of up to 0.2 mA/1x result for CdS cells.

Photoconductors respond to radiation ranging from long thermal radiation through the infrared, visible, and ultraviolet regions of the electromagnetic spectrum. The sensitivity S changes significantly with wavelength and drops sharply at both short and long wavelengths; consequently, photoconductive cells exhibit the same disadvantage as photoconduction tubes (photoemissive cells) in that the calibration constant depends on wavelength.

The photocurrent requires some time to develop after the excitation is applied and some time to decay after the excitation is removed. The rise and fall times for commercially available photoconductors are usually several seconds. Because of these delays, the CdS and CdSe photoconductors are not suitable for dynamic measurements. Instead, their simplicity, high sensitivity, and low cost lend them to applications involving counting and switching based on a slowly varying light intensity.

5.9.3 Semiconductor Photodiodes

Although semiconductor photodiodes may be classified as photoconductors, photodiodes fabricated from silicon and incorporating a P/N junction differ so markedly in both construction and performance that they are described separately here. The

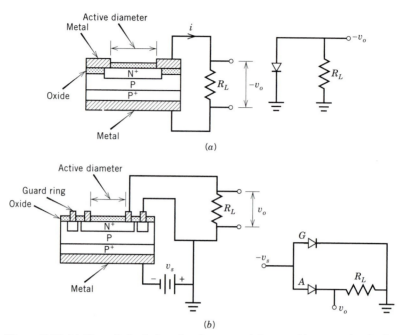

Figure 5.23 (*a*) Photodiode designed and connected for zero-bias operation in the photovoltaic mode. (*b*) Photodiode designed with a guard ring and connected in a circuit with negative-bias for operation in the photoconduction mode.

semiconductor photodiode may be operated as a photoconductor (changing resistance) or as a photovoltaic device (a voltage source) depending on the external circuit. When the P/N junction of the photodiode shown in Fig. 5.23*a* is illuminated and a zero-bias connection is made to both sides of the junction (no externally applied voltage), a current flows that is proportional to the light intensity. This behavior is the well-known photovoltaic effect used in solar cells to produce light-to-voltage converters for solar power supplies. High-performance photovoltaic detectors have recently been developed that have low noise, extended linearity, and a wide dynamic range and are thus suitable for the detection and measurement of light.

If the P/N junction is placed in a circuit with an externally applied reverse bias, as illustrated in Fig.5.23*b,* and illuminated, then a current will flow. The current is composed of two parts: a dark current caused by reverse leakage, which remains constant, and the photocurrent, which varies linearly with the intensity of the incident light.

The choice of using these semiconductor diodes as a photovoltaic detector or as a photoconductive detector depends primarily on the frequency response required in the measurement. For light-intensity fluctuations at the lower frequencies (less than 100 kHz), the photovoltaic circuit in Fig. 5.23*a* gives a lower NEP than the photoconduction circuit in Fig. 5.23*b*. The circuit shown in Fig. 5.23*b* is used for measuring high-speed light pulses or high-frequency modulation of a continuous light beam. The reverse bias serves to accelerate the electron/hole transition times and improve the frequency response. Photoconductive diodes operate over a frequency range from dc to 100 MHz and are capable of measuring the intensity of light pulses with rise times in the range of 3 to 12 ns.

Semiconductor photodiodes are small, rugged, and inexpensive, and because of these advantages they have replaced the vacuum-tube detector (see Section 5.9.1) in

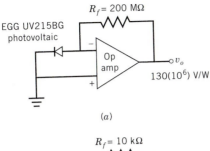

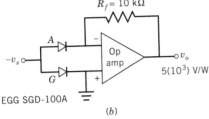

Figure 5.24 Operational amplifiers used with photodiodes to increase their responsivity. (*a*) Photovoltaic arrangement, and (*b*) photoconductive arrangement.

many applications. The photodiodes may be used in either mode with operational amplifiers to give an exceptionally high responsivity. For example, the photovoltaic detector shown in Fig. 5.24*a* gives $R = 130 \times 10^6$ V/W with a gain–bandwidth product of 100 kHz when the signal is amplified with an operational amplifier. The photoconductive detector shown in Fig. 5.24*b* gives a lower responsivity $R = 5 \times 10^3$ V/W, but the gain-bandwidth product is 20 MHz.

Continued improvements in semiconductor photodiodes expand the sensors available. Space limitations, however, preclude a more thorough treatment in this book. Placement of large numbers of photodiodes along a line or over an area to create line or area arrays of sensors has greatly expanded the number of applications for photodiodes in field measurements of light patterns.

5.10 RESISTANCE TEMPERATURE DETECTORS

The change in resistance of metals with temperature provides the basis for a family of temperature sensors known as resistance temperature detectors (RTDs). The sensor is simply a conductor fabricated either as a wire-wound coil or as a film or foil grid. The change in resistance of the conductor with temperature is given by the expression

$$\frac{\Delta R}{R_0} = \gamma_1(T - T_0) + \gamma_2(T - T_0)^2 + \cdots + \gamma_n(T - T_0)^n \qquad (5.30)$$

where

T_0 is a reference temperature
R_0 is a reference resistance at temperature T_0
$\gamma_1, \gamma_2, \ldots, \gamma_n$ are temperature coefficients of resistance

Resistance temperature detectors are often used in ovens and furnaces where inexpensive but accurate and stable temperature measurements and controls are required.

Platinum is widely used for sensor fabrication because it is the most stable of all the metals, it is the least sensitive to contamination, and it is capable of operating over a wide range of temperatures (4 K to 1064°C). Because platinum provides an extremely reproducible output, it has been selected for use in establishing the International Temperature Scale (1990) over the range from 13.8033 K to 961.78°C.

The performance of platinum RTDs depends strongly on the design and construction of the package containing the sensitive wire coil. The most precise sensors are fabricated with a minimum amount of support; they are fragile and often fail if subjected to rough handling, shock, or vibration. Most sensors used in transducers for industrial applications have platinum coils supported on ceramic or glass tubes. These fully supported sensors are quite rugged and will withstand shock levels up to 100 g. Unfortunately, the range and accuracy of such sensors are limited to some degree by the influence of the constraining package. More information on the performance characteristics of packaged RTDs and the circuits used for their application is presented in Chapter 11.

The sensitivity S of a platinum RTD is relatively high ($S = 0.390 \ \Omega/°C$ for an RTD with a resistance of 100 Ω at 0°C); however, the sensitivity varies with temperature because of the nonlinear response of the sensor to temperature, as indicated by Eq. 5.30. The sensitivity S decreases to 0.378, 0.367, 0.355, 0.344, and 0.332 $\Omega/°C$ at temperatures of 100, 200, 300, 400, and 500°C, respectively.

The dynamic response of an RTD depends almost entirely on construction details. For large coils mounted on heavy ceramic cores and sheathed in stainless steel tubes, the response time may be several seconds or more. For film or foil elements mounted on thin polymide substrates, the response time can be less than 0.1 s.

5.11 THERMISTORS

A second type of temperature sensor based on resistance change of the sensing element with temperature is known as a *thermistor*. A thermistor differs from a resistance temperature detector because the sensing element is fabricated from a semiconducting material instead of a metal. The semiconducting materials, which include oxides of copper, cobalt, manganese, nickel, and titanium, exhibit very large changes in resistance with temperature. As a result, thermistors can be fabricated in the form of extremely small beads, as shown in Fig. 5.25.

Resistance change with temperature for a thermistor can be expressed by an equation of the form

$$\ln \rho = A_0 + \frac{A_1}{\theta} + \frac{A_2}{\theta^2} + \cdots + \frac{A_n}{\theta^n} \tag{5.31}$$

where

ρ is the specific resistance of the material
$A_1, A_2, \ldots, A_n$ are material constants
θ is the absolute temperature

The temperature–resistance relationship as expressed by Eq. 5.31 is usually approximated by retaining only the first two terms. The simplified equation is then expressed as

$$\ln \rho = A_0 + \frac{\beta}{\theta} \tag{5.32}$$

Figure 5.25 Miniature bead-type thermistors.
(Courtesy of Dale Electronics, Inc.)

Use of Eq. 5.32 is convenient and acceptable when the temperature range is small and the higher order terms of Eq. 5.31 are negligible. Differentiation of Eq. 5.32 shows that the slope of the specific resistance curve is negative.

Thermistors have many advantages over other temperature sensors and are widely used in industry. They can be small (0.005-in. diameter) and consequently permit point sensing and rapid response to temperature change. Their high resistance minimizes lead-wire problems, and their output is more than 10 times that of an RTD, as shown in Fig. 5.26. Finally, thermistors are very rugged, which permits use in those industrial environments where shock and vibration occur. The disadvantages of thermistors include nonlinear output with temperature, as indicated by Eqs. 5.31 and 5.32, and a limited range. Significant advances have been made in thermistor fabrication methods, and it is possible to obtain stable, reproducible, interchangeable thermistors that are accurate to 0.5 percent over a specified temperature range.

5.12   THERMOCOUPLES

A thermocouple is a temperature sensor consisting of two dissimilar materials that are in thermal and electrical contact. A potential develops at the interface of the two materials as the temperature changes. This thermoelectric phenomenon is known as the *Seebeck effect*. Thermoelectric sensitivities (μV/°C) for a number of different materials in combination with platinum are listed in Table 5.4. The data of Table 5.4 can be used to determine the sensitivity of any thermocouple junction by noting, for example, that

$$S_{\text{Chromel/Alumel}} = S_{\text{Chromel/platinum}} - S_{\text{Alumel/platinum}}$$
$$= +25.8 - (-13.6) = 39.4 \ \mu\text{V/}^\circ\text{C}$$

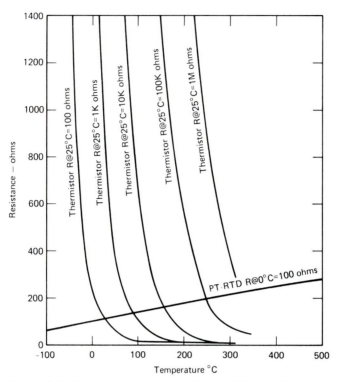

Figure 5.26 Resistance–temperature characteristics of thermistors and RTDs.

Common thermocouple material combinations include iron/constantan, Chromel/Alumel, Chromel/constantan, copper/constantan, and platinum/platinum–rhodium.

The output voltage from a thermocouple junction is measured by connecting two identical thermocouples into a circuit, as shown in Fig. 5.27. The output voltage v_o from this circuit is related to the temperature at each junction by an expression of the form

$$v_o = S_{A/B}(T_1 - T_2) \tag{5.33}$$

where

$S_{A/B}$ is the sensitivity of material combination A and B
T_1 is the temperature at junction 1
T_2 is the temperature at junction 2

Table 5.4 Thermoelectric Sensitivities for Different Materials in Contact with Platinum

Material	Sensitivity (μV/°C)	Material	Sensitivity (μV/°C)
Constantan	-35	Copper	$+\ 6.5$
Nickel	-15	Gold	$+\ 6.5$
Alumel	-13.6	Tungsten	$+\ 7.5$
Carbon	$+\ 3$	Iron	$+\ 18.5$
Aluminum	$+\ 3.5$	Chromel	$+\ 25.8$
Silver	$+\ 6.5$	Silicon	$+440$

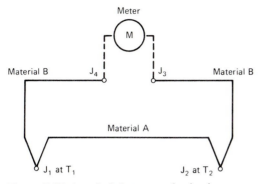

Figure 5.27 A typical thermocouple circuit.

In practice, junction J_2 is a reference junction, which is maintained at a carefully controlled reference temperature T_2. Junction J_1 is placed in contact with the body at the point where a temperature is to be measured. When a meter is inserted into the thermocouple circuit, junctions J_3 and J_4 are created at the contact between the wires (material B) and the terminals of the meter. If these terminals are at the same temperature ($T_3 = T_4$), the added junctions (J_3 and J_4) do not affect the output voltage v_o.

The use of thermocouples in temperature measurement is covered in much more detail in Chapter 11. As a sensor, the thermocouple can be made quite small (0.0005-in.-diameter wire is available) to reduce its mass and size. These miniaturized thermocouples have a rapid response time (milliseconds) and provide essentially point measurements of temperature. A thermocouple can cover a wide range of temperatures; however, the output is nonlinear, and linearizing circuits are required to relate the output voltage v_o to the temperature T. In addition to nonlinear output, thermocouple sensors suffer the disadvantage of very low signal output and the need to carefully control the reference temperature at junction J_2.

5.13 CRYSTAL OSCILLATORS

Time is defined as the interval between two events, and measurements of this interval are made by making comparisons with some reproducible event. For example, the time required for the earth to orbit the sun gives the measurement of a year, and the time required for the earth to rotate on its axis gives the measurement of a day. *Emphemeris time* is based on astronomical measurements of the time required for the earth to orbit the sun. *Sidereal time* is earth rotation time measured with respect to a distant star. *Solar time* is earth rotation time measured with respect to the sun. Sidereal time is used primarily in astronomical laboratories. Solar time is used for navigation on earth and in daily life.

The fundamental unit of time in both the English and International (SI) systems of units is the second. Prior to 1956, the second was defined as 1/86,400 of a mean solar day (average period of revolution of the earth on its axis). In the late 1950s, atomic research revealed that certain atomic transitions can be measured with excellent repeatability. As a result, the Thirteenth General Conference on Weights and Measures, which was held in Paris in 1967, redefined the second as the duration of 9,192,631,770 periods of the radiation corresponding to the transition between the two hyperfine levels of the fundamental state of the atom of cesium 133. The estimated accuracy of this standard is two parts in 10^9. Currently, standards laboratories throughout the world have cesium-beam oscillators that agree in frequency to within a few parts in 10^{11}.

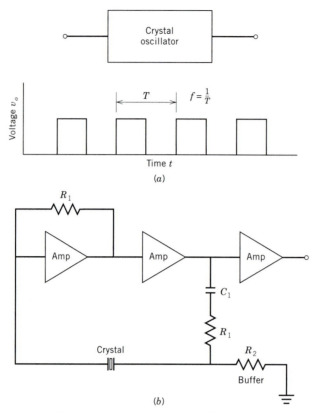

Figure 5.28 (a) The pulse train generated by a crystal oscillator operating at a frequency f. (b) Feedback from the crystal in a series-resonance oscillator operating at the natural frequency of the crystal maintains the stability of the oscillator.

Electronic counters, which are multipurpose instruments based on digital circuits, are utilized in engineering laboratories to measure time and several other time-related quantities. Components used in these instruments include an internal clock, gates, comparators, scalers, and counters.

The internal clock, which provides the time base for the digital electronic counter, is the time sensor. The clock is a crystal oscillator that issues a train of pulses, as illustrated in Fig. 5.28a. The oscillator circuit, shown in Fig. 5.28b, incorporates a piezoelectric crystal, usually quartz, to provide stability to the circuit. The oscillator is driven at the natural frequency of the crystal, and the feedback from the crystal maintains the oscillator frequency equal to the natural frequency for extended periods of time. Long-term frequency stability varies, depending on the quality of the oscillator, from one part in 10^5 to one part in 10^8. Short-term stability, measured in hours, is about one part in 10^9. Small frequency changes occur owing to aging of the crystal and are usually of the order of 1 to 10 ppm/yr.

Crystal oscillators are available with frequencies that cover a very wide range (1 to 150 MHz). At the lower frequencies (1 to 20 MHz), the quartz crystal is usually driven at its fundamental frequency, and at the higher frequencies, overtone modes of resonance are employed to stabilize the oscillator.

Most counters used in engineering laboratories utilize a 10 MHz crystal oscillator, which is housed in a small oven to maintain a constant temperature and improve stability.

5.14   SUMMARY

A wide variety of basic sensors have been described in this chapter. These sensors are sometimes used directly to measure an unknown quantity (such as use of a thermocouple to measure temperature); however, in many other instances, the sensors are one component of a more involved transducer, such as a piezoelectric crystal in an accelerometer. Important characteristics that must be considered in the selection process of a sensor include the following:

1. *Size.* Smaller is better, because of enhanced dynamic response and minimum interference with the process or event.
2. *Range.* Extended range is preferred so as to increase the latitude of operation.
3. *Sensitivity.* Higher output devices have the advantage of requiring less amplification.
4. *Accuracy.* Devices that exhibit errors of 1 percent or less after considering zero shift, linearity, and hysteresis are preferred.
5. *Frequency response.* Wide-response sensors that permit application in both static and dynamic loading situations are preferred.
6. *Stability.* Low drift in output over extended periods of time and with changes in temperature and humidity is essential.
7. *Temperature limits.* The ability to operate from cryogenic to elevated temperatures is considered important.
8. *Economy.* Reasonable costs are preferred.
9. *Ease of application.* Reliability and simplicity are significant factors.

REFERENCES

1. Brindley, K.: *Sensors and Transducers,* Heinemann, London, 1988.
2. Foster, R. L., and S. P. Wnuk, Jr.: "High Temperature Capacitive Displacement Sensing," Instrument Society of America, Paper No. 85-0123, 1985, pp. 245–252.
3. Geyling, F. T., and J. J. Forst: "Semiconductor Strain Transducers," *Bell Syst. Tech. J.,* vol. 39, 1960.
4. Herceg, E. E.: *Handbook of Measurement and Control,* Schaevitz Engineering, Pennsauken, N.J., 1972.
5. Kinzie, P. A.: *Thermocouple Temperature Measurements,* Wiley, New York, 1973.
6. Macklen, E.: *Thermistors,* Electrochemical Publications, Ayr, Scotland, 1979.
7. Mason, W. P., and R. N. Thurston: "Piezoresistive Materials in Measuring Displacement, Force, and Torque," *J. Acoust. Soc. Am.,* vol. 29, no. 10, 1957, pp. 1096–1101.
8. Neubert, H. K. P.: *Instrument Transducers: An Introduction to Their Performance and Design,* 2nd ed., Oxford Univ. Press (Clarendon), Oxford, 1975.
9. Sachse, H.: *Semiconducting Temperature Sensors and Their Applications,* Wiley, New York, 1975.
10. Sanchez, J. C., and W. V. Wright: "Recent Developments in Flexible Silicon Strain Gages," in M. Dean and R. D. Douglas (eds.), *Semiconductor and Conventional Strain Gages,* Academic Press, New York, 1962, pp. 307–346.
11. Schaevitz, H.: "The Linear Variable Differential Transformer," *Proc. Soc. Exp. Stress Anal.,* vol. IV, no. 2, 1947, pp. 79–88.
12. Sedra, A. S., and K. C. Smith: *Microelectronic Circuits,* 3rd ed., Holt, Rinehart & Winston, New York, 1991.
13. Smith, C. S.: "Piezoresistive Effect in Germanium and Silicon," *Phys. Rev.,* vol. 94, 1954, pp. 42–49.
14. Yang, E. S.: *Microelectronic Devices,* McGraw–Hill, New York, 1988.

EXERCISES

5.1 Describe the differences between a sensor and a transducer. Give an example of a transducer that incorporates a displacement sensor. Give another example of a sensor that is also a transducer.

5.2 A slide-wire potentiometer with a length of 100 mm is fabricated by winding wire with a diameter of 0.10 mm around a cylindrical insulating core. Determine the resolution limit of this potentiometer.

5.3 If the potentiometer of Exercise 5.2 has a resistance of 2000 Ω and can dissipate 2 W of power, determine the voltage required to maximize the sensitivity. What voltage change corresponds to the resolution limit?

5.4 A 20-turn potentiometer with a calibrated dial (100 divisions per turn) is used as a balance resistor in a Wheatstone bridge. If the potentiometer has a resistance of 20 kΩ and a resolution of 0.05 percent, what is the minimum incremental change in resistance ΔR that can be read from the calibrated dial?

5.5 Why are potentiometers limited to static or quasi-static applications?

5.6 List several advantages of the conductive-film type of potentiometer.

5.7 A new elevator must be tested to determine its performance characteristics. Design a displacement transducer that utilizes a 20-turn potentiometer to monitor the position of the elevator over its 100-m range of travel.

5.8 Compare the potentiometer and LVDT as displacement sensors with regard to the following characteristics: range, accuracy, resolution, frequency response, reliability, complexity, and cost.

5.9 Prepare a block diagram representing the electronic components in an LVDT. Describe the function of each component.

5.10 Prepare a sketch of the output signal as a function of time for an LVDT with its core located in a fixed off-center position if the demodulator is
(a) functioning
(b) removed from the circuit

5.11 Prepare a sketch of the output signal as a function of time for an LVDT with its core moving at constant velocity from one end of the LVDT through the center to the other end if the demodulator is
(a) functioning
(b) removed from the circuit

5.12 Describe the basic differences between an LVDT and a DCDT.

5.13 Design a 50-mm strain extensometer to be used for a simple tension test of mild steel. If the strain extensometer is to be used only in the elastic region and is to detect the onset of yielding, specify the maximum range. What is the advantage of limiting the range?

5.14 Compare the cylindrical potentiometer (helipotentiometer) and the RVDT as sensors for measuring angular displacement.

5.15 What two factors are responsible for the resistance change dR/R in an electrical resistance strain gage? Which is the most important for gages fabricated from constantan?

5.16 What function does the thin plastic film serve for an electrical resistance strain gage?

5.17   A strain gage with an initial resistance R_0 and a gage factor S_g is subjected to a strain ε. Determine ΔR and $\Delta R/R$ for the conditions listed below:

	$R_0(\Omega)$	S_g	$\varepsilon(\mu\text{m/m})$
(a)	120	2.02	1600
(b)	350	3.47	650
(c)	350	2.07	650
(d)	1000	2.06	200

5.18   For the conditions described in Exercise 5.17, determine the output voltage v_o for an initially balanced bridge if the input voltage v_i is
(a)   2 V   (c)   7 V
(b)   4 V   (d)   10 V

5.19   Exercise 5.18 shows that increasing v_i increases v_o. What happens if v_i is increased to 50 V to improve the output v_o?

5.20   For the gages specified in Exercise 5.17, monitored in a single-arm Wheatstone bridge with $v_s = 5$ V, determine the strain ε if the output voltage v_o is
(a)   1.5 mV   (c)   4.8 mV
(b)   3.3 mV   (d)   5.7 mV

5.21   A short-range displacement transducer utilizes a cantilever beam as the mechanical element and a strain gage as the sensor. Derive an expression for the displacement δ of the end of the beam in terms of the output voltage v_o.

5.22   Verify Eqs. 5.11 and 5.12.

5.23   Prepare a graph of the sensitivity S of a capacitance sensor as a function of frequency ω. Assume the dielectric in the gap is air and consider probe diameters of 1, 2, 5, and 10 mm.

5.24   Write an engineering brief describing the advantages and disadvantages of capacitance sensors.

5.25   Describe the operating principles of an instrumentation system, as shown in Fig. 5.12, used to monitor the output of a capacitance sensor.

5.26   An eddy-current sensor is calibrated for use on an aluminum target material. The gage is then used to monitor the displacement of a specimen fabricated from 304 stainless steel. Is an error produced? Is so, estimate the magnitude of the error.

5.27   Can eddy-current sensors be employed with the following target materials?
(a)   magnetic materials
(b)   polymers
(c)   nonmagnetic metallic foils
Indicate procedures that permit usage of the sensor in these three cases.

5.28   Determine the charge q developed when a piezoelectric crystal with $A = 15$ mm^2 and $h = 8$ mm is subjected to a pressure $p = 2$ MPa if the crystal is
(a)   X-cut, length-longitudinal quartz
(b)   parallel-to-polarization barium titanate

5.29   Determine the output voltages for the piezoelectric crystals described in Exercise 5.28.

5.30   Compare the use of quartz and barium titanate as materials for
(a)   piezoelectric sensors
(b)   ultrasonic signal sources

5.31   The equivalent circuit (see Fig. 5.16) for a measuring system incorporating a piezoelectric crystal consists of the following: $R_p = 10$ TΩ, $R_A = 1$ GΩ, $C_p = 20$ pF, and $C_A = 15$ pF. Let C_L be a variable ranging from 10 pF to 1000 pF. Determine the effective time constant τ_e for the circuit. If the error must be limited to 5 percent, determine the time available for measurement of the magnitude of a step pulse of unit magnitude.

5.32   Compare the characteristics of a piezoresistive sensor with those of a piezo-electric sensor.

5.33   What advantages does the piezoresistive sensor have over the common (metal) electrical resistance strain gage? What are some disadvantages?

5.34   For the silicon photodetector with the output current–light intensity specification shown in Fig. 5.18, determine the responsivity R. Assume that the output voltage is caused by the voltage drop across resistances of 10, 100, 1000, and 10,000 Ω.

5.35   The sensitivity of a photoconducting tube employing Sb Cs as the photoemissive material is 50 mA/W. Determine the responsivity of the circuit defined in Fig. 5.20 if the area of the detector is 8 mm^2 and $R_l = 100,000$ Ω.

5.36   Write an engineering brief describing a photomultiplier tube. Cite advantages and disadvantages of using these tubes.

5.37   Describe the difference between a photoconducting tube and a photoconduction cell.

5.38   Design a circuit to turn on outside lights at your home as it begins to get dark. Use a photoconduction cell in the circuit and provide for an adjustment to control the intensity level for switch activation.

5.39   Sketch the circuit for a photovoltaic cell and write a paragraph explaining its operation. Write another paragraph stating the advantages and disadvantages of this light sensor.

5.40   Sketch the circuit for a photodiode used in the photoconduction mode and write a paragraph explaining its operation. Write another paragraph stating the advantages and disadvantages of this light sensor.

5.41   Prepare a graph showing the sensitivity S of a platinum RTD with a 100 Ω resistance as a function of temperature T. For the three circuits shown in Fig. E5.41, plot a graph of v_o versus T.

5.42   Write an engineering brief describing the effects of the three circuits in Exercise 5.41 on the v_o versus T relationship.

5.43   Compare resistance temperature detectors and thermistors as temperature sensors. Give an example that is best suited for each sensor.

5.44   Determine the sensitivity of the following thermocouples:

(a)   chromel–Alumel   (c)   iron–constantan   (e)   gold–silver
(b)   copper–constantan   (d)   iron–nickel

5.45   Compute the voltage output (approximate) at the meter in Fig. 5.27 for the five thermocouples defined in Exercise 5.44 if

	$T_1(°C)$	$T_2(°C)$
(a)	300	0
(b)	200	0
(c)	250	10
(d)	−100	100

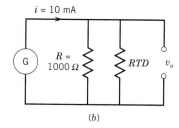

$i = 10$ mA

$R = 500\Omega$

G

RTD v_o

(a)

$i = 10$ mA

G $R = 1000\,\Omega$ RTD v_o

(b)

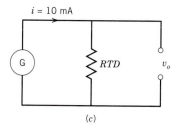

$i = 10$ mA

G RTD v_o

(c)

Figure E5.41

5.46 Describe the crystal oscillator that is used as a clock in almost all instruments today. Indicate the stability that can be anticipated.

5.47 If you designed a clock using a crystal oscillator, what error would you expect in one year. State all of your assumptions in your error calculation.

5.48 Write your own summary of the important topics in this chapter.

Chapter **6**

Signal Conditioning Circuits

6.1 INTRODUCTION

An instrumentation system, as noted in Chapter 1, contains many elements that are used either to supply power to the transducer or to condition the output from the transducer so that it can be displayed by a voltage-measuring instrument. Signal conditioning circuits and power supplies are common to instrumentation systems designed to measure acceleration, displacement, flow, force, strain, and so on. For this reason each type of power supply and circuit will be described independent of its application in a particular measuring system.

Since a very large number of signal conditioning circuits are available today, a complete coverage of the subject is beyond the scope of this book. However, the general characteristics of circuits frequently encountered in engineering measurements will be covered.

6.2 POWER SUPPLIES

With few exceptions, transducers are driven (provided the energy needed for their operation) with either a constant-voltage or a constant-current power supply.

6.2.1 Battery Supplies

The simplest and least expensive constant-voltage power supply is the common battery, which can provide a reasonably constant voltage with large current flow for short periods of time. The difficulty experienced with batteries is that the voltage decays with time under load, and they must be replaced or recharged periodically.

The problem of voltage decay from the battery supply can be eliminated by using a simple regulating circuit incorporating a Zener diode, as shown in Fig. 6.1a. The Zener diode is operated in reverse bias and only leakage current flows until v_i exceeds the breakdown voltage v_z. When v_i exceeds v_z, in reverse bias, the Zener diode breaks down and high currents flow, as indicated in Fig. 6.1b. The resistor is used to limit the current flow and to protect the Zener diode. As long as v_i exceeds v_z, the output voltage v_o equals the breakdown voltage v_z and the breakdown voltage controls the output and not the condition of the battery.

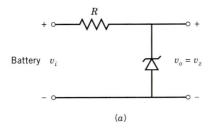

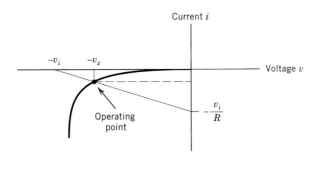

(a)

(b)

Figure 6.1 (a) Voltage regulation with a Zener diode operating in reverse bias. (b) Voltage-current characteristics of a Zener diode showing breakdown at $v_z = v_o$.

Zener diodes are available with breakdown voltages v_z ranging from 2 to 200 V with tolerances of 10 percent and power capabilities to 50 W. Good regulation can be achieved by using the type of circuit shown in Fig 6.1. Better regulation, with temperature compensation and voltage control, can be achieved with the type of circuit shown in Fig. 6.2. This circuit incorporates an operational amplifier that provides a constant current i_z to the Zener diode and provides a gain to increase the output voltage above v_z. This circuit provides a reference-quality voltage source, which can be used as a power supply with batteries providing the required power.

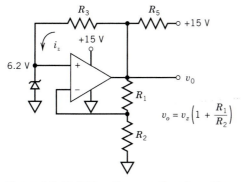

Figure 6.2 Variable voltage supply using a Zener diode and an operational amplifier. Note: The ±15-V supply may be provided by batteries.

The type of battery used depends on the requirements of the instrument. For relatively large power requirements, the lead-calcium rechargeable (LCR) battery is used. These batteries are available in many different sizes and with voltages ranging from 4 to 12 V. The discharge rate is quite high at 20 A/h and capacity in A·h is a function of size. At 12 V, these batteries exhibit a capacity of about 16 V·A·h/lb. Storage life is temperature dependent and will exceed 6 months at normal room temperature (70–75°F). They can be recharged about 1000 times if the extent of the discharge is limited to 50 percent of the capacity. Since this type of battery is completely sealed, it can be used in any position and treated as a dry cell.

Nickel-cadmium batteries are also used when recharging is an important requirement. Nickel-cadmium cells are much smaller than lead-calcium cells and exhibit lower capacities. Individual cell voltage is 1.2 V, and capacities range from 100 to 4000 mA·h. At 1.2 V, these batteries also exhibit a capacity of 16 V·A·h/lb. The discharge characteristics of a Ni-Cd battery are superior to those of ordinary dry cells, and they can be recharged about 500 times before replacement is required.

In the past decade, primary lithium batteries with a number of different cell chemistries have been developed. Some common compositions include lithium-iodine, $Li-MnO_2$, and Li-CF. Lithium batteries exhibit outstanding performance characteristics with high individual cell voltages, low decay of voltage under load, extremely long shelf life, and high energy density. The performance of a lithium-iodine battery (shown in Fig. 6.3) indicates continuous operation with negligible voltage decay for 3 to 9 years, depending on load resistance. The capacity is 110 V·A·h/lb, which is eight times better than that of the Ni-Cd or the LCR battery. With proper load resistances to limit current flow, the life is so long that recharging (which is impossible) is not required.

In some instruments, where the current requirements are low and replacement is convenient, more conventional alkaline and carbon-zinc dry cells are used. As these are primary cells, which cannot be recharged, they are replaced when the supply voltage drops below specified values.

Properly regulated battery power supplies, which can be recharged, are often superior to much more expensive and complex power supplies that convert an ac line voltage to a dc output voltage, since ripple and noise is eliminated. The portability

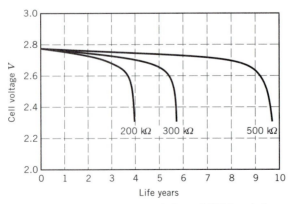

Figure 6.3 Voltage-life characteristics at 25°C for a primary lithium-iodine cell as a function of resistive load.

offered by a battery supply is also an advantage in field measurements, where ac power is not available.

6.2.2 Line Voltage Supplies

The use of general-purpose power supplies that convert an ac line voltage (either 110 V or 220 V) to a lower dc output voltage (often variable) is quite common. A typical example of a high-performance dc power supply, which is capable of delivering nearly constant voltage or constant current, is shown in Fig. 6.4. The power supply uses a rectifier to convert the ac line voltage to a dc output voltage and a filter to reduce the ripple resulting from the rectification. The ripple and regulation are further improved by incorporating a transistor series regulator between the filter and the output. Performance characteristics of the unit, which is capable of providing 0 to 40 V dc and 0 to 3 A, are described in the following paragraphs as they are important features of a power supply.

The load effect, which is the voltage drop from an initial setting as the current is increased from zero to the maximum rated value, is 0.01 percent plus 200 μV when the unit is operated as a constant-voltage source. When operated as a constant-current source, the current increases 0.02 percent plus 500 μA as the voltage is increased from zero to its maximum rated value.

The source effect, which is the change in output for a change in line voltage (between 104 and 127 V for 110-V units), is 0.01 percent plus 200 μV for the voltage and 0.02 percent plus 500 μA for the current. The ripple and noise, which is a small ac signal superimposed on the dc output, is 10 mV peak to peak and 3 mA rms.

The temperature-effect coefficient, which is the change in output voltage or current per degree Celsius following a warm-up period of 30 min, is 0.01 percent plus 200 μV for the voltage and 0.01 percent plus 1 mA for the current.

The drift stability, which is the change in output under constant load over an 8-h period following a 30-min warm-up period, is 0.03 percent plus 500 μV and 0.03 percent plus 3 mA for the voltage and current, respectively.

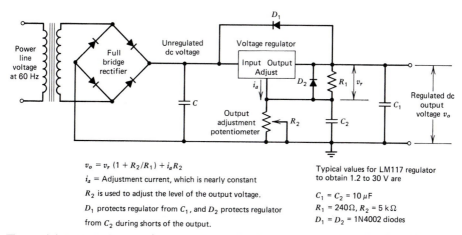

$v_o = v_r (1 + R_2/R_1) + i_a R_2$

i_a = Adjustment current, which is nearly constant

R_2 is used to adjust the level of the output voltage.

D_1 protects regulator from C_1, and D_2 protects regulator from C_2 during shorts of the output.

Typical values for LM117 regulator to obtain 1.2 to 30 V are

$C_1 = C_2 = 10\,\mu$F
$R_1 = 240\Omega$, $R_2 = 5\,$kΩ
$D_1 = D_2 = $ 1N4002 diodes

Figure 6.4 Typical elements for a simple, regulated dc power supply with adjustable output voltage.

The output impedance of the power supply can be represented by a resistor and an inductor in series. The output impedance is usually low for voltage supplies because values for R and L of 2 mΩ and 1 μH are typical.

6.3   POTENTIOMETER CIRCUIT (CONSTANT VOLTAGE)

The potentiometer circuit employed with resistance-type transducers to convert the transducer output $\Delta R/R$ to a voltage signal Δv is shown in Fig. 6.5. With fixed-value resistors in the circuit, the open-circuit output voltage v_o can be expressed as

$$v_o = \frac{R_1}{R_1 + R_2}v_s = \frac{1}{1 + r}v_s \tag{6.1}$$

where

  v_s is the input voltage
  r is the resistance ratio R_2/R_1

If the resistors R_1 and R_2 are varied by ΔR_1 and ΔR_2, the change Δv_o in the output voltage can be determined from Eq. 6.1 as

$$v_o + \Delta v_o = \frac{R_1 + \Delta R_1}{R_1 + \Delta R_1 + R_2 + \Delta R_2}v_s \tag{a}$$

Solving for Δv_o gives

$$\Delta v_o = \left(\frac{R_1 + \Delta R_1}{R_1 + \Delta R_1 + R_2 + \Delta R_2} - \frac{R_1}{R_1 + R_2}\right)v_s \tag{b}$$

Equation b can be reduced and expressed in a more useful form by introducing the resistance ratio r. Thus

$$\Delta v_o = \frac{\dfrac{r}{(1 + r)^2}\left(\dfrac{\Delta R_1}{R_1} - \dfrac{\Delta R_2}{R_2}\right)}{1 + \dfrac{1}{1 + r}\left(\dfrac{\Delta R_1}{R_1} + r\dfrac{\Delta R_2}{R_2}\right)}v_s \tag{6.2}$$

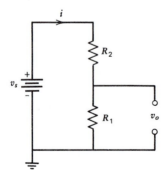

Figure 6.5 The constant-voltage potentiometer circuit.

Equation 6.2 indicates that the change in output voltage Δv_o for the potentiometer circuit is a nonlinear function of the inputs $\Delta R_1/R_1$ and $\Delta R_2/R_2$. The nonlinear effects associated with the circuit can be expressed as a nonlinear term η, where

$$\eta = 1 - \cfrac{1}{1 + \cfrac{1}{1+r}\left(\cfrac{\Delta R_1}{R_1} + r\cfrac{\Delta R_2}{R_2}\right)} \tag{6.3}$$

Equation 6.2 then becomes

$$\Delta v_o = \frac{r}{(1+r)^2}\left(\frac{\Delta R_1}{R_1} - \frac{\Delta R_2}{R_2}\right)(1-\eta)v_s \tag{6.4}$$

The nonlinear effects of the potentiometer circuit can be evaluated by considering a situation that is typical of many applications ($r = 9$ and $\Delta R_2 = 0$). For this simplified case, the nonlinear term η can be expressed as

$$\eta = 1 - \cfrac{1}{1 + \left(0.1\cfrac{\Delta R_1}{R_1}\right)}$$

$$= \left(0.1\frac{\Delta R_1}{R_1}\right) - \left(0.1\frac{\Delta R_1}{R_1}\right)^2 + \left(0.1\frac{\Delta R_1}{R_1}\right)^3 - \cdots \tag{6.5}$$

Results from Eq. 6.5 are shown in Fig. 6.6. Note that linearity within 1 percent can be obtained if $\Delta R_1/R_1 < 0.1$.

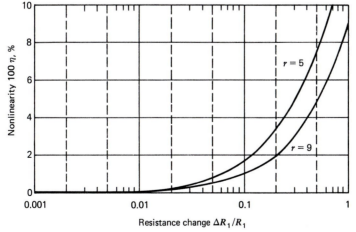

Figure 6.6 Nonlinear term η as a function of resistance change $\Delta R_1/R_1$ for a constant-voltage potentiometer circuit with $r = 5$ or 9 ($\Delta R_2 = 0$).

The range of the potentiometer circuit is defined as the maximum $\Delta R_1/R_1$ that can be recorded without exceeding some specified value of the nonlinear term (usually 1 or 2 percent). In the special case with $r = 9$ and $\Delta R_2 = 0$, the range is $(\Delta R_1/R_1)_{max} = 0.101$ for linearity within 1 percent and $(\Delta R_1/R_1)_{max} = 0.204$ for linearity within 2 percent.

The sensitivity of the potentiometer circuit is defined for a case where $\Delta R_2 = 0$ as

$$S_c = \frac{\Delta v_o}{\Delta R_1/R_1} = \frac{r}{(1+r)^2} v_s \tag{6.6}$$

Equation 6.6 indicates that the sensitivity can be increased without limit simply by increasing the input voltage v_s; however, all transducers have power-dissipation capabilities that limit the input voltage. The power p_T dissipated by a transducer in a potentiometer circuit is given by the expression

$$p_T = \frac{v_T^2}{R_T} \tag{6.7}$$

where

v_T is the voltage across the transducer
R_T is the transducer resistance

From Eq. 6.1

$$v_T = \frac{v_s}{1+r} \tag{6.8}$$

The upper limit of the voltage that can be applied to the potentiometer circuit as obtained from Eqs. 6.7 and 6.8 is

$$v_{s_{max}} = (1+r)\sqrt{p_T R_T} \tag{6.9}$$

A realistic expression for the sensitivity of the constant-voltage potentiometer circuit S_{cv} is obtained by substituting Eq. 6.9 into Eq. 6.6. Thus

$$S_{cv} = \frac{r}{1+r}\sqrt{p_T R_T} \qquad \text{with} \qquad \Delta R_2 = 0 \tag{6.10}$$

It is clear from Eq. 6.10 that maximum sensitivity is achieved with large r, with a high-resistance transducer, and with a transducer capable of dissipating a large amount of power. In practice, sensitivity is usually limited by voltage requirements. For r greater than 5 or 6, the higher voltages required cannot be justified by the small additional gains in sensitivity.

The preceding equations for the potentiometer circuit have been based on the assumption that the input impedance of the voltage recording instrument is infinite (open-circuit voltage) and that no power is required to measure $v_o + \Delta v_o$. In practice, recording instruments have a finite resistance, some power is drawn from the circuit, and there is an error owing to the load of the recorder. Reference to Fig. 3.2 shows the load error as a function of R_s/R_m and indicates that error will be small provided that the recorder impedance $R_m > 100 R_s$.

6.4 POTENTIOMETER CIRCUIT (CONSTANT CURRENT)

The potentiometer circuit described in Section 6.3, which was driven wit stant-voltage power supply, exhibited a nonlinear output voltage Δv_o when $\Delta R/R$ exceeded certain limits. In many applications, this nonlinear behavior limits the usefulness of the circuit; therefore, means are sought to extend the linear range of operation.

Constant-current power supplies with sufficient regulation for instrumentation systems have been made possible by recent advances in solid-state electronics. The constant-current power supply automatically adjusts its output voltage to compensate for a changing resistive load in order to maintain the current at a constant value.

A potentiometer circuit with a constant-current power supply is shown schematically in Fig. 6.7a. The open-circuit output voltage v_o (measured with a very high impedance recording instrument so that loading errors are negligible) is

$$v_o = iR_1 \tag{6.11}$$

When the resistances R_1 and R_2 are changed by the amounts ΔR_1 and ΔR_2, the output voltage becomes

$$v_o + \Delta v_o = i(R_1 + \Delta R_1) \tag{a}$$

From Eqs. 6.11 and a,

$$\Delta v_o = i\,\Delta R_1 = iR_1\frac{\Delta R_1}{R_1} \tag{6.12}$$

Equation 6.12 indicates that neither R_2 nor ΔR_2 influences the output of the constant-current potentiometer circuit; therefore, it is possible to eliminate R_2 and use the simple circuit shown in Fig. 6.7b. It should also be observed that the change in output voltage Δv_o is a linear function of the input $\Delta R_1/R_1$, regardless of the magnitude of ΔR_1. This linear behavior extends the usefulness of the potentiometer circuit for many applications.

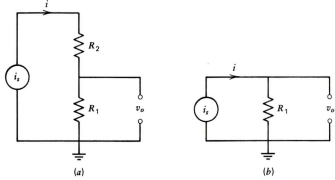

(a)   (b)

Figure 6.7 Constant-current potentiometer circuits. (a) Two-element circuit, and (b) single-element circuit.

The circuit sensitivity S_{cc} for the constant-current potentiometer circuit is

$$S_{cc} = \frac{\Delta v_o}{\Delta R_1 / R_1} = iR_1 \tag{6.13}$$

If the constant-current source is adjustable so that the current i can be increased to the maximum power that can be dissipated by the transducer, then

$$i = \sqrt{\frac{p_T}{R_1}} \tag{b}$$

Substituting Eq. b into Eq. 6.13 yields

$$S_{cc} = \sqrt{p_T R_T} \tag{6.14}$$

Equations 6.10 and 6.14 indicate that the sensitivity of the potentiometer circuit is improved by a factor of $(1 + r)/r$ by using the constant-current source.

6.5   WHEATSTONE BRIDGE (CONSTANT VOLTAGE)

The Wheatstone bridge, shown in Fig. 6.8, is another circuit commonly used to convert a change in resistance to an output voltage. The output voltage v_o of the bridge can be determined by treating the top and bottom parts of the bridge as individual voltage dividers. Thus,

$$v_{AB} = \frac{R_1}{R_1 + R_2} v_s \tag{a}$$

$$v_{AD} = \frac{R_4}{R_3 + R_4} v_s \tag{b}$$

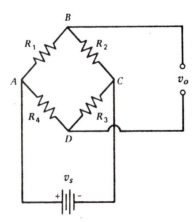

Figure 6.8 The constant-voltage Wheatstone bridge circuit.

The output voltage v_o of the bridge is

$$v_o = v_{BD} = v_{AB} - v_{AD}$$

Substituting Eqs. a and b into c yields

$$v_o = \frac{R_1 R_3 - R_2 R_4}{(R_1 + R_2)(R_3 + R_4)} v_s \qquad (6.15)$$

Equation 6.15 indicates that the initial output voltage will vanish ($v_o = 0$) if

$$R_1 R_3 = R_2 R_4 \qquad (6.16)$$

When Eq. 6.16 is satisfied, the bridge is said to be *balanced*. The ability to balance the bridge (and zero v_o) represents a significant advantage, since it is much easier to measure small changes in voltage Δv_o from a null voltage than from an elevated voltage v_o, which may be as much as 1000 times greater than Δv_o.

With an initially balanced bridge, an output voltage Δv_o develops when resistances R_1, R_2, R_3, and R_4 are varied by amounts ΔR_1, ΔR_2, ΔR_3, and ΔR_4, respectively. From Eq. 6.15, with these new values of resistance,

$$\Delta v_o = \frac{(R_1 + \Delta R_1)(R_3 + \Delta R_3) - (R_2 + \Delta R_2)(R_4 + \Delta R_4)}{(R_1 + \Delta R_1 + R_2 + \Delta R_2)(R_3 + \Delta R_3 + R_4 + \Delta R_4)} v_s \qquad (d)$$

Expanding, neglecting higher order terms, and substituting Eq. 6.15 yields

$$\Delta v_o = \frac{R_1 R_2}{(R_1 + R_2)^2} \left(\frac{\Delta R_1}{R_1} - \frac{\Delta R_2}{R_2} + \frac{\Delta R_3}{R_3} - \frac{\Delta R_4}{R_4} \right) v_s \qquad (6.17)$$

An equivalent form of this equation is obtained by substituting $r = R_2/R_1$ in Eq. 6.17 to give

$$\Delta v_o = \frac{r}{(1 + r)^2} \left(\frac{\Delta R_1}{R_1} - \frac{\Delta R_2}{R_2} + \frac{\Delta R_3}{R_3} - \frac{\Delta R_4}{R_4} \right) v_s \qquad (6.18)$$

Equations 6.17 and 6.18 indicate that the output voltage from the bridge is a linear function of the resistance changes. This apparent linearity results from the fact that the higher order terms in Eq. d were neglected. If the higher order terms are retained, the output voltage Δv_o is a nonlinear function of the $\Delta R/R$'s, which can be expressed as

$$\Delta v_o = \frac{r}{(1 + r)^2} \left(\frac{\Delta R_1}{R_1} - \frac{\Delta R_2}{R_2} + \frac{\Delta R_3}{R_3} - \frac{\Delta R_4}{R_4} \right)(1 - \eta) v_s \qquad (6.19)$$

where

$$\eta = \cfrac{1}{1 + \cfrac{r + 1}{\cfrac{\Delta R_1}{R_1} + \cfrac{\Delta R_4}{R_4} + r \left(\cfrac{\Delta R_2}{R_2} + \cfrac{\Delta R_3}{R_3} \right)}} \qquad (6.20)$$

In a widely used form of the bridge, $R_1 = R_2 = R_3 = R_4$. In this case, Eq. 6.20 reduces to

$$\eta = \frac{\sum\limits_{i=1}^{4}\left(\dfrac{\Delta R_i}{R_i}\right)}{\sum\limits_{i=1}^{4}\left(\dfrac{\Delta R_i}{R_i}\right) + 2} \tag{6.21}$$

The error owing to the nonlinear effect is shown as a function of $\Delta R_1/R_1$ and r in Fig. 6.9 for a bridge with one active transducer in arm R_1 and fixed-value resistors in the other three arms. These results show that $\Delta R_1/R_1$ must be less than 0.02 if the error owing to the nonlinear effect is not to exceed 1 percent. Although this limit may appear quite restrictive, the Wheatstone bridge is usually employed with transducers that exhibit very small changes in $\Delta R_1/R_1$.

The sensitivity S_c of a Wheatstone bridge with a constant-voltage power supply and a single active arm is determined from Eq. 6.18 as

$$S_c = \frac{\Delta v_o}{\Delta R_1/R_1} = \frac{r}{(1+r)^2}v_s \tag{6.22}$$

Again it is clear that increasing v_s produces an increase in sensitivity; however, the power p_T that can be dissipated by the transducer limits the supply voltage v_s to

$$v_s = i_T(R_1 + R_2)$$
$$= i_T R_T (1 + r) = (1 + r)\sqrt{p_T R_T} \tag{6.23}$$

Substituting Eq. 6.23 into Eq. 6.22 gives

$$S_{cv} = \frac{r}{1+r}\sqrt{p_T R_T} \tag{6.24}$$

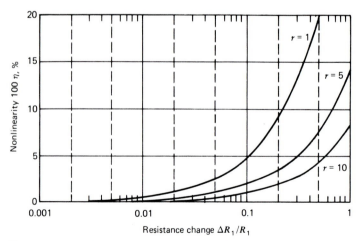

Figure 6.9 Nonlinear term η as a function of resistance change $\Delta R_1/R_1$ for a constant-voltage Wheatstone bridge circuit with one active gage.

Equation 6.24 indicates that the circuit sensitivity of the constant-voltage Wheatstone bridge is due to two factors: (1) the circuit efficiency $r/(1 + r)$, and (2) the characteristics of the transducer as indicated by p_T and R_T. Increasing r increases circuit efficiency; however, r should not be so high as to require unusually large supply voltages. For example, a 500-Ω sensor capable of dissipating 0.2 W in a bridge with $r = 4$ (80 percent circuit efficiency) will require a supply voltage $v_s = 50$ V, which is higher than the capacity of most highly regulated power supplies.

The selection of a sensor with a high resistance and a high heat-dissipating capability is much more effective in maximizing circuit sensitivity than increasing the circuit efficiency beyond 70 or 80 percent. The product $p_T R_T$ for commercially available sensors can range from about 1 W·Ω to 1000 W·Ω; therefore, much more latitude exists for increasing circuit sensitivity S_{cv} by transducer selection than by increasing circuit efficiency.

Circuit sensitivity S_{cv} can also be increased, as indicated by Eq. 6.18, by using multiple sensors (one in each arm of the bridge). In most cases, however, the cost of the additional sensors is not warranted. Instead, it is usually more economical to use a high-gain differential amplifier to increase the output signal Δv_o from the Wheatstone bridge.

Load effects in a Wheatstone bridge are negligible if a high-impedance voltage-measuring instrument (such as a DVM for static signals or an oscilloscope for dynamic signals) is used with the bridge. The output impedance Z_B of the bridge can be determined by using Thevenin's theorem. Thus,

$$Z_B = R_B = \frac{R_1 R_2}{R_1 + R_2} + \frac{R_3 R_4}{R_3 + R_4} \qquad (6.25)$$

In most bridge arrangements, R_B is less than 10^4 Ω. Since the input impedance Z_M of most modern voltage recording devices exceeds 10^6 Ω, the ratio $Z_B/Z_M < 0.01$ and loading errors are usually less than 1 percent.

6.6   WHEATSTONE BRIDGE (CONSTANT CURRENT)

Use of a constant-current power supply with the potentiometer circuit improves the circuit sensitivity and eliminates nonlinear effects. Advantages of using a constant-current power supply with the Wheatstone bridge can be determined by considering the circuit shown in Fig. 6.10. The current i_s delivered to the bridge by the power supply divides at point A into currents i_1 and i_2, where

$$i_s = i_1 + i_2 \qquad \text{(a)}$$

The voltage drop across resistance R_1 is

$$v_{AB} = i_1 R_1 \qquad \text{(b)}$$

Similarly, the voltage drop across resistance R_4 is

$$v_{AD} = i_2 R_4 \qquad \text{(c)}$$

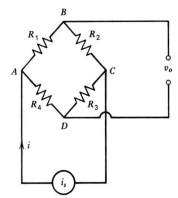

Figure 6.10 The constant-current Wheatstone bridge circuit.

Thus, the output voltage v_o from the bridge is

$$v_o = v_{BD} = v_{AB} - v_{AD} = i_1 R_1 - i_2 R_4 \qquad \textbf{(6.26)}$$

From Eq. 6.26 it is clear that the bridge will be balanced ($v_o = 0$) if

$$i_1 R_1 = i_2 R_4 \qquad \textbf{(d)}$$

This balance equation is not in a useful form, because the currents i_1 and i_2 are unknowns. The magnitudes of these currents can be determined by observing that the voltage v_{AC} can be expressed in terms of i_1 and i_2 as

$$v_{AC} = i_1(R_1 + R_2) = i_2(R_3 + R_4) \qquad \textbf{(e)}$$

From Eqs. a, d, and e,

$$i_1 = \frac{R_3 + R_4}{R_1 + R_2 + R_3 + R_4} i_s$$

$$\qquad \textbf{(f)}$$

$$i_2 = \frac{R_1 + R_2}{R_1 + R_2 + R_3 + R_4} i_s$$

Substituting these equations into Eq. 6.26 gives

$$v_o = \frac{i_s}{R_1 + R_2 + R_3 + R_4}(R_1 R_3 - R_2 R_4) \qquad \textbf{(6.27)}$$

This equation shows that the balance requirement for the constant-current Wheatstone bridge,

$$R_1 R_3 = R_2 R_4 \qquad \textbf{(6.16)}$$

is the same condition as that for the constant-voltage Wheatstone bridge.

The open-circuit output voltage Δv_o, from an initially balanced bridge ($v_o = 0$), owing to resistance changes ΔR_1, ΔR_2, ΔR_3, and ΔR_4, is determined from Eq. 6.27 as

$$\Delta v_o = \frac{i_s}{\Sigma R + \Sigma \Delta R}[(R_1 + \Delta R_1)(R_3 + \Delta R_3) - (R_2 + \Delta R_2)(R_4 + \Delta R_4)]$$

$$= \frac{i_s R_1 R_3}{\Sigma R + \Sigma \Delta R}\left(\frac{\Delta R_1}{R_1} - \frac{\Delta R_2}{R_2} + \frac{\Delta R_3}{R_3} - \frac{\Delta R_4}{R_4} + \frac{\Delta R_1}{R_1}\frac{\Delta R_3}{R_3} - \frac{\Delta R_2}{R_2}\frac{\Delta R_4}{R_4}\right)$$

(6.28)

where

$$\Sigma R = R_1 + R_2 + R_3 + R_4$$
$$\Sigma \Delta R = \Delta R_1 + \Delta R_2 + \Delta R_3 + \Delta R_4$$

Equation 6.28 shows that the constant-current Wheatstone bridge exhibits a nonlinear output voltage Δv_o. The nonlinearity is due to the $\Sigma \Delta_R$ term in the denominator and to the two second-order terms within the bracketed quantity. Consider a typical application with a transducer in arm R_1 and fixed-value resistors in the other three arms of the bridge such that

$$R_1 = R_4 = R_T, \qquad R_2 = R_3 = rR_T, \qquad \Delta R_2 = \Delta R_3 = \Delta R_4 = 0 \qquad \textbf{(g)}$$

For this example, Eq. 6.28 reduces to

$$\Delta v_o = \frac{i_s R_T r}{2(1 + r) + (\Delta R_T / R_T)}\left(\frac{\Delta R_T}{R_T}\right)$$

(6.29)

which can also be expressed as

$$\Delta v_o = \frac{i_s R_T r}{2(1 + r)}\frac{\Delta R_T}{R_T}(1 - \eta)$$

(6.30)

where

$$\eta = \frac{\Delta R_T / R_T}{2(1 + r) + (\Delta R_T / R_T)}$$

(6.31)

It is evident from Eq. 6.31 that the nonlinear effect can be reduced by increasing r. The percent error ($100\,\eta$) as a function of $\Delta R_t / R_T$ and r is shown in Fig. 6.11. Comparing the errors illustrated in Figs. 6.9 and 6.11 clearly shows the advantage of the constant-current power supply in extending the range of the Wheatstone bridge circuit.

The circuit sensitivity S_{cc} as obtained from Eq. 6.30 is

$$S_{cc} = \frac{\Delta v_o}{\Delta R_T / R_T} = \frac{i_s R_T r}{2(1 + r)}.$$

(6.32)

For the example being considered, the bridge is symmetric; therefore, the current $i_T = i_s / 2$. The power dissipated by the transducer is

$$p_T = i_T^2 R_T = \frac{1}{4}i_s^2 R_T$$

(h)

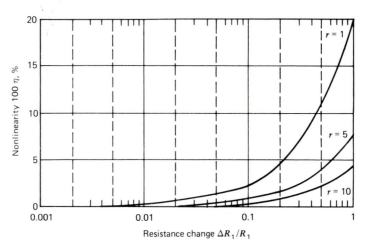

Figure 6.11 Nonlinear term η as a function of resistance change $\Delta R_1/R_1$ for a constant-current Wheatstone bridge circuit with one active gage.

Substituting Eq. h into Eq. 6.32 yields

$$S_{cc} = \frac{r}{1 + r} \sqrt{p_T R_T} \qquad \textbf{(6.24)}$$

which indicates that the circuit sensitivity is the same for constant-voltage and constant-current Wheatstone bridges.

The principal advantage of a Wheatstone bridge, when compared with a potentiometer, is the initial balancing of the Wheatstone bridge to produce a zero output voltage ($v_o = 0$). A second advantage is the ability to use the bridge in a null-balance mode (see Section 7.5), which eliminates the need for a precise voltage-measuring instrument.

6.7   AMPLIFIERS

An amplifier is one of the most important components in an instrumentation system. An amplifier is used in nearly every system to increase low-level signals from a transducer to a level sufficiently high for recording with a voltage-measuring instrument. An amplifier is represented in schematic diagrams of instrumentation systems by the triangular symbol shown in Fig. 6.12. The voltage input to the amplifier is v_i; the voltage output is v_o. The ratio v_o/v_i is the gain G of the amplifier. As the input voltage is increased, the output voltage increases in the linear range of the amplifier according to the relationship

$$v_o = G v_i \qquad \textbf{(6.33)}$$

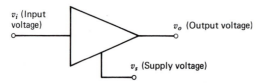

v_i (Input voltage)

v_o (Output voltage)

v_s (Supply voltage)

Figure 6.12 Symbol for an amplifier.

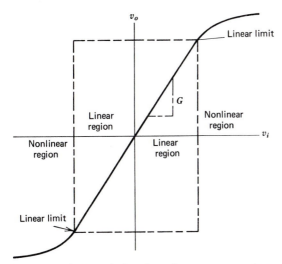

Figure 6.13 A typical voltage-input versus voltage-output curve for an amplifier.

The linear range of an amplifier is finite, since the output voltage is limited by the supply voltage and the characteristics of the amplifier components. A typical input–output graph for an amplifier is shown in Fig. 6.13. If the amplifier is driven beyond the linear range (overdriven), serious errors can result if the gain G is treated as a constant.

If the gain from a single amplifier is not sufficient, two or more amplifiers can be series connected (cascaded), as shown in Fig. 6.14. Such an amplifier system has an output voltage v_o given by the expression

$$v_o = G^3 \left(\frac{Z_i}{Z_i + Z_1} \right) \left(\frac{Z_i}{Z_i + Z_o} \right)^2 \left(\frac{Z_2}{Z_o + Z_2} \right) v_i \qquad (6.34)$$

where the impedances Z_i, Z_o, Z_1, and Z_2 are defined in Fig. 6.14. Each bracketed term in Eq. 6.34 represents the voltage attenuation owing to the current required to drive the next stage. By maintaining $Z_i \gg Z_o$ and using a recorder with a high-input impedance so that $Z_2 \gg Z_o$, these attenuation terms approach unity and

$$v_o = G^3 v_i \qquad (6.35)$$

With proper selection of input and output impedances, the overall gain of a cascaded amplifier system equals the product of the gains of the individual stages.

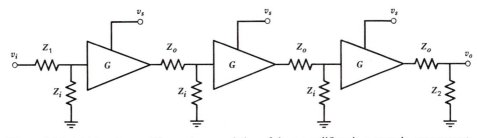

Figure 6.14 A high-gain amplifier system consisting of three amplifiers in a cascade arrangement.

Frequency response of an amplifier must also be given careful consideration when designing an instrumentation system. The gain is a function of the frequency of the input signal; there is always a high frequency at which the gain of the amplifier will be less than its value at the lower frequencies. This frequency effect on amplifier gain is similar to inertia effects in a mechanical system. A finite time (transit time) is required for the input current to pass through all the components in the amplifier and reach the output terminal. Time is also required for the output voltage to develop, since some capacitance is always present in the input impedance of the recording instrument.

The frequency response of an amplifier–recorder system can be illustrated in two different ways. First, the output voltage can be described as a function of time for a step input as shown in Fig. 6.15a. The rise in output voltage for this representation can be approximated by an exponential function of the form.

$$v_o = G(1 - e^{-t/\tau})v_i \qquad (6.36)$$

where τ is the time constant for the amplifier.

The second method of illustrating frequency effects utilizes a graph showing gain plotted as a function of frequency, as shown in Fig. 6.15b. The output of the amplifier

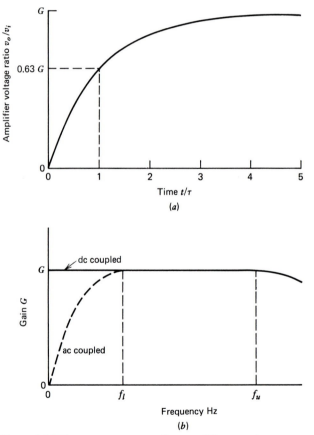

Figure 6.15 Frequency response of an amplifier-recorder system. (a) Amplifier response to a step-input voltage. (b) Gain as a function of frequency of the input voltage.

is flat between the lower and upper frequency limits f_l and f_u. Thus, a dynamic signal with all frequency components within the band between f_l and f_u will be amplified with a constant gain. *DC* or *dc-coupled amplifiers* are designed with input circuits that maintain a constant gain down to zero frequency. However, if a capacitor is placed in series with the input to the amplifier to block the dc components of the input signal, the gain G goes to zero as the frequency of the input signal decreases toward zero. The addition of the series capacitor produces an ac-coupled amplifier, with a variable gain for frequencies between zero and f_1 (see Fig. 6.15*b*).

Amplifiers are classified as either single ended or dual input. With *single-ended amplifiers* both the input and output voltages are referenced to ground, as indicated in Fig. 6.16*a*. Single-ended types can be used only when the output from the signal conditioning circuit is referenced to ground, as is the case for the potentiometer circuit described in Section 6.3. The output from a Wheatstone bridge is not referenced to ground; therefore, single-ended amplifiers cannot be used with this signal conditioning circuit. *Dual-input* or *differential amplifiers* must be employed (see Fig. 6.16*b*) where two separate voltages, each referenced to ground, are connected to the inputs. The output is single ended and referenced to ground. The ideal output voltage from a differential amplifier is

$$v_o = G(v_{i1} - v_{i2}) \tag{6.37}$$

Generally, these input voltages can be expressed as

$$
\begin{aligned}
v_{i1} &= v + \Delta v \\
v_{i2} &= v
\end{aligned}
\tag{a}
$$

where v is the common-mode voltage and Δv is the small difference voltage that is to be amplified. Unfortunately, owing to slight differences in the amplifier's components, the output voltage v_o is not zero, as indicated by Eq. 6.37, when Δv is zero. It is more accurate to write Eq. 6.37 as

$$v_o = G_d \Delta v + G_c v \tag{6.38}$$

where

G_d is the gain for the difference voltage Δv
G_c is the gain for the common-mode voltage v

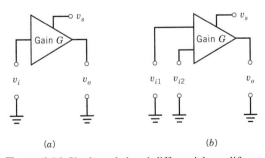

(a)   (b)

Figure 6.16 Single-ended and differential amplifiers.
(*a*) Single-ended input and output. (*b*) Double-ended input and single-ended output.

One measure of the quality of a differential amplifier is the *common-mode rejection ratio* (CMRR), where

$$CMRR = \frac{G_d}{G_c} \tag{6.39}$$

To minimize G_c relative to G_d, a very high value of the CMRR is preferred. Values of CMRR ranging from 1000 to 20,000 are typical for differential amplifiers, with the lower values occurring at the higher frequencies. With CMRR $\geq$ 1000, Eq. 6.38 is closely approximated by Eq. 6.37.

A high value for the CMRR in the difference mode is important because it implies that spurious signals common to both inputs v_{i1} and v_{i2}, such as noise, power-supply ripple, and temperature-induced drift, are canceled. The ability of the differential amplifier to eliminate these undesirable components of the input signal is a significant advantage. Differential amplifiers can be used with all types of signal conditioning circuits.

Another measure of the quality of an amplifier is related to the signal-to-noise ratio (S/N), which is written as

$$(S/N)_i = \left(\frac{v_i}{v_{ni}}\right)^2 \tag{6.40}$$

where v_{ni} is voltage superimposed on the input signal by noise. Note that the voltage ratio v_i/v_{ni} in Eq. 6.40 is squared because the signal-to-noise ratio is defined as the ratio of the signal power to the noise power.

To evaluate the quality of an amplifier in limiting noise on the output signal, consider the input/output noise figure F_n

$$F_n = 10\log\left[\frac{(S/N)_i}{(S/N)_o}\right] = 10\log(NF) \tag{6.41a}$$

where NF is the noise factor given by

$$NF = \frac{(S/N)_i}{(S/N)_o} > 1.0 \tag{6.41b}$$

Note that the signal-to-noise ratio on the output is

$$(S/N)_o = \left(\frac{v_o}{v_{no}}\right)^2 = \frac{G_p v_i^2}{G_p v_{ni}^2 + v_{nA}^2} \tag{6.42}$$

where

v_{nA} is the noise introduced by the amplifier
G_p is the power gain of the amplifier

Substituting Eqs. 6.40 and 6.42 into Eq. 6.41b gives

$$NF = 1 + \frac{v_{nA}^2}{G_p v_{ni}^2} \tag{6.43}$$

We seek to minimize the term $(v_{nA}^2 / G_p v_{ni}^2)$ so that the noise Factor NF is close to one. If NF = 2, it is evident from Eq. 6.43 that the amplifier and input source are adding a noise signal equal to the noise in the input signal, which is clearly unacceptable.

Another measure of the quality of an amplifier is its dynamic range R_d, which is defined as

$$R_d = 20 \log \left(\frac{v_m}{v_n} \right) \tag{6.44}$$

where v_m is the maximum input signal before the amplifier becomes nonlinear and $v_n = v_{nA}/G$. A high dynamic range is desired to permit extended usage in the linear range of the amplifier.

6.8 OPERATIONAL AMPLIFIERS

An *operational amplifier* (op-amp) is an integrated circuit consisting of miniaturized transistors, diodes, resistors, and capacitors, which have been placed on a small silicon chip to form the complete amplifier circuit. Operational amplifiers serve many functions because they can easily be adapted to perform several mathematical operations by adding a small number of external passive components, such as resistors or capacitors. Operational amplifiers have an extremely high gain ($G = 10^5$ is a typical value), and G is usually considered infinite in the analysis and design of circuits containing the op-amp. The input impedance (typically $R = 4$ MΩ and $C = 8$ pF) is so high that circuit loading usually is not a consideration. Output resistance (on the order of 100 Ω) is sufficiently low to be considered negligible in most applications.

Figure 6.17 shows the symbols used to represent the internal op-amp circuit in schematic diagrams, a pin-connection diagram, and the physical size of a typical op-amp. The two input terminals are identified as the *inverting* (−) *terminal* and the *noninverting* (+) *terminal*. The output voltage v_o of an op-amp is given by the expression

$$v_o = G(v_{i1} - v_{i2}) \tag{6.37}$$

It is evident from Eq. 6.37 that the op-amp is a differential amplifier; however, it is not used as an instrument differential amplifier because of its extremely high gain and very poor stability. The op-amp can be used effectively, however, as a part of a larger circuit (with more accurate and more stable passive elements) for many applications. Several applications of the op-amp, including inverting amplifiers, voltage followers, summing amplifiers, integrating amplifiers, and differentiating amplifiers, will be covered in subsequent subsections.

6.8.1 Inverting Amplifier

An *inverting amplifier* with a single-ended input and output can be assembled from an op-amp and resistors, as shown in Fig. 6.18. In this circuit, the input voltage v_i is applied to the negative input terminal of the op-amp through an input resistor R_1. The positive input terminal of the op-amp is connected to the common ground wire. The output voltage v_o is fed back to the negative terminal of the op-amp through a feedback resistor R_f.

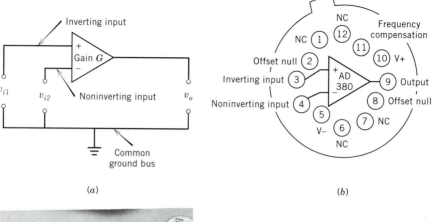

(a)

(b)

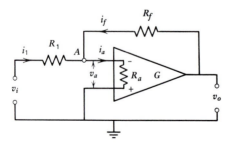

(c)

Figure 6.17 Characteristics of an operational amplifier. (a) Circuit diagram. (b) Pin-connection diagram. (c) Photograph of one type of op-amp.

The gain of the inverting amplifier can be determined by considering the sum of the currents at point A in Fig. 6.18. Thus,

$$i_1 + i_f = i_a \tag{a}$$

If v_a is the voltage drop across the input terminals of the op-amp,

$$i_1 = \frac{v_i - v_a}{R_1} \qquad i_f = \frac{v_o - v_a}{R_f} \qquad i_a = \frac{v_a}{R_a} \tag{b}$$

Figure 6.18 An inverting amplifier with single-ended input and output.

The voltage drop across the op-amp v_a is related to the output voltage v_o by the gain relation. Therefore,

$$v_a = -\frac{v_o}{G} \tag{c}$$

From Eqs. a, b, and c,

$$\frac{v_o}{v_i} = -\frac{R_f}{R_1}\left[\frac{1}{1 + \dfrac{1}{G}\left(1 + \dfrac{R_f}{R_1} + \dfrac{R_f}{R_a}\right)}\right] \tag{6.45}$$

As an example, consider an op-amp with a gain $G = 200{,}000$ and $R_a = 4\ \mathrm{M\Omega}$, with $R_1 = 0.1\ \mathrm{M\Omega}$ and $R_f = 1\ \mathrm{M\Omega}$. Substituting these values into Eq. 6.45 yields

$$\frac{v_o}{v_i} = -10\left[\frac{1}{1 + 5.6(10^{-5})}\right] = -9.9994 \approx -10$$

Thus, it is obvious that the gain term in Eq. 6.45 can be neglected without introducing appreciable error (0.0056 percent in this example), and the gain of the circuit G_c can be accurately approximated by

$$G_c = \frac{v_o}{v_i} \approx -\frac{R_f}{R_1} \tag{6.46}$$

Op-amps can also be used in noninverting amplifiers and differential amplifiers in addition to inverting amplifiers. The circuits for each of these amplifiers are shown in Fig. 6.19, and the governing equations for these circuits are given in the following sections.

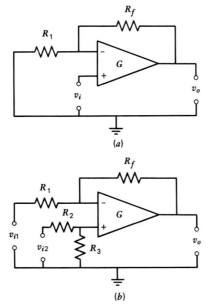

(a)

(b)

Figure 6.19 Instrument amplifiers that use operational amplifiers as the central circuit element. (a) Noninverting amplifier. (b) Differential amplifier.

Noninverting Amplifier

$$G_c = \frac{v_o}{v_i} = \frac{G}{1 + \dfrac{GR_1}{R_1 + R_f}} \tag{6.47}$$

$$G_c \approx 1 + \frac{R_f}{R_1} \tag{6.48}$$

6.8.2 Differential Amplifier

$$v_o \approx \frac{R_3}{R_2}\left(\frac{1 + \dfrac{R_f}{R_1}}{1 + \dfrac{R_3}{R_2}}\right)v_{i2} - \left(\frac{R_f}{R_1}\right)v_{i1} \tag{6.49}$$

If $R_f/R_1 = R_3/R_2$,

$$G_c \approx \frac{v_o}{v_{i2} - v_{i1}} \approx \frac{R_f}{R_1} \tag{6.50}$$

The circuits shown in Fig. 6.19 have been simplified to illustrate the concept of developing an amplifier with a gain G_c that is essentially independent of the op-amp gain G. In practice, these circuits must be modified to account for zero-offset voltages since, ideally, the output voltage v_o of the amplifier should be zero when the inputs (+) and (−) of the op-amp are connected to the common ground wire. In practice, this zero voltage is not achieved automatically, since the op-amps exhibit a zero-offset voltage. To avoid serious measurement errors, a biasing circuit that can be adjusted to restore the output voltage to zero must be added to the amplifier. Since the magnitude of the offset voltage changes (drifts) as a result of temperature, time, and power-supply voltage variations, it is advisable to adjust the bias circuit periodically to restore the zero output conditions.

A biasing circuit for an inverting amplifier with single-ended input and output is shown in Fig. 6.20. Common values of resistances R_2, R_3, and R_4 are $R_3 = R_1$, $R_2 = 10\ \Omega$, and $R_4 = 25\ \text{k}\Omega$. Voltages $v_1 = \pm15$ V are often used since the zero-

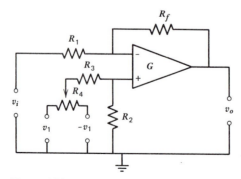

Figure 6.20 Biasing circuit for an amplifier with single-ended input and output.

offset voltage of the op-amp can be either positive or negative. The magnitude of the bias voltage that must be applied to the op-amp seldom exceeds a few millivolts.

The frequency response of instrument amplifiers constructed with op-amps depends on the frequency response of the op-amp and the circuit gain G_c. Since the gain G of an op-amp decreases with increasing frequency, the gain G_c of the circuit also decreases with increasing frequency. The frequency responses of op-amps vary appreciably, depending on design characteristics. Typical curves showing open-loop gain G and closed-loop gain G_c with frequency are presented in Fig. 6.21a. The decrease in the common-mode rejection ratio (CMRR) with frequency is shown in Fig. 6.21b.

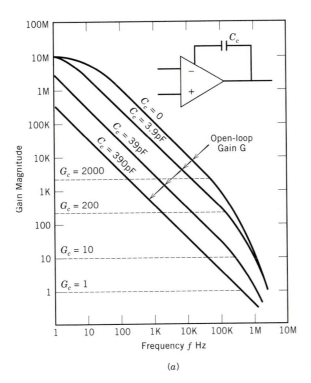

(a)

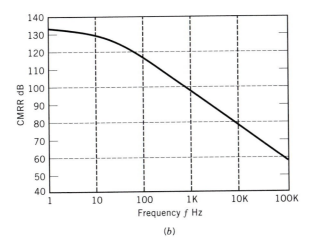

(b)

Figure 6.21 Characteristics of the AD 504 operational amplifier. (a) Gain G and G_c variations with frequency. (b) CMRR variations with frequency. (Courtesy of Analog Devices.)

6.8.3 Voltage Follower

An op-amp can also be used to construct an instrument with a very high input impedance for use with transducers such as piezoelectric sensors which require high input impedances. The high-impedance circuit, shown in Fig. 6.22, is know as a *voltage follower* and has a circuit gain of unity ($G_c = 1$). The voltage follower serves to adjust the impedance between the transducer and the voltage recording instrument. The voltage follower is also known as a unity-gain buffer amplifier.

The gain G_c of the voltage follower can be determined form Eq. 6.37. Thus,

$$v_o = G(v_{i1} - v_{i2}) = G(v_i - v_o) \tag{a}$$

Solving Eq. a for the circuit gain G_c gives

$$G_c = \frac{v_o}{v_i} = \frac{G}{1 + G} \tag{6.51}$$

When the gain G of the op-amp is very large, the gain G_c of the circuit approaches unity.

The input resistance of the voltage-follower circuit is given by Ohm's law as

$$R_{ci} = \frac{v_i}{i_a} \tag{b}$$

The input current i_a can be expressed in terms of the input and output voltages of the op-amp and the input resistance of the op-amp as

$$i_a = \frac{v_i - v_o}{R_a} \tag{c}$$

Combining Eqs. a, b, and c gives

$$R_{ci} = \frac{v_i R_a}{v_i - v_o} = \frac{v_i R_a}{v_i - \dfrac{G v_i}{1 + G}} = (1 + G)R_a \tag{6.52}$$

Since both G and R_a are very large for op-amps, the input impedance R_{ci} of the voltage-follower circuit can be made quite large. In practice, field-effect transistor (FET)–electrometer op-amps are used because they exhibit $G = 50 \times 10^3$ and $R_a = 10^{13}\,\Omega$. This combination will give an input impedance $R_{ci} = 5 \times 10^{17}\,\Omega$, which is

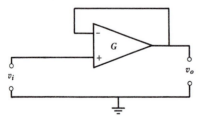

Figure 6.22 A high-impedance voltage-follower circuit.

sufficient to minimize any drain of charge from a piezoelectric transducer during static readout of short duration.

The output resistance R_{co} of the voltage-follower circuit is extremely low and is given by the expression

$$R_{co} = \frac{R_o}{1 + G} \tag{6.53}$$

where R_o is the output resistance of the op-amp, which is typically less than $100 \ \Omega$.

6.8.4 Summing Amplifier

In some data-analysis applications, signals from two or more sources must be added to obtain an output signal that is proportional to the sum of the input signals. Adding can be accomplished with the op-amp circuit known as a *summing amplifier,* shown in Fig. 6.23.

Operation of the summing amplifier can be established by considering current flow at point A of Fig. 6.23, which can be expressed as

$$i_1 + i_2 + i_3 + i_f = 0 \tag{a}$$

Applying Ohm's law to Eq. a yields

$$\frac{v_{i1}}{R_1} + \frac{v_{i2}}{R_2} + \frac{v_{i3}}{R_3} + \frac{v_o}{R_f} = 0 \tag{b}$$

Solving Eq. b for v_o gives

$$v_o = -R_f \left(\frac{v_{i1}}{R_1} + \frac{v_{i2}}{R_2} + \frac{v_{i3}}{R_3} \right) \tag{6.54}$$

Equation 6.54 indicates that the input signals v_{i1}, v_{i2}, and v_{i3} are scaled by ratios R_f/R_1, R_f/R_2, and R_f/R_3, respectively, and then summed. If $R_1 = R_2 = R_3 = R_f$, the inputs sum without scaling and Eq. 6.54 reduces to

$$v_o = -(v_{i1} + v_{i2} + v_{i3}) \tag{6.55}$$

If the gain G and the input impedance R_a are finite, the analysis of the circuit illustrated in Fig. 6.23 is more involved; however, it can be shown that

$$v_o = - \frac{\dfrac{v_{i1}}{R_1} + \dfrac{v_{i2}}{R_2} + \dfrac{v_{i3}}{R_3}}{\dfrac{1}{R_f} + \dfrac{1}{G} \left(\dfrac{1}{R_1} + \dfrac{1}{R_2} + \dfrac{1}{R_3} + \dfrac{1}{R_f} + \dfrac{1}{R_a} \right)} \tag{6.56}$$

Equations 6.54 and 6.56 indicate that the term

$$\frac{1}{G} \left(\frac{1}{R_1} + \frac{1}{R_2} + \frac{1}{R_3} + \frac{1}{R_f} + \frac{1}{R_a} \right)$$

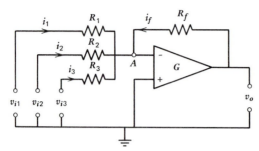

Figure 6.23 A summing amplifier incorporating an op-amp circuit.

represents an error in the scaling and summing operation. The magnitude of this error is small if G and R_a are large.

The circuit shown in Fig. 6.23 can be modified to produce an adding–subtracting amplifier if the positive terminal of the op-amp is used. The adding–subtracting amplifier circuit is shown in the figure of Exercise 6.39.

6.8.5 Integrating Amplifier

An *integrating amplifier* utilizes a capacitor in the feedback loop, as shown in Fig. 6.24. An expression for the output voltage from the integrating amplifier can be established by following the procedure used for the summing amplifier. Thus, by considering current flow at point A of Fig. 6.24 and taking $R_a \to \infty$,

$$i_1 + i_f = 0 \tag{a}$$

If Ohm's law is applied, Eq. a can be written as

$$\frac{v_i - v_i'}{R_1} + i_f = 0 \tag{b}$$

where

$$v_i' = -\frac{v_o}{G} \tag{c}$$

The voltage $v_i' \to 0$ when the gain G is large, and Eq. b becomes

$$\frac{v_i}{R_i} + i_f = 0 \tag{d}$$

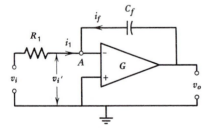

Figure 6.24 An integrating amplifier incorporating an op-amp circuit.

The charge q on the capacitor is given by the expression

$$q = \int_0^t i_f \, dt = C_f v_o \tag{e}$$

Substituting Eq. d into Eq. e and solving for the output voltage v_o yields

$$v_o = -\frac{1}{R_1 C_f} \int_0^t v_i \, dt \tag{6.57}$$

It is clear from Eq. 6.57 that the output voltage v_o from the circuit of Fig. 6.24 is the integral of the input voltage v_i with respect to time multiplied by the constant $-1/R_1 C_f$.

6.8.6 Differentiating Amplifier

The *differentiating amplifier* is similar to the integrating amplifier except that the positions of the resistor and capacitor of Fig. 6.24 are interchanged, as shown in Fig. 6.25. An expression for the output voltage v_o of the differentiating amplifier can be developed by following the procedure outlined for the integrating amplifier. The results are

$$v_o = -R_f C_1 \frac{dv_i}{dt} \tag{6.58}$$

Considerable care must be exercised to minimize noise on the input signal when the differentiation amplifier is used. Noise superimposed on the input voltage is differentiated and contributes significantly to the output voltage, thereby producing large error. The effects of high-frequency noise can be suppressed by placing a capacitor across resistance R_f; however, the presence of this capacitor affects the differentiating process, and Eq. 6.58 must be modified to account for its effects.

6.9 FILTERS

In many instrumentation applications, the signal from the transducer is combined with noise or some other parasitic signal. These parasitic voltages can often be eliminated with a filter that is designed to attenuate the undesirable noise signals but transmit the transducer signal without significant attenuation or distortion. Filtering of the signal

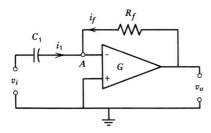

Figure 6.25 A differentiating amplifier incorporating an op-amp circuit.

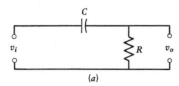

(a)

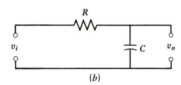

(b)

Figure 6.26 Filters commonly used for signal conditioning. (*a*) High-pass RC filter, and (*b*) low-pass RC filter.

is possible if the frequencies of the parasitic and transducer signals are sufficiently different. Two filters that utilize passive components and are commonly employed in signal conditioning include: the high-pass RC filter and the low-pass RC filter. Schematic diagrams of these filters are shown in Fig. 6.26.

6.9.1 High-Pass RC Filter

A simple yet effective high-pass resistance–capacitance (RC) filter is illustrated in Fig. 6.26*a*. The behavior of this filter in response to a sinusoidal input voltage of the form

$$v_i = v_a e^{j\omega t} \tag{a}$$

can be determined by summing the voltage drops around the loop of Fig. 6.26*a*. Thus,

$$v_i - \frac{q}{C} - Ri = 0 \tag{b}$$

where q is the charge on the capacitor. Equation b can be expressed in a more useful form by differentiating with respect to time to obtain

$$RC\frac{di}{dt} + i = j\omega C v_a e^{j\omega t} \tag{c}$$

Solving Eq. c by letting $i = i_a e^{j\omega t}$ and using Eq. 2.34 yields

$$i = \frac{j\omega C v_a}{1 + j\omega RC}e^{j\omega t} \tag{d}$$

The output voltage v_o is the voltage drop across the resistance R; therefore, from Eq. d

$$v_o = iR = \frac{\omega RC v_a}{\sqrt{1 + (\omega RC)^2}}e^{(j\omega t + \phi)} \tag{e}$$

where

$$\phi = \frac{\pi}{2} - \tan^{-1} \omega RC$$

The ratio of the amplitudes of the output and input voltages v_o/v_i obtained from Eqs. a and f gives the frequency response function for the high-pass filter as

$$\frac{v_o}{v_i} = H(\omega) = \frac{j\omega RC}{1 + j\omega RC} = |H(\omega)|e^{j\phi}$$

where

$$|H(\omega)| = \frac{\omega RC}{\sqrt{1 + (\omega RC)^2}} \tag{6.59}$$

Equation 6.59 indicates that $v_o/v_i \rightarrow 1$ as the frequency becomes large; thus, this filter is known as a high-pass filter. The response curve for a high-pass RC filter is shown in Fig. 6.27. At zero frequency (dc), the voltage ratio v_o/v_i vanishes, which indicates that the filter completely blocks any dc component of the output voltage. This dc-blocking capability of the high-pass RC filter can be used to great advantage when a low-amplitude transducer signal is superimposed on a large dc output voltage (see, for example, the potentiometer circuit of Section 6.3). Since the RC filter eliminates the dc voltage, a low-magnitude but frequency-dependent signal from a transducer can be amplified to produce a satisfactory display.

When a high-pass RC filter is used, $\omega RC \geq 5$ is needed to ensure that the input signal is transmitted through the filter with an attenuation that is less than 2 percent. The phase angle ϕ for $\omega RC \geq 5$ leads by less than 11 deg.

6.9.2 Low-Pass RC Filter

A low-pass RC filter is produced by interchanging the position of the resistor and capacitor of the high-pass RC filter. This modified RC circuit, shown in Fig. 6.26b, has transmission characteristics opposite to those of the high-pass RC filter; namely, it transmits low-frequency signals and attenuates high-frequency signals.

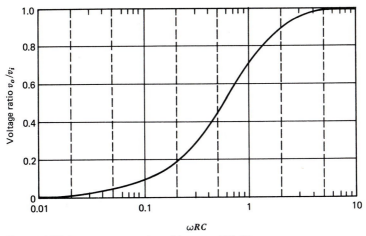

Figure 6.27 Response curve for a high-pass RC filter.

Using the methods of circuit analysis described previously, it is easy to show that the ratio of the amplitudes of the output and input voltages for this filter is

$$\frac{v_o}{v_i} = H(\omega) = \frac{1}{1 + j\omega RC} = |H(\omega)|e^{-j\phi} \tag{6.60}$$

where

$$|H(\omega)| = \frac{1}{\sqrt{1 + (\omega RC)^2}}$$

$$\phi = \tan^{-1}(\omega RC)$$

The frequency response curve for this filter is shown in Fig. 6.28. These results indicate that $|H(\omega)|$ varies from 1 to 0 as ωRC changes from 0.1 to 100. Both ends of this response curve are important. The low-frequency end, presented in expanded form in Fig. 6.28*b*, is important because it controls the attenuation of the input signal. Note that a 2 percent attenuation occurs when $\omega RC = 0.203$. To avoid errors greater than 2 percent when designing the low-pass filter, values for R and C must be selected so that $\omega RC < 0.203$.

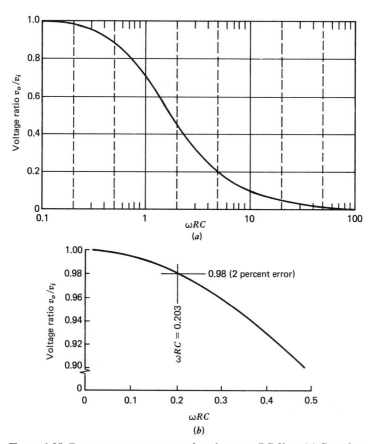

Figure 6.28 Frequency response curve for a low-pass RC filter. (*a*) Complete response curve. (*b*) Low-frequency end of the response curve.

The high-frequency response of the filter is also important, since it controls the attenuation of the parasitic or noise signal. A reduction of 90 percent in the noise signal can be achieved if $\omega_p RC = 10$ (ω_p is the circular frequency of the parasitic signal). It is not always possible with this passive filter to simultaneously limit the attenuation of the input signal to 2 percent while reducing the parasitic voltages by 90 percent, since to do so requires $\omega_p/\omega_i \geq 20$. If $\omega_p/\omega_i < 20$, it will be necessary to accept a higher ratio of parasitic signal or to accept a higher loss of the input signal.

6.9.3 Active Filter

Operational amplifiers are employed to construct active filters where select frequencies can be attenuated and the signal amplified during the filtering process. Several active filters are illustrated in Fig. 6.29. The filter shown in Fig. 6.29a is a low-pass filter

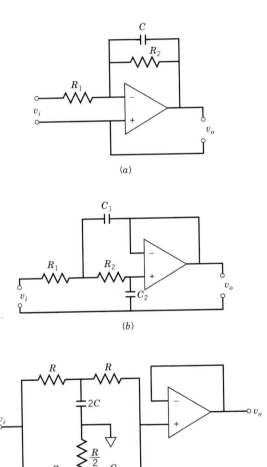

Figure 6.29 Active filter circuits that use operational amplifiers. (*a*) Inverting op-amps act as an RC filter with independent gain and frequency adjustments. (*b*) A second-order critically damped low-pass filter. (*c*) A twin-tee notch filter.

identical to that presented in Fig. 6.26b. The ratio of the amplitudes of the input and output voltages for this filter is

$$\frac{v_o}{v_i} = -\frac{R_2}{R_1} \frac{1}{\sqrt{1 + (\omega/\omega_c)}}$$

$$\omega_c = \frac{1}{R_2 C}$$

(6.61)

The addition of the operational amplifier permits the gain of the filter to be adjusted (R_2/R_1) independently of the critical frequency ω_c, which is varied by changing $R_2 C$. For independent settings of G_c and ω_c, R_2 is fixed, and C is adjusted to vary ω_c while R_1 is adjusted to vary G_c.

A two-pole Bessel filter is presented in Fig. 6.29b. The ratio of the amplitudes of the input and output voltages for this filter is

$$\frac{v_o}{v_i} = \frac{1}{\sqrt{[1 - (\omega/\omega_c)^2]^2 + 4(\omega/\omega_c)^2}}$$

(6.62)

$$\omega_c = \frac{2}{3RC} \quad \text{when} \quad R_1 = R_2 = R \quad \text{and} \quad C_1 = C_2 = C$$

The addition of the second pole (the use of a second capacitor) increases the rate of roll-off in attenuating the higher frequency signals and improves filter performance. Exercises concerning filters are given at the end of the chapter to enable comparison of the performances of different types of filters.

Noise at 60 Hz, caused by the presence of motors and lights operating at line frequency, is extremely common and annoying in most laboratories. If proper shielding does not eliminate these noise signals, a notch filter that attenuates signals at 60 Hz can be employed. A twin-tee notched filter, shown in Fig. 6.29c, has an attenuating notch that depends on the critical frequency $f_c = 1/(2\pi RC)$. Selecting $R = 41.9$ kΩ and $C = 0.0633$ μF places f_c at 60 Hz. The frequencies of the input signal should be considerably lower or higher than f_c since the notch is broad.

Many other filters, such as the fourth-order low-pass Butterworth filter and the Chebyshev filter, are often used in signal-processing applications. An excellent and extensive treatment of active filters is given in Reference 6.

6.10 AMPLITUDE MODULATION AND DEMODULATION

Amplitude modulation is a signal conditioning process in which the signal from a transducer is multiplied by a carrier signal of constant frequency and amplitude. The carrier signal can have any periodic form, such as a sinusoid, square wave, sawtooth, or triangle. The transducer signal can be sinusoidal, transient, or random. The only requirement for mixing carrier and transducer signals is that the frequency ω_c of the carrier signal must be much higher than the frequency ω_i of the transducer signal.

The significant aspects of data transmission with amplitude modulation can be illustrated by considering a case in which both the carrier and transducer signals are sinusoidal. The output voltage v_o is then given by the expression

$$v_o = (v_i \sin \omega_i t)(v_c \sin \omega_c t)$$

(a)

where

v_i is the amplitude of the transducer signal
v_c is the amplitude of the carrier signal

Next recall the trigonometric identity

$$\sin A \sin B \;=\; \frac{1}{2}\cos (A - B) - \frac{1}{2}\cos (A + B) \qquad\qquad \textbf{(b)}$$

Substituting Eq. b into Eq. a gives

$$v_o \;=\; \frac{v_i v_c}{2}\left[\cos (\omega_c - \omega_i)t - \cos (\omega_c + \omega_i)t\right] \qquad\qquad \textbf{(6.63)}$$

The results of Eq. 6.63 give an amplitude-modulated output signal v_o that is illustrated in Fig. 6.30.

Equation 6.63 indicates that the output signal is being transmitted at two discrete frequencies $(\omega_c - \omega_i)$ and $(\omega_c + \omega_i)$. The amplitude $(\frac{1}{2})$ associated with each frequency is the same. The transmission of data at the higher frequencies permits use of high-pass filters to eliminate noise signals that usually occur at much lower frequencies. For example, consider use of a carrier frequency of 4000 Hz with a transducer-signal frequency of 60 Hz. Normally any 60-Hz noise would be difficult to eliminate because of the coincidence of the frequencies of the transducer signal and the noise. However, with amplitude modulation, the data (in this example) are transmitted at frequencies of 3940 and 4060 Hz and the 60-Hz noise can be eliminated with a high-pass filter.

Although amplitude modulation offers several advantages in data transmission (stability, low power, and noise suppression), the output signal is not suitable for display or interpretation until the input signal is separated from the carrier signal. The process of separating the transducer signal from the carrier signal is known as *demodulation*. The demodulation process is illustrated in a block diagram in Fig. 6.31*a*.

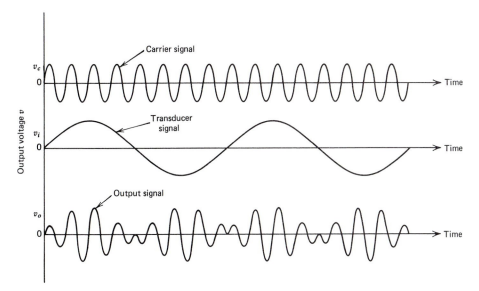

Figure 6.30 Carrier, transducer, and amplitude-modulated output signals.

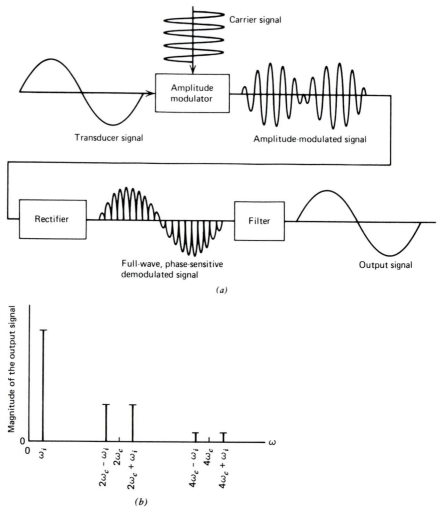

Figure 6.31 (*a*) Amplitude modulation and demodulation. (*b*) Frequency spectrum for the output from a full-wave, phase-sensitive rectifier.

The first step in the demodulation process involves rectifying the signal. This is usually accomplished with a full-wave, phase-sensitive rectifier. The output from this type of rectifier is a series of half-sine waves with amplitude and sense corresponding to the output signal from the transducer. The frequency spectrum associated with a rectified signal is shown in Fig. 6.31*b*. The frequency spectrum contains the input-signal frequency as a single line at ω_i and four other lines that depend strongly on the carrier frequency. There are many other carrier lines at higher frequencies; however, these have been omitted since they can be easily eliminated by filtering. The transducer signal is separated with a low-pass filter that transmits ω_i and severely attenuates the frequencies $2\omega_c \pm \omega_i$, $4\omega_c \pm \omega_i$, and so on.

Carrier frequencies of 10 to 100 times the transducer frequencies are required to eliminate the carrier signal. Multipole low-pass filters with sharp roll-off at the higher frequencies are usually employed in demodulation processes.

6.11   TIME-MEASURING CIRCUITS

Several different time measurements, including frequency, time interval, and event counting, are made with digital electronic counters. All these measurements are based on crystal oscillators (see Section 5.13), which are extremely accurate over short counting periods (1 part in 10^9). These crystal oscillators, known as clocks, provide a train of square-wave pulses within a precisely controlled period that are counted to determine the time interval.

In addition to the clock, time-measuring instruments incorporate digital counting units, gates, triggers, and liquid-crystal displays. Since the operation of each of these units is important in understanding methods of measuring time, each unit will be described.

6.11.1 Binary Counting Unit

The binary counting unit, shown in Fig. 6.32, contains a counting register consisting of four interconnected flip-flops. Each flip-flop stores one piece of binary information (a bit) and is either high (1) or low (0). Registers have many designs to accommodate different means of loading, reading, and resetting. The unit in Fig. 6.32 involves toggle flip-flops, which change state when the T input goes from 0 to 1. The register stores a digital code (a binary number) representing the number of input pulses. The counting process begins with a clearing signal, which sets the output Q of each flip-flop at 0 and the output register reads 0000. On the first counting pulse, the input T of flip-flop zero goes from 0 to 1 and the output Q goes from 0 to 1. The register now reads 0001, which gives a count of 1. On the next pulse, flip-flop zero returns to 0, which toggles flip-flop one and the register reads 0010, which gives a count of 2. After the third pulse, flip-flop zero goes high (1), flip-flop one is not changed, and the register reads 0011, which gives a count of 3. It is clear from this sequence of switching that the output Q of flip-flop zero represents 2^0, flip-flop one represents 2^1, flip-flop two represents 2^2, and flip-flop three represents 2^3. The output Q of the flip-flops that can be on or off gives a binary number representing the instantaneous count.

The counter of Fig. 6.32 can be extended by adding more flip-flops in the line. Each additional flip-flop gives another binary digit and increases the number of counts possible. The four flip-flop counter described here counts from 0 to 15 and on the 16th pulse the counter returns to 0000. For n flip-flops the max count C is

$$C = 2^n - 1 \qquad\qquad (4.1)$$

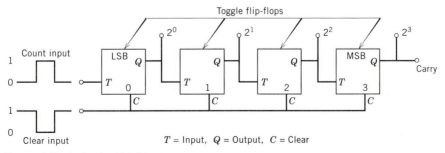

Figure 6.32 A simple 4-bit binary counter.

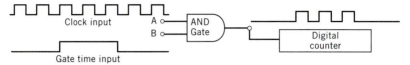

Figure 6.33 Gating a digital counter with an AND logic gate.

From Eq. 4.1 it is evident that a binary counter capable of a count $C = 1,048,575$ contains a line of 20 flip-flops (i.e., $n = 20$).

Other types of counters that use different digital codes, such as binary coded decimal (BCD) and decade scaling, are available; however, they are similar in concept to the binary counters and will not be described here. The reader is referred to Reference 10 for additional information on digital counters.

6.11.2   Gates in Counter Applications

Gates are used to control the digital counters by turning them on or off at appropriate times. The simplest counter-input gate, illustrated in Fig. 6.33, is the AND gate, which was described in Section 2.4.

The AND gate serves to transmit the clock pulses on terminal A to the digital counter only during the interval of time when the input signal from the gate time pulse to terminal B is high. Thus the gate time input signal controls the duration of the count.

In Fig. 6.33, the gate time input pulse was high for three clock pulses and these three pulses were transmitted as the AND gate turned on and off with the clock input. The digital counter recorded the count of 3 during the counting interval.

6.11.3 Triggers

Triggers are used to monitor input signals. At a preselected level, the trigger circuit issues a sharp-front output pulse. The Schmitt trigger, shown in Fig. 6.34, incorporates two transistors, which can be adjusted to provide upper- and lower-level thresholds for switching. The output v_o is high when transistor Q_1 is open and low when Q_1

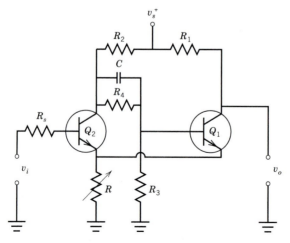

Figure 6.34 A Schmitt trigger circuit.

conducts. The status of transistor Q_1 depends on whether the trigger voltage v_i is above or below its upper and lower threshold values. For example, the output v_o of a Schmitt trigger with an oscillating input voltage is presented in Fig. 6.35. It is evident that v_o goes high when the input signal exceeds the upper threshold voltage and stays high until the input signal becomes less than the lower threshold voltage. The output is in the form of a sharp (square-wave) pulse with an amplitude suitable for logic-level transitions. The output amplitude is independent of the amplitude of the input pulse if the thresholds are exceeded. It should be noted that double minimums associated with valleys do not actuate the trigger. Both v_{iU} and v_{iL} must be crossed to change the output state of the trigger circuit.

The difference between v_{iU} and v_{iL} is known as a hysteresis zone. Hysteresis is essential for the operation of the trigger circuit and is useful in the use of a digital electronic counter with a noisy analog input. If the hysteresis zone is adjusted to be relatively large, the noise superimposed on the analog input signal v_i will produce multiple valleys or multiple peaks but will not cause triggering. On the other hand, if the hysteresis zone is too large, the trigger may not change state on a significant change of the input voltage. In this situation a real signal is not detected and a count is missed. The effect of the width of the hysteresis zone is illustrated in Fig. 6.36. Note that the output voltage corresponding to a small and a large hysteresis band is illustrated for an oscillating analog input signal.

6.11.4 Counting Instruments

A typical instrument for counting performs three different functions by utilizing clocks, gates, triggers, and counters arranged to perform EPUT, TIM, and GATE measurements.

A. EPUT

EPUT is an acronym for events per unit time. EPUT is the measurement of the number of events that occur in a precisely controlled interval of time. Signal frequency is the most common EPUT measurement, and it is a special case because the signal is

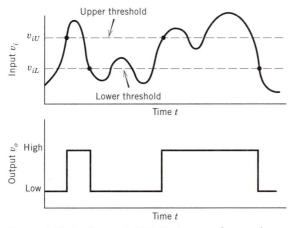

Figure 6.35 Logic-signal (high-low) output from an irregular analog signal input to a Schmitt trigger.

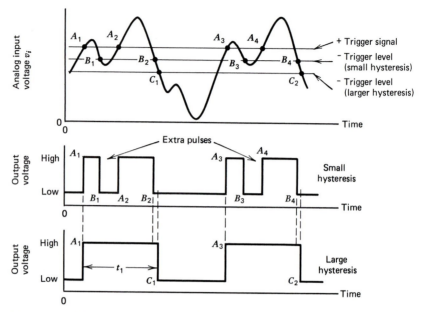

Figure 6.36 Effect of trigger level and hysteresis on the conversion of analog signals to digital pulses.

periodic. In the more general case of an arbitrary waveform, the signal need not be periodic.

The arrangement for a counter configured to measure EPUT is shown in Fig. 6.37. The input signal is amplified and then converted to a train of square-wave logic-level pulses by the Schmitt trigger. The gate controls the time interval over which the pulses are counted. A clock and decade counter switch the gate on (start) and off (stop) for precise intervals of time. These intervals can usually be selected as 0.001, 0.01, 0.1, 1, or 10 s. The count is accurate to ± 1.

B. TIM

TIM is the acronym for time-interval measurement and represents the time between two events. The configuration of a digital counter to measure TIM is shown in Fig. 6.38. The time between events is measured by taking the output from a crystal oscillator through a gate to a digital counter. The critical element is the gate, which is controlled by two event lines. The signal from event 1 is amplified and then passed through a Schmitt trigger, which issues a square-wave signal to start the gate. The signal from event 2 is also amplified to activate the second trigger, which produces the stop pulse.

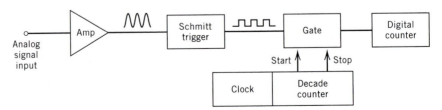

Figure 6.37 Counter arrangement for EPUT measurement.

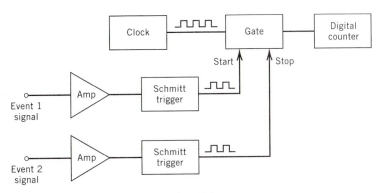

Figure 6.38 Counter arrangement for TIM measurement.

The count is in μs when the clock operates at 1 MHz or in steps of 100 ns when the clock operates at 10 MHz.

When the period of a repetitive signal is measured, only one of the amplifier-trigger lines is necessary. However, in this measurement the trigger is permitted to fire repeatedly. The sequencing for the stop and start signals on the gate is controlled by a logic circuit that senses the periodic input signal. At the initiation of the input signal, the gate goes high and the count starts. At the end of the period, the gate goes low and the count stops. The digital count is displayed, and the entire process is repeated at intervals depending on the rate of sampling. The period or frequency ($1/T$) displayed is updated after each measurement.

The count is extremely accurate because of the excellent quality of the crystal oscillator. However, errors are introduced by the triggering. Noise signals superimposed on the analog input signal can cause either a premature triggering or a delay in the triggering and in the generation of the start or the stop signals. The magnitude of the error depends on the precise characters of the analog signal and the noise signal.

C. GATE

The GATE function involves counting pulses associated with event 1 during a time interval that is started by event 2 and stopped by event 3. RATIO is a special case of GATE, where the ratio of two frequencies is measured.

Component arrangements for GATE measurements and the two simpler cases of RATIO and COUNT are shown in Fig. 6.39. In the GATE arrangement, illustrated in Fig. 6.39a, signal 2 through the amplifier-trigger generates the start pulse, and signal 3 through the other amplifier-trigger generates the stop pulse. Signal 1, through the first amplifier-trigger, provides the pulses to be counted. A precision time base is not necessary for any of the GATE measurements. The start and stop trigger levels must be set with precision to obtain an accurate GATE time interval.

The component arrangement for the simpler case of RATIO measurement is shown in Fig. 6.39b. For this measurement, signal 2 through the amplifier-trigger provides both the start pulse and the stop pulse. Signal 1 through the first amplifier-trigger provides the pulses to be counted. Thus, the digital counting unit measures the number of A (signal 1) events per B (signal 2) event (i.e., the ratio A/B). This measurement can be used for monitoring digital data transmission and for reliability assurance studies. For example, during digital data transmission, sending and receiving stations

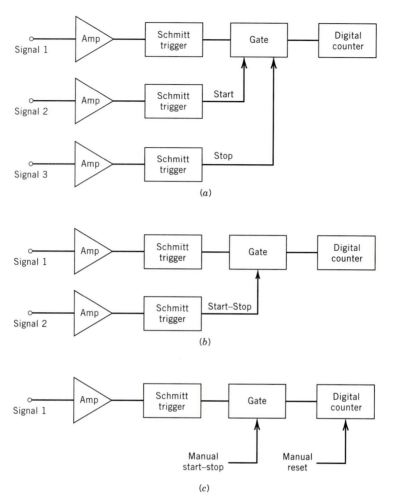

Figure 6.39 Component arrangements for GATE, RATIO, and COUNT measurements. (*a*) General GATE arrangement, (*b*) RATIO arrangement, and (*c*) COUNT arrangement.

can each count the number of bits sent and received. Agreement in the counts indicates a high probability that the data were properly received.

The COUNT mode of operation, shown in Fig. 6.39*c*, is a completely manual mode of operation. A manual start–stop switch both initiates and terminates the counting process. The digital counting unit is also manually reset to zero.

6.12 SUMMARY

Different signal conditioning circuits are employed in instrumentation systems, and the performance of the entire system can be markedly affected by the behavior of any of these circuits. Power supplies, both constant-voltage and constant-current, are commonly used in several different elements of an instrumentation system. It is imperative that the power supplies be stable over long periods of time and that noise and ripple be suppressed. Many portable instruments are equipped with long-life batteries using Zener diodes to stabilize the supply voltage.

Both the potentiometer circuit and the Wheatstone bridge circuit convert resistance change to a voltage variation. The significant characteristics of each type of circuit are described by the following equations:

Constant-Voltage Potentiometer Circuit

Output voltage:

$$\Delta v_o = \frac{r}{(1 + r)^2} \left(\frac{\Delta R_1}{R_1} - \frac{\Delta R_2}{R_2} \right)(1 - \eta)v_s \tag{6.4}$$

Nonlinear term:

$$\eta = 1 - \frac{1}{1 + \dfrac{1}{1 + r} \left(\dfrac{\Delta R_1}{R_1} + r \dfrac{\Delta R_2}{R_2} \right)} \tag{6.3}$$

Circuit sensitivity:

$$S_{cv} = \frac{r}{1 + r} \sqrt{p_T R_T} \tag{6.10}$$

Constant-Current Potentiometer Circuit

Output voltage:

$$v_o = i R_1 \frac{\Delta R_1}{R_1} \tag{6.12}$$

Circuit sensitivity:

$$S_{cc} = \sqrt{p_T R_T} \tag{6.14}$$

The Wheatstone bridge is widely used for converting resistance change to voltage, since it can be employed for both static and dynamic measurements. The Wheatstone bridge can also be driven with either a constant-voltage or a constant-current power supply. The equations describing the behavior of the Wheatstone bridge are as follows:

Constant-Voltage Wheatstone Bridge

Output voltage:

$$\Delta v_o = \frac{r}{(1 + r)^2} \left(\frac{\Delta R_1}{R_1} - \frac{\Delta R_2}{R_2} + \frac{\Delta R_3}{R_3} - \frac{\Delta R_4}{R_4} \right)(1 - \eta)v_s \tag{6.19}$$

Nonlinear term:

$$\eta = \frac{1}{1 + \dfrac{r + 1}{\dfrac{\Delta R_1}{R_1} + \dfrac{\Delta R_4}{R_4} + r \left(\dfrac{\Delta R_2}{R_2} + \dfrac{\Delta R_3}{R_3} \right)}} \tag{6.20}$$

Circuit sensitivity:

$$S_{cv} = \frac{r}{1+r} \sqrt{p_T R_T} \tag{6.24}$$

Constant-Current Wheatstone Bridge
Output voltage:

$$\Delta v_o = \frac{i_s R_T r}{2(1+r)} \frac{\Delta R_T}{R_T}(1 - \eta) \tag{6.30}$$

Nonlinear term:

$$\eta = \frac{\Delta R_T / R_T}{2(1+r) + (\Delta R_T / R_T)} \tag{6.31}$$

Circuit sensitivity:

$$S_{cv} = \frac{r}{1+r} \sqrt{p_T R_T} \tag{6.24}$$

The constant-current Wheatstone bridge is superior because of its extended range.

Amplifiers are used in most instrumentation systems to increase the output signal from a transducer to a level sufficiently large for recording with a voltage-measuring instrument. Ideally, the output and input signals for the amplifier are related by the expression

$$v_o = G v_i \tag{6.36}$$

Frequency response and linearity are two important characteristics of instrument amplifiers that must be adequate if signal distortion is to be avoided. A popular amplifier in instrumentation systems is the differential amplifier, because it rejects common-mode signals. Instrument amplifiers that employ op-amps with resistor feedback are commonly used because of their stability, low cost, and favorable operating characteristics.

Operational amplifiers (op-amps), such as voltage followers, summing amplifiers, integrating amplifiers, and differentiating amplifiers, are the active circuit element in signal conditioning circuits. The voltage follower exhibits a gain of unity and is used because of its high input impedance. Summing, integrating, and differentiating amplifiers, as the names imply, are used to add (or subtract) two or more input signals, integrate an input signal with respect to time, or differentiate an input signal with respect to time. The important characteristics of each are summarized by the following equations:

Voltage Follower
Input resistance:

$$R_{ci} = (1 + G)R_a \tag{6.52}$$

Summing Amplifier
Output voltage:

$$v_o = -R_f \left(\frac{v_{i1}}{R_1} + \frac{v_{i2}}{R_2} + \frac{v_{i3}}{R_3} \right) \tag{6.54}$$

Integrating Amplifier
Output voltage:

$$v_o = -\frac{1}{R_1 C_f} \int_0^t v_i \, dt \qquad (6.57)$$

Differentiating Amplifier
Output voltage:

$$v_o = -R_f C_1 \frac{dv_i}{dt} \qquad (6.58)$$

Filters are used to eliminate undesirable signals such as noise, a dc signal, or a high-frequency carrier signal. Voltage ratios v_o/v_i for two commonly used filters are given by the following expressions:

High-Pass RC Filter
Voltage ratio:

$$\frac{v_o}{v_i} = |H(\omega)| = \frac{\omega RC}{\sqrt{1 + (\omega RC)^2}} \qquad (6.59)$$

Low-Pass RC Filter
Voltage ratio:

$$\frac{v_o}{v_i} = |H(\omega)| = \frac{1}{\sqrt{1 + (\omega RC)^2}} \qquad (6.60)$$

Filters must be selected carefully; otherwise, the filter may attenuate both the noise signal and the input signal (if the frequencies are similar) and produce serious error. Active filters, which employ operational amplifiers, combine amplification and filtering functions. They can also be employed to produce notched filters (see Fig. 6.29), which are useful in eliminating 60-Hz noise signals.

Amplitude modulation is a signal conditioning process in which the input signal is multiplied by a carrier signal of a much higher frequency. The resulting output voltage is given by the expression

$$v_o = \frac{v_i v_c}{2} [\cos(\omega_c - \omega_i)t - \cos(\omega_c + \omega_i)t] \qquad (6.63)$$

The higher frequencies $(\omega_c - \omega_i)$ and $(\omega_c + \omega_i)$ associated with Eq. 6.63 permit the use of high-pass filters to eliminate low-frequency noise.

Time-measuring circuits employ clocks, digital counting units, gates, Schmitt triggers with hysteresis, and displays to perform frequency (EPUT), time-interval (TIM), and GATE measurements.

REFERENCES

1. Ahmed, H., and P. J. Spreadbury: *Analogue and Digital Electronics for Engineers,* 2nd ed., Cambridge Univ. Press, London/New York, 1984.
2. Barna, A., and D. I. Porat: *Operational Amplifiers,* 2nd ed., Wiley, New York, 1989.

3. Brophy, J. J.: *Basic Electronics for Scientists,* 5th ed., McGraw–Hill, New York, 1990.

4. Doebelin, E. O.: *Measurement Systems,* 4th ed., McGraw–Hill, New York, 1990.

5. Herpy, M., and J. C. Berka: *Active RC Filter Design,* Elsevier, Amsterdam/New York, 1986.

6. Hilburn, J. L., and D. E. Johnson: *Manual of Active Filter Design,* 2nd ed., McGraw–Hill, New York, 1982.

7. Hughes, F. W.: *Op-Amp Handbook,* 2nd ed., Prentice–Hall, Englewood Cliffs, N.J., 1986.

8. Irvine, R. G.: *Operational Amplifier Characteristics and Applications,* 2nd ed., Prentice–Hall, Englewood Cliffs, N.J., 1987.

9. Lenk, J. D.: *Handbook of Modern Solid-State Amplifiers,* Prentice–Hall, Englewood Cliffs, N.J., 1974.

10. Malmstadt, H. V., C. G. Enke, and S. R. Crouch: *Electronics and Instrumentation for Scientists,* Benjamin/Cummings, Menlo Park, Calif., 1981.

11. Meiksin, Z. H.: *Complete Guide to Active Filter Design, Op Amps, and Passive Components,* Prentice–Hall, Englewood Cliffs, N.J., 1989.

12. Parks, T. W., and C. S. Burrus: *Digital Filter Design,* Wiley, New York, 1987.

13. Stephenson, F. W.: *RC Active Filter Design Handbook,* Wiley, New York, 1985.

14. Van Valkenburg, M. E.: *Analog Filter Design,* Holt, Rinehart & Winston, New York, 1982.

EXERCISES

6.1 Describe the operation of the Zener-controlled battery power supply shown in Fig. 6.1. If the battery supply voltage is $v_s = 9$ V and the Zener breakdown voltage is $v_z = 5.2$ V, select R to limit the current flow to 100 mA. If a resistive load of 1000 Ω is placed across the output terminals, describe the effect on the voltage v_o.

6.2 How and when will the power supply of Exercise 6.1 fail?

6.3 Compare the characteristics of the LCR, Ni-Cd, and Li-I batteries.

6.4 Sketch a circuit showing the output impedance of a dc power supply.

6.5 Prepare a graph showing the sensitivity S_{cv} versus r for a potentiometer circuit. Consider the product $p_T R_T$ as a variable equal to 100, 200, 500, and 1000 W·Ω.

6.6 A strain gage with $R_g = 350$ Ω and $S_g = 2.00$ is used to monitor a sinusoidal signal with an amplitude of 1200 μin./in. and a frequency of 200 Hz. Determine the output voltage v_o if a constant-voltage potentiometer circuit is used to convert the resistance change to voltage. Assume $v_s = 8$ V and $r = 2$.

6.7 Determine the magnitude of the nonlinear term η for the data of Exercise 6.6

6.8 If the strain gage described in Exercise 6.6 can dissipate 0.25 W, determine the input voltage v_s required to maximize the output voltage v_o.

6.9 Determine the circuit sensitivity S_{cv} for the constant-voltage potentiometer circuit described in Exercise 6.6.

6.10 Determine the load error $\mathscr{E}$ if the output voltage v_o of Exercise 6.6 is monitored with:

(a) an oscilloscope having an input impedance of 10^6 Ω

(b) an oscillograph having an input impedance of 350 Ω

6.11 If a constant-current potentiometer circuit was used in Exercise 6.6 in place of the constant-voltage potentiometer circuit, determine the output voltage v_o if $i = 3$ mA.

6.12 Determine the magnitude of the nonlinear term η for the data of Exercise 6.11.

6.13 If the strain gage described in Exercise 6.6 can dissipate 0.25 W, determine the current i that should be used with a constant-current potentiometer circuit to maximize the output voltage v_o.

6.14 Determine the circuit sensitivity S_{cc} for the constant-current potentiometer circuit of Exercise 6.11.

6.15 Determine the circuit sensitivity S_{cc} for the constant-current potentiometer circuit of Exercise 6.13.

6.16 Prepare a graph showing S_{cc} as a function of R_T from 100 to 10,000 Ω for a constant-current potentiometer circuit. Let $p_T = 0.1, 0.2, 0.5,$ and 1 W.

6.17 A constant-voltage Wheatstone bridge circuit is employed with a displacement transducer (potentiometer type) to convert resistance change to output voltage. If the displacement transducer has a total resistance of 2000 Ω, then $\Delta R = \pm 1000\ \Omega$ if the wiper is moved from the center position to either end. If the transducer is placed in arm R_1 of the bridge and if $R_1 = R_2 = R_3 = R_4 = 1000\ \Omega$, determine the magnitude of the nonlinear term η as a function of ΔR. Prepare a graph of η versus ΔR as ΔR varies from $-1000\ \Omega$ to $+1000\ \Omega$.

6.18 Determine the output voltage v_o as a function of ΔR for the displacement transducer and Wheatstone bridge described in Exercise 6.17 if $v_s = 8$ V.

6.19 The nonlinear output voltage of Exercise 6.18 makes data interpretation difficult. How can the Wheatstone bridge circuit be modified to improve the linearity of the output voltage v_o?

6.20 A strain gage with $R_g = 350\ \Omega$, $p_T = 0.25$ W, and $S_g = 2.05$ is used in arm R_1 of a constant-voltage Wheatstone bridge. If the available power supply is limited to 28 V, determine:
   (a)   the values of R_2, R_3, and R_4 needed to maximize v_o
   (b)   the circuit sensitivity of the bridge

6.21 If the strain gage of Exercise 6.20 is subjected to strain of 1200 μin./in., determine the output voltage v_o.

6.22 Four strain gages are installed on a cantilever beam as shown in Fig. E6.22 to produce a displacement transducer.
   (a)   Indicate how the gages should be wired into a Wheatstone bridge to produce maximum signal output.
   (b)   Determine the circuit sensitivity if $R_g = 350\ \Omega$, $p_T = 0.15$ W, and $S_g = 2.00$.
   (c)   Determine the calibration constant $c = \delta/v_o$ for the transducer.

6.23 If the cantilever beam of Exercise 6.22 is used as a load transducer, determine the calibration constant $C = P/v_o$.

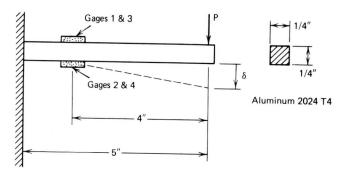

Figure E6.22

6.24   A strain gage with $R_g = 350\ \Omega$, $p_T = 0.10$ W, and $S_g = 2.00$ is used in arm R_1 of a constant-current Wheatstone bridge. Determine:
  (a)   the values of R_2, R_3, and R_4 needed to maximize v_o if the available power supply can deliver a maximum of 10 mA.
  (b)   the circuit sensitivity of the bridge of part a
  (c)   the output voltage v_o if the gage is subjected to a strain of 1500 μin./in.

6.25   If the displacement transducer of Exercise 6.17 is used with a constant-current Wheatstone bridge, determine the magnitude of the nonlinear term η as a function of ΔR. Prepare a graph of η versus ΔR as ΔR varies from $-1000\ \Omega$ to $+1000\ \Omega$.

6.26   Determine the output voltage v_o as a function of ΔR for the displacement transducer and Wheatstone bridge described in Exercise 6.25 if $i = 20$ mA.

6.27   Prepare a graph showing v_o/v_i for an amplifier responding to a step-input voltage. Let $\tau = 10\ \mu$s and consider gains of 10, 100, and 1000.

6.28   Sketch simple circuits showing the difference between single-ended and differential amplifiers.

6.29   If we have a common voltage of 0.1 V on the input to a differential amplifier with a gain $G_d = 500$ and we measure a voltage difference $\Delta v = 10$ mV, find the output voltage v_o if the common-mode rejection ratio is as follows:
  (a)   1000   (c)   10,000
  (b)   5000   (d)   20,000

6.30   Prepare a graph showing the dynamic range R as a function of gain G for an amplifier with a maximum input voltage $v_i = 500$ mV. Assume that the amplifier noise v_{nA} is 5 μV.

6.31   Use an op-amp with a gain of 100 dB and $R_a = 7\ M\Omega$ to design an inverting amplifier with a gain of
  (a)   10   (c)   50
  (b)   20   (d)   100

6.32   Use an op-amp with a gain of 100 dB and $R_a = 7\ M\Omega$ to design a noninverting amplifier with a gain of
  (a)   10   (c)   50
  (b)   20   (d)   100

6.33   Use an op-amp with a gain of 100 dB and $R_a = 7\ M\Omega$ to design a differential amplifier with a gain of
  (a)   10   (c)   50
  (b)   20   (d)   100

6.34   Determine the input and output impedances for a voltage follower that incorporates an op-amp having a gain of 120 dB and $R_a = 10\ M\Omega$.

6.35   Verify Eq. 6.52.

6.36   Verify Eq. 6.56.

6.37   Verify Eq. 6.57.

6.38   Three signals v_{i1}, v_{i2}, and v_{i3} are to be summed so that the output voltage v_o is proportional to $v_{i1} + 3v_{i2} + \frac{1}{3}v_{i3}$. Select resistances R_1, R_2, R_3, and R_f to accomplish this operation.

6.39   Show that the op-amp circuit shown in Fig. E6.39. is a compound adding/scaling and subtracting/scaling amplifier by deriving the following equation for the output voltage v_o:

$$v_o = \frac{R_f^*}{R_4}v_{i4} + \frac{R_f^*}{R_5}v_{i5} - \frac{R_f}{R_1}v_{i1} - \frac{R_f}{R_2}v_{i2} - \frac{R_f}{R_3}v_{i3}$$

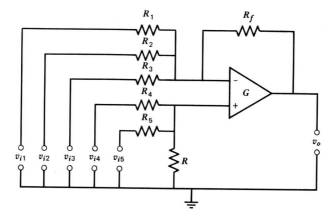

Figure E6.39

where

$$R_f^* = R_f \left(\frac{\dfrac{1}{R_1} + \dfrac{1}{R_2} + \dfrac{1}{R_3} + \dfrac{1}{R_f}}{\dfrac{1}{R_4} + \dfrac{1}{R_5} + \dfrac{1}{R}} \right)$$

6.40 The signals shown in Fig. E6.40 are to be used as input to an integrating amplifier having $R_1 = 1$ MΩ and $C_f = 0.5$ μF. Sketch the output signal corresponding to each of the input signals.

6.41 Discuss potential problems associated with the output voltages from the signals a and c of Exercise 6.40.

6.42 Repeat Exercise 6.40 with a differentiating amplifier in place of the integrating amplifier.

6.43 Draw circuits for the simple high-pass and low-pass RC filters. Sketch response curves for these filters.

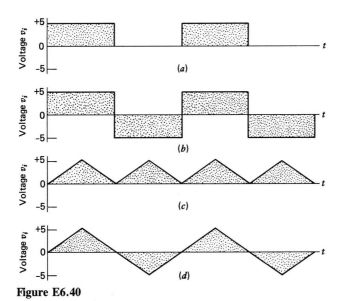

Figure E6.40

6.44   Verify Eq. 6.59.

6.45   Verify Eq. 6.60.

6.46   Select R and C for a high-pass filter so that the dc component of the output from a potentiometer circuit will be blocked while the ac signal from a transducer will be transmitted with less than 2 percent attenuation, if the transducer signal is

(a)   10 Hz   (c)   30 Hz

(b)   20 Hz

6.47   Select R and C for a low-pass filter that will block 60-Hz noise but transmit the following low-frequency signals with less than 1 percent loss:

(a)   5 Hz   (c)   20 Hz

(b)   10 Hz

6.48   Select R and C in Fig. 6.29 to give a notch filter with a critical frequency f_c of

(a)   60 Hz   (c)   10,000 Hz

(b)   1200 Hz

6.49   If the data signal is a triangular wave and the carrier signal is a square wave, sketch the amplitude-modulated signal.

6.50   Write an engineering brief that can be understood by a business school graduate, which explains amplitude modulation and demodulation.

6.51   Describe the operation of the 4-bit binary counter shown in Fig. 6.32.

6.52   Describe the operation of a Schmitt trigger and indicate why it is important in measuring time.

6.53   Sketch a block diagram representing the EPUT measurement and describe the operating principle of this counter arrangement.

6.54   Sketch a block diagram representing the TIM measurement and describe the operating principle of this counter arrangement.

6.55   Sketch a block diagram representing the GATE measurement and describe the operating principle of this counter arrangement.

6.56   Sketch a block diagram representing the RATIO measurement and describe the operating principle of this counter arrangement.

6.57   Sketch a block diagram representing the COUNT measurement and describe the operating principle of this counter arrangement.

Chapter 7

Resistance-Type Strain Gages

7.1 INTRODUCTION

Historically, the development of strain gages has followed many different approaches, and gages have been developed based on mechanical, optical, electrical, acoustical, and even pneumatic principles. A strain gage has several characteristics that should be considered in judging its adequacy for a particular application. These characteristics are as follows:

1. The calibration constant for the gage should be stable; it should not vary with either time, temperature, or other environmental factors.
2. The gage should be able to measure strains with an accuracy of ± 1 μin./in. (μm/m) over a large strain range (± 10 percent).
3. The gage size (the gage length ℓ_0 and width w_0) should be small so that strain (a point quantity) is approximated with small error.
4. The response of the gage, largely controlled by its inertia, should be sufficient to permit recording of dynamic strains with frequency components exceeding 100 kHz.
5. The gage system should permit both on-location and remote readout.
6. The output from the gage during the readout period should be independent of temperature and other environmental parameters.
7. The gage and the associated auxiliary equipment should both be low in cost to permit wide usage.
8. The gage system should be easy to install and operate.
9. The gage should exhibit a linear response to strain over a wide range.
10. The gage should be suitable for use as the sensing element in other transducer systems where an unknown quantity such as pressure is measured in terms of strain.

Although no single gage system can be considered optimum, the electrical resistance strain gage meets nearly all of the required characteristics listed above.

7.2   ETCHED-FOIL STRAIN GAGES

The sensitivity S_A of a metallic conductor to strain was developed in Section 5.4, and it is evident from Eq. 5.4

$$S_A = \frac{dR/R}{\epsilon_a} = \frac{d\rho/\rho}{\epsilon_a} + (1 + 2\nu) \tag{5.4}$$

that it is possible to measure strain with a straight length of wire if the change in resistance is monitored as the wire is subjected to a strain. The circuits required to measure dR (in practice ΔR), however, have power supplies with limited current capabilities and the power dissipated by the gage itself must be limited. As a result, strain gages are usually manufactured with a resistance of 120 Ω or, preferably, more. These high values of gage resistance in most cases preclude fabrication from a straight length of wire, since the length of the gage becomes excessive.

When electrical resistance strain gages were first introduced (1936–1956), the gage element was produced by forming a grid configuration with very fine diameter wire. Since the late 1950s, most gages have been fabricated from ultrathin metal foil by using a precise photoetching process. Since this process is quite versatile, a wide variety of gage sizes and grid shapes are produced (see Fig. 5.9). Gages as small as 0.20 mm in length are commercially available. Standard gage resistances are 120 Ω and 350 Ω; but in some configurations, resistances of 500 Ω, 1000 Ω, and 5000 Ω are available. The foil gages are normally fabricated from Advance, Karma, or Isoelastic alloys (see Table 5.2). In addition, high-temperature gages are available in several heat-resistant alloys.

The etched metal–film grids are very fragile and easy to distort or tear. To avoid these difficulties, the metal film is bonded to a thin sheet of plastic (see Fig. 5.7), which serves as a backing material and carrier before the photoetching process is performed. The carrier contains markings for the centerlines of the gage length and width to facilitate installation and serves to electrically insulate the metal grid from the specimen once it is installed.

For general-purpose strain-gage applications, a polyimide plastic that is tough and flexible is used for the carrier. For transducer applications, where precision and linearity are extremely important, a very thin, brittle, high-modulus epoxy is used for the carrier. Glass-reinforced epoxy is used when the gage will be exposed to high-level cyclic strains or when the gage will be employed at temperatures as high as 750°F (400°C).

For very high temperature applications, a gage with a strippable carrier is available. The carrier is removed during installation of the gage. A ceramic adhesive is used to maintain the grid configuration and to electrically insulate the grid from the specimen.

7.3   STRAIN-GAGE INSTALLATION

The bonded type of electrical resistance strain gage is a high-quality precision resistor that must be attached to a specimen with the correct adhesive and proper bonding procedures. The adhesive serves a vital function in the strain-measuring system, because it must transmit the surface displacement from the specimen to the gage grid without distortion. Initially it appears that this function can be accomplished with almost any strong adhesive; however, experience has shown that improperly selected and cured adhesives can seriously degrade a gage installation by changing the gage factor or the initial resistance of the gage. Improperly cured or viscoelastic adhesives also produce hysteresis and signal loss owing to stress relaxation. Best results are obtained with a strong, low-viscous, well-cured adhesive that forms a very thin elastic bond line.

The surface of the component in areas where gages are to be positioned must be carefully prepared before the gages are installed. This preparation consists of removal of paint and rust followed by sanding to obtain a smooth, but not polished, surface. Solvents are then used to eliminate all traces of grease and oil. Finally, the surface should be treated with a basic solution to give it the proper chemical affinity for the adhesive.

Next, the gage location is lightly scribed on the specimen and the gage, without adhesive, is positioned by using a rigid transparent tape in the manner illustrated in Fig. 7.1. The position and orientation of the gage are maintained by the tape as the adhesive is applied. The gage is pressed into place using the tape as a carrier, and then the excess adhesive is squeezed out from under the gage to produce a very thin bond line.

Once the gage is positioned, the adhesive must be subjected to a proper combination of pressure and temperature for the time required to ensure a complete cure. The curing process is quite critical since the adhesive will expand during heating, experience a

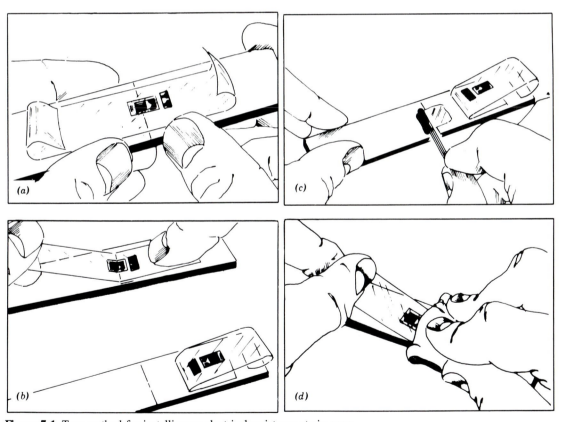

Figure 7.1 Tape method for installing an electrical resistance strain gage.
(*a*) Position the gage and terminal patch with transparent tape.
(*b*) Role back the gage and the terminal patch.
(*c*) Apply adhesive over the bonding area.
(*d*) Reposition the gage and terminal patch with tape, using finger pressure to force out the excessive adhesive.
(Courtesy of Micro-Measurements Division, Measurements Group, Inc., USA.)

volume change during polymerization, contract while cooling, and sometimes exhibit a postcure shrinkage. Since the adhesive is rigid enough to affect deformation of the gage, changes in the volume of the adhesive influence the resistance of the gage. Of particular importance is postcure shrinkage, which can influence the gage resistance long after the adhesive is supposed to be completely cured. If a long-term measurement of strain is made with a gage having an adhesive that has not completely polymerized, the signal from the gage will drift with time and accuracy of the data will be seriously impaired.

For most strain-gage applications, either cyanoacrylate or epoxy adhesives are used. The cyanoacrylate adhesive has the advantage of being easier to apply, since it requires no heat, cures with a gentle pressure that can be applied with one's thumb, and requires only about 10 min for complete polymerization. Its disadvantages include deterioration of strength with time, with water absorption, and with elevated temperatures. Epoxy adhesives are superior to cyanoacrylates; however, they are more difficult to apply since they require a pressure of 5 to 20 psi (35 to 140 kPa) and often need to be heated for an hour or more while the pressure is applied. After the adhesive is completely cured, the gage should be waterproofed with a light overcoating of crystalline wax or a polyurethane varnish.

Lead wires are attached to the terminals of the gage so that the change in resistance can be monitored with a suitable instrumentation system. Since the foil strain gages are fragile even when bonded to a structure, care must be exercised when the lead wires are attached to the soldering tabs. Intermediate anchor terminals (see Fig. 7.2), which are much more rugged than the strain-gage tabs, are used to protect the gage from damage. A small-diameter wire (32 to 36 gage) with a strain relief loop is used to connect the gage terminal to the anchor terminal. Three lead wires are soldered to the anchor terminal, as shown in Fig. 7.2, to provide for temperature compensation of the lead wires used to connect the strain gages to the Wheatstone bridge (see Section 7.4).

7.4 WHEATSTONE BRIDGE SIGNAL CONDITIONING

The basic equations governing the balance condition, output voltage, nonlinearity, and sensitivity of Wheatstone bridges with constant-voltage and constant-current power supplies were developed in Sections 6.5 and 6.6. Since the Wheatstone bridge is the circuit most commonly employed to convert the resistance change $\Delta R / R$ from a strain gage to an output voltage v_o, its application for this purpose is considered in

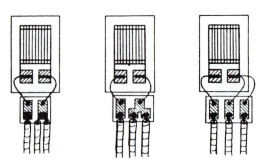

Figure 7.2 A strain-gage installation with anchor terminals. (Courtesy of Micro-Measurements Division, Measurements Group, Inc., USA.)

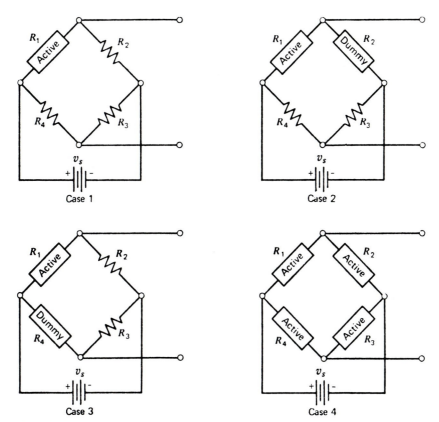

Figure 7.3 Four common strain-gage arrangements in a Wheatstone bridge.

detail in this section. One of the first questions that arises pertaining to use of the Wheatstone bridge for strain measurements concerns location of the gage or gages within the bridge. This question is answered by considering the four common bridge arrangements shown in Fig. 7.3.

Case 1. This bridge arrangement utilizes a single active gage in position R_1 and is often employed for both static and dynamic strain-gage measurements if temperature compensation is not required. The resistance $R_1 = R_g$ and the other three resistances are selected to maximize the circuit sensitivity while maintaining the balance condition $R_1R_3 = R_2R_4$.

The sensitivity S_s of the strain-gage–Wheatstone bridge system is defined as the product of the sensitivity of the gage S_g and the sensitivity of the bridge circuit S_c. Thus,

$$S_s = S_g S_c = \frac{\Delta R_g / R_g}{\epsilon} \left(\frac{\Delta v_o}{\Delta R_g / R_g} \right) = \frac{\Delta v_o}{\epsilon} \tag{7.1}$$

From Eqs. 5.5 and 6.24,

$$S_s = \frac{r}{1 + r} S_g \sqrt{p_g R_g} \tag{7.2}$$

Equation 7.2 indicates that the sensitivity of the system is controlled by the circuit efficiency $r/(1 + r)$ and the characteristics of the strain gage S_g, p_g, and R_g. Most important are the characteristics of the strain gage, which vary widely with gage selection. The gage factor S_g is about 2 for gages fabricated from Advance or Karma alloys and about 3.6 for gages fabricated from Isoelastic alloy. Resistances of 120 and 350 Ω are available for most grid configurations; resistances of 500, 1000, and 5000 Ω can be obtained for a few configurations. Power dissipation p_g is more difficult to specify, since it depends on the conductivity and heat-sink capacity of the specimen to which the gage is bonded. Power density p_D is defined as

$$p_D = \frac{p_g}{A} \tag{7.3}$$

where

p_g is the power that can be dissipated by the gage
A is the area of the grid of the gage

Recommended power densities for different materials and different types of test specimens are given in Table 7.1.

A graph showing bridge supply or input voltage v_s as a function of grid area for a large number of different gage configurations is shown in Fig. 7.4. The bridge voltage v_s, specified in Fig. 7.4, is for a four-equal-arm bridge with $r = 1$. In this case, the bridge voltage is given by

$$v_s = 2\sqrt{A p_D R_g} \tag{7.4}$$

When $r \neq 1$, the bridge voltage is given by

$$v_s = (1 + r)\sqrt{A p_D R_g} \tag{7.5}$$

The power that can be dissipated by a gage will vary over very wide limits. A small gage with a grid area of 0.001 in.^{2} bonded to an insulating material such as a ceramic ($p_D = 0.2$ W/in.^{2}) can dissipate 0.2 mW. On the other hand, a large strain gage with $A = 0.2$ in.^{2} mounted on a heavy aluminum section ($p_D = 10$ W/in.^{2}) can dissipate 2 W.

Table 7.1 Recommended Power Densities

Power Density p_D		Specimen Conditions
W/in^2	W/mm^2	
5–10	0.008–0.016	Heavy aluminum or copper sections
2–5	0.003–0.008	Heavy steel sections
1–2	0.0015–0.003	Thin steel sections
0.2–0.5	0.0003–0.0008	Fiberglass, glass, ceramics
0.02–0.05	0.00003–0.00008	Unfilled plastics

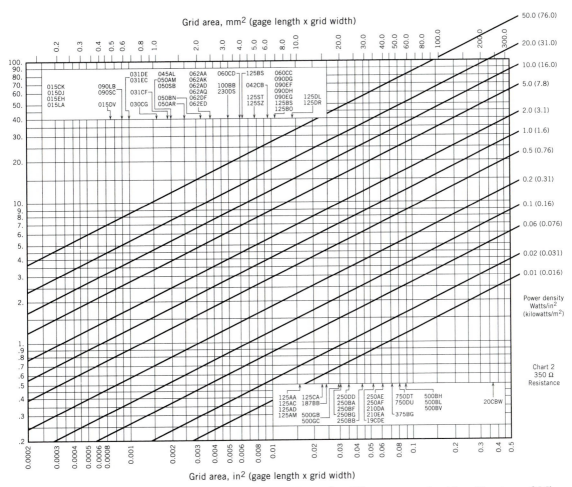

Figure 7.4 Allowable bridge voltage as a function of grid area for different power densities. (Courtesy of Micro-Measurements Division, Measurements Group, Inc., USA.)

System sensitivity can be maximized by selecting high-resistance gages with the largest grid area consistent with allowable error from the effects of gage length and width. Specification of Isoelastic alloy to obtain $S_g = 3.6$ should be limited to dynamic strain measurements where temperature stability of the gage is not a consideration.

The second factor controlling system sensitivity is circuit efficiency $r/(1 + r)$. The value of r should be selected to increase circuit efficiency, but should not be so high that the bridge voltage given by Eq. 7.5 increases beyond reasonable limits. Values of r between 3 and 5 give circuit efficiencies between 75 and 83 percent while maintaining v_s at reasonable values. For this reason most bridges are designed with r in this range.

Case 2. This bridge arrangement contains a single active gage in arm R_1, a dummy gage in arm R_2, and fixed-value resistors in arms R_3 and R_4. The active gage and the dummy gage must be identical (preferably two gages from the same lot of production), must be applied with the same adhesive, and must be subjected to the same curing cycle. The dummy gage can be mounted in a stress-free region of the specimen or on a small block of specimen material that is placed in the same thermal environment as the specimen. In the Wheatstone bridge, the dummy gage output serves to cancel any active gage output caused by temperature fluctuations during the test interval. The manner in which this bridge arrangement compensates for temperature changes can be illustrated by considering the resistance changes experienced by the active and dummy gages during a test. Thus

$$\left(\frac{\Delta R_g}{R_g}\right)_a = \left(\frac{\Delta R_g}{R_g}\right)_\epsilon + \left(\frac{\Delta R_g}{R_g}\right)_{\Delta T} \tag{a}$$

$$\left(\frac{\Delta R_g}{R_g}\right)_d = \left(\frac{\Delta R_g}{R_g}\right)_{\Delta T} \tag{b}$$

In Eqs. a and b the subscripts a and d refer to the active and dummy gages, respectively, and the subscripts ϵ and ΔT refer to the effects of strain and temperature. Substituting Eqs. a and b into Eq. 6.18 and noting that $\Delta R_3 = \Delta R_4 = 0$ (fixed-value resistors) gives

$$\Delta v_o = v_s \frac{r}{(1 + r)^2} \left[\left(\frac{\Delta R_g}{R_g}\right)_\epsilon + \left(\frac{\Delta R_g}{R_g}\right)_{\Delta T} - \left(\frac{\Delta R_g}{R_g}\right)_{\Delta T}\right] \tag{7.6}$$

Since the last two terms in the bracketed quantity cancel, the output Δv_o is caused only by the strain applied to the active gage, and temperature compensation is achieved.

With this bridge arrangement, r must equal 1 to satisfy the bridge balance requirement; therefore, the system sensitivity obtained from Eq. 7.2 is

$$S_s = \frac{1}{2} S_g \sqrt{p_g R_g} \tag{7.7}$$

Equation 7.7 indicates that placement of a dummy gage in arm R_2 of the Wheatstone bridge to effect temperature compensation reduces the circuit efficiency to 50 percent. This undesirable feature can be avoided by use of the bridge arrangement described under Case 3.

Case 3. In this bridge arrangement, the dummy gage is inserted in arm R_4 of the bridge instead of arm R_2. The active gage remains in arm R_1, and fixed-value resistors are used in arms R_2 and R_3. With this positioning of the dummy gage, r is not restricted by the balance condition and the system sensitivity is the same as that given by Eq. 7.2. Temperature compensation is achieved in the same manner that was illustrated in Case 2, but without loss of circuit efficiency. When a dummy gage is to be used to effect temperature compensation, arm R_4 of the bridge is the preferred location for the dummy gage.

Case 4. Four active gages are used in this Wheatstone bridge arrangement: one active gage in each arm of the bridge (thus, $r = 1$). When the gages are placed on a specimen, such as a cantilever beam in bending, with tensile strains on gages 1 and 3 (top surface of the beam) and compressive strains on gages 2 and 4 (bottom surface of the beam), then

$$\frac{\Delta R_1}{R_1} = \frac{\Delta R_3}{R_3} = -\frac{\Delta R_2}{R_2} = -\frac{\Delta R_4}{R_4} \tag{c}$$

Substituting these equations into Eq. 6.18 gives

$$\Delta v_o = \frac{1}{4} v_s \left(4 \frac{\Delta R_g}{R_g} \right) = v_s \frac{\Delta R_g}{R_g} \tag{7.8}$$

The Wheatstone bridge has added the four resistance changes to increase the output voltage; therefore, the system sensitivity is

$$S_c = \frac{1}{2} \left(4 S_g \sqrt{p_g R_g} \right) = 2 S_g \sqrt{p_g R_g} \tag{7.9}$$

This arrangement (with four active gages) has doubled the system sensitivity of Cases 1 and 3 and has quadrupled the sensitivity of Case 2. This bridge arrangement also provides temperature compensation. The use of multiple gages to gain sensitivity is not usually recommended because of the costs involved in the installation of the extra gages. High-quality, high-gain differential amplifiers usually can be used more economically to increase the output signal.

Examination of the four bridge arrangements shows that the system sensitivity can be varied from one half to two times $S_g \sqrt{p_g R_g}$. Temperature compensation is best achieved by placing the dummy gage in position R_4 to avoid loss of system sensitivity. System sensitivity can be improved by using multiple gages; however, the costs involved for the added gages is usually not warranted except for transducer applications, where the additional gages serve other purposes (see Chapter 8).

7.5 RECORDING INSTRUMENTS FOR STRAIN GAGES

The selection of a recording system for strain-gage applications depends primarily on the nature of the strain to be measured (static or dynamic) and the number of strain gages to be monitored. Static recording is far easier and less expensive than dynamic recording. Noise problems arise as a result of the higher level of amplification needed for dynamic recording, and the complex multichannel dynamic recorders significantly increase the costs of making the measurements.

Many different recording instruments can be used to monitor the output of the Wheatstone bridge. These recorders are described in considerable detail in Chapter 3. This section describes four different instrumentation systems that have been specifically adapted for strain-gage applications and are widely used in industry.

7.5.1 Direct-Reading Strain Indicator

A direct-reading strain indicator uses an integrating digital voltmeter to record the system output. This system also contains a Wheatstone bridge that is initially balanced by a potentiometer connected parallel to the output terminals. The voltage output from the bridge is amplified and then displayed on a $4\frac{1}{2}$-digit DVM, which gives a range of 19,999 $\mu\epsilon$. The bridge excitation of 2 V dc is low enough to avoid gage heating in most applications. System calibration is performed with a shunt resistor switched across either 120 or 350 dummy gages to yield an output signal equivalent to 5000 $\mu\epsilon$. Another potentiometer is used to set the gage factor. The gage factor adjustment attenuates the output of the instrument amplifier so that the DVM reads directly in units of strain ($\mu\epsilon$).

This strain indicator is also equipped with an analog output, which can be used with an oscilloscope for dynamic recording. In this application, the strain indicator serves as a bridge and a preamplifier for the oscilloscope. Note that the frequency response of this amplifier is limited and the output is down 3 dB at 4 kHz.

Another useful feature is a transducer connector, which permits the strain indicator to be used for monitoring and displaying the output signals from transducers that incorporate strain gages as the sensing elements. Amplifier-gain and gage-factor adjustments are made to permit direct reading of the transducer variable (load, pressure, or torque) on the DVM display.

The resolution of the unit is ±1 $\mu\epsilon$. The accuracy is 0.05 percent of the strain magnitude, or ±3 $\mu\epsilon$.

Several strain gages can be monitored with this indicator if a separate parallel-balance potentiometer is provided for each gage and if a multiple switch is provided so that each gage and balance resistor can be switched, in sequence, into the bridge.

7.5.2 Null-Balance Bridges

The direct-reading strain indicator described in Section 7.5.1 uses a DVM to measure the output voltage Δv_o of the bridge. For static measurements of strain, it is possible to employ a null-balance bridge, where the resistance in a nonactive arm is changed to match the resistance change $\Delta R/R$ of the active gage. The null-balance approach is much slower than direct-reading methods because of the time required to balance the bridge for each reading; however, it provides accurate strain measurements with low-cost instruments.

The reference bridge, illustrated in Fig. 7.5, is a strain indicator that utilizes the null-balance principle. Here, two bridges are used in combination to achieve null balance. The strain-gage bridge on the left is used for the active gages (one to four), and the reference bridge on the right is used for the variable resistors needed to affect balance. When gages and fixed-value bridge-completion resistors are inserted in the strain-gage bridge, an initial unbalance occurs between the two bridges. This unbalance produces a voltage difference that, after amplification, causes the galvanometer to deflect. The variable resistor in the reference bridge is then adjusted to balance the voltages at points A and B on the bridges (indicated by a zero or null reading on the galvanometer). A reading on the scale associated with the variable resistor (after initial balance is achieved) provides a datum reading for subsequent strain measurements. If strains are applied to the gages, an unbalance again occurs and is eliminated by changing the

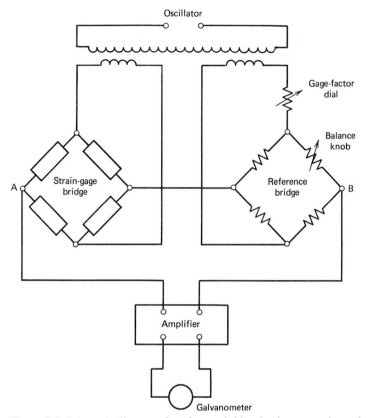

Figure 7.5 Schematic diagram of a reference bridge that is commonly used for the measurement of static strains.

setting of the variable resistor. The difference between the two readings provides a measure of the resistance change $\Delta R/R$ experienced by the active gage as a result of the applied strain. The scale on the variable resistor can be constructed to read strain directly.

In actual instruments, the circuits are more complex than the schematic arrangement shown in Fig. 7.5. The bridges are powered by a common oscillator with a 1000-Hz square-wave output of 1.5 V (rms). The voltage to the reference bridge is adjusted with a potentiometer (calibrated by using the gage-factor setting) so that readout from the scale of the variable resistor is direct in terms of strain.

The null-balance strain indicator functions over a range of gage resistances from 50 Ω to 2000 Ω. With gage resistances less than 50 Ω, the oscillator is overloaded; with gage resistances greater than 2000 Ω, the load on the amplifier becomes excessive. The gage-factor adjustment accommodates gages within the range $1.5 \leq S_g \leq 4.5$. The scale on the variable resistor can be read to ± 2 $\mu\epsilon$ and is accurate to ± 0.1 percent of the reading, or 5 $\mu\epsilon$, whichever is greater. The range of strain that can be measured is $\pm 50,000$ $\mu\epsilon$. This strain indicator, together with a switch and balance unit, provides a small, lightweight, and portable strain-measurement system. It is easy

Figure 7.6 An integrated strain-gage signal conditioner, which includes a power supply, half bridge, amplifier, and low-pass filter. (Courtesy of Analog Devices.)

to operate and is adequate for all static measurements of strain except those requiring a very large number of gages with extensive data analysis.

7.5.3 Strain-Gage Signal Conditioners

Complete signal conditioners, which include an internal half bridge, an instrument amplifier, an adjustable-output regulated power supply, and a low-pass filter, have recently become available. A typical example, presented in Fig. 7.6, shows the compactness of this integrated unit.

The instrument amplifier has an adjustable gain that can be changed from 2 to 5000 by varying the resistance across two accessible terminals. The excitation can be adjusted with external resistors to give bridge supply voltages between 4 and 15 V. The low-pass filter is set for a cutoff frequency of 1 kHz, although it can be varied between 10 Hz and 20 kHz with external resistors or capacitors.

These integrated signal conditioners are recommended because they provide four well-matched circuits (bridge, power supply, amplifier, and filter) at remarkably low cost ($45 in lots of 100) for the complete pin-mounted package.

7.5.4 Wheatstone Bridge and Oscilloscope

When strain gages are used to measure high-frequency dynamic strains at only a few locations, the oscilloscope is probably the best recording instrument. A typical Wheatstone bridge–oscilloscope arrangement is shown schematically in Fig. 7.7. The connection from the bridge to the oscilloscope can be direct if the oscilloscope has

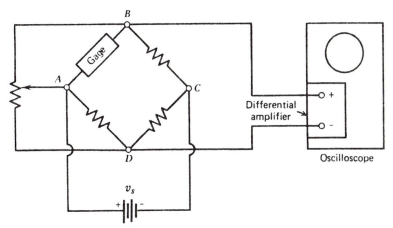

Figure 7.7 A Wheatstone bridge–oscilloscope strain-measurement system.

a differential amplifier with sufficient gain. Some single-ended amplifiers and power supplies cannot be used, since they ground point D of the bridge. This grounding seriously affects the output voltage of the bridge and introduces major errors in the strain measurements.

The input impedance of an oscilloscope is quite high (1 to 10 MΩ); as a consequence, loading errors are negligible for the Wheatstone bridge–oscilloscope combination because $R_s/R_m < 0.001$. The frequency response of an oscilloscope is extremely high, and even low-frequency models with a 10-MHz bandwidth greatly exceed the requirements for mechanical strain measurements, which rarely exceed 100 kHz. The observation interval depends on the sweep rate and can range from about 1 μs to 50 s. With digital oscilloscopes, the observation period depends on the number of words that can be stored and the sampling rate.

Strain as a function of time is displayed as a trace on the face of the cathode-ray tube (CRT). The trace can be photographed or, if a digital oscilloscope is used, readings of voltage and strain can be taken directly from the CRT. With an analog oscilloscope, strain ϵ is computed from the height d_s of the strain–time pulse, as illustrated in Fig. 7.8, and the distance d_c between two calibration lines produced by shunt calibration (see Section 7.6) of the bridge. The strain ϵ_g indicated by the gage is

$$\epsilon_g = \frac{d_s}{d_c}\epsilon_c \tag{7.10}$$

where ϵ_c is the equivalent strain produced by shunt calibration.

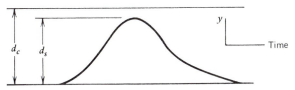

Figure 7.8 Determining strain from an oscilloscope trace.

If the shunt calibration technique is not used, the strain must be computed by using the output voltage from the bridge. For example, consider the output voltage Δv_o from a single gage (see Fig. 7.8) that is given by using Eqs. 6.18 and 5.5 as

$$\Delta v_o = \frac{r}{(1 + r)^2} v_s S_g \epsilon \tag{a}$$

This output voltage can be expressed in terms of oscilloscope parameters as

$$\Delta v_o = S_R d_s \tag{b}$$

where

d_s is the height of the strain–time pulse in CRT divisions
S_R is the sensitivity of the oscilloscope in volts per division

Substituting Eq. b into Eq. a and solving for the strain gives

$$\epsilon_g = \frac{(1 + r)^2}{r} \frac{S_R d_s}{v_s S_g} \tag{7.11}$$

With a digital oscilloscope the strain is determined using the on-board microprocessor to perform the multiplication indicated in Eq. 7.11.

7.5.5 Wheatstone Bridge and Oscillograph

The oscillograph (see Section 3.5) is also used with the Wheatstone bridge for dynamic strain measurements. The oscillograph is preferred over the oscilloscope[1] when large numbers of strain gages must be monitored and when the observation period is relatively long. When an oscillograph is used for dynamic strain measurements, care must be exercised in selection of the galvanometers, since many galvanometers are satisfactory only for low-frequency signals.

When an oscillographic galvanometer is connected directly to a Wheatstone bridge, the output voltage of the bridge is seriously affected. These galvanometers have a very low input impedance (between 30 and 300 Ω) and draw significant current from the output of the bridge. The equations developed previously were based on use of a high-impedance recorder and must be modified to account for the characteristics of the galvanometer. A circuit diagram for an oscillograph connected directly to a Wheatstone bridge is shown in Fig. 7.9a. An equivalent circuit, obtained by using Thevenin's theorem and consisting of an equivalent resistance R_B and a voltage source Δv_o replacing the bridge, is shown in Fig. 7.9b. The Wheatstone bridge–oscillograph circuit must be designed such that the equivalent resistance R_B provides the external damping resistance R_x required by the galvanometer to maintain frequency response. Also, the bridge should be designed to produce the largest possible output current i_G per unit strain in order to maximize the deflection θ of the galvanometer.

The equivalent circuit shown in Fig. 7.9b can be used to determine the current i_G passing through the galvanometer in terms of strain gage and circuit parameters.

---

[1] Waveform recorders with large RAM memories also provide very long observation periods together with a capability for expanding the time scale to improve resolution.

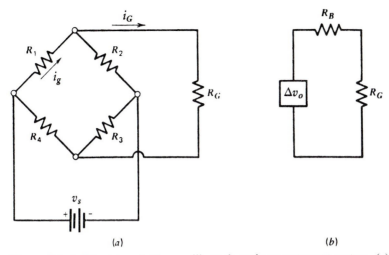

Figure 7.9 A Wheatstone bridge–oscillograph strain-measurement system. (*a*) Schematic diagram with R_G representing the input impedance of the galvanometer. (*b*) Equivalent circuit for determining i_G.

Consider an initially balanced bridge and define resistance ratios r and q as

$$r = \frac{R_2}{R_1} \quad \text{and} \quad q = \frac{R_2}{R_3} \tag{a}$$

The value of the equivalent resistance R_B can be obtained from Eq. 6.25 and Eq. a as

$$R_B = \frac{r}{1+r} \frac{1+q}{q} R_1 \tag{7.12}$$

If the bridge contains one active gage in arm R_1, Eqs. 6.18 and 5.5 give the output voltage as

$$\Delta v_o = \frac{R}{(1+r)^2} v_s S_g \epsilon \tag{7.13}$$

Since the voltage v_s is limited by the power p_g that can be dissipated by the gage,

$$v_s = (1+r) \sqrt{p_g R_g} \tag{b}$$

Substituting Eq. b into Eq. 7.13 gives

$$\Delta v_o = \frac{r}{1+r} \sqrt{p_g R_g} S_g \epsilon \tag{7.14}$$

For the equivalent circuit of Fig. 7.9*b*, the current i_G is given by the expression

$$i_G = \frac{q}{1+q} \sqrt{\frac{p_g}{R_g}} S_g \epsilon \left(\frac{R_B}{R_B + R_G} \right) \tag{7.15}$$

The deflection θ of the galvanometer is given by Eq. 3.13 as

$$\theta = S_G i_G \tag{3.13}$$

where S_G is the galvanometer sensitivity. System sensitivity is defined as

$$S_s = \frac{\theta}{\epsilon} \tag{7.16}$$

Therefore,

$$S_s = \frac{q}{1+q} \sqrt{\frac{p_g}{R_g}} S_g \left(\frac{1}{1+(R_G/R_B)} \right) S_G \tag{7.17}$$

Equation 7.17 shows that the system sensitivity is controlled by

1. Circuit efficiency $[q/(1+q)]$.
2. Gage selection ($\sqrt{p_g/R_g S_g}$).
3. The ratio of galvanometer resistance to equivalent bridge resistance (R_G/R_B).
4. Galvanometer sensitivity (S_G).

The constraint that R_B must provide the required external resistance for proper damping of the galvanometer limits the options available for maximizing system sensitivity. Most galvanometers used for strain measurements are designed so that the required external resistance R_x equals 120 or 350 Ω (see Table 2.2). The equivalent resistance of a four-equal-arm bridge ($R_1 = R_2 = R_3 = R_4$) with 120- or 350-Ω gages is 120 and 350 Ω, respectively. Thus, $R_B = R_g$ and $q = 1$. Under these conditions, Eq. 7.17 reduces to

$$S_s = \frac{1}{2} \sqrt{\frac{p_g}{R_g}} S_g \left(\frac{1}{1+(R_G/R_g)} \right) S_G \tag{7.18}$$

If the equivalent resistance of the bridge is less than the required external resistance for the galvanometer, which is often the case, a series resistor R_s must be added to the circuit between the bridge and the galvanometer. The value of the series resistor is given by

$$R_s = R_x - R_B \tag{d}$$

When the series resistor is added, the circuit sensitivity given by Eq. 7.17 becomes

$$S_s = \frac{q}{1+q} \sqrt{\frac{p_g}{R_g}} S_g \left(\frac{R_B}{R_B + R_G + R_s} \right) S_G \tag{7.19}$$

If the external resistance is maintained at the value specified in Table 3.2, the frequency response of the system will be flat within a ± 5 percent accuracy band over the frequency range $0 \leq \omega \leq 0.87\omega_n$.

In many cases, the system sensitivity achieved by connecting a galvanometer with adequate frequency response to the Wheatstone bridge is too low to provide sufficient galvanometer deflection. In these instances, a current amplifier must be inserted between the bridge and the galvanometer to provide the current required for adequate response. Such amplifiers are designed to provide an output impedance equal to or less than the required external resistance for most galvanometers. When the output impedance is less than R_x, a series resistor is used on the output of the amplifier for proper matching.

Strain is determined from the oscillograph record by using Eq. 7.10. In all instances, the system is calibrated either by applying a known strain to a calibration gage placed in arm R_1 of the Wheatstone bridge or by applying an equivalent strain with a shunt resistor across arm R_2 of the bridge. Calibration methods are described in Section 7.6.

7.6 CALIBRATION METHODS

A strain-measurement system (see Fig. 7.10) usually includes one or more strain gages, a power supply, circuit-completion resistors, an amplifier, and a voltage- or current-measuring instrument. It is possible to calibrate such a system by precisely measuring R_1, R_2, R_3, R_4, v_s; the gain G of the amplifier; and the sensitivity S_R of the recorder. The system calibration constant C for the entire system is then given by

$$C = \frac{(1 + r)^2 S_A S_R}{r v_s S_g} \tag{7.20}$$

where

S_A is the amplifier sensitivity
S_R is the recorder sensitivity (volts per division)

The strain recorded with the system is given in terms of the system calibration constant as

$$\epsilon = C d_s \tag{7.21}$$

where d_s is the deflection of the recorder in divisions. This procedure is time-consuming and is subject to errors in measuring each of the quantities in Eq. 7.20. A more direct, less time-consuming, and more accurate procedure is to calibrate the

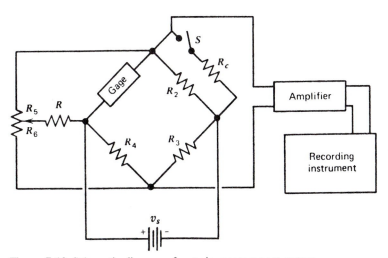

Figure 7.10 Schematic diagram of a strain-measurement system.

complete system. This may be accomplished by introducing a known strain in the bridge (either mechanically or electrically), measuring d_s resulting from this strain, and determining the system calibration constant C from Eq. 7.21.

Mechanical calibration is performed by mounting a strain gage (which must have the same gage factor S_g as the gages being employed for the measurements) on a cantilever beam, connecting this calibration gage into arm R_1 of the Wheatstone bridge, and observing the deflection of the trace on the recorder as a known strain is applied to the gage. If the free end of a cantilever beam is deflected a distance δ, the calibration strain ϵ_c induced in the calibration gage is

$$\epsilon_c = \frac{3hx}{2\ell^3}\delta = k\delta \tag{7.22}$$

where

    h is the depth of the cantilever beam
    ℓ is the length of the cantilever beam
    x is the distance from the load point to the center of the gage
    k is a constant defined in Eq. 7.22

The voltage output from an initially balanced bridge is recorded before and after the beam is deflected (for example, note the two horizontal traces shown in Fig. 7.8). The distance between these two lines d_c is used with the calibration strain ϵ_c to determine the calibration constant C. Thus,

$$C = \frac{\epsilon_c}{d_c} \tag{7.23}$$

Electrical calibration is performed in a similar manner, except that the calibration strain is induced by shunting a calibration resistor R_c across arm R_2 of the Wheatstone bridge, as shown in Fig. 7.10. The effective resistance of arm R_2 with R_c in place is

$$R_{2e} = \frac{R_2 R_c}{R_2 + R_c} \tag{7.24}$$

The change of resistance $\Delta R_2 / R_2$ is then given by

$$\frac{\Delta R_2}{R_2} = \frac{R_{2e} - R_2}{R_2} = -\frac{R_2}{R_2 + R_c} \tag{7.25}$$

The output voltage produced by shunting R_c across R_2 is obtained by substituting Eq. 7.25 into Eq. 6.17 to give

$$\Delta v_o = v_s \frac{R_1 R_2}{(R_1 + R_2)^2}\left(\frac{R_2}{R_2 + R_c}\right) \tag{a}$$

The output from a single active gage in arm R_1 of a bridge caused by a strain equal to the calibration strain is given by Eq. 6.17 as

$$\Delta v_o = v_s \frac{R_1 R_2}{(R_1 + R_2)^2} S_g \epsilon_c \tag{b}$$

Equating Eqs. a and b and solving for ϵ_c gives

$$\epsilon_c = \frac{R_2}{S_g(R_2 + R_c)} \tag{7.26}$$

After ϵ_c is determined, the calibration constant C is found by using Eq. 7.23. This technique of shunt calibration is accurate and simple to use. It provides a single calibration constant for the complete system that incorporates the sensitivities of all components. Unfortunately, the effect of lead-wire resistance is not accounted for when the calibration resistor is shunted on resistance R_2 (see Section 7.7.1 for details).

7.7   EFFECTS OF LEAD WIRES, SWITCHES, AND SLIP RINGS

The resistance change from a strain gage is very small; therefore, any disturbance that produces a resistance change within the bridge circuit is extremely important, because it also affects the output voltage. Components within the bridge include gages, soldered joints, terminals, lead wires, and binding posts. Frequently, switches and slip rings are also included. Since the effects of lead wires, switches, and slip rings are the most important, they will be covered in this section. The effects of soldered joints, terminals, and binding posts must not be neglected because they can also produce significant errors; however, if cold-soldered connections are avoided and if binding posts are clean and tight, then joint resistance will be constant and negligibly small.

7.7.1 Lead Wires

Frequently, a strain gage is mounted on a component that is located a significant distance from the bridge and recording system. The gage must be connected to the bridge with two long lead wires, as shown in Fig. 7.11a. With this arrangement, two detrimental effects occur: (1) signal attenuation, and (2) loss of temperature compensation. Both effects seriously compromise the accuracy of the measurements.

Signal attenuation or loss owing to the resistance of the two lead wires can be determined by noting in Fig 7.11a that

$$R_1 = R_g + 2R_L \tag{a}$$

where R_L is the resistance of a single lead wire. The added resistance in arm R_1 of the bridge (owing to the lead wires) leads to the expression

$$\frac{\Delta R_1}{R_1} = \frac{\Delta R_g}{R_g + 2R_L} = \frac{\Delta R_g}{R_g}\left[\frac{1}{(1 + (2R_L/R_g))}\right] \tag{b}$$

Equation b can be rewritten in terms of a signal loss factor $\mathscr{L}$ as

$$\frac{\Delta R_1}{R_1} = \frac{\Delta R_g}{R_g}(1 - \mathscr{L}) \tag{c}$$

where

$$\mathscr{L} = \frac{2R_L/R_g}{1 + (2R_L/R_g)} \tag{7.27}$$

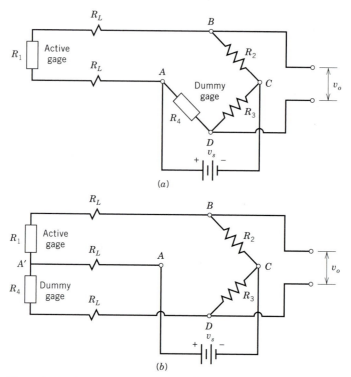

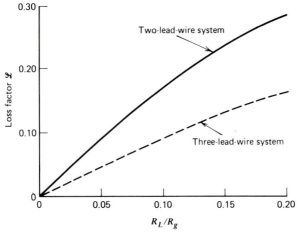

Figure 7.11 Gage connections to the Wheatstone bridge. (*a*) Two-lead-wire system. (*b*) Three-lead-wire system.

Signal loss factor $\mathscr{L}$ is shown as a function of resistance ratio R_L/R_g in Fig. 7.12. Error caused by lead wires can be reduced to less than 1 percent if $R_L/R_g \leq 0.005$. The resistance of a 100-ft (30.5-m) length of solid-copper lead wire and the associated signal loss factor as a function of gage size is listed in Table 7.2. It is obvious from the data in Table 7.2 that long lengths of small-diameter wire must be avoided in strain-gage measurements.

Figure 7.12 Loss factor $\mathscr{L}$ as a function of resistance ratio R_L/R_g for two- and-three-lead-wire systems.

Table 7.2 Resistance (Ω per 100 ft or 30.5 m)
of Solid-Conductor Copper Wire and Signal Loss Factor
$\mathcal{L}$ for Gages with $R_g = 120 \ \Omega$

Gage Size	R_L	$2R_L/R_g$	$\mathcal{L}(\%)$
12	0.159	0.00265	0.26
14	0.253	0.00422	0.42
16	0.402	0.00670	0.67
18	0.639	0.01065	1.05
20	1.015	0.01692	1.67
22	1.614	0.0269	2.62
24	2.567	0.0428	4.10
26	4.081	0.0670	6.28
28	6.490	0.1082	9.76
30	10.310	0.1718	14.67
32	16.41	0.2735	21.5
34	26.09	0.4348	30.3
36	41.48	0.6913	40.9
38	65.96	1.0993	52.4
40	104.90	1.7483	63.6

The second detrimental effect resulting from lead wires is loss of temperature compensation. As an example, consider a Wheatstone bridge with an active gage and two long lead wires in arm R_1 and a dummy gage with two short lead wires in arm R_4. If both gages and all lead wires are subjected to the same temperature change ΔT during the time interval when strain is being monitored, the output of the bridge is given by Eq. 6.18 as

$$\Delta v_o = v_s \frac{r}{(1+r)^2}\left[\left(\frac{\Delta R_g}{R_g + 2R_L}\right)_\epsilon + \left(\frac{\Delta R_g}{R_g + 2R_L}\right)_{\Delta T}\right.$$
$$\left.+\left(\frac{2\Delta R_L}{R_g + 2R_L}\right)_{\Delta T} - \left(\frac{\Delta R_g}{R_g}\right)_{\Delta T}\right] \tag{7.28}$$

The first and second terms in the brackets are the resistance changes in the active gage caused by the strain and temperature changes, respectively. The third term is the resistance change in the lead wires of arm R_1 caused by the temperature change. The fourth term is the resistance change in the dummy gage resulting from the temperature change. The resistance change in the short lead wires to arm R_4 is considered negligible. In this example, temperature compensation is not achieved, since the second and fourth terms do not cancel. Additional error as a result of resistance changes in the lead wires is represented by the third term in Eq. 7.28.

The detrimental effects of long lead wires can be reduced by employing the simple three-wire system illustrated in Fig. 7.11b. With this three-wire arrangement, both the active gage and the dummy gage are located at the remote site. One of the three wires is not considered a lead wire, since it is not within the bridge (not in either arm R_1 or R_4) and serves only to transfer point A of the bridge to the remote location. The active and dummy gages each have one long lead wire with resistance R_L connecting to points B and D, respectively, and one short lead wire with negligible resistance connecting to point A'. The signal loss factor for the three-wire system is

$$\mathscr{L} = \frac{R_L/R_g}{1 + (R_L/R_g)} \tag{7.29}$$

A comparison of Eqs. 7.27 and 7.29 indicates that signal attenuation as a result of lead-wire resistance is reduced by a factor of nearly 2 by using the three-wire system (also see Fig. 7.12).

The temperature-compensating feature of the Wheatstone bridge is retained when the three-wire system is used. In this case, Eq. 7.28 is modified to read

$$
\begin{aligned}
\Delta v_o = v_s \frac{r}{(1 + r)^2} &\left[\left(\frac{\Delta R_g}{R_g + R_L} \right)_\epsilon + \left(\frac{\Delta R_g}{R_g + R_L} \right)_{\Delta T} + \left(\frac{\Delta R_L}{R_g + R_L} \right)_{\Delta T} \right.\\
&\left. - \left(\frac{\Delta R_g}{R_g + R_L} \right)_{\Delta T} - \left(\frac{\Delta R_L}{R_g + R_L} \right)_{\Delta T} \right]
\end{aligned}
\tag{7.30}
$$

It is clear from Eq. 7.30 that temperature compensation is achieved because all the temperature-dependent terms in the bracketed quantity cancel.

In all cases where lead-wire resistance causes appreciable signal attenuation, the calibration resistor should be placed across the remote dummy gage to include the effects of the lead wires in the system calibration constant.

7.7.2 Switches

Frequently, many gages are necessary to evaluate a structure, and the output of each gage is read several times during a typical test. In this case, the number of gages is too large to employ a separate recording system for each gage. Instead, a single instrument system is used and the gages are switched in and out of the system according to some schedule. Two different switching arrangements are commonly employed with multiple-gage installations.

The most common and least expensive arrangement is shown in Fig. 7.13. One side of each active gage is switched, in turn, into arm R_1 of the bridge, while the other side of each active gage is connected to the terminals of the bridge with a common lead wire. This arrangement places the switch in arm R_1 of the bridge;

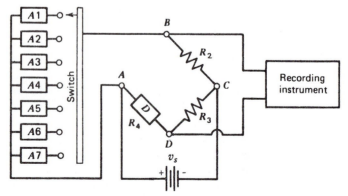

Figure 7.13 Switching a large number of individual gages into arm R_1 of the Wheatstone bridge with a single-pole switch.

therefore, a high-quality switch with a small and reproducible contact resistance (less than 500 $\mu\Omega$) must be employed. Low resistance is achieved by using silver-tipped contacts and two or more parallel contacts per switch. If the switch resistance is not reproducible, the change in switch resistance ΔR_s adds to the strain-induced change in gage resistance ΔR_g to produce an apparent strain ϵ', which can be expressed as

$$\epsilon' = \frac{\Delta R_s / R_g}{S_g} \tag{7.31}$$

The quality of a switch can be easily checked, since a nonreproducible switch resistance results in a shifting of the zero reading. Switches must be cleaned regularly, because even high-quality switches will begin to perform erratically when the contacts become dirty or when surface films develop from chemical reactions.

A second switching arrangement, shown in Fig. 7.14, employs a three-pole switch to transfer terminals A, B, and D to the power supply and the recording instrument. Terminal C of each bridge is grounded in common with the power supply with a single

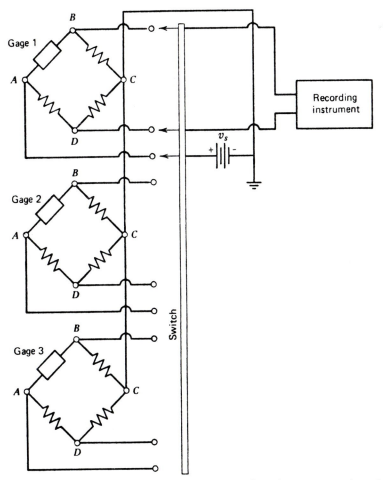

Figure 7.14 Switching several complete bridges into the power supply and recording instrument with a three-pole switch.

common lead wire. Because the switches are not located within the bridge, switch resistance is much less important. However, switching the complete bridge requires separate dummy gages and two bridge-completion resistors for each active gage.

A major disadvantage of all switching schemes is the thermal drift caused by heating of the gages and resistors when power is suddenly applied to the bridge as it is switched into the instrument system. Depending on the thermal capacity of the specimen, this drift may continue for a minute or more before thermal equilibrium at the gage site is achieved.

7.7.3 Slip Rings

When strain gages are used on rotating members, slip rings are often used to complete the lead-wire connections, as shown in Fig. 7.15a. The slip rings are usually mounted on a shaft that can be attached to the rotating member so that the axes of rotation of the shaft and member coincide. The outer shell of the slip-ring assembly is stationary and carries several brushes per ring to transfer the signal from the rotating rings to terminals on the stationary shell. Satisfactory operation up to speeds of 24,000 rpm is possible with a properly designed slip-ring assembly.

Brush contact and dirt collecting on the slip rings cause brush wear and tend to produce significant fluctuations in resistance. It is possible to reduce these fluctuations by using multiple brushes in parallel for each lead wire. However, even with multiple brushes, fluctuation in resistance between rings and brushes tends to be so large that slip rings are not placed within the arms of the bridge. Instead, a complete bridge is assembled for each active gage on the rotating member, as shown in Fig. 7.15b. The slip rings are used only to connect the bridge to the power supply and the instrument system. This arrangement minimizes the effect of resistance fluctuations produced by the slip rings and provides a means for accurately recording strain-gage signals from rotating members.

7.8 ELECTRICAL NOISE

The output voltage from a Wheatstone bridge caused by the resistance change $\Delta R / R$ of a strain gage is usually only a few millivolts. Because of this very small output voltage Δv_o, electrical noise is frequently a problem. Electrical noise occurs as a result of magnetic fields generated by current flow in wires in close proximity to the lead wires or bridge, as shown in Fig. 7.16. When common line current flows in an adjacent wire, a 60-Hz magnetic field is produced, which cuts both wires of the signal circuit and induces a voltage (noise) in the signal loop. The magnitude of this noise signal is proportional to the current i flowing in the disturbing wire and the area enclosed by the signal loop, and is inversely proportional to the distance between the disturbing wire and the strain-gage lead wires (see Fig. 7.16). In some instances, the voltage induced by the magnetic field is so large that the signal-to-noise ratio becomes excessive and it is difficult to separate the noise from the strain-gage signal.

Three precautions can be taken to minimize noise. First, all lead wires should be twisted or arranged in a ribbon conductor to minimize the area of the signal loop. Second, only shielded cables should be used, and the shields should be grounded only at the negative terminal of the power supply to the bridge, as shown in Fig. 7.17. With this arrangement, the shield is grounded without forming a ground loop and any noise voltage generated in the shield is maintained at ground potential. The power

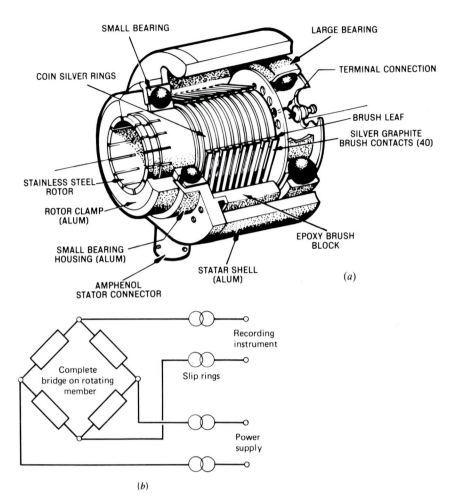

SMALL BEARING

LARGE BEARING

COIN SILVER RINGS

TERMINAL CONNECTION

BRUSH LEAF

SILVER GRAPHITE
BRUSH CONTACTS (40)

STAINLESS STEEL
ROTOR

ROTOR CLAMP
(ALUM)

SMALL BEARING
HOUSING (ALUM)

AMPHENOL
STATOR CONNECTOR

STATAR SHELL
(ALUM)

EPOXY BRUSH
BLOCK

(a)

Recording
instrument

Complete
bridge on rotating
member

Slip rings

Power
supply

(b)

(c)

Figure 7.15 Signal transfer from rotating members with slip rings. (a) Construction details of a slip-ring unit. (b) Schematic diagram of a strain-measurement system with slip rings. (c) Several slip-ring units. (Courtesy of Michigan Scientific Corp.)

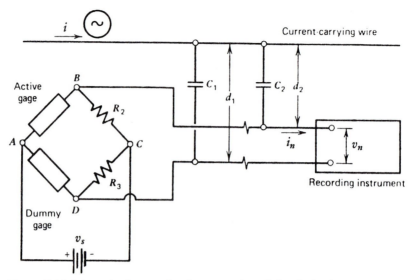

Figure 7.16 Schematic diagram showing generation of electrical noise.

supply is floated relative to the system ground (the third conductor in the power cord) to avoid a ground loop at the supply. Third, differential amplifiers should be used to reduce noise by common-mode rejection. If the lead wires are twisted, the noise signal developed on both lead wires will be equal and will occur simultaneously. A differential amplifier rejects these noise signals and only the strain signal is amplified. Common-mode rejection for good-quality instrumentation amplifiers is about 10^6 to 1 at 60 Hz; therefore, most of the noise is suppressed.

If these three techniques for suppressing noise are employed, the signal-to-noise ratio can be maximized and low-magnitude strain signals can be recorded even at locations with adverse electrical conditions.

7.9 TEMPERATURE-COMPENSATED GAGES

Temperature compensation within the Wheatstone bridge was discussed in Section 7.4; however, temperature compensation of the gage itself is possible. In static applications, both the bridge and the gage should be compensated to nullify any signal resulting

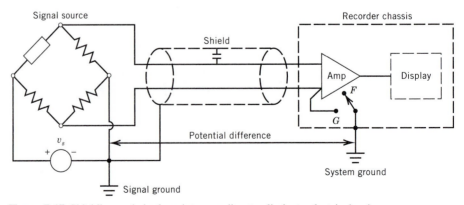

Figure 7.17 Shielding and single-point grounding to eliminate electrical noise.

from temperature variations during the test period. When the ambient temperature changes, four effects occur that influence the signal $\Delta R/R$ from the gage:

1. The gage factor S_g changes with temperature.
2. The grid undergoes an elongation or contraction ($\Delta \ell/\ell = \alpha \Delta T$).
3. The specimen elongates or contracts ($\Delta \ell/\ell = \beta \Delta T$).
4. The resistance of the gage changes ($\Delta R/R = \gamma \Delta T$).

The strain sensitivities S_A of the two most commonly used alloys (Advance and Karma) are linear functions of temperature, as shown in Fig. 7.18. The slope of the $S_A - T$ line indicates that $\Delta S_A/\Delta T$ equals 0.0000735/°C and −0.0000975/°C for Advance and Karma alloys, respectively. Since these changes are small (less than 1 percent for $\Delta T = 100$°C), variations in S_A with temperature are usually neglected when temperature variations are less than 50°C. However, in thermal stress studies where temperature variations of several hundred degrees are common, changes in S_A become significant and must be considered.

The remaining three effects are much more significant and combine to produce a change in resistance of the gage that can be expressed as

$$\left(\frac{\Delta R}{R}\right)_{\Delta T} = (\beta - \alpha)S_g \Delta T + \gamma \Delta T \tag{7.32}$$

where

α is the thermal coefficient of expansion of the gage alloy
β is the thermal coefficient of expansion of the specimen material
γ is the temperature coefficient of resistivity of the gage alloy

A differential expansion between the gage grid and the specimen owing to a temperature change ($\alpha \neq \beta$) subjects the gage to a thermally induced mechanical strain $\epsilon_T = (\beta - \alpha)\Delta T$, which does not occur in the specimen. The gage responds to the strain ϵ_T in the same way in which it responds to a load-induced strain ϵ in the specimen. Unfortunately, it is impossible to separate the component of the response caused by temperature change from that caused by load.

If the gage alloy is matched to the specimen ($\alpha = \beta$), the first term in Eq. 7.32 does not produce a response; however, the second term causes a response and indicates an apparent strain that does not exist in the specimen. A temperature-compensated gage is obtained only if both terms in Eq. 7.32 are zero or if they cancel.

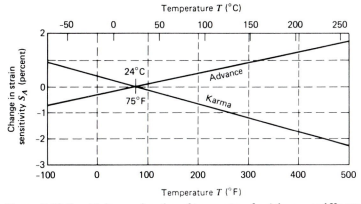

Figure 7.18 Sensitivity as a function of temperature for Advance and Karma alloys.

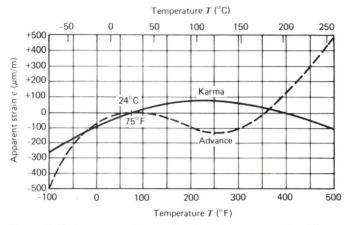

Figure 7.19 Apparent strain as a function of temperature for Advance and Karma alloys.

The values of α and γ are quite sensitive to the composition of the alloy and to the degree of cold working imparted during the rolling of the foil. It is common practice for strain-gage manufacturers to measure the thermal-response characteristics of sample gages made from each roll of foil in their inventories. Because of variations in α and γ between melts and rolls of foil, it is possible to select gage alloys from inventory that are temperature compensated for almost any specimen material. These gages are known as selected-melt or temperature-compensated gages.

Unfortunately, selected-melt gages are not perfectly compensated over a wide range of temperature because of nonlinear terms that were omitted in Eq. 7.32. A typical selected-melt strain gage exhibits an apparent strain with temperature as shown in Fig. 7.19. The apparent strain produced by a temperature change of a few degrees in the neighborhood of 75°F (24°C) is quite small (less than 0.5 $\mu\epsilon/°C$); however, when the temperature change is large, the apparent strain generated by the gage becomes significant and corrections to account for this apparent strain are necessary.

7.10   ALLOY SENSITIVITY, GAGE FACTOR, AND CROSS-SENSITIVITY FACTOR

The sensitivity of a single, uniform length of conductor to strain was defined as

$$S_A = \frac{dR/R}{\epsilon} \approx \frac{\Delta R/R}{\epsilon} \tag{5.4}$$

where S_A is the alloy sensitivity (see Section 5.4). In a typical strain gage, the conductor is formed into a pattern (commonly referred to as the grid) to keep the gage length short. Also, the conductor is usually not uniform over its entire length. As a result, the alloy sensitivity S_A is not a true calibration constant for a strain gage.

A better understanding of the response of a grid-type strain gage can be obtained by considering a gage mounted on a specimen that is subjected to a biaxial strain field. For this situation,

$$\frac{\Delta R}{R} = S_a \epsilon_a + S_t \epsilon_t + S_s \gamma_{at} \tag{7.33}$$

where

ϵ_a is the normal strain along the axial direction of the gage
ϵ_t is the normal strain along the transverse direction of the gage
γ_{at} is the shearing strain associated with the a and t directions
S_a is the sensitivity of the gage to axial strain
S_t is the sensitivity of the gage to transverse strain
S_s is the sensitivity of the gage to shearing strain

The gage sensitivity to shearing strain is believed to be small and is neglected. The gage sensitivity to transverse strain is usually not small enough to neglect; therefore, manufacturers provide a transverse sensitivity factor K_t for each gage, which is defined as

$$K_t = \frac{S_t}{S_a} \tag{7.34}$$

If Eq. 7.34 is substituted into Eq. 7.33 with $S_s = 0$,

$$\frac{\Delta R}{R} = S_a(\epsilon_a + K_t\epsilon_t) \tag{7.35}$$

The sensitivity of strain gages is usually expressed in terms of a gage factor S_g previously indicated as

$$\frac{\Delta R}{R} = S_g\epsilon_a \tag{5.5}$$

The gage factor S_g is determined by the manufacturer by measuring $\Delta R/R$ for a sample of gages drawn from each production lot. In calibration, the sample gages are mounted on a beam with a Poisson's ratio $\nu_0 = 0.285$. A known axial strain ϵ_a is applied to the beam which produces a transverse strain ϵ_t given by

$$\epsilon_t = -\nu_0\epsilon_a \tag{7.36}$$

The response of the gage in calibration is obtained by substituting Eq. 7.36 into Eq. 7.35 to give

$$\frac{\Delta R}{R} = S_a\epsilon_a(1 - \nu_0 K_t) \tag{7.37}$$

A comparison of Eqs. 7.37 and 5.5 indicates that the gage factor S_g can be expressed in terms of S_a and K_t as

$$S_g = S_a(1 - \nu_0 K_t) \tag{7.38}$$

The simplified form of the $\Delta R/R$ versus ϵ_a relationship given by Eq. 5.5 is usually used to interpret strain-gage response. It is very important to recognize that this equation is approximate unless either $K_t = 0$ or $\epsilon_t = 0$. The magnitude of the error incurred by using Eq. 5.5 can be determined by considering the response of a gage in a general biaxial strain field. If Eq. 7.38 is substituted into Eq. 7.35, the gage response is given as

$$\frac{\Delta R}{R} = \frac{S_g \epsilon_a}{1 - \nu_0 K_t}\left(1 + K_t \frac{\epsilon_t}{\epsilon_a}\right) \tag{7.39}$$

The true value of strain ϵ_a can then be written as

$$\epsilon_a = \frac{\Delta R/R}{S_g}\left[\frac{1 - \nu_0 K_t}{1 + K_t(\epsilon_t/\epsilon_a)}\right] \tag{7.40}$$

The apparent strain ϵ_a', obtained by using Eq. 5.5, is

$$\epsilon_a' = \frac{\Delta R/R}{S_g} \tag{a}$$

Substituting Eq. a into Eq. 7.40 gives

$$\epsilon_a = \epsilon_a'\left[\frac{1 - \nu_0 K_t}{1 + K_t(\epsilon_t/\epsilon_a)}\right] \tag{7.41}$$

The percent error $\mathscr{E}$ incurred by neglecting the transverse sensitivity of a strain gage in a general biaxial strain field is obtained from Eqs. 7.40 and a as

$$\mathscr{E} = \frac{\epsilon_a' - \epsilon_a}{\epsilon_a}(100) = \frac{K_t(\epsilon_t/\epsilon_a + \nu_0)}{1 - \nu_0 K_t}(100) \tag{7.42}$$

Some representative values of $\mathscr{E}$ as a function of K_t for different biaxial ratios ϵ_t/ϵ_a are illustrated in Fig. 7.20.

Two different procedures are used to correct for the error involved with the use of Eq. 5.5. First, if the biaxiality ratio ϵ_t/ϵ_a is known (in a thin-walled cylinder under

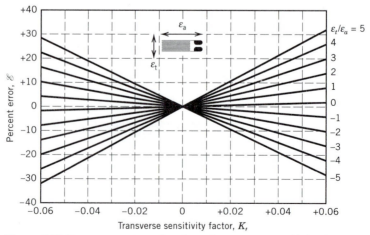

Figure 7.20 Percent error as a function of transverse sensitivity factor for several different ratios of transverse to axial strain.

internal pressure, for example), the bracketed term in Eq. 7.41 can be viewed as a correction factor C_f that modifies the apparent strain ϵ_a' to give the true strain ϵ_a. The factor by which all apparent strain values must be multiplied to give true strains is

$$C_f = \frac{1 - \nu_0 K_t}{1 + K_t(\epsilon_t/\epsilon_a)} \tag{7.43}$$

Alternatively, a corrected gage factor S_g^* can be used in place of S_g to adjust the bridge before readings are taken. The corrected gage factor as determined from Eqs. 5.5 and 7.41 is

$$S_g^* = S_g \left[\frac{1 + K_t(\epsilon_t/\epsilon_a)}{1 - \nu_0 K_t} \right] = \frac{S_g}{C_f} \tag{7.44}$$

The second correction procedure is used when the biaxiality ratio ϵ_t/ϵ_a is not known. If apparent strains ϵ_{xx}' and ϵ_{yy}' are recorded in orthogonal directions, then from Eqs. 7.39 and 5.5

$$\epsilon_{xx}' = \frac{1}{1 - \nu_0 K_t}(\epsilon_{xx} + K_t\epsilon_{yy})$$

$$\epsilon_{yy}' = \frac{1}{1 - \nu_0 K_t}(\epsilon_{yy} + K_t\epsilon_{xx}) \tag{7.45}$$

Solving for the true strains ϵ_{xx} and ϵ_{yy} yields

$$\epsilon_{xx} = \frac{1 - \nu_0 K_t}{1 - K_t^2}(\epsilon_{xx}' - K_t\epsilon_{yy}')$$

$$\epsilon_{yy} = \frac{1 - \nu_0 K_t}{1 - K_t^2}(\epsilon_{yy}' - K_t\epsilon_{xx}') \tag{7.46}$$

7.11   DATA-REDUCTION METHODS

Strain gages are normally bonded on the free surface of a specimen ($\sigma_{zz} = \tau_{zx} = \tau_{zy} = 0$) to determine the stresses at specific points when the specimen is subjected to a system of loads. The conversion from strains to stresses requires knowledge of the elastic constants E and ν of the material and, depending on the state of stress at the point, from one to three normal strains. Three different stress states are considered in the following subsections. Data-analysis methods and special-purpose strain gages are described for each stress state.

7.11.1 The Uniaxial State of Stress ($\sigma_{xx} \neq 0$, $\sigma_{yy} = \tau_{xy} = 0$)

In a uniaxial state of stress (encountered in tension members and in beams in pure bending, for example), the stress σ_{xx} is the only nonzero component and its direction is known. In this case, a single-element strain gage (see Fig. 5.9a, b, and c) oriented with its axis in the x direction is used to determine the strain ϵ_{xx}. The stress is then given by the uniaxial form of Hooke's law as

$$\sigma_{xx} = E\epsilon_{xx} \tag{7.47}$$

One principal direction coincides with the x axis; and all directions perpendicular to the x axis are also principal directions.

7.11.2 The Biaxial State of Stress ($\sigma_{xx} \neq 0$, $\sigma_{yy} \neq 0$, $\tau_{xy} = 0$)

If the directions of the principal stresses are known at the gage location (on an axis of load and geometric symmetry, for example), two strain measurements in perpendicular directions provide sufficient data to determine the stresses at the point. Special strain gages known as two-element rectangular rosettes (see Fig. 5.9d, e, and f) are available for this purpose. The two-element rosette should be oriented on the specimen with its axes coincident with the principal stress directions in order to determine the two principal strains ϵ_1 and ϵ_2. The stresses are then given by the biaxial form of generalized Hooke's law as

$$\sigma_1 = \frac{E}{1 - \nu^2}(\epsilon_1 + \nu\epsilon_2)$$

$$\tag{7.48}$$

$$\sigma_2 = \frac{E}{1 - \nu^2}(\epsilon_2 + \nu\epsilon_1)$$

7.11.3 The General State of Stress ($\sigma_{xx} \neq 0$, $\sigma_{yy} \neq 0$, $\tau_{xy} \neq 0$)

In the most general case, the principal stress directions are not known; therefore, three unknowns σ_1, σ_2, and the principal angle ϕ_1 must be determined in order to specify the state of stress at the point. Three-element rosettes (see Fig. 5.9g, h, and i) are used in these cases to obtain the required strain data. The fact that three strain measurements are sufficient to determine the state of strain at a point on a free surface is demonstrated by considering three gages aligned along axes A, B, and C, as shown in Fig. 7.21. Using one of the equations of strain transformation, we write

$$\epsilon_A = \epsilon_{xx} \cos^2 \theta_A + \epsilon_{yy} \sin^2 \theta_A + \gamma_{xy} \sin \theta_A \cos \theta_A$$

$$\epsilon_B = \epsilon_{xx} \cos^2 \theta_B + \epsilon_{yy} \sin^2 \theta_B + \gamma_{xy} \sin \theta_B \cos \theta_B \tag{7.49}$$

$$\epsilon_C = \epsilon_{xx} \cos^2 \theta_C + \epsilon_{yy} \sin^2 \theta_C + \gamma_{xy} \sin \theta_C \cos \theta_C$$

The Cartesian components of strain ϵ_{xx}, ϵ_{yy}, and τ_{xy} are determined by solving Eqs. 7.49 if ϵ_A, ϵ_B, and ϵ_C are known. The principal strains ϵ_1 and ϵ_2 and the principal direction ϕ_1 are then determined from

$$\epsilon_1 = \frac{1}{2}(\epsilon_{xx} + \epsilon_{yy}) + \frac{1}{2}\sqrt{(\epsilon_{xx} - \epsilon_{yy})^2 + \gamma_{xy}^2}$$

$$\epsilon_2 = \frac{1}{2}(\epsilon_{xx} + \epsilon_{yy}) - \frac{1}{2}\sqrt{(\epsilon_{xx} - \epsilon_{yy})^2 + \gamma_{xy}^2}$$

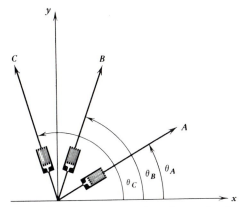

Figure 7.21 Three gages oriented at angles θ_A, θ_B, θ_C with respect to the x axis.

$$\phi_1 = \frac{1}{2} \tan^{-1} \frac{\gamma_{xy}}{\epsilon_{xx} - \epsilon_{yy}} \tag{7.50}$$

where ϕ_1 is the angle between the principal direction for ϵ_1 and the x axis.

Two of the most commonly employed rosettes are the delta rosette and the three-element rectangular rosette. The three-element rectangular rosette is discussed here; analysis of the delta rosette is left as an exercise (see Exercise 7.41). The three-element rectangular rosette is designed with $\theta_A = 0°$, $\theta_B = 45°$, $\theta_C = 90°$. With these fixed angles, Eqs. 7.49 reduce to

$$\epsilon_A = \epsilon_{xx}$$

$$\epsilon_B = \frac{1}{2}(\epsilon_{xx} + \epsilon_{yy} + \gamma_{xy}) \tag{a}$$

$$\epsilon_C = \epsilon_{yy}$$

From Eqs. a it is clear that

$$\gamma_{xy} = 2\epsilon_B - \epsilon_A - \epsilon_C \tag{b}$$

The principal strains ϵ_1 and ϵ_2 and the principal angle θ_1 are obtained in terms of ϵ_A, ϵ_B, and ϵ_C from Eqs. a, b, and 7.50 as

$$\epsilon_1 = \frac{1}{2}(\epsilon_A + \epsilon_C) + \frac{1}{2}\sqrt{(\epsilon_A - \epsilon_C)^2 + (2\epsilon_B - \epsilon_A - \epsilon_C)^2}$$

$$\epsilon_2 = \frac{1}{2}(\epsilon_A + \epsilon_C) - \frac{1}{2}\sqrt{(\epsilon_A - \epsilon_C)^2 + (2\epsilon_B - \epsilon_A - \epsilon_C)^2} \tag{7.51}$$

$$\phi_1 = \frac{1}{2} \tan^{-1} \frac{2\epsilon_B - \epsilon_A - \epsilon_C}{\epsilon_A - \epsilon_C}$$

Equation 7.51 yields two values for the angle ϕ. One value, ϕ_1, refers to the angle between the x axis and the axis of ϵ_1; the second value, ϕ_2, refers to the angle between the x axis and the axis of ϵ_2. A classification procedure, shown in Eq. 7.52 (see Exercise 7.43), is employed to define the angle ϕ_1:

$$0° < \phi_1 < 90° \quad \text{when} \quad \epsilon_B > \frac{1}{2}(\epsilon_A + \epsilon_C)$$

$$-90° < \phi_1 < 0° \quad \text{when} \quad \epsilon_B < \frac{1}{2}(\epsilon_A + \epsilon_C) \tag{7.52}$$

$$\phi_1 = 0 \quad \text{when} \quad \epsilon_A > \epsilon_C \quad \text{and} \quad \epsilon_A = \epsilon_1$$

$$\phi_1 = \pm 90 \quad \text{when} \quad \epsilon_A < \epsilon_C \quad \text{and} \quad \epsilon_A = \epsilon_2$$

Finally, the principal stresses are determined in terms of ϵ_A, ϵ_B, and ϵ_C by substituting Eqs. 7.51 into Eqs. 7.48 to give

$$\sigma_1 = E\left[\frac{\epsilon_A + \epsilon_C}{2(1 - \nu)} + \frac{1}{2(1 + \nu)} \sqrt{(\epsilon_A - \epsilon_C)^2 + (2\epsilon_B - \epsilon_A - \epsilon_C)^2} \right]$$

$$\sigma_2 = E\left[\frac{\epsilon_A + \epsilon_C}{2(1 - \nu)} - \frac{1}{2(1 + \nu)} \sqrt{(\epsilon_A - \epsilon_C)^2 + (2\epsilon_B - \epsilon_A - \epsilon_C)^2} \right] \tag{7.53}$$

Although derivation of these equations is tedious, they are easy to employ. As a result, rosettes are widely used to establish the complete state of stress at a point on the free surface of a general three-dimensional body.

7.12 HIGH-TEMPERATURE STRAIN MEASUREMENTS

The behavior of strain gages under a wide range of temperature was described in Section 7.9, and results showing changes in alloy sensitivity and apparent strain with temperature were presented in Figs. 7.18 and 7.19. It is important to note that these results are not given for temperatures above 500°F (260°C). As the temperature increases beyond this limit, the use of strain gages becomes more difficult. At elevated temperatures, the response of a resistance strain gage must be considered to be a function of temperature T and time t in addition to strain. Recasting Eq. 3.4 to accommodate time and temperature gives

$$\frac{\Delta R}{R} = S_g \epsilon + S_T \Delta T + S_\tau \Delta t \tag{7.54}$$

where

  S_T is the gage sensitivity to temperature
  S_τ is the gage sensitivity to time

Examination of Fig. 7.18 indicates that S_g changes with temperature for both Advance and Karma alloys, and Fig. 7.19 shows that S_T becomes larger for both of these alloys at higher temperatures. Indeed, S_T becomes so large for Advance that its use is not recommended above 180°C.

The gage sensitivity with time S_τ, which is quite small at or near room temperature, becomes significant at higher temperature. This term $S_\tau \Delta t$, often termed strain-gage

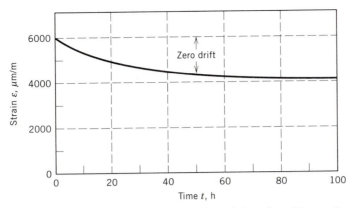

Figure 7.22 Strain-gage drift as a function of time for a Karma-alloy strain gage with a glass-fiber epoxy-phenolic carrier at 560°F (from data by Hayes).

drift, must be taken into account in all static strain measurements at elevated temperatures. An example of strain-gage drift for a Karma-alloy gage at 560°F (293°C) is presented in Fig. 7.22.

Stability or drift is affected by stress relaxation in the adhesive bond and carrier material and by metallurgical changes (phase transformations and annealing) in the strain-gage alloy. Drift rates, or the magnitude of S_τ, depend on the gage alloy, adhesive and carriers materials, strain level, and temperature. For high-temperature strain measurements, it is suggested that a series of strain–time calibration curves (like the one shown in Fig. 7.22) be developed to cover the range of strains and temperatures to be encountered in the actual experiment. Corrections can then be taken from the appropriate curve to eliminate the $S_\tau \Delta t$ term in Eq. 7.54.

The stability problem becomes much less critical as the observation period decreases, and for times of a few seconds or less, zero drift resulting from time can usually be neglected. Also, for short observation times, the temperature changes are small and apparent strains as a result of the $S_T \Delta T$ term tend to vanish. Thus, dynamic measurement of strain, even at very high temperatures (to 1600°F), is possible. For the dynamic measurement, $\Delta T = \Delta \tau \approx 0$ and Eq. 7.54 reduces to $\Delta R/R = S_g \epsilon$.

Strain measurements above 500°F (260°C) are usually performed by modifying the gage and installation procedures. Polymeric carriers are not satisfactory and are stripped from the gage after a facedown installation. Polymeric adhesives are replaced with ceramic adhesives to avoid the degradation that occurs when polymers are exposed to high temperatures. Metallurgically stable alloys, such as Armour D (70 Fe, 20 Cr, and 10 Al) or alloy 479 (92 Pt and 8 W), are used in fabricating the gage grids to avoid phase changes or oxidation at the higher temperatures.

7.13 SUMMARY

The electrical resistance strain gage nearly satisfies all of the optimum requirements for a strain gage; therefore, it is widely employed in stress analysis and as the sensing element in other transducers. Although the strain gage is inexpensive and relatively easy to use, care must be exercised in its installation to ensure that it is properly bonded to the specimen, properly waterproofed, and correctly connected into the Wheatstone bridge.

The voltage that can be applied to a Wheatstone bridge with a single active gage is limited by the power the gage can dissipate. The maximum supply voltage is

$$v_s = (1 + r) \sqrt{A p_D R_g} \tag{7.5}$$

With this voltage applied, the system sensitivity is

$$S_s = \frac{r}{1 + r} S_g \sqrt{p_g R_g} \tag{7.2}$$

The bridge can provide temperature compensation if a temperature-compensating gage (dummy gage) is used in arm R_4 of the bridge.

Digital voltmeters and oscilloscopes are high-impedance recording instruments that can be used with the Wheatstone bridge to measure the output voltage Δv_o without introducing appreciable loading errors. When the Wheatstone bridge is used with a low-impedance oscillograph recording system, however, loading effects are significant and system sensitivity is reduced to

$$S_s = \frac{1}{2} \sqrt{\frac{p_g}{R_g}} \left(\frac{S_g S_G}{1 + (R_G / R_g)} \right) \tag{7.18}$$

Also, care must be exercised to ensure that the required external resistance is used in the circuit in order for the galvanometer to maintain its range of frequency response.

Both electrical and mechanical procedures are used to calibrate a strain-measuring system. For electrical calibration, the calibration strain ϵ_c is simulated by shunting a calibration resistor R_c across arm R_2 of the bridge. The magnitude of the calibration strain is given by

$$\epsilon_c = \frac{R_2}{S_g (R_2 + R_c)} \tag{7.26}$$

The strain is obtained by comparing deflections of the recorder trace induced by the calibration strain and the load-induced strain. Thus,

$$\epsilon = \frac{d_s}{d_c} \epsilon_c = C d_s \tag{7.21}$$

Lead wires, slip rings, and switches, commonly employed with strain gages, can in some cases seriously degrade the instrumentation system. The detrimental effects of long lead wires are reduced appreciably by using a three-wire system. Signal loss $\mathcal{L}$, owing to long lead wires in a three-wire system, is

$$\mathcal{L} = \frac{R_L / R_g}{1 + (R_L / R_g)} \tag{7.29}$$

The effects of long lead wires on calibration can be eliminated by shunt calibration at the remote gage location. Only high-quality switches should be used in the arms

of a Wheatstone bridge; otherwise, errors will result from variations in switch resistance. Slip rings should not be used within the arms of the bridge to transmit
signals from rotating members. Instead, a complete bridge is assembled on the rotating member, and the supply voltage and output voltage are transmitted with the slip
rings.

Noise in strain-gage circuits is common and can be minimized by employing twisted
leads with a properly grounded shield. Also, the use of differential amplifiers with
common-mode rejection further reduces the noise-to-signal ratio.

Temperature-compensating strain gages are available for a wide range of specimen
materials and should be used for all tests where large temperature changes are expected
to occur.

Strain gages exhibit a sensitivity to both axial and transverse strains given by

$$\frac{\Delta R}{R} = \frac{S_g \epsilon_a}{1 - \nu_0 K_t} \left(1 + K_t \frac{\epsilon_t}{\epsilon_a} \right) \tag{7.39}$$

If the transverse sensitivity of the gage is neglected, the response from the gage is
related to the axial strain by the simple but approximate expression

$$\frac{\Delta R}{R} = S_g \epsilon_a \tag{5.5}$$

The percent error resulting from use of this approximate relation is

$$\mathscr{E} = \frac{K_t(\epsilon_t/\epsilon_a + \nu_0)}{1 - \nu_0 K_t}(100) \tag{7.42}$$

If the ratio ϵ_t/ϵ_a is known, a corrected gage factor S_g^* is used to eliminate error
resulting from the transverse sensitivity of the gage. The corrected gage factor S_g^* is

$$S_g^* = \left[\frac{1 + K_t(\epsilon_t/\epsilon_a)}{1 - \nu_0 K_t} \right] S_g \tag{7.44}$$

If the ratio ϵ_t/ϵ_a is not known, transverse sensitivity errors are eliminated by measuring two orthogonal apparent strains ϵ_{xx}' and ϵ_{yy}' with a two-element rectangular
rosette and computing true strains by using the equations

$$\epsilon_{xx} = \frac{1 - \nu_0 K_t}{1 - K_t^2}(\epsilon_{xx}' - K_t \epsilon_{yy}')$$

$$\epsilon_{yy} = \frac{1 - \nu_0 K_t}{1 - K_t^2}(\epsilon_{yy}' - K_t \epsilon_{xx}') \tag{7.46}$$

Strain measurements can be converted to stresses for the uniaxial state of stress by
using the simple expression

$$\sigma_{xx} = E \epsilon_{xx}$$

In more complex strain fields, where a three-element rectangular rosette is used to record strains ϵ_A, ϵ_B, and ϵ_C, the principal stresses σ_1 and σ_2 and their directions ϕ are obtained from these three strains by using the equations

$$\sigma_1 = E\left[\frac{\epsilon_A + \epsilon_C}{2(1 - \nu)} + \frac{1}{2(1 + \nu)}\sqrt{(\epsilon_A - \epsilon_C)^2 + (2\epsilon_B - \epsilon_A - \epsilon_C)^2}\right]$$

$$\sigma_2 = E\left[\frac{\epsilon_A + \epsilon_C}{2(1 - \nu)} - \frac{1}{2(1 + \nu)}\sqrt{(\epsilon_A - \epsilon_C)^2 + (2\epsilon_B - \epsilon_A - \epsilon_C)^2}\right] \quad \textbf{(7.53)}$$

$$\phi = \frac{1}{2}\tan^{-1}\frac{2\epsilon_B - \epsilon_A - \epsilon_C}{\epsilon_A - \epsilon_C} \quad \textbf{(7.51)}$$

To employ strain gages at elevated temperatures, the relation for $\Delta R/R$ must be modified to account for sensitivity to temperature changes and instabilities with time to

$$\frac{\Delta R}{R} = S_g\epsilon + S_T\Delta T + S_\tau\Delta t \quad \textbf{(7.54)}$$

REFERENCES

1. Brace, W. F.: "Effect of Pressure on Electrical-Resistance Strain Gages," Exp. Mech., vol. 4, no. 7, 1964, pp. 212–216.
2. Dally, J. W., and W. F. Riley: *Experimental Stress Analysis,* 3rd ed., McGraw–Hill, New York, 1991.
3. Freynik, H. S., and G. R. Dittbenner: "Strain Gage Stability Measurements for a Year at 75°C in Air," University of California Radiation Laboratory Report 76039, 1975.
4. Maslen, K. R., and I. G. Scott: "Some Characteristics of Foil Strain Gauges," Royal Aircraft Establishment Technical Note Instruction 134, 1953.
5. McClintock, F. A.: "On Determining Principal Strains from Strain Rosettes with Arbitrary Angles," *Proc. Soc. Exp. Stress Anal.,* vol. IX, no. 1, 1951, pp. 209–210.
6. Measurements Group, Inc., "Strain Gage Installations," *Instruction Bulletin* B-130-2, July 1972.
7. Measurements Group, Inc., "Errors Due to Wheatstone Bridge Nonlinearity," *Technical Note* 139, 1974, pp. 1–4.
8. Measurements Group, Inc., "Temperature-Induced Apparent Strain and Gage Factor Variation in Strain Gages," *Technical Note* 504, 1976, pp. 1–9.
9. Measurements Group, Inc., "Strain Gage Selection Criteria, Procedures, & Recommendations," *Technical Note* 505, 1976, pp. 1–12.
10. Measurements Group, Inc., "Errors Due to Misalignment of Strain Gages," *Technical Note* 138, 1979, pp. 1–7.
11. Measurements Group, Inc., "Optimizing Strain Gage Excitation Levels," *Technical Note* 502, 1979, pp. 1–5.
12. Measurements Group, Inc., "Noise Control in Strain Gage Measurements," *Technical Note* 501, 1980, pp. 1–5.
13. Measurements Group, Inc., "Fatigue Characteristics of Micro-Measurements Strain Gages," *Technical Note* 508, 1982, pp. 1–4.
14. Measurements Group, Inc., "Errors Due to Transverse Sensitivity in Strain Gages," *Technical Note* 509, 1982, pp. 1–8.

15. Murray, W. M.: "Some Simplifications in Rosette Analysis," *Proc. Soc. Exp. Stress Anal.,* vol. XV, no. 2, 1958, pp. 39–52.
16. Palermo, P. M.: "Methods of Waterproofing SR-4 Strain Gages," *Proc. Soc. Exp. Stress Anal.,* vol. XIII, no. 2, 1956, pp. 79–88.
17. Perry, C. C., and H. R. Lissner: *The Strain Gage Primer,* 2nd ed., McGraw–Hill, New York, 1962.
18. Simmons, E. E., Jr.: Material Testing Apparatus, U. S. Patent 2,292,549, February 23, 1940.
19. Stein, P. K.: "A Simplified Method of Obtaining Principal Stress Information from Strain Gage Rosettes," *Proc. Soc. Exp. Stress Anal.,* vol. XV, no. 2, 1958, pp. 21–38.
20. Stein, P. K.: *Advanced Strain Gage Techniques,* Chapter 2, Stein Engineering Services, Phoenix, Ariz., 1962.
21. Telinde, J. C.: "Strain Gages in Cryogenics and Hard Vacuum," *Proc. West. Regional Strain Gage Committee,* 1968, pp. 45–54.
22. Telinde, J. C.: "Strain Gages in Cryogenic Environment," *Exp. Mech.,* vol. 10, no. 9, 1970, pp. 394–400.
23. Thompson, W. (Lord Kelvin): "On the Electrodynamic Qualities of Metals," *Philos. Trans. R. Soc. London,* vol. 146, 1856, pp. 649–751.
24. Tomlinson, H.: "The Influence of Stress and Strain on the Action of Physical Forces," *Philos. Trans. R. Soc. London,* vol. 174, 1883, pp. 1–172.
25. Weymouth, L. J.: "Strain Measurement in Hostile Environment," *Appl. Mech. Rev.,* vol. 18, no. 1, 1965, pp. 1–4.

EXERCISES

7.1   A strain gage is to be fabricated from Advance wire having a diameter of 0.001 in. and a resistance of 25 Ω/in. The gage is to have a gage length of 2 in. and a resistance of 500 Ω. Design a grid configuration.

7.2   Write an engineering brief describing the characteristics of an optimum strain gage.

7.3   Write a concise description of a foil-type resistance strain gage. Include the alloys employed and cover the role of the carrier and the tabs.

7.4   Write a specification describing the procedure to be followed by a laboratory technician in installing strain gages on metallic components.

7.5   Write a specification describing the procedure to be followed by a laboratory technician in installing strain gages on components fabricated from engineering polymers.

7.6   Prepare a graph showing system sensitivity S_s as a function of the power p_g dissipated by the gage. Let $r = 3$, $S_g = 2$, and consider $R = 120, 350, 500,$ and $1000 \ \Omega$.

7.7   Determine the system sensitivity for a bridge with a single active gage having $R_g = 350 \ \Omega$ and $S_g = 2.05$ if $r = 3$ and if the bridge voltage is 5 V.

7.8   If the gage in Exercise 7.7 can dissipate 0.01 W, is the bridge voltage correct? If not, what is the correct voltage?

7.9   Determine the voltage output from a Wheatstone bridge if a single active gage is used in an initially balanced bridge to measure a strain of 1200 μm/m. Assume that a digital voltmeter will be used to measure the voltage and that $S_g = 2.06$, $r = 1$, and $v_s = 6$ V.

7.10   Prepare a graph showing bridge voltage v_s as a function of gage area A if $r = 1$ and $R_g = 350 \ \Omega$. Consider $p_D = 0.1, 0.2, 0.5, 1, 2, 5,$ and 10 W/in.^{2}

7.11 A 350-Ω strain gage with $S_g = 2.07$ is employed in a single-arm Wheatstone bridge with $r = 1$. If the gage is subjected to a strain of 1600 μin./in., determine the reading on a $4\frac{1}{2}$-digit DVM if

	v_s (V)	Amplifier Gain G
(a)	2	10
(b)	4	10
(c)	6	100
(d)	5	50

7.12 Write an engineering brief describing the difference between a direct-reading and a null-balance bridge. Cite advantages and disadvantages of each type.

7.13 Four strain gages ($R_g = 500$ Ω, $S_g = 2.06$) are attached to a bar to produce a load cell. The supply voltage $v_s = 4$ V and the amplifier (with a variable gain 10–100) is set at $G = 20$. When a load of 10,000 lb is applied to the bar the $4\frac{1}{2}$-digit DVM provides a count of 6280. What adjustments are required to make the count correspond to the applied load?

7.14 An oscilloscope with an input impedance of 10^6 Ω is connected to a Wheatstone bridge with one active gage ($R_g = 350$ Ω, $S_g = 3.35$, and $r = 5$). The bridge is powered with a 9-V constant-voltage supply. If the gage responds to a dynamic strain pulse having a magnitude of 900 μm/m, determine the sensitivity setting on the oscilloscope that will give a trace deflection of four divisions.

7.15 If the bridge and gage of Exercise 7.14 respond to a strain of 1400 μm/m, determine the trace deflection if an oscilloscope having a sensitivity of 1 mV per division is used for the measurement.

7.16 If the bridge, gage, and oscilloscope of Exercise 7.15 record a trace deflection of 5.2 divisions, determine the strain at the gage location.

7.17 Determine the sensitivity of a gage–bridge–galvanometer system if a four-equal-arm bridge ($R_1 = R_2 = R_3 = R_4 = R_g$) is used and if $R_g = 350$ Ω, $S_g = 2.07$, $p_g = 0.25$ W, $R_G = 100$ Ω, and $S_G = 0.003$ mm/μA.

7.18 Would the sensitivity of the system described in Exercise 7.17 be improved by replacing the 350-Ω gage with a 500-Ω gage? Explain.

7.19 An oscillograph chart shows a calibration displacement $d_c = 30$ mm. Determine the strain corresponding to a trace displacement $d_s = 26$ mm if the calibration constant $C = 40$ μm/m per millimeter of displacement.

7.20 Determine the value of the calibration constant C for a gage–bridge–amplifier–recorder system if $S_g = 2.04$, $r = 2$, $v_s = 4$ V, $G = 50$, and $S_R = 10$ mV per division.

7.21 Determine the resistance R_c that must be shunted across arm R_2 of the listed Wheatstone bridges to produce the given calibration strains

	ϵ_c (μm/m)	r	R_g (Ω)	S_g
(a)	600	3	350	2.06
(b)	1000	2	500	2.07
(c)	900	1	120	2.05
(c)	2000	2	350	2.09

7.22 Design a displacement fixture to be used with a strain gage mounted on a cantilever beam to mechanically produce a calibration strain ϵ_c ranging from 0 to 2000 μm/m in 500 μm/m increments.

7.23 A 120-Ω strain gage is connected into a single-arm bridge using 190 ft of twin lead wire. What is the signal loss factor if the wire is gage
   (a) 30   (c) 20
   (b) 26   (d) 14

7.24 Solve Exercise 7.23 if a 350-Ω gage is used in place of the 120-Ω gage.

7.25 Solve Exercise 7.23 if a three-wire system is employed to connect the gage into the bridge.

7.26 A switch similar to the one shown in Fig. 7.13 is employed to switch a series of 120-Ω gages ($S_g = 2.09$) in and out of a single-arm bridge. Prepare a graph showing the apparent strain ϵ' as a function of switch resistance ΔR_s. Let ΔR_s range from 100 $\mu\Omega$ to 100 mΩ.

7.27 Write an engineering brief describing the operation of a set of slip rings to be employed with a rotating shaft.

7.28 Write an engineering specification describing grounding procedures to minimize noise pickup on strain-gage leads.

7.29 What three precautions used to minimize noise must be included in the specification of Exercise 7.28?

7.30 Determine the error that results if a strain gage compensated for aluminum [$\alpha = 130(10^{-6})/°\text{F}$] is used on steel [$\alpha = 6(10^{-6})/°\text{F}$]. The total response of the gage was 200 μm/m and the temperature change between the zero and final reading was 30°F. Assume that a dummy gage was not used in the bridge.

7.31 Determine the axial sensitivity S_a of a strain gage if $S_g = 2.04$ and $K_t = 0.02$.

7.32 Determine the error involved if transverse sensitivity is neglected in a measurement of hoop strain in a thin-walled steel ($E = 29,000$ ksi and $v = 0.29$) cylindrical pressure vessel when K_t for the gage is 0.03.

7.33 If strain gages with $S_g = 2.03$ and $K_t = 0.04$ are used to determine the apparent strains ϵ'_{xx} and ϵ'_{yy}, determine the true strains ϵ_{xx} and ϵ_{yy} if:

Case	ϵ'_{xx} (μm/m)	ϵ'_{yy} (μm/m)
1	800	1200
2	640	−720
3	1120	−240
4	−560	2400
5	240	1440

7.34 Determine the error produced by ignoring transverse sensitivity effects if strain gages with $K_t = 0.03$ are employed in a simple tension test to measure Poisson's ratio of a material.

7.35 Assume that K_t for an ordinary strain gage is zero. Show how this gage could be used to measure stress in a specified direction. Clearly indicate all assumptions made in your derivation.

7.36 A stress gage is fabricated with two grids, each of which exhibits an axial sensitivity $S_a = 2.00$. Determine the output from the gage in terms of $\Delta R/R$ if it is mounted on a steel specimen and subjected to a stress $\sigma_a = 50,000$ psi.

7.37 Determine the uniaxial state of stress associated with the following strain measurements:

ϵ (μin./in.)	Material
800	Steel ($E = 29,000$ ksi, $v = 0.29$)
1100	Aluminum ($E = 10,000$ ksi, $v = 0.33$)
1620	Titanium ($E = 14,000$ ksi, $v = 0.25$)

7.38   Determine the biaxial state of stress associated with the following strain measurements:

ϵ_1 (μin./in.)	ϵ_2 (μin./in.)	Material
-600	-900	Aluminum
1220	-470	Titanium
1115	820	Steel

See Exercise 7.37 for material properties.

7.39   Determine the general state of stress associated with the following strain measurements made with a 0°, 45°, and 90° rosette.

ϵ_A (μin./in.)	ϵ_B (μin./in.)	ϵ_C (μin./in.)	Material
600	1200	-300	Aluminum
1050	1050	1050	Steel
-450	-900	1350	Titanium

See Exercise 7.37 for material properties.

7.40   Beginning with Eqs. 7.49 verify Eqs. 7.51 for the three-element rectangular rosette.

7.41   Derive relations of the same form as Eq. 7.51 that give ϵ_1, ϵ_2, and ϕ in terms of ϵ_A, ϵ_B, and ϵ_C for a three-element delta rosette.

7.42   Continue Exercise 7.41 and expand the solution to show:
   (a)   the classification procedure for the principal angles in terms of ϵ_A, ϵ_B, and ϵ_C — similar to Eqs. 7.52
   (b)   the relations for the principal stresses in terms of ϵ_A, ϵ_B, and ϵ_C — similar to Eqs. 7.53

7.43   Use Mohr's strain cycle to verify the classification procedure for a three-element rectangular rosette described by Eqs. 7.52.

7.44   Write an engineering brief that describes the capabilities of measuring strain at high temperatures by employing electrical resistance strain gages.

Chapter **8**

Force, Torque, and Pressure Measurements

8.1 INTRODUCTION

Transducers that measure force, torque, or pressure usually contain an elastic member that converts the mechanical quantity to a deflection or strain. A deflection sensor or a set of strain gages is then used to give an electrical signal proportional to the quantity of interest (force, torque, or pressure). Characteristics of the transducer, such as range, linearity, and sensitivity, are determined by the size and shape of the elastic member, the material used in its fabrication, and the sensor.

A wide variety of transducers are commercially available for measuring force (load cells), torque (torque cells), and pressure. The different elastic members employed in the design of these transducers include links, columns, rings, beams, cylinders, tubes, washers, diaphragms, shear webs, and numerous other shapes for special-purpose applications. Strain gages are commonly used as sensors; however, linear potentiometers and linear variable differential transformers (LVDTs) are sometimes used for static or quasi-static measurements. A selection of force transducers (load cells) are shown in Fig. 8.1.

In quasi-static measurements the loads are applied slowly to the elastic member contained in the transducer. If the time required for the load to reach its maximum value exceeds the period of the natural frequency of the elastic element by a factor of 10, then the dynamic response of the transducer is not a serious consideration.

A spring-mass-dashpot representation of a load cell permits examination of the response of a load cell to terminated ramp and periodic input functions (see Section 8.7). A seismic transducer model is introduced in Chapter 9. The seismic transducer model is important because it can be applied to describe dynamic behavior of transducers in measuring displacement, velocity, and acceleration as well as force and pressure.

8.2 FORCE MEASUREMENTS (LOAD CELLS)

The elastic members commonly used in load cells are links, beams, rings, and shear webs. The operating characteristics for several transducers with these elastic members are developed in the following subsections.

Figure 8.1 A selection of force transducers. (Courtesy of Hottinger–Baldwin Measurements, Inc.)

8.2.1 Link-Type Load Cell

A simple link-type load cell with strain gages constituting the sensor is shown in Fig. 8.2a. The load P can be either tensile or compressive. The four strain gages are bonded to the link such that two are in the axial direction and two are in the transverse direction. The four gages are wired into a Wheatstone bridge with the axial gages in arms 1 and 3 of the bridge and the transverse gages in arms 2 and 4, as shown in Fig. 8.2b.

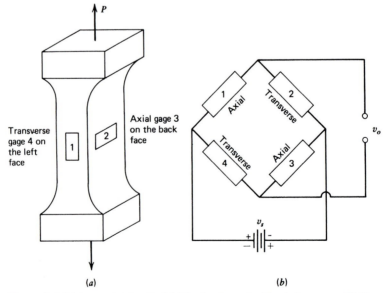

Figure 8.2 Link-type load cell. (a) Elastic element with strain gages. (b) Gage positions in the Wheatstone bridge.

When the load P is applied to the link, axial and transverse strains ϵ_a and ϵ_t, respectively, develop in the link and are related to the load by the expressions

$$\epsilon_a = \frac{P}{AE} \qquad \epsilon_t = -\frac{\nu P}{AE} \tag{a}$$

where

A is the cross-sectional area of the link
E is the modulus of elasticity of the link material
ν is Poisson's ratio of the link material

The response of the gages to the applied load P is given by Eqs. 5.5 and a as

$$\frac{\Delta R_1}{R_1} = \frac{\Delta R_3}{R_3} = S_g \epsilon_a = \frac{S_g P}{AE}$$

$$\frac{\Delta R_2}{R_2} = \frac{\Delta R_4}{R_4} = S_g \epsilon_t = -\frac{\nu S_g P}{AE} \tag{b}$$

The output voltage v_o from the Wheatstone bridge is expressed in terms of the load P by substituting Eqs. b into Eq. 6.18. If the four strain gages on the link are identical, then $R_1 = R_2$ and Eq. 6.18 yields

$$v_o = \frac{S_g P(1 + \nu)v_s}{2AE} \tag{8.1}$$

or

$$P = \frac{2AE}{S_g(1 + \nu)v_s} v_o = Cv_o \tag{8.2}$$

Equation 8.2 indicates that the load P is linearly proportional to the output voltage v_o and that the constant of proportionality or calibration constant C is

$$C = \frac{2AE}{S_g(1 + \nu)v_s} \tag{8.3}$$

The sensitivity of the load cell–Wheatstone bridge combination is given by Eq. 1.6 as $S = v_o/P$; therefore, from Eq. 8.3

$$S = \frac{v_o}{P} = \frac{1}{C} = \frac{S_g(1 + \nu)v_s}{2AE} \tag{8.4}$$

Clearly, the sensitivity of the link-type load cell depends on the cross-sectional area of the link (A), the elastic constants of the material used in fabricating the link (E and ν), the gage factor of the gages (S_g), and the input voltage applied to the Wheatstone bridge (v_s).

The range of a link-type load cell is determined by the cross-sectional area of the link and by the fatigue strength S_f of the material used in its fabrication. Thus,

$$P_{max} = S_f A \tag{8.5}$$

Since both sensitivity and range depend on the cross-sectional area A of the link, high sensitivities are associated with low-capacity load cells and low sensitivities are associated with high-capacity load cells.

The voltage ratio at maximum load $(v_o/v_s)_{max}$ for the link-type load cell is obtained from Eqs. 8.5 and 8.1 as

$$\left(\frac{v_o}{v_s}\right)_{max} = \frac{S_g S_f (1 + \nu)}{2E} \tag{8.6}$$

Most load-cell links are fabricated from AISI 4340 steel ($E = 30,000,000$ psi and $\nu = 0.30$) that is heat treated to give a fatigue strength $S_f \approx 80,000$ psi. Since $S_g \approx 2$ for the strain gages, Eq. 8.6 gives $(v_o/v_s)_{max} = 3.47$ mV/V. Many link-type load cells are rated at $(v_o/v_s)^* = 3$ mV/V at the full-scale value of the load ($P = P_{max}$). With this full-scale specification of voltage ratio $(v_o/v_s)^*$, the load P on the load cell is given by

$$P = \frac{v_o/v_s}{(v_o/v_s)^*} P_{max} \tag{8.7}$$

The supply voltage applied to the bridge in the load cell is typically about 10 V, which gives an output voltage at the maximum rated load of approximately 30 mV. This output can be monitored with a digital voltmeter or, if the signal is dynamic, it can be displayed on an oscilloscope.

The link-type load cell has a high spring rate and a high natural frequency that depends on the mass at each end of the link.

8.2.2 Beam-Type Load Cell

Beam-type load cells are commonly employed for measuring low-level loads where the link-type load cell is too stiff to be effective. A simple cantilever beam serves as the elastic member (see Fig. 8.3a). Two strain gages on the top surface and two strain gages on the bottom surface (all oriented along the axis of the beam) act as the sensor. The gages are connected into a Wheatstone bridge as shown in Fig. 8.3b.

The load P produces a moment $M = Px$ at the gate location x, which causes equal and opposite strains

$$\epsilon_1 = -\epsilon_2 = \epsilon_3 = -\epsilon_4 = \frac{6M}{Ebh^2} = \frac{6Px}{Ebh^2} \tag{a}$$

where

b is the width of the cross section of the beam
h is the height of the cross section of the beam

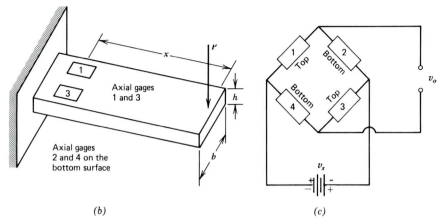

(b)   (c)

Figure 8.3 (*a*) Beam-type load cells incorporate an elastic element with strain gages. (*b*) Gage positions in the Wheatstone bridge.

The response of the strain gages is obtained from Eqs. 5.5 and a. Thus,

$$\frac{\Delta R_1}{R_1} = -\frac{\Delta R_2}{R_2} = \frac{\Delta R_3}{R_3} = -\frac{\Delta R_4}{R_4} = \frac{6S_g Px}{Ebh^2} \qquad \textbf{(b)}$$

The output voltage v_o from the Wheatstone bridge, owing to the load P, is obtained by substituting Eq. b into Eq. 6.18. If the four strain gages on the beam are identical, the bridge output voltage v_o becomes

$$v_o = \frac{6S_g Pxv_s}{Ebh^2} \qquad \textbf{(8.8)}$$

or

$$P = \frac{Ebh^2}{6S_g xv_s}v_o = Cv_o \qquad \textbf{(8.9)}$$

and the sensitivity is

$$S = \frac{v_o}{P} = \frac{1}{C} = \frac{6S_g xv_s}{Ebh^2} \qquad \textbf{(8.10)}$$

The sensitivity of the beam-type load cell depends on the shape of the beam cross section (b and h), the modulus of elasticity of the material used in fabricating the beam (E), the location of the load with respect to the gages (x), the strain gages (S_g), and the input voltage applied to the Wheatstone bridge (v_s).

The range of a beam-type load cell depends on the shape of the cross section of the beam, the location of the point of application of the load, and the fatigue strength of the material from which the beam is fabricated. If the gages are located at or near the beam support, then $M_{\text{gage}} \approx M_{\text{max}}$ and

$$P_{\text{max}} = \frac{S_f bh^2}{6x} \qquad \textbf{(8.11)}$$

Equations 8.10 and 8.11 indicate that both the range and the sensitivity of a beam-type load cell can be changed by varying the point of application of the load. Maximum sensitivity and minimum range occur as x approaches the length of the beam. The sensitivity decreases and the range increases as the point of load application moves nearer the gages.

The voltage ratio at maximum load $(v_o/v_s)_{max}$ is obtained from Eqs. 8.11 and 8.8 as

$$\left(\frac{v_o}{v_s}\right)_{max} = \frac{S_g S_f}{E} \tag{8.12}$$

A comparison of Eq. 8.12 with Eq. 8.6 shows that the beam-type load cell is approximately 50 percent more sensitive than the link-type load cell. Beam-type load cells are commercially available with ratings of $(v_o/v_s)^*$ between 4 and 5 mV/V at full-scale load.

8.2.3 Ring-Type Load Cell

Ring-type load cells, illustrated in Fig. 8.4, incorporate a proving ring as the elastic element. The ring element can be designed to cover a very wide range of loads by varying the radius R, the thickness t, or the depth w of the ring. Either strain gages or an LVDT are used as sensors.

If an LVDT is used to measure the diametric compression or extension δ of the ring, the relationship between displacement δ and load P is given by the following approximate expression:

$$\delta = 1.79 \frac{PR^3}{Ewt^3} \tag{8.13}$$

Equation 8.13 is approximate, since the reinforced areas at the top and bottom of the ring that accommodate the loading attachments have not been considered in its development. The output voltage v_o of an LVDT can be expressed as

$$v_o = S\delta v_s \tag{8.14}$$

where

S is the sensitivity of the LVDT
v_s is the voltage applied to the primary winding of the LVDT

Expressions relating output voltage v_o and load P are obtained by substituting Eq. 8.13 into Eq. 8.14. Thus,

$$v_o = 1.79 \frac{SPR^3 v_s}{Ewt^3} \tag{8.15}$$

or

$$P = 0.56 \frac{Ewt^3}{SR^3 v_s} v_o = Cv_o \tag{8.16}$$

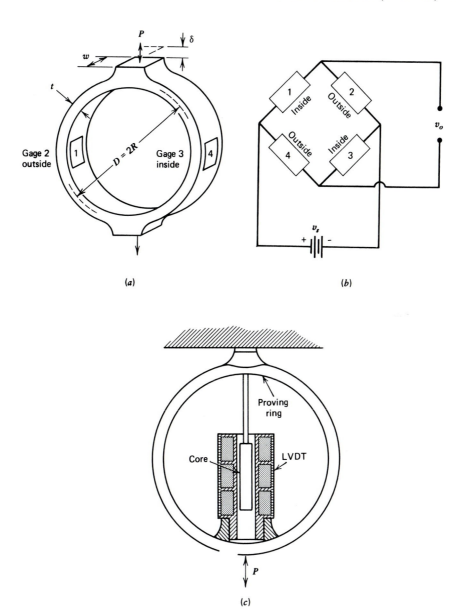

Figure 8.4 Ring-type load cell. (*a*) Elastic element with strain-gage sensors. (*b*) Gage positions in the Wheatstone bridge. (*c*) Elastic element with an LVDT sensor.

where

$$C = 0.56\frac{Ewt^3}{SR^3 v_s} \tag{a}$$

The sensitivity of the ring–LVDT combination S_t is

$$S_t = \frac{v_o}{P} = \frac{1}{C} = 1.79\frac{SR^3 v_s}{Ewt^3} \tag{8.17}$$

The sensitivity of the ring-type load cell with an LVDT sensor depends on the geometry of the ring (R, t, and w), the ring material (E), and the characteristics of the LVDT (S and v_s).

The range of a ring-type load cell is controlled by the strength of the material used in fabricating the ring. If the load cell is used to measure cyclic loads, the fatigue strength S_f is important. If the load cell is used only to measure static loads, the proportional limit S_{pl} of the material establishes the range of the load cell. The maximum stress in a ring element, reinforced at the top and the bottom, is highest on the inside surface of the ring on a diameter perpendicular to the line of the loads. The approximate value for the stress at this location is

$$\sigma_\theta = 1.09 \frac{PR}{wt^2} \tag{8.18}$$

From Eq. 8.18, for cyclic load measurements, the maximum load is

$$P_{max} = 0.92 \frac{wt^2 S_f}{R} \tag{8.19}$$

The voltage ratio at maximum load $(v_o/v_s)_{max}$ is obtained from Eqs. 8.19 and 8.15 as

$$\left(\frac{v_o}{v_s}\right)_{max} = 1.64 \frac{SR^2 S_f}{Et} \tag{8.20}$$

The rated voltage ratio $(v_o/v_s)^*$ for most ring-type load cells will be slightly less than that predicted by Eq. 8.20, because the ring will not be operated at a stress level equal to the fatigue strength of the material. Once the rated voltage ratio $(v_o/v_s)^*$ and maximum load P_{max} for a particular load cell are known, Eq. 8.7 is used to determine the load from a measured output voltage.

A typical short-range LVDT (± 1.25 mm) used for the load-cell sensor exhibits a sensitivity of 250 (mV/V)/mm. If the ring element of the load cell is designed with a maximum deflection $\delta_{max} = 1.25$ mm at P_{max}, then Eq. 8.14 gives

$$\left(\frac{v_o}{v_s}\right)_{max} = S\delta_{max} = 313 \text{ mV/V} \tag{b}$$

Ring-type load cells rated at $(v_o/v_s)^* \approx 300$ mV/V are available and have the capability of measuring both tensile and compressive loads (universal load cells). The rated output of a ring-type load cell with an LVDT sensor is higher (by about a factor of 10) than the output achieved with strain-gage sensors.

8.2.4 Shear-Web-Type Load Cell

The shear-web-type load cell (also known as a low-profile or flat load cell) is useful for applications where space is limited along the line of action of the load. The flat load cell consists of an inner loading hub and an outer supporting flange connected

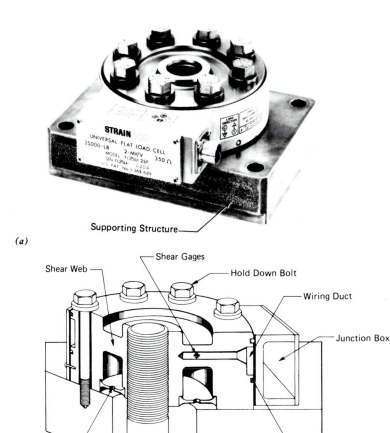

(a)

(b)

Figure 8.5 (a) Universal flat load cell. (b) Construction details. (Courtesy of Strainsert Company.)

by a continuous shear web (see Fig. 8.5). Strain gages that respond to shear strain are used as the sensor. The strain gages are installed in small holes drilled into the neutral surface of the web (see Fig. 8.5b). Some characteristics of flat load cells are shown in Table 8.1.

The shear-web load cell is compact and stiff; therefore, it can be used to measure dynamic loads at higher frequencies than the beam-type load cell. The effective weight w_e in Table 8.1 consists of the inner hub and a portion of the shear web. The listed natural frequencies are for a rigidly mounted load cell with no mass attached; therefore, they are the upper frequency limit. The attachment of any additional mass will reduce this natural frequency and will also reduce the frequency range of the transducer. The

Table 8.1 Mechanical Properties of Flat Load Cells.[a]

Force Capacity P (lb)	Spring Rate k (lb/in.)	Effective Weight w_e (lb)	Natural Frequency f_n (kHz)
250	920,000	0.028	18
1,000	1,220,000	0.023	22.8
5,000	6,600,000	0.135	22.0
10,000	8,500,000	0.34	15.7
50,000	20,400,000	1.58	11.3
100,000	28,600,000	4.50	7.9
500,000	65,000,000	33.0	4.4

[a]Courtesy of Strainsert Company (from Technical Bulletin No. 365-4MP).

manufacturer recommends the following equation for estimating the natural frequency of these load cells:

$$f_n = 3.13\sqrt{\frac{k}{w_e + w_x}} \qquad (8.21)$$

where

f_n is the transducer natural frequency (Hz)
k is the spring rate (lb/in.)
w_e is the effective weight of the hub assembly (lb)
w_x is the externally attached weight at the hub (lb)

The natural frequencies listed in Table 8.1 were calculated assuming that no external weight ($w_x = 0$) was attached to the load cell. With external weight attached to the load cell, Eq. 8.21 indicates that the frequency limit will be lowered. A complete discussion of the performance of force transducers under dynamic loading is presented in Section. 8.7.

It should be recognized that the frequency response of the load cells described in this section is limited. The exact natural frequency is difficult to predict because of the effect of the mass that is added to the load cell in connecting it to the load train. Piezoelectric force transducers with significantly better frequency response will be described in Section 9.5.

8.3 TORQUE MEASUREMENT (TORQUE CELLS)

Torque cells are transducers that convert an applied torque to an electrical output signal. The two types of torque cells in common usage are those installed on fixed shafts and those installed on rotating shafts. The latter type is more difficult to utilize, because the electrical signal must be transmitted from the rotating shaft to a stationary instrument station. The problem of signal transmission is treated after design concepts associated with torque cells are discussed.

8.3.1 Torque Cells—Design Concepts

Torque cells are similar to load cells; they contain a mechanical element (usually a shaft with a circular cross section) and a sensor (usually electrical resistance strain

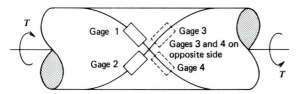

Figure 8.6 A circular shaft with strain gages used as a torque cell.

gages). A circular shaft with four strain gages mounted on two perpendicular 45-deg helices that are diametrically opposite one another is shown in Fig. 8.6. Gages 1 and 3, mounted on the right-hand helix, sense a positive strain, and gages 2 and 4, mounted on the left-hand helix, sense a negative strain. The two 45-deg helices define the principal stress and strain directions for a circular shaft subjected to pure torsion.

The shearing stress τ in the circular shaft is related to the applied torque T by the equation

$$\tau_{xz} = \frac{TD}{2J} = \frac{16T}{\pi D^3} \tag{8.22}$$

where

D is the diameter of the shaft

J is the polar moment of inertia of the circular cross section

Since the normal stresses $\sigma_x = \sigma_y = \sigma_z = 0$ for a circular shaft subjected to pure torsion, it is easy to show that

$$\sigma_1 = -\sigma_2 = \tau_{xz} = \frac{16T}{\pi D^3} \tag{8.23}$$

Principal strains ϵ_1 and ϵ_2 are obtained by using Eqs. 8.23 and Hooke's law for the plane state of stress. Thus,

$$\epsilon_1 = \frac{16T}{\pi D^3} \left(\frac{1 + \nu}{E} \right) \qquad \epsilon_2 = -\frac{16T}{\pi D^3} \left(\frac{1 + \nu}{E} \right) \tag{8.24}$$

The response of the strain gages is obtained by substituting these equations into Eq. 5.5

$$\frac{\Delta R_1}{R_1} = -\frac{\Delta R_2}{R_2} = \frac{\Delta R_3}{R_3} = -\frac{\Delta R_4}{R_4} = \frac{16T}{\pi D^3} \left(\frac{1 + \nu}{E} \right) S_g \tag{a}$$

If the gages are connected into a Wheatstone bridge, as illustrated in Fig 8.4b, the relationship between output voltage v_o and torque T is obtained by substituting Eq. a into Eq. 6.18 to give

$$v_o = \frac{16T}{\pi D^3} \left(\frac{1 + \nu}{E} \right) S_g v_s \tag{8.25}$$

or

$$T = \frac{\pi D^3 E}{16(1 + \nu)S_g v_s} v_o = C v_o \qquad (8.26)$$

where

$$C = \frac{\pi D^3 E}{16(1 + \nu)S_g v_s} \qquad (b)$$

The sensitivity is

$$S = \frac{v_o}{T} = \frac{1}{C} = \frac{16(1 + \nu)S_g v_s}{\pi D^3 E} \qquad (8.27)$$

The sensitivity of a torque cell depends on the diameter of the shaft (D), the shaft material (E and ν), the gage factor (S_g), and the voltage applied to the Wheatstone bridge (v_s).

The range of the torque cell depends on the diameter D of the shaft and the proportional limit S_τ of the material in torsion. For static applications, the range is given by Eq. 8.22 as

$$T_{max} = \frac{\pi D^3 S_\tau}{16} \qquad (8.28)$$

The voltage ratio at maximum torque $(v_o/v_s)_{max}$ is obtained from Eqs. 8.27 and 8.28 as

$$\left(\frac{v_o}{v_s}\right)_{max} = \frac{S_\tau S_g (1 + \nu)}{E} \qquad (8.29)$$

If the torque cell is fabricated from heat-treated steel ($S_\tau \approx 60{,}000$ psi), then $(v_o/v_s)_{max} = 5.2$ mV/V. Typically, torque cells are rated at values of $(v_o/v_s)^*$ between 4 and 5 mV/V. The torque T corresponding to an output voltage v_o is then given by Eq. 8.7.

8.3.2 Torque Cells—Data Transmission

Frequently, torque is measured on a rotating shaft, which necessitates signal transmission between a Wheatstone bridge on the rotating shaft and a stationary instrumentation center. Signal transmission between a rotating body and a fixed instrument is accomplished with either slip rings or telemetry.

A. Signal Transmission with Slip Rings

A schematic illustration of a slip-ring connection between a Wheatstone bridge on a rotating shaft and a recording instrument at a stationary location is shown in Fig. 8.7. The slip-ring assembly contains a series of insulated rings mounted on a shaft and a companion series of insulated brushes mounted in the case. High-speed bearings

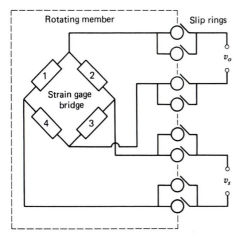

Figure 8.7 Schematic illustration of a slip-ring connection between a rotating member and a fixed instrumentation station.

between the shaft and the case enable the case to remain stationary while the shaft rotates with the torque cell. A commercial slip-ring assembly is shown in Fig. 8.8.

The major problem in using slip rings is noise, generated by variations in contact resistance between the rings and brushes. These variations can be maintained within acceptable limits by fabricating the rings from monel metal (a copper-nickel alloy), and the brushes from a silver-graphite mixture. Also, it is important to maintain the ring-brush contact pressure between 50 and 100 psi. Rotational speed limits of slip-ring assemblies are determined by the concentricity that exists between the shaft and the case and by the quality of the bearings. Slip-ring units with speed ratings of 6000 rpm are commercially available.

B. Signal Transmission with Telemetry

In many applications, a shaft end is not available for mounting the slip-ring assembly and telemetry is used to transmit the bridge signal from the rotating shaft to a record-

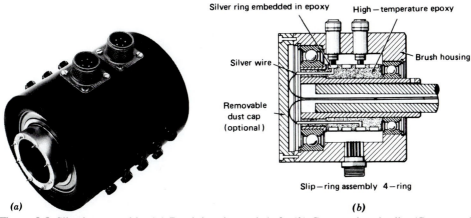

(a)

(b)

Figure 8.8 Slip-ring assembly. (*a*) Brush housing and shaft. (*b*) Construction details. (Courtesy of Lebow Products, Eaton Corp.)

ing instrument. In a simple telemetry system, the Wheatstone bridge output voltage is used to modulate a radio signal. The strain gages, bridge, power supply, and radio transmitter are mounted on the rotating shaft, and the receiver and recorder are stationary nearby. Usually, the signal is transmitted only a few feet; therefore, low-power (and unlicensed) transmitters are used.

A commercial short-range telemetry system, designed to measure rotating-shaft torques, is shown in Fig. 8.9. A split collar that fits over the shaft contains a power supply, a modulator, a voltage-controlled oscillator (VCO), and an antenna. The bridge signal modulates the pulse width of a constant-amplitude 5-kHz square wave (i.e., the time width of the positive part of the square wave is proportional to bridge output while the square-wave period remains the same). The square wave is used to vary the VCO frequency, which is centered at 10.7 MHz. The VCO signal is transmitted at low power by an antenna in the split collar. This signal is received by a stationary loop antenna that encircles the split collar, as shown in Fig. 8.9. The transmitting unit is completely self-contained and receives its power through inductive coupling of a 160-kHz signal from the stationary loop antenna.

A more complex telemetry system is required for longer range multiple-transducer applications. Licensing is required for these systems, because the transmitted signals have far greater power and the available transmission frequencies are limited and crowded. In the United States, telemetry transmissions are limited to only two bands: 1435 to 1535 MHz, and 2200 to 2300 MHz. As the number of transducers increases in an application requiring telemetry, it becomes less practical to use a separate transmitter and receiver for each signal. Instead, several transducer signals are combined into a single, composite transmission signal in a *multiplexing* process. Two types of multiplexing are: frequency-division multiplexing and time-division multiplexing. The radio receiver contains circuitry to separate the composite signal into the individual signals that were recorded. A *frequency-division multiplexing* system is illustrated

Figure 8.9 Telemetry system for data transmission. Stationary loop antenna and rotating collar on a shaft. (Courtesy of Wireless Data Corporation.)

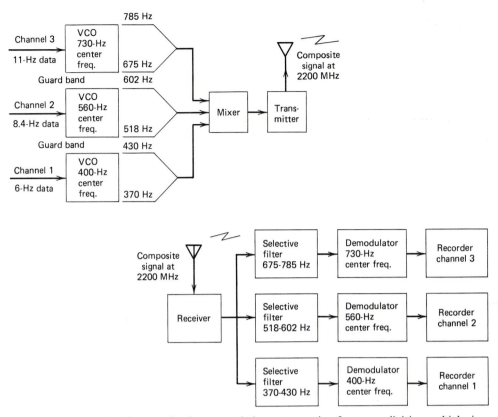

Figure 8.10 Schematic diagram of a data-transmission system using frequency-division multiplexing.

in Fig. 8.10. Three transducer output voltages modulate their respective subcarrier frequencies. These signals are combined and transmitted as a single signal. As shown, the first transducer oscillator is centered at 400 Hz and the transducer signal produces a maximum frequency deviation of ±30 Hz (±7.5 percent). Similarly, the second transducer oscillator is centered at 560 Hz with a maximum deviation of ±42 Hz, and the third transducer oscillator is centered at 730 Hz with a maximum deviation of ±55 Hz. There is no overlapping between channels, and guard bands are placed between channels to ensure separation. The three distinct channel signals are mixed together to form a composite signal ranging from 370 to 785 Hz. This composite signal is transmitted at 2200 MHz over the radio link. The receiving station has three band-pass filters, which separate the composite signal into three recovery band signals. These separated signals are connected to individual discriminator circuits, which demodulate and recover the original transducer signals for recording. It is clear that each channel must have the same phase-shift characteristics in order to maintain relative phase integrity.

With *time-division multiplexing,* all channels are transmitted at the same transmission frequency, one channel at a time. Each channel is sampled in a repeated sequence to give a composite signal consisting of time-spaced segments of each transducer signal. Since each channel is not monitored continuously, the sampling rate must be sufficient to ensure that individual signal amplitudes do not change significantly during the time between samples. Sampling rates must be at least five times greater than the highest frequency component in any signal for time-division multiplexing to be acceptable.

8.4 COMBINED MEASUREMENTS OF FORCE AND MOMENTS OR TORQUES

Some experiments require that two or more quantities be measured simultaneously. For example, a six-component wind-tunnel support string must simultaneously measure three orthogonal forces and three orthogonal moments. These combined measurements are accomplished by using two or more separate strain-gage bridges mounted on an elastic element or by using selected gage combinations to form a single bridge. Many different gage and bridge combinations can be designed. To illustrate the concepts while limiting the length of this section, only two combined-measurement transducers are discussed: a force–moment transducer and a force-torque transducer.

8.4.1 Force–Moment Measurements

A simple force–moment transducer uses an elastic link, as shown in Fig. 8.11. For simplicity, assume that the link has a square cross section ($A = h^2$) with strain gages bonded on the centerline of each side and aligned with the longitudinal (P_z) direction. An axial force P_z is measured by connecting gages A and C into Wheatstone bridge arm positions 1 and 3, as shown in Fig. 8.12a. Resistances R_2 and R_4 are equal fixed-value resistors with $R_2 = R_4 = R_g$. Under these conditions, Eq. 6.17 reduces to

$$v_o = \frac{1}{4}\left(\frac{\Delta R_1}{R_1} + \frac{\Delta R_3}{R_3}\right)v_s \qquad \textbf{(a)}$$

The corresponding strain-gage response is given by Eq. 5.5 as

$$\frac{\Delta R_1}{R_1} = \frac{\Delta R_3}{R_3} = S_g\epsilon = \frac{S_g P_z}{AE} \qquad \textbf{(b)}$$

Substitution of Eq. b into Eq. a gives an input–output of

$$v_o = \left(\frac{S_g v_s}{2AE}\right)P_z = S_a P_z \qquad \textbf{(8.30)}$$

where the term in parentheses is the force–voltage sensitivity S_a. A sensitivity comparison between this two-gage load cell and a four-gage load cell (see Eq. 8.4) shows a sensitivity loss of $1/(1 + \nu)$, or about 25 percent. This sensitivity loss is the price paid for having gages B and D available for moment M_x measurements.

Moment M_x is measured by connecting gages B and D into the Wheatstone bridge as shown in Fig. 8.12b. With gages B and D in bridge arms R_1 and R_4 and equal

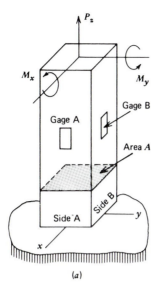

(a)

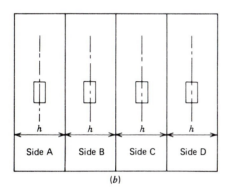

(b)

Figure 8.11 Combined-measurement transducer used to measure axial load P_z and moments M_x and M_y. (a) Elastic element with strain gages. (b) Developed surface showing strain-gage orientations.

fixed-value resistors in arms R_2 and R_3, the bridge output voltage is given by Eq. 6.17 as

$$v_o = \frac{1}{4}\left(\frac{\Delta R_1}{R_1} - \frac{\Delta R_4}{R_4}\right)v_s \tag{c}$$

The corresponding strain-gage response is given by Eq. 5.5 as

$$\frac{\Delta R_1}{R_1} = -\frac{\Delta R_4}{R_4} = S_g\epsilon = \frac{6S_g M_x}{Eh^3} \tag{d}$$

Substitution of Eq. d into Eq. c gives an input–output of

$$v_o = \left(\frac{3S_g v_s}{Eh^3}\right)M_x = S_x M_x \tag{8.31}$$

where the term in parentheses is the moment–voltage sensitivity S_x.

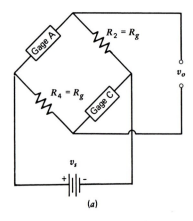

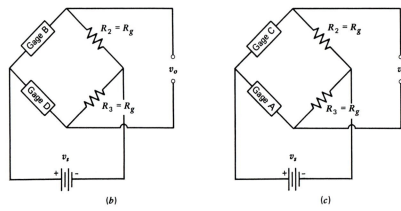

Figure 8.12 Wheatstone bridge arrangements used with a combined-measurement transducer to measure an axial force P_z and moments M_x and M_y. (*a*) Bridge arrangement for measuring force P_z. (*b*) Bridge arrangement for measuring moment M_x. (*c*) Bridge arrangement for measuring moment M_y.

Similarly, moment M_y can be measured by connecting gages C and A into bridge arms R_1 and R_4, respectively, and using equal fixed-value resistors in arms R_2 and R_3, as indicated in Fig. 8.12*c*. Thus,

$$v_o = \left(\frac{3S_g v_s}{E h^3}\right) M_y = S_y M_y \tag{8.32}$$

where the term in parentheses is the moment-voltage sensitivity S_y. A comparison of Eqs. 8.31 and 8.32 shows equal voltage sensitivities for the x and y directions because the cross section used in designing the link was square.

Transducers designed for combined measurements are usually equipped with switch boxes, which contain the required bridge-completion resistors and the connections needed to switch the gages into the correct bridge locations. Care must be exercised during certain measurements, since temperature compensation is not maintained within the Wheatstone bridge.

8.4.2 Force–Torque Measurements

A transducer designed to measure both axial force P_z and torque M_z is shown in Fig. 8.13. The load link has a circular cross section. For measuring axial force P_z, gages A and C are connected to bridge arms R_1 and R_3, respectively, and equal fixed-value resistors are used in the other two arms. This arrangement is identical to that shown in Fig. 8.12a; therefore, Eq. 8.30 applies to this case as well.

Torque is measured by connecting gages B and D into bridge arms R_1 and R_4, as shown in Fig. 8.12b. With equal fixed-value resistors in the other two bridge arms, Eq. 6.17 reduces to Eq. c. The strain-gage response for this case becomes

$$\frac{\Delta R_1}{R_1} = -\frac{\Delta R_4}{R_4} S_g \epsilon = \frac{16(1 + \nu)S_g M_z}{\pi D^3 E} \tag{e}$$

Substitution of Eq. e into Eq. c gives

$$v_o = \left[\frac{8(1 + \nu)S_g v_s}{\pi D^3 E} \right] M_z = S_z M_z \tag{8.33}$$

A comparison of Eq. 8.33 for two active strain gages with Eq. 8.27 for a standard four-gage torque cell indicates that the combined transducer has one-half the sensitivity of the standard cell. This loss in sensitivity results from using only half as many strain gages in the Wheatstone bridge.

8.5   PRESSURE MEASUREMENTS (PRESSURE TRANSDUCERS)

Pressure transducers are devices that convert an applied pressure into an electrical signal through a measurement of displacement, strain, or piezoelectric response. The

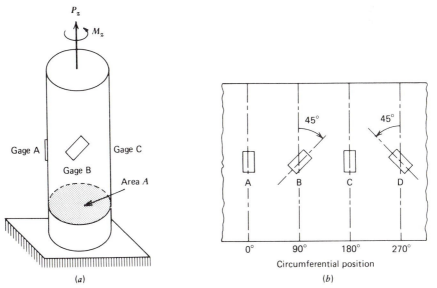

Figure 8.13 Combined-measurement transducer used to measure axial load P_z and moment M_z. (a) Elastic element with strain gages. (b) Developed surface showing strain-gage orientation.

quasi-static operating characteristics for each of these transducer types are covered in the following subsections. The dynamic operating characteristics of pressure transducers will be covered in Section 8.7.

8.5.1 Displacement-Type Pressure Transducer

A pressure transducer that employs a displacement sensor to produce an output voltage is illustrated in Fig. 8.14. This transducer utilizes a bourdon tube as the elastic element and a linear variable differential transformer (LVDT) as the sensor. The bourdon tube is a C-shaped conductor with a flat-oval cross section that tends to straighten as internal pressure is applied. In this displacement-type pressure transducer, one end of the bourdon tube is fixed, while the other end is free to displace. The core of the LVDT is attached to the free end of the bourdon tube and to a small cantilever spring that maintains tension on the core assembly. The coil of the LVDT is attached to the housing that anchors the fixed end of the tube (see Fig. 8.14).

As pressure is applied to the bourdon tube, it pulls the core of the LVDT through the coil and an output voltage develops. The output voltage is a linear function of the pressure, provided that the displacement of the bourdon tube is small. Displacement-type transducers of this type provide stable and reliable measurements of pressure over extended periods of time. Such transducers are excellent for static or quasi-static applications; however, they are not suitable for dynamic measurements of pressure, since the masses of the tube, the fluid in the tube, and the core limit frequency response to 10 Hz or less.

8.5.2 Diaphragm-Type Pressure Transducer

A second general type of pressure transducer utilizes either a clamped circular plate (diaphragm) or a hollow cylinder as the elastic element and electrical resistance strain gages as the sensor. Diaphragms are used for low- and middle-pressure ranges (0 to

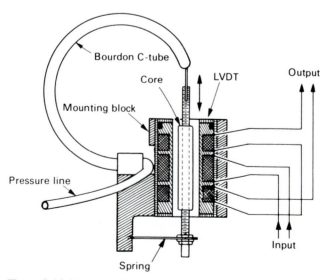

Figure 8.14 Pressure transducer that utilizes a bourdon tube as the elastic element and a linear variable differential transformer as the sensor.

30,000 psi), and cylinders are primarily used for the high- and very high-pressure ranges (30,000 to 100,000 psi). The strain distribution resulting from a uniform pressure on the face of a clamped circular plate of constant thickness is given in reference [5] as

$$\epsilon_{rr} = \frac{3p(1 - \nu^2)}{8Et^2}(R_o^2 - 3r^2)$$

$$\epsilon_{\theta\theta} = \frac{3p(1 - \nu^2)}{8Et^2}(R_o^2 - r^2) \tag{8.34}$$

where

p is the pressure
t is the thickness of the diaphragm
R_o is the outside radius of the diaphragm
r is a position parameter

Examination of these equations indicates that the circumferential strain $\epsilon_{\theta\theta}$ is always positive and is a maximum at $r = 0$. The radial strain ϵ_{rr} is positive in some regions, but negative in others, and assumes its maximum negative value at $r = R_o$. Both distributions are shown in Fig. 8.15.

A special-purpose diaphragm strain gage, which has been designed to take advantage of this strain distribution, is widely used in diaphragm-type pressure transducers. Circumferential elements are employed in the central region of the diaphragm, where $\epsilon_{\theta\theta}$ is a maximum. Similarly, radial elements are employed near the edge of the diaphragm, where ϵ_{rr} is a maximum. Also, the circumferential and radial elements are

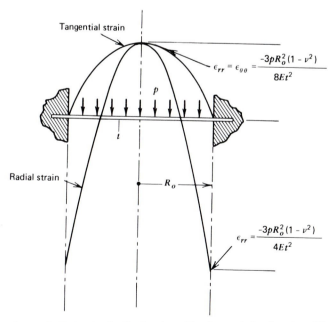

Figure 8.15 Strain distribution in a thin clamped circular plate (diaphragm) owing to a uniform lateral pressure.

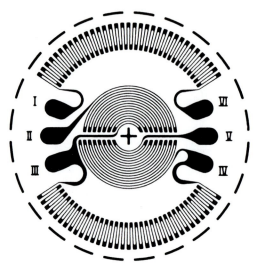

Figure 8.16 Special-purpose four-element strain gage for diaphragm-type pressure transducers. (Courtesy of Micro-Measurements Division, Measurements Group, Inc., USA.)

each divided into two parts, as shown in Figure 8.16, so that the special-purpose gage actually consists of four separate gages. The individual gages are connected into a Wheatstone bridge with the circumferential elements in arms R_1 and R_3 and the radial elements in arms R_2 and R_4. If the strains are averaged over the areas of the circumferential and radial elements and if the average values of $\Delta R/R$ (with a gage factor $S_g = 2$) obtained from Eq. 5.5 are substituted into Eq. 6.18, the output voltage v_o is given by

$$v_o = 0.82 \frac{p R_o^2 (1 - \nu^2)}{E t^2} v_s = S_p p \tag{8.35}$$

where the pressure-voltage sensitivity S_p depends on the geometry (R_o and t), material properties (E and ν), and supply voltage (v_s). The power supplied to a Wheatstone bridge is controlled by the power p_T that can be dissipated by the gage elements. The voltage–power relationship of a four-arm Wheatstone bridge is given by Eq. 6.24 as

$$v_s = 2 \sqrt{p_T R_T} \tag{a}$$

Substituting Eq. a into Eq. 8.35 and solving for S_p gives

$$S_p = 1.64 \frac{R_o^2 (1 - \nu^2) \sqrt{p_T R_T}}{E t^2} \tag{8.36}$$

Clearly, Eq. 8.36 shows that a diaphragm-type pressure gage can be designed with a wide range of sensitivities by varying the (R_o/t) ratio. Unlike most other transducers, diaphragm deflection rather than yield strength controls the limit of (R_o/t). The relationship between pressure and output voltage will be linear within 0.3 percent if

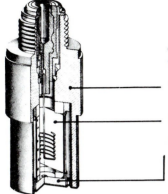

Quartz pressure transducers consist basically of three parts (see sectional drawing):

1. Transducer housing, which serves for
   mounting and encloses the quartz elements hermetically at the same time.
2. Quartz elements, yielding an electrical
   charge proportional to the pressure.
3. Diaphragm, welded tightly with the transducer housing and transmitting to the
   quartz elements the pressure exerted by
   the medium.

Figure 8.17 Design details for a piezoelectric pressure transducer. (Courtesy of
Kistler Instrument Corp.)

the center deflection is limited to be less than $t/4$. Using this deflection criterion, one
can show that the maximum sensitivity becomes

$$S_{p_{max}} = 2.19 \left(\frac{t}{R_o} \right)^2 \frac{\sqrt{p_T R_T}}{p_{max}} \tag{8.37}$$

The amount of damping in a diaphragm pressure transducer is highly dependent on
the fluid in contact with the diaphragm surface. Also, the effective seismic mass of
the diaphragm is dependent on the density of the contacting fluid. The upper resonant
natural frequency can be estimated by using the expression

$$f_r = 0.471 \frac{t}{R_o^2} \sqrt{\frac{E g}{w(1 - \nu^2)}} \tag{8.38}$$

where

f_r is the resonant frequency in Hz
g is the gravitational constant (386 in./s^2 or 9.81 m/s^2)
w is the specific weight of the diaphragm material (lb/in.^{3} or N/mm^3)

Since typical values of f_r range from 10 to 50 kHz, the diaphragm pressure transducer
can be used over a range of frequencies from dc (static) to dynamic measurements
involving frequencies as high as $f_r/5$ (2 to 10 kHz).

8.5.3 Piezoelectric-Type Pressure Transducer

The piezoelectric type of pressure transducer uses a piezoelectric crystal (see Section
5.7) as both the elastic element and the sensor. Quartz is the most widely used piezoelectric material because of its high modulus of elasticity, high resonant frequency,
linearity over a wide range, and very low hysteresis. Resonant frequencies of 0.25 to
0.50 MHz can be achieved while maintaining relatively high sensitivity.

Design details of a piezoelectric pressure transducer are shown in Fig. 8.17. The
quartz crystal is enclosed in a cylindrical shell, which has a thin pressure-transmitting

diaphragm on one end and a rigid support base for the crystal on the other end. As pressure is applied to the face of the crystal in contact with the diaphragm, an electrostatic charge is generated. The magnitude of the charge depends on the pressure, the size of the crystal, and the orientation of the crystal axes, as indicated by Eq. 5.16. Miniature pressure transducers that utilize a quartz crystal having a diameter of 6 mm and a length of 6 mm exhibit a sensitivity of approximately 1 pC/psi, and those with larger crystals (11-mm diameter and 12-mm length) have sensitivities of 5 pC/psi. Piezoelectric transducers can be used for very high pressure (up to 100,000 psi) measurements and for pressure measurements at temperatures as high as 350°C. If water cooling is used to protect the crystal and its insulation, the temperature range can be extended.

Piezoelectric transducers all exhibit an extremely high output impedance that depends on the frequency of the applied pressure, as indicated by Eq. 2.42. Because of this high output impedance, a charge amplifier or cathode follower must be inserted between the transducer and any conventional voltage-measuring instrument. A charge amplifier converts the charge to a voltage, amplifies the voltage, and provides an output impedance of approximately 100 Ω, which is satisfactory for most voltage-measuring instruments. A complete discussion of the circuits used for dynamic applications of pressure transducers is presented in Section. 9.6.

The low-frequency response characteristics of piezoelectric sensors depend on the effective time constant of the measuring circuit (see Eq. 5.17). The time constant depends on the input impedance of the charge amplifiers or built-in voltage followers that are employed with the sensor. The performance characteristics of several pressure transducers are given in Table 8.2.

8.6   MINIMIZING ERRORS IN TRANSDUCERS

Most transducers designed to measure force, torque, or pressure utilize electrical resistance strain gages as sensors because they are inexpensive, easy to install, and provide an output voltage v_o (when used as elements of a Wheatstone bridge) that can be related easily to the load, torque, or pressure. In applications of strain gages to

Table 8.2 Characteristics of Piezoelectric Pressure Transducers

	Manufacturer		
	Endevco	PCB	Kistler
Model	2501-2000	109A	601B1
Crystal material	Piezite P-1	Quartz	Quartz
Charge sensitivity (pC/psi)	14	—	1.0
Voltage sensitivity (mV/psi)	35	0.1	—
Capacitance (pF)	300	—	20
Internal time constant (s)	—	2000	—
Frequency response (Hz)	2 to 10,000	0 to 100,000	0 to 100,000
Mounted resonant frequency (Hz)	50,000	500,000	250,000
Range (psi)	0 to 2000	0 to 80,000	0 to 15,000
Amplitude linearity	±3%	±2%	±1%
Weight (grams)	21	—	7
Acceleration sensitivity (psi/g)	0.03	0.004	0.002
Temperature range (°F)	−22 to +230	−100 to +275	−450 to +500

stress analysis, accuracies of ± 2 percent are usually acceptable. When strain gages are used as sensors in transducers, accuracy requirements are an order of magnitude more stringent; therefore, more care must be exercised in the selection and installation of the gages and in the design of the Wheatstone bridge. Typical performance specifications for general-purpose, improved-accuracy, and high-accuracy load cells are listed in Table 8.3.

Errors that degrade the accuracy of a transducer include dual sensitivity, zero shift with temperature change, bridge balance, span adjust, and span change with temperature change. Each of these sources of error is discussed in the following subsections together with procedures for minimizing error.

8.6.1 Dual Sensitivity

All transducers exhibit a *dual sensitivity* to some degree, which means that the output voltage is the result of both a primary quantity, such as load, torque, or pressure, and a secondary quantity, such as temperature or a secondary load. Provision must be made during design of the transducer to minimize these secondary sensitivities.

A. Dual Sensitivity — Temperature

As an example of dual sensitivity owing to temperature, consider a link-type load cell subjected to a load P_z and a temperature change ΔT that occurs during the measurement period. The strain gages on the link will respond to both the strain ϵ produced by the load and the apparent strain ϵ' caused by the temperature change. The total response of each gage will appear as

$$\frac{\Delta R_1}{R_1} = \frac{\Delta R_g}{R_g}\bigg]_{P_z} + \frac{\Delta R_g}{R_g}\bigg]_{\Delta T} = S_g(\epsilon_1 + \epsilon_1') \qquad \textbf{(a)}$$

Table 8.3 Specifications for Load-Cell Accuracies

Characteristic	General-Purpose Load Cell	Improved-Accuracy Load Cell	High-Accuracy Load Cell
Calibration inaccuracy	0.5% FS[a]	0.25% FS	0.1% FS
Temperature effect on zero	$\pm 0.005\%/°F$ FS	$\pm 0.0025\%/°F$ FS	$\pm 0.0015\%/°F$ FS
Zero-balance error	$\pm 5\%$ FS	$\pm 2.5\%$ FS	$\pm 1\%$ FS
Temperature effect on span	$\pm 0.01\%/°F$ OL[b]	$\pm 0.005/°F$ OL	$\pm 0.008\%/°F$ OL
Nonlinearity	0.25% FS	0.1% FS	0.05% FS
Hysteresis	0.1% FS	0.05% FS	0.02% FS
Nonrepeatability	0.1% FS	0.05% FS	0.02% FS
System inaccuracy[c]	1% FS	0.5% FS	0.15% FS

[a]Full scale.
[b]Of load.
[c]Combined effects, but not including temperature.

Similar expressions will apply for the other three gages. If the strain gages are identical and if the temperature change for each gage is the same, Eq. 6.18 indicates that the response of the gages as a result of temperature will cancel and the output of the Wheatstone bridge will be a function only of the load-induced strains in the elastic element (link) of the transducer. In this example, the signal-summing property of the Wheatstone bridge provides temperature compensation of the load cell.

B. Dual Sensitivity—Secondary Load

When link-type load cells are used for force measurements, it is usually difficult to apply the load P_z coincident with the centroidal axis of the link. As a consequence, both the load P_z and a bending moment M are imposed on the link and it is necessary to design the transducer with a very low sensitivity to M while maintaining a high sensitivity to P_z. This objective is accomplished in the link-type load cell by proper placement of the strain gages.

As an example, consider that an arbitrary moment M is being applied to the cross section of the elastic element of the load cell, as shown in Fig. 8.18. If the moment M is resolved into Cartesian components M_x and M_y, the effect of M_x will be to bend the element about the x axis such that

$$\epsilon_{a1} = -\epsilon_{a3} \tag{a}$$

Since the transverse gages are on the neutral axis for bending about the x axis,

$$\epsilon_{t2} = \epsilon_{t4} = 0 \tag{b}$$

Equations a and b indicate that the response of the gages to a moment M_x will be

$$\frac{\Delta R_1}{R_1} = -\frac{\Delta R_3}{R_3} \tag{c}$$

$$\frac{\Delta R_2}{R_2} = \frac{\Delta R_4}{R_4} = 0 \tag{d}$$

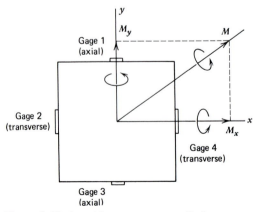

Figure 8.18 An arbitrary moment applied to a cross section of the elastic element in a link-type load cell.

When Eqs. c and d are substituted into Eq. 6.18, the output resulting from the strain-gage response to moment M_x vanishes. Since neither M_x nor M_y produces an output, any arbitrary moment M can be applied to the load cell without influencing the measurement of the load P_z. In this example, proper placement of the strain gages eliminates any sensitivity to the secondary load.

8.6.2 Zero Shift with Temperature Change

It was shown in Section 7.9 that some electrical resistance strain gages are temperature compensated (resistance changes owing to temperature change are minimized through proper selection of the gage alloy) over a limited range of temperature. When temperature variations are large, changes in resistance occur and zero output under zero load is not maintained.

Zero shift with temperature change is reduced by using either half or full Wheatstone bridges, as discussed previously. Temperature-induced resistance changes are partially canceled by the summing properties of the Wheatstone bridge. However, since the strain gages are never identical, some zero shift persists.

A third compensation procedure for reducing zero shift in transducers is illustrated in Fig. 8.19, where a low-resistance copper ladder gage is inserted between arms 3 and 4 of the Wheatstone bridge. Since the ladder gage is a part of both arms, it increases both ΔR_3 and ΔR_4 when the temperature is increased. During calibration (which involves temperature cycling of the transducer over its specified range of operation), the ladder gages are trimmed by cutting one or more rungs (ΔR_3 and ΔR_4 caused by temperature change are adjusted) until the zero shift for this particular transducer is within acceptable limits.

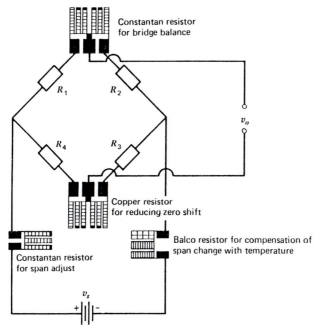

Figure 8.19 Compensation resistors introduced into the Wheatstone bridge of a transducer to minimize the effects of temperature change. (After Dorsey.)

8.6.3 Bridge Balance

In general, transducers should exhibit zero output under no-load conditions. Unfortunately, the strain gages employed as sensors do not have exactly the same resistance; therefore, the Wheatstone bridge is usually out of balance under the no-load condition. Balance can be achieved by inserting a compensation resistor between arms 1 and 2 of the bridge, as shown in Fig. 8.19. The compensation resistor is a double-ladder gage that can be trimmed to add either ΔR_1 or ΔR_2 until nearly perfect balance is achieved.

8.6.4 Span Adjust

Span refers to the sensitivity of the transducer. In instrumentation systems in which transducers are often interchanged or replaced, it is important to have transducers with the span adjusted to a preselected level. The span is usually adjusted by using a temperature-insensitive resistor (ladder gage) in series with the voltage supply, as indicated in Fig. 8.19. As the ladder gage reduces the supply voltage applied to the bridge, the span of the transducer is adjusted to the specified value (usually 3 mV/V full scale).

8.6.5 Span Change with Temperature

Compensation of span change with temperature is difficult because the procedure involves simultaneous application of the load and cycling of the temperature. Usually, temperature compensation of span is achieved by inserting a resistor that changes with temperature (a nickel-iron alloy known as Balco) in the second lead from the voltage supply, as shown in Fig. 8.19. This ladder resistor is trimmed to give a resistance change with temperature that compensates for changes in sensitivity with temperature.

Considerable fine-tuning of a transducer is required to achieve the accuracies specified in Table 8.3. Also periodic recalibration is needed to ensure that the transducer is operating within specified limits of accuracy. If these high accuracies are not required, then some of the compensation procedures will not be necessary and lower cost transducers can be specified.

8.7 FREQUENCY RESPONSE OF TRANSDUCERS

There are two different approaches employed in describing the frequency response of transducers in applications involving measurements of mechanical quantities that vary with time. One approach assumes that a fixed point exists where the base of the transducer is attached. The load or pressure is applied to the elastic member, which deforms relative to this fixed point. The second approach does not utilize a fixed reference point. Instead, the base and a representative mass in the transducer both move in response to the dynamic input. The relative motion between the base and the mass is used to sense kinematic quantities, such as displacement, velocity, and acceleration, when a fixed reference point is not available.

This section employs the first approach and assumes that the base of the transducer is mounted to a rigid surface that remains stationary during the dynamic application of force to the elastic member. In Chapter 9 the seismic transducer model is introduced. This model accommodates the motion of the base of the transducer and must be used if it is not possible to attach the base of the transducer to a fixed reference surface. The dynamic response of a galvanometer was considered previously in Section 3.5 to illustrate the importance of frequency response characteristics to the recording process.

The response characteristics of a transducer are equally important, since serious errors can be introduced in dynamic measurements if the frequency response of the transducer is not adequate.

Transducers for measuring load, torque, and pressure are all second-order systems; therefore, their dynamic behavior can be described by a second-order differential equation similar to Eq. 3.33. The application of second-order theory to transducers will be illustrated by considering a link-type load cell with a uniaxial elastic member, such as the one illustrated in Fig. 8.2. The dynamic response of this load cell can be described by the differential equation of motion for the mass-spring-dashpot combination shown in Fig. 8.20. The elastic member of the load cell is represented by the spring. The spring modulus or spring constant k is given by the expression

$$k = \frac{P}{\delta} = \frac{AE}{L} \tag{8.39}$$

where

P is the load
δ is the extension or contraction of the elastic member
A is the cross-sectional area of the elastic member
E is the modulus of elasticity
L is the length of the link

The dashpot represents the parameters in the transducer system (such as internal friction) that produce damping. In load and torque cells, damping is usually a very small quantity; however, in diaphragm pressure transducers, damping is larger because the fluid interacts with the diaphragm. The mass is a lumped mass consisting of the mass of the object to which the transducer is attached plus the effective mass of the elastic element of the transducer. The force $F(t)$ acts on the moving object and is the quantity being measured. The position parameter x describes the motion (displacement) of the mass m as a function of time. The dynamic response of this second-order system is described by the differential equation

$$\frac{1}{\omega_n^2} \frac{d^2 x}{dt^2} + \frac{2d}{\omega_n} \frac{dx}{dt} + x = \frac{F(t)}{k} \tag{8.40}$$

where ω_n is the natural frequency of the system ($\omega_n = \sqrt{k/m}$).

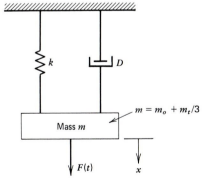

Figure 8.20 A spring-mass-dashpot representation of a link-type load cell.

The natural frequency ω_n of the system depends on the mass of the object and the effective mass of the elastic element. For the link-type load cell, the effective mass m is approximated as

$$m = m_o + \left(\frac{m_t}{3}\right)$$

where

m_o is the mass of the object
m_t is the mass of the elastic element in the transducer

Thus, both the mass of the elastic element and the mass of the object affect the fidelity of the measurement of $F(t)$.

8.7.1 Response of a Force Transducer to a Terminated Ramp Function

The fidelity of a force measurement depends primarily on the rise time associated with $F(t)$. In the treatment of galvanometers and other electronic recording instruments, $F(t)$ is usually considered a step function, since electrical signals can be applied almost instantaneously. In mechanical systems, however, it is not realistic to consider $F(t)$ a step function, since application of $F(t)$ requires some finite time even in the most severe dynamic application. For this reason, a more realistic forcing function for mechanical systems is the terminated ramp function (also called the ramp-hold function) shown in Fig. 8.21. This function can be expressed in equation form as

$$F(t) = \frac{F_0 t}{t_0} \qquad \text{for } 0 \le t \le t_0$$

$$F(t) = F_0 \qquad \text{for } t > t_0 \tag{8.41}$$

Since the degree of damping in the link-type load cell is very small, only the underdamped solution to Eq. 8.40 needs to be considered. After Eq. 8.41 is substituted into Eq. 8.40, the differential equation is solved to yield:

Homogeneous solution

$$x_h = e^{-d\omega_n t}\left(C_1 \sin \sqrt{1 - d^2}\, \omega_n t + C_2 \cos \sqrt{1 - d^2} \omega_n t\right) \tag{a}$$

Particular solution

$$x_p = \frac{F_0}{k t_0}\left(t - \frac{2d}{\omega_n}\right) \qquad \text{for } 0 \le t \le t_0$$

$$x_p = \frac{F_0}{k} \qquad \text{for } t > t_0 \tag{b}$$

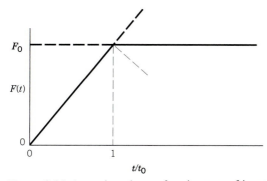

Figure 8.21 A terminated ramp function type of input to a load cell.

The coefficients C_1 and C_2 in Eq. a are obtained by using the initial conditions for the system. These conditions are

$$x = 0 \quad \text{and} \quad \frac{dx}{dt} = 0 \quad \text{at } t = 0 \tag{c}$$

The general solution for the ramp region ($x_r = x_p + x_h$) is obtained from Eqs. a, b, and c as

$$\frac{x_r(t)}{x_0} = \left(\frac{t}{t_0} - \frac{\tau}{t_0} \right) - \frac{e^{-at}}{bt_0} \sin (bt - \phi) \qquad \text{for } t < t_0 \tag{8.42}$$

where

$$\tau = \frac{2d}{\omega_n} \qquad x_0 = \frac{F_0}{k}$$

$$a = d\omega_n \qquad \phi = \tan^{-1} \left(\frac{2d\sqrt{1 - d^2}}{1 - 2d^2} \right) \tag{8.43}$$

$$b = \omega_n \sqrt{1 - d^2}$$

The hold portion of the input in Fig. 8.21 is composed of the initial ramp with a second ramp subtracted starting at $t = t_0$, as shown by the dashed lines. Thus

$$x_h(t) = x_r(t) - x_r(t - t_0) \tag{d}$$

Substituting Eq. 8.42 into Eq. d gives

$$\frac{x_h(t)}{x_0} = 1 - \frac{e^{-at}}{bt_0} \{ \sin (bt - \phi) - e^{at_0} \sin [b(t - t_0) - \phi] \}. \qquad \text{for } t > t_0 \tag{8.44}$$

Equations 8.42, 8.43, and 8.44 give the information needed to determine the error introduced by a load cell while tracking a terminated ramp function if the degree of damping d is known.

In most force transducers, the degree of damping d is very low (less than 0.02); therefore, $\phi \cong 0$ and Eqs. 8.42, 8.43, and 8.44 reduce to

$$\frac{x_r(t)}{x_0} = \frac{t}{t_0} - \frac{\sin \omega_n t}{\omega_n t_0} \qquad \text{for } t < t_0 \tag{8.45}$$

$$\frac{x_h(t)}{x_0} = 1 - \frac{1}{\omega_n t_0} [\sin \omega_n t - \sin \omega_n (t - t_0)] \qquad \text{for } t > t_0$$

$$= 1 + \frac{\sqrt{2(1 - \cos \omega_n t_0)}}{\omega_n t_0} \sin(\omega_n t + \theta) \tag{8.46}$$

where

$$\theta = \tan^{-1}\left(\frac{\sin \omega_n t_0}{1 - \cos \omega_n t_0}\right) \tag{8.47}$$

The first term in Eq. 8.45 represents the ramp, and the second term represents an oscillation about the ramp that has an amplitude of $1/\omega_n t_0$. This amplitude is the maximum deviation from the ramp, as illustrated in Fig. 8.22. The deviation is minimized by ensuring that $1/\omega_n t_0$ is small compared with the peak response of 1. The maximum error that can occur in the measurement after the peak response is given by Eq. 8.46 as

$$\mathcal{E} = \frac{2.0}{\omega_n t_0} \tag{8.48}$$

where $\omega_n t_0$ is an odd multiple of π. To limit the error to a specified amount $\mathcal{E}$, the natural frequency of the transducer must be selected such that

$$\omega_n \geq \frac{2.0}{\mathcal{E} t_0} \tag{8.49}$$

The error may be less than that given by Eq. 8.48 if $\omega_n t_0$ is near to an even multiple of π, such as 2π, 4π, and so on. The precise error at the peak response is obtained from Eq. 8.46 as

$$\mathcal{E} = \frac{\sqrt{2(1 - \cos \omega_n t_0)}}{\omega_n t_0} \tag{8.50}$$

Since ω_n and t_0 are not always known prior to the measurement, Eq. 8.48 is usually employed to judge the adequacy of a transducer for measuring the peak of a terminated ramp function with a rise time of t_0. Consider, for example, a load cell and mass system with a natural frequency of 5000 Hz intended to measure a dynamic load with a rise time of 1 ms. The maximum error in the peak response that can occur is given by Eq. 8.48 as 0.0636, or 6.4 percent. If this error is acceptable, the measurement can be made with confidence. However, if the error is too high, a transducer with a higher natural frequency must be utilized for the measurement.

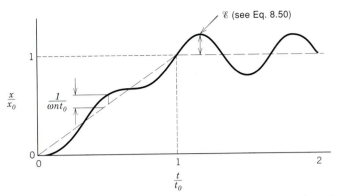

Figure 8.22 Response of a load cell with $d = 0$ to the terminated ramp function type of input.

8.7.2 Response of a Force Transducer to a Sinusoidal Forcing Function

The dynamic response of transducers to a periodic forcing function can be studied by letting $F(t) = F_0 \sin \omega t$ in Eq. 8.40. Thus,

$$\frac{1}{\omega_n^2} \frac{d^2 x}{dt^2} + \frac{2d}{\omega_n} \frac{dx}{dt} + x = \frac{F_0}{k} \sin \omega t \tag{8.51}$$

where ω is the circular frequency of the applied force.

If the damping coefficient d is small (as it is in most transducers), Eq. 8.51 reduces to

$$\frac{1}{\omega_n^2} \frac{d^2 x}{dt^2} + x = \frac{F_0}{k} \sin \omega t \tag{8.52}$$

Since the forcing function is periodic, the complementary solution to Eq. 8.52 has no significance and the particular solution gives the steady-state response of the transducer as

$$x_p = \frac{1}{1 - (\omega/\omega_n)^2} \frac{F_0}{k} \sin \omega t \tag{8.53}$$

Equation 8.53 can be expressed in terms of the periodic forcing function $F(t)$ as

$$k x_p = \frac{1}{1 - (\omega/\omega_n)^2} F_0 \sin \omega t = A \sin \omega t \tag{8.54}$$

where $A = F_0 / [1 - (\omega/\omega_n)^2]$ is an amplification factor that relates the steady-state response to the amplitude of the forcing function.

The error $\mathcal{E}$ associated with a measurement of F_0 is determined from Eq. 8.54 in terms of the frequency ratio as

$$\mathcal{E} = \frac{A - F_0}{F_0} = \frac{(\omega/\omega_n)^2}{1 - (\omega/\omega_n^2)} \tag{8.55}$$

Equation 8.55 indicates that substantial error can occur in a measurement of force, torque, or pressure unless the frequency ratio ω/ω_n is very small. For example, if $\omega/\omega_n = 0.142$, an error of 2 percent occurs. Similarly, if $\omega/\omega_n = 0.229$, the error is 5 percent. If the error is to be kept within reasonable limits, the natural frequency of the transducer system (including all attached mass) must be 5 to 10 times higher than the frequency of the forcing function.

The dynamic response of transducers to other common inputs, such as the impulse function and the ramp function, should also be studied. These cases are covered in the exercises at the end of the chapter.

8.8   CALIBRATION OF TRANSDUCERS

All transducers must be calibrated periodically to ensure that the sensitivity has not changed with time or misuse. Load cells and torque cells are usually calibrated using a testing machine with a scale certified to be accurate within specified limits. After the transducer is mounted in the testing machine, load is applied in increments that cover the range of the transducer. The output from the transducer is compared with the load indicated by the testing machine at each level of load, and differences (errors) are recorded. If the error is small, the calibration constant for the transducer is verified and the transducer can be used with confidence. If the error is excessive but consistent (i.e., the response is linear, but the slope is not correct), the calibration constant can be adjusted to correct the error. If the calibration constant requires correction, the calibration test should be repeated to ensure that the new calibration constant is reproducible and correct. In some cases the error is not consistent, and the output from the transducer is erratic. If the instrumentation system is checked and found to be operating properly, then the transducer is malfunctioning and cannot be calibrated. Such transducers should be removed from service immediately and returned to the manufacturer for repair.

A second method of calibrating load cells and torque cells is commonly employed in instances where a testing machine certified to the required limits of accuracy is not available. This method utilizes two transducers connected in series. One transducer is a standard that is used only for calibration purposes; the other is the working transducer. With this method, the calibration loads are given by the standard transducer instead of the testing machine.

For low-capacity load cells and torque cells, deadweight loads are frequently used in the calibration process. The standard weights (traceable to the National Institute of Standards and Technology or certified by weighing with a calibrated scale) are applied directly to the transducer to provide the known input. The transducer output is compared with the known input, as discussed previously. Although deadweight loading in calibration has many advantages and is usually preferred, it is not practical when the range or capacity of the load cell exceeds a few hundred pounds (about 1 kN). Also, deadweight calibration is dependent on the local acceleration resulting from gravity, a quantity that varies with location and altitude.

Pressure transducers are usually calibrated with a deadweight pressure source, such as the one illustrated schematically in Fig. 8.23. The calibration pressures are generated in the deadweight tester by adding standard weights to the piston tray. The calibration pressure is related to the weight by the expression

$$p = \frac{W}{A} \tag{8.56}$$

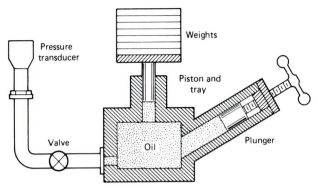

Figure 8.23 Schematic representation of a deadweight calibration system for pressure transducers.

where

    W is the total weight of the piston, tray, and standard weights

    A is the cross-sectional area of the piston

After the weights are placed on the piston tray, a screw-driven plunger is forced into the hydraulic oil chamber to reduce its volume and to lift the piston-weight assembly. The piston-weight assembly is then rotated to eliminate frictional forces between the piston and the cylinder. By adding weights incrementally to the piston tray, it is possible to generate 10 to 12 calibration pressures that cover the operating range of the transducer. Comparisons are made between the calculated calibration pressures and the pressures indicated by the transducer in order to certify the calibration constant. Since deadweight testers are relatively inexpensive to operate over a very wide range of pressures, they are preferred in calibrating pressure gages over methods that utilize a standard transducer. The cost associated with the purchase of a number of standard transducers to cover a wide range of pressures exceeds the cost of a deadweight tester to cover the same range.

Dynamic calibration of pressure transducers is usually accomplished with a shock tube. A shock tube is simply a closed section of smooth-walled tubing that is divided by a diaphragm into a short high-pressure chamber and a long low-pressure chamber. When the diaphragm is ruptured, a shock wave propagates into the low-pressure chamber, as illustrated schematically in Fig. 8.24. The pressure associated with the shock wave (the dynamic calibration pressure p_c), with air as the gas in the shock tube, is given by the expression

$$p_c = p_h \left(\frac{5}{6} - \frac{v}{c} \right) \tag{8.57}$$

where

    p_h is the static pressure in the high-pressure chamber

    v is the velocity of the shock wave in the low-pressure chamber

    c is the velocity of sound at the static pressure p_l in the low-pressure chamber

The velocity v of the shock wave is determined by placing a number of pressure transducers along the length of the low-pressure chamber and measuring the time of arrival of the shock front at the various locations. With a shock tube it is possible to

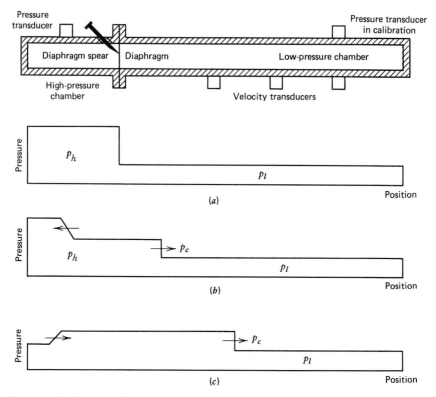

Figure 8.24 Schematic illustration of the use of a shock tube to generate dynamic pressure pulses for transducer calibration. (*a*) Pressure distribution with the diaphragm intact. (*b*) Pressure distribution before reflection of the rarefaction wave. (*c*) Pressure distribution after reflection of the rarefaction wave.

apply a sharp-fronted pressure pulse to a transducer so that its dynamic response can be characterized.

8.9   SUMMARY

A wide variety of transducers are commercially available for measuring load, torque, and pressure. These transducers incorporate an elastic member and a sensor to convert the deformation of the elastic member into an electrical signal. Accuracies of 0.1 to 0.2 percent are often specified.

Most load cells use strain gages as the sensor; however, for static measurements, where long-term stability is important, the linear variable differential transformer sensors are more suitable. For dynamic measurements, where a very high natural frequency is required, piezoelectric sensors are recommended. Load cells are covered in sufficient detail in Section 8.2 to give the reader adequate background to design special-purpose transducers and to thoroughly understand the sensitivities that can be achieved with many of the commonly employed elastic elements. It should be noted, however, that it is usually less expensive to buy a transducer than to build one, and the reader is encouraged to design and fabricate transducers only in those instances when it is not possible to purchase one with suitable characteristics.

Torque cells are similar to load cells. Strain gages are the most common sensors and a simple circular shaft is the most common elastic member. The most significant difference between load and torque measurements arises when torque measurements must be made on a rotating shaft. In these measurements, either slip rings or telemetry must be used to transmit the signal from the rotating member to the stationary instrumentation station. Slip rings are usually preferred if the end of the shaft is accessible for mounting of the slip-ring assembly. If the shaft ends are not accessible, telemetry is usually employed for signal transmission.

Pressure transducers are available in a wide variety of designs and capacities. The diaphragm-type pressure transducer, with electrical resistance strain gage sensors, is probably the most common type because of ease in manufacturing. Selection of a pressure transducer for a given application is usually made on the basis of stability and frequency response. For long-term stability, transducers with linear variable differential transformer sensors are usually preferred. For quasi-static and medium-frequency measurements, the diaphragm-type pressure transducer with strain-gage sensors has advantages. For extremely high frequency measurements, transducers with piezoelectric sensors are used.

Considerable care must be exercised in all dynamic measurements to ensure that the desired quantities are recorded with the required accuracy. The capability of a transducer to record a dynamic signal depends primarily on the natural frequency of the transducer since damping is very small. Because error increases in proportion to the frequency ratio ω/ω_n, piezoelectric sensors with natural frequencies of 100 kHz have significant advantages. Also, the form of the dynamic signal is important. The errors occurring in measuring a terminated ramp function and a periodic-input function are:

For the terminated ramp input

$$\mathscr{E} = \frac{2.0}{\omega_n t_0} \tag{8.48}$$

For the sinusoidal input

$$\mathscr{E} = \frac{(\omega/\omega_n)^2}{1 - (\omega/\omega_n)^2} \tag{8.55}$$

Periodic transducer calibration and complete system calibration must be performed to ensure continued satisfactory performance. Calibration must be regarded as the most important step in the measurement process.

REFERENCES

1. Beckwith, T. G., and R. D. Marangoni: *Mechanical Measurements*, 4th ed., Addison–Wesley, Reading, Mass., 1990.
2. Brindley, K.: *Sensors and Transducers*, Heinemann, London, 1988.
3. Dally, J. W., and W. F. Riley: *Experimental Stress Analysis*, 3rd ed., McGraw–Hill, New York, 1991.
4. Measurements Group, Inc.: "Design Considerations for Diaphragm Pressure Transducers," *Technical Note* 105, 1982.
5. Measurements Group, Inc.: "Strain Gage Based Transducers," 1988.

6. Neubert, H. K. P.: *Instrument Transducers: An Introduction to Their Performance and Design*, 2nd ed., Univ. Press (Clarendon Press), London/New York, 1975.

7. Norton, H. N.: *Handbook of Transducers*, Prentice–Hall, Englewood Cliffs, N.J., 1989.

8. Perry, C. C., and H. R. Lissner: *The Strain Gage Primer*, 2nd ed., McGraw–Hill, New York, 1962.

9. Tandeske, D.: *Pressure Sensors: Selection and Application*, Dekker, New York, 1991.

EXERCISES

8.1 Write an engineering brief describing a
   (a) force transducer
   (b) torque transducer
   (c) pressure transducer

8.2 Prepare a list of sensors used in fabricating the transducers listed in Exercise 8.1

8.3 Prepare a graph showing the sensitivity of a steel ($E = 29,000$ ksi and $\nu = 0.29$) link-type load cell–bridge combination. Use $S_g = 2$ and $v_s = 10$ V. Let the cross-sectional area A of the link vary from .02 in.^{2} to 20 in.^{2}

8.4 The sensitivity of the transducer of Exercise 8.3 can be increased if the input voltage v_s is increased. If each gage in the bridge can dissipate 0.5 W of power, determine the maximum sensitivity that can be achieved without endangering the strain-gage ($R_g = 350\ \Omega$) sensors.

8.5 Determine the voltage ratio v_o/v_s for the load cell of Exercise 8.3 if the fatigue strength S_f of the elastic member is 90,000 psi.

8.6 If the load cell of Exercise 8.3 is used in a static load application, what maximum load can be placed on the transducer? What voltage ratio v_o/v_s would result?

8.7 The calibration constant of a transducer procured from a commercial supplier is listed as 2 mV/V. Determine the sensitivity S of the transducer if $P_{max} = 100,000$ lb and $v_s = 10$ V.

8.8 Design a beam-type load cell with variable range and sensitivity. Use aluminum ($E = 10,000,000$ psi, $\nu = 0.33$, and $S_f = 20,000$ psi) as the beam material and four electrical resistance strain gages ($S_g = 2.00$ and $R_g = 350\ \Omega$) as the sensors. Design the load cell to give the following sensitivities and corresponding range:

$(v_o/v_s)^*$ (mV/V)	Range (lb)
2	1000
3	500
4	200

8.9 Design a ring-type load cell with an LVDT sensor. The load cell should have a capacity of 10,000 lb. The radius-to-thickness ratio of the ring R/t should be 10. Select an LVDT for this application from Table 5.1. Use steel ($E = 30,000,000$ psi and $\nu = 0.30$) for the ring. Determine the sensitivity S_t for your transducer.

8.10 For the transducer designed in Exercise 8.9, determine $(v_o/v_s)^*$ if the fatigue strength S_f of the steel is 85,000 psi.

8.11 Show that the torque cell shown in Fig. 8.6 is insensitive to both axial load P and moments M_x and M_y.

8.12 Determine the sensitivity of a torque cell if $E = 30,000,000$ psi, $\nu = 0.30$, $v_s = 8$ V, $D = 1$ in., $S_g = 2.00$, and $R_g = 350$ Ω.

8.13 The sensitivity of the torque cell described in Exercise 8.12 can be increased if the input voltage v_s is increased. If each gage in the bridge can dissipate 0.5 W of power, determine the maximum sensitivity that can be achieved without endangering the strain-gage sensors.

8.14 Determine the sensitivity of the torque cell of Exercise 8.13 if strain gages having $R_g = 1000$ Ω are used in place of the 350-Ω gages.

8.15 A torque cell with a capacity of 500 ft·lb is supplied with a calibration constant of $(v_o/v_s)^* = 4$ mV/V and a recommendation that the input voltage $v_s = 10$ V. If the cell is used with $v_s = 8$ V and a measurement of v_o yields 18 mV, determine the torque T.

8.16 Determine the sensitivity of the torque cell described in Exercise 8.15.

8.17 Why are at least four slip rings used to transmit the voltages associated with a torque cell on a rotating shaft?

8.18 Outline the advantages associated with the use of telemetry for data transmission from a rotating shaft.

8.19 A solid circular shaft having a diameter of 3 in. is rotating at 800 rpm and is transmitting 200 hp. Show how four strain gages can be used to convert the shaft itself into a torque cell. Determine the sensitivity of this shaft-torque transducer if the shaft is made of steel having $E = 30,000,000$ psi and $\nu = 0.30$.

8.20 Design a static pressure transducer having a 1.5-in. diameter diaphragm fabricated from steel with a fatigue strength of 75,000 psi. Select an LVDT from Table 5.1 to use as a sensor to convert the center point deflection of the diaphragm to an output voltage v_o. The capacity of the transducer is to be 2000 psi and linearity must be maintained within 0.3 percent. Determine the sensitivity of your transducer.

8.21 Repeat Exercise 8.20 by using a special-purpose four-element diaphragm strain gage as the sensor in place of the LVDT. Assume that for each element of the strain gage $p_g = 1$ W, $R_g = 350$ Ω, and $S_g = 2.00$.

8.22 Determine the natural frequency of the pressure transducer of Exercise 8.20.

8.23 A cylindrical elastic element for a very high pressure transducer is shown in Fig. E8.23. If the capacity of the pressure transducer is to be 60,000 psi and if the fatigue strength of the steel used in fabricating the cylindrical elastic element is 80,000 psi, determine the diameters D_i and D_o.

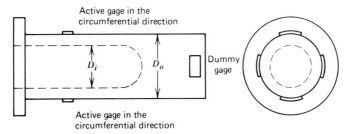

Figure E8.23

8.24 Determine the sensitivity of the pressure transducer described in Exercise 8.23 if the strain gages used as sensors have $p_g = 0.5$ W, $S_g = 2.00$, and $R_g = 350\ \Omega$.

8.25 Explain why dummy gages are mounted at the positions shown in E8.23.

8.26 The load cell depicted in Fig. 8.11 was shown to be insensitive to moments M_x and M_y. Show that it is also insensitive to torque M_z.

8.27 A load cell with a natural frequency $f_n = 10$ kHz is to be used to measure a terminated ramp function that exhibits a rise time of 1 ms. Determine the maximum error resulting from the response characteristics of the transducer.

8.28 Prepare a graph showing the maximum error of a load cell with a natural frequency of 20 kHz as a function of rise time if the loading function is a terminated ramp.

8.29 A transducer having a natural frequency ω_n will be used to monitor a sinusoidal forcing function having a frequency ω. Determine the error if ω/ω_n equals:

(a) 0.05   (c) 0.20
(b) 0.10   (d) 0.50

Plot a curve showing error as a function of frequency ratio ω/ω_n.

8.30 Derive the response equation for a transducer with $d = 0$ if the input function is a ramp that can be expressed as $F(t) = \dot{F}_0 t$.

8.31 Interpret the results of Exercise 8.30 and determine the magnitude of any error that may result.

8.32 Derive the response equation for a transducer with damping $(d \neq 0)$ if the input function is an impulse I_0 as shown in Fig. E8.32.

8.33 Use the results of Exercise 8.32 to show the response of the transducer if d is

(a) 0.01   (c) 0.20   (e) 0.70
(b) 0.10   (d) 0.50

8.34 Write an engineering brief outlining factors that produce errors in transducers.

8.35 Write a specification describing a calibration procedure for static application of

(a) pressure transducers   (c) torque cells
(b) load cells

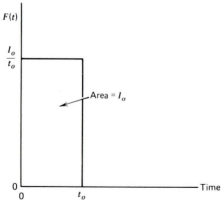

Figure E8.32

Chapter 9

Displacement, Velocity, and Acceleration Measurements

9.1 INTRODUCTION

Many methods have been developed to measure linear and angular displacements (s and θ), velocities (v and ω), and accelerations (a and α). Displacements and accelerations are usually measured directly, whereas velocities are sometimes obtained by integrating acceleration signals. The definitions of velocity ($v = ds/dt$ or $\omega = d\theta/dt$) and acceleration ($a = dv/dt = d^2s/dt^2$ or $\alpha = d\omega/dt = d^2\theta/dt^2$) suggest that any convenient quantity can be measured and the others can be obtained by integrating or differentiating the recorded signal. Since the differentiation process amplifies errors resulting from noise and other signal disturbances, it is rarely used to determine either v or a. The integration process reduces error and can be employed to determine v by integrating a or to determine s by integrating v. However, it is important to account for the initial conditions (the lower limit in the integration process) to avoid errors. Displacement measurements are most frequently made in manufacturing and process-control applications, and acceleration measurements are made in vibration, shock, or motion measurement situations.

Measurements of kinematic quantities, such as displacement, velocity, and acceleration, must be made with respect to a system of reference axes. The basic frame of reference in mechanics is the *primary inertial system,* which consists of an imaginary set of rectangular axes fixed in space. Measurements made with respect to this primary inertial system are absolute, and the laws of Newtonian mechanics are valid as long as velocities are small compared with the speed of light (300,000 km/s or 186,000 mi/s).

A reference frame attached to the surface of the earth exhibits motion in the primary inertial system; therefore, corrections to the basic equations of mechanics may be required when measurements are made relative to an earth-based reference frame.

For example, the absolute motion of the earth must be considered in calculations related to rocket-flight trajectories. However, for engineering measurements involving machines and structures that remain on the surface of the earth, corrections are extremely small and are usually neglected. Thus, measurements made relative to the earth on earthbound engineering applications are considered absolute.

Measurements of quantities that describe motion (displacement, velocity, and acceleration) are made using two significantly different approaches. One approach incorporates a fixed reference plane to which the base of the transducer is attached. The other approach involves seismic transducers and is employed when the use of a fixed reference plane is not possible. We will describe the principles of seismic transducers in the next section and cover motion measurements relative to a fixed plane in Section 9.12.

9.2   THE SEISMIC TRANSDUCER MODEL

The mechanical behavior of seismic transducers is characterized by considering the single-degree-of-freedom dynamic model shown in Fig. 9.1a. It is assumed that $\dot{y} > \dot{x}$ and $y > x$ in constructing the free-body diagram of the seismic mass shown in Fig. 9.1b. The equation of motion of the seismic mass, obtained from Newton's second law, is

$$m\ddot{y} + C(\dot{y} - \dot{x}) + k(y - x) = F(t) \tag{9.1}$$

where

m is the seismic mass
k is the spring constant
C is the viscous damping constant
x, $\dot{x}$, and $\ddot{x}$ are the displacement, velocity, and acceleration of the base plane
y, $\dot{y}$, and $\ddot{y}$ are the displacement, velocity, and acceleration of the seismic mass
$F(t)$ is a time-dependent external forcing function that results from either a force or a pressure [$F(t) = Ap(t)$]

The force between the transducer base and the support structure is F_b. The transducer is not affected by this attachment force; however, the structure is affected.

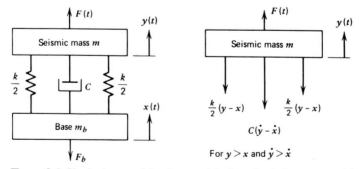

Figure 9.1 Single-degree-of-freedom model of a seismic instrument. (a) Schematic of a seismic transducer. (b) Free-body diagram of the seismic mass.

To design a transducer using the seismic model, a sensor is placed between the base and the seismic mass and the relative displacement z is measured. For the coordinates shown in Fig. 9.1a, the relative displacement and its derivatives are

$$z = y - x$$
$$\dot{z} = \dot{y} - \dot{x}$$
$$\ddot{z} = \ddot{y} - \ddot{x} \qquad \textbf{(9.2)}$$

Substituting these equations into Eq. 9.1 gives the differential equation of motion

$$m\ddot{z} + C\dot{z} + kz = F(t) - m\ddot{x} = R(t) \qquad \textbf{(9.3)}$$

where $R(t) = F(t) - m\ddot{x}$ represents the transducer excitation owing to the external force and the inertia force produced by base acceleration. Equation 9.3 shows that the relative motion z depends on the excitation, the damping C, and the frequency of the excitation $R(t)$ relative to the natural frequency ($\omega_n = \sqrt{k/m}$) of the transducer.

The seismic transducer model, illustrated in Fig. 9.1, is employed as a seismic reference frame to support accelerometers and dynamic force transducers. In the design of motion or force transducers, a sensor is placed between the base and the seismic mass to measure the relative displacement z or its derivatives $\dot{z}$ and $\ddot{z}$. Since z, $\dot{z}$ and $\ddot{z}$ are related to $R(t)$ by Eq. 9.3, the unknown quantity $\ddot{x}$ or F(t) can be determined.

9.3   DYNAMIC RESPONSE OF THE SEISMIC MODEL

The dynamic response of the seismic model is given by Eq. 9.3. A general solution of this second-order differential equation is important in order to be able to characterize the response of specific transducers to different transient excitations.

Before beginning the solution of Eq. 9.3, note that the transducer excitation $R(t)$ is

$$R(t) = F(t) - m\ddot{x} \qquad \textbf{(9.4)}$$

When the seismic frame is used in the design of motion transducers to measure x, $\dot{x}$, or $\ddot{x}$, the applied force $F(t) = 0$ and $R(t) = -m\ddot{x}$. When the seismic frame is used in the design of a dynamic force transducer and the base of the transducer is fixed to a reference plane, then $\ddot{x} = 0$ and $R(t) = F(t)$. When the seismic frame of a force transducer is attached to an accelerating base, however, the output is due to both $F(t)$ and $\ddot{x}$, as indicated in Eq. 9.4. Under these conditions, significant force-measurement errors can occur when $F(t)$ is small and $\ddot{x}$ is large. This condition occurs when the structure under test is in a resonating condition. Also, in some situations the gravity force mg causes acceleration-measurement errors (see Exercise 9.4).

9.3.1 Sinusoidal Excitation

The phasor method of analysis described in Section 2.6 is employed to obtain the steady-state response. The excitation has magnitude R_0 and frequency ω and is written as

$$R(t) = R_0 e^{j\omega t} \qquad \textbf{(a)}$$

The corresponding relative motion and its derivatives are written as

$$z = z_0 e^{j\omega t} \qquad \dot{z} = j\omega z_0 e^{j\omega t} \qquad \ddot{z} = -\omega^2 z_0 e^{j\omega t} \qquad \textbf{(b)}$$

where z_0 is the complex amplitude (magnitude and phase) of the response phasor z_0 relative to the excitation phasor R_0. Substituting Eqs. a and b into Eq. 9.3 yields

$$(k - m\omega^2 + jC\omega)z_0 e^{j\omega t} = R_0 e^{j\omega t} \qquad \textbf{(9.5)}$$

The frequency response function $H(\omega)$ of the output phasor z_0 relative to the input excitation phasor R_0 is

$$H(\omega) = \frac{z_0}{R_0} = \frac{1}{k - m\omega^2 + jC\omega} = \frac{1}{[(k - m\omega^2)^2 + (C\omega)^2]^{1/2}} e^{-j\phi} \qquad \textbf{(9.6)}$$

Both magnitude and phase information are given by Eq. 9.6. The magnitude $|H(\omega)|$ is

$$|H(\omega)| = \frac{1}{k[(1 - r^2)^2 + (2rd)^2]^{1/2}} \qquad \textbf{(9.7)}$$

and the phase angle ϕ is

$$\phi = \tan^{-1} \frac{C\omega}{k - m\omega^2} = \tan^{-1} \frac{2rd}{1 - r^2} \qquad \textbf{(9.8)}$$

where

$d = C/(2\sqrt{km})$ is a dimensionless damping ratio
$r = \omega/\omega_n = f/f_n$ is a dimensionless frequency ratio

The particular solution of Eq. 9.3 for sinusoidal excitation is

$$z = \frac{R_0}{[(k - m\omega^2)^2 + (C\omega)^2]^{1/2}} e^{j(\omega t - \phi)}$$

$$= \frac{R_0}{k[(1 - r^2)^2 + (2rd)^2]^{1/2}} e^{j(\omega t - \phi)}$$

$$= H(\omega)R_0 e^{j\omega t} \qquad \textbf{(9.9)}$$

The angle between the rotating phasors R_0 and z_0 is illustrated in Fig. 9.2. Phase angles associated with the response phasor z_0 are positive when counterclockwise and negative when clockwise with respect to the excitation phasor R_0.

The frequency response function $|H(\omega)|$ for mechanical transducers, Eq. 9.7, is identical to that obtained for galvanometers (Eq. 3.44). The magnitude $|H(\omega)|$ and phase ϕ for the seismic model are shown in Fig. 9.3 as a function of the dimensionless frequency ratio r for various damping ratios d. The magnitude and phase vary considerably with input frequency. Chapter 3 noted the need to use approximately

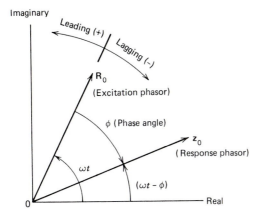

Figure 9.2 Rotating phasor representation of steady-state instrument response.

64 percent damping for galvanometers because this amount of damping provides a nearly linear phase shift. A linear phase shift is essential to avoid signal distortion for frequency components below the galvanometer natural frequency. The 64 percent damping is satisfactory for use in either transducers or voltage recorders since signals of different frequencies are shifted the same amount in time. Unfortunately, transducers exhibit very low damping ratios (0.1 to about 4 percent) and the nearly linear phase shift with excitation frequency cannot be achieved. To avoid signal distortion owing to nonlinear phase shift, the frequency range of seismic transducers is limited to 0.2 f_n. The phase shift is nearly zero in this low-frequency range for such light damping. Also, an examination of Fig. 9.3 shows the dynamic response to be the same as that for static loads when $(f/f_n) < 0.2$, because $|H(\omega)| \cong 1$ and $kz \cong R_0$.

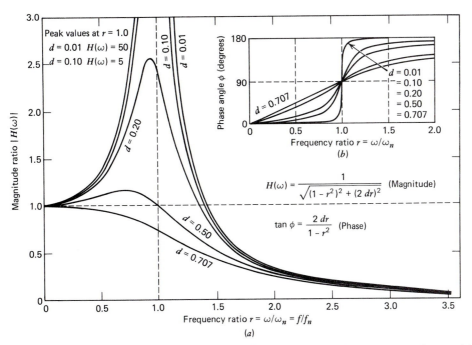

Figure 9.3 Frequency response function for acceleration, force, and pressure transducers. (*a*) Magnitude. (*b*) Phase.

9.3.2 Transient Excitations

A basic concern when seismic transducers are used to make a measurement is the response of the mechanical sensing structure to transient events. A *transient input* involves rapid changes that occur over a short period of time. The rapid change is preceded and followed by essentially a constant input for long periods of time compared with system response times. Step, ramp-hold, half-sine, and triangular inputs are classic transient signals used to develop transducer response characteristics.

A. Step Excitation

The characteristic response of a second-order system to a step input was discussed in Chapter 3 for three different cases of damping: (1) overdamped, (2) critically damped, and (3) underdamped. The corresponding solutions for these three cases are given by Eqs. 3.36, 3.37, and 3.38. Figure 3.15 shows the typical response for each case. For lightly damped transducers ($d < 2$ percent), Eq. 3.38 indicates that the initial cycle of response is approximated by $[1 - \cos(\omega_n t)]$. A graph of this response, shown in Fig. 9.4, demonstrates that it bears little resemblance to the step input. This example is indicative of the type of measurement error that occurs when the transducer is forced to "ring" at its natural frequency. The step input produces changes that are much too rapid for a mechanical system to respond to.

B. Ramp-Hold Excitation

The terminated ramp-hold function described in Section 8.7 applies to seismic transducers as well, since the governing differential equations are identical in form. Thus, the deviation of $1/\omega_n t_0$ for the ramp portion and the maximum error of $2/\omega_n t_0$ for the hold portion apply. Generally, $\omega_n t_0 > 10\pi$ is required in order to obtain satisfactory measurements.

C. Half-Sine and Triangle Pulses

Two common transients for impact loading are the half-sine and triangle pulses shown in Fig. 9.5. The figure shows the response of the transducer for pulse durations of T_n and $5T_n$, where T_n is the natural period of the transducer. It is evident from these graphs that a reasonable response from the transducer requires a pulse duration that

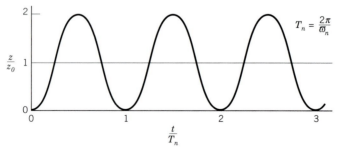

Figure 9.4 Transducer ringing at its natural frequency when subjected to a step excitation.

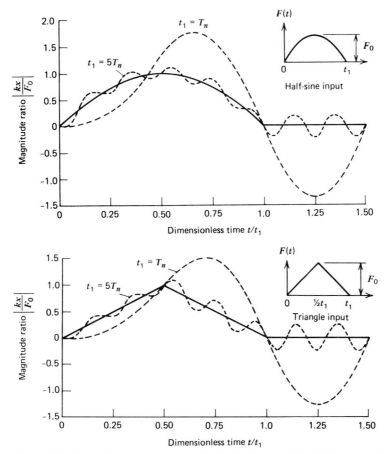

Figure 9.5 Mechanical response of a seismic transducer to (*a*) a half-sine input and (*b*) a triangular input.

is at least five times the natural period of the transducer and preferably larger. For pulse durations equal to the natural period of the transducer T_n, the output is severely distorted, as shown in Fig. 9.5. The conclusion drawn from this transient analysis is that either the transient rise time (or fall time) or the duration time must be greater than five times the period of the transducer. Transducer ringing at its natural frequency, which severely distorts the input signal, is a clear indication that this rule is being violated.

9.4 SEISMIC MOTION TRANSDUCERS

The theoretical basis for seismic motion instruments is the seismic mechanical model, which leads to the equation of motion given by Eq. 9.3. The external force $F(t)$, which acts on the seismic mass m, is zero in motion-measurement applications (see Exercise 9.4 for a special case) because the seismic mass is isolated in a protective case. The transducer is excited by an inertia force $m\ddot{x}$ applied through base motion. In this treatment, the base motion is considered to be sinusoidal so that

$$x = x_0 e^{j\omega t} \qquad \dot{x} = j\omega x_0 e^{j\omega t} \qquad \ddot{x} = -\omega^2 x_0 e^{j\omega t} \tag{9.10}$$

Thus, the excitation force $R(t)$ in Eq. 9.3 becomes

$$R(t) = R_0 e^{j\omega t} = m\omega^2 x_0^{j\omega t} \tag{9.11}$$

These equations are essential in describing seismic displacement, velocity, and acceleration measuring instruments.

9.4.1 Seismic Displacement Transducers

The characteristic behavior of a seismic displacement transducer is obtained by substituting Eq. 9.11 into the steady-state solution of Eq. 9.3 as given by Eq. 9.9. The result for the relative displacement z is

$$z = \frac{r^2 x_0}{[(1 - r^2)^2 + (2rd)^2]^{1/2}} e^{j(\omega t - \phi)} = H(\omega) x_0 e^{j\omega t} \tag{9.12}$$

The frequency response function $H(\omega)$ is shown in Fig. 9.6 in the form of a magnitude ratio z/x_0 and a phase angle ϕ versus frequency ratio $r = \omega/\omega_n$ for specific amounts of damping. It is evident from these curves that ratio z/x_0 approaches unity and phase angle ϕ approaches 180 deg for frequency ratios greater than 4, irrespective of the amount of damping. The peak of the response curve decreases and the peaks occur at higher frequency ratios as the damping ratio is increased from 0 to 0.707.

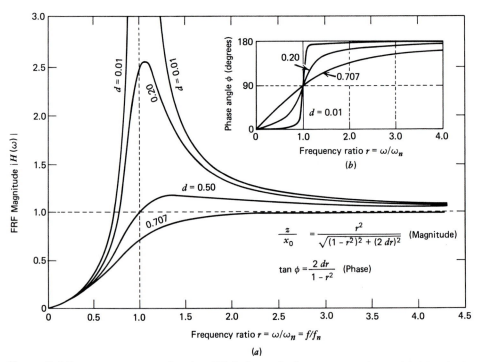

Figure 9.6 Frequency response function $H(\omega)$ for a displacement transducer. (*a*) Magnitude. (*b*) Phase.

The results of Fig. 9.6 indicate that seismic displacement transducers must have a very low natural frequency so that the frequency ratio ω/ω_n is large in most situations. With the constraints $r \gg 1$ and $\phi \cong \pi$, Eq. 9.12 reduces to

$$z = \frac{r^2 x_0}{[(-r^2)^2]^{1/2}} e^{j(\omega t - \pi)} \cong -x_0 e^{j\omega t} = -x \tag{9.13}$$

Equation 9.13 indicates that the seismic mass does not move; therefore, the transducer output response z is equal in magnitude to the input base motion x. The amount of damping used is not critical, since the response curves for both $|H(\omega)|$ and ϕ converge for $r > 4$.

Seismic displacement transducers are designed with soft springs and a relatively large seismic mass to give an instrument with a very low natural frequency. As a result, the common sensors used to measure the relative displacement z are linear variable differential transformers (LVDTs) or electrical resistance strain gages mounted on elastic supports that also serve as the soft springs.

9.4.2 Seismic Velocity Transducers

A seismic displacement transducer becomes a seismic velocity transducer when a magnetic velocity sensor is used to measure the relative velocity $\dot{z}$ between the seismic mass and the instrument base. The governing equation that describes this type of instrument is obtained by differentiating Eq. 9.13 with respect to time. Thus,

$$\dot{z} = \frac{r^2(j\omega x_0)}{[(1 - r^2)^2 + (2rd)^2]^{1/2}} e^{j(\omega t - \phi)} = H(\omega)(j\omega x_0)e^{j\omega t} \tag{9.14}$$

The frequency response function $H(\omega)$ is the same as that of a displacement transducer. Thus, for the conditions that $r \gg 1.0$ and $\phi \cong \pi$, Eq. 9.14 reduces to

$$\dot{z} \cong -j\omega x_0 e^{j\omega t} = -\dot{x} \tag{9.15}$$

Equation 9.15 indicates that transducer output motion $\dot{z}$ is the same as base input velocity $\dot{x}$ when the excitation frequency is much higher than the natural frequency of the transducer.

9.4.3 Seismic Acceleration Transducers

Linear seismic accelerometers have been used for many years; however, angular seismic accelerometers are a recent development. Both types are described in this subsection.

A. Linear Seismic Accelerometers

In seismic displacement and velocity transducers, soft springs are used to achieve a low natural frequency so that the instruments can utilize conventional displacement and velocity sensors. For accelerometers, a stiff spring and a high natural frequency are required, since the inertia force $m\ddot{x}$ is to be measured. Equation 9.12 gives the basic response for sinusoidal input motion and Eq. 9.10 indicates that $\ddot{x} = -\omega^2 x_0 e^{j\omega t}$. Thus,

$$z = -\frac{e^{-j\phi}}{\omega_n^2[(1 - r^2)^2 + (2rd)^2]^{1/2}}\ddot{x} = -H(\omega)\frac{1}{\omega_n^2}\ddot{x} \qquad (9.16)$$

The frequency response function $H(\omega)$ in Eq. 9.16 is the same as that given in Eq. 9.6 and illustrated in Fig. 9.3. It is evident from these curves that the magnitude of $H(\omega)$ approaches unity and the phase angle ϕ approaches zero when r is very small, irrespective of the amount of damping. Under the condition that $r \ll 1$, Eq. 9.16 reduces to

$$z \cong -\frac{1}{\omega_n^2}\ddot{x} = -\frac{m\ddot{x}}{k} \qquad (9.17)$$

Equation 9.17 shows that in a seismic acceleration transducer the inertia force $m\ddot{x}$ is resisted by the spring force kz. This type of instrument requires a very stiff spring and a very small seismic mass so that it will have a very high natural frequency. The piezoelectric sensor is ideal for this application, because it serves as a very stiff spring while providing an electrical output with a high sensitivity. Consequently, seismic transducers are designed to be small and light, so that the presence of the transducer has a nearly insignificant effect on most vibrating structures.

Two common accelerometer designs, where the piezoelectric crystal is subjected to either compression or shear, are shown in Fig. 9.7. The single-ended compression type consists of a base, a threaded center post, a flat, washer-shaped piezoelectric element, a seismic mass, and a nut. The piezoelectric element is prestressed by torquing the nut onto the center post. This unit has two spring elements, center post k_1 and piezoelectric element k_2, which act in series. The seismic mass is protected from the environment by the cover. This design is somewhat susceptible to base bending.

The shear design consists of a base with a center post, a cylindrical piezoelectric element, and a seismic mass. The piezoelectric element is bonded to both the center post and the seismic mass. The entire assembly is sealed in a protective case. This design is less sensitive to base bending than is the single-ended compression type. If the center post is hollow and extends through the case at the top and bottom, the transducer can be mounted by using a bolt that passes through the hole. This design

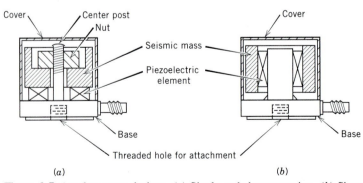

Figure 9.7 Accelerometer designs. (*a*) Single-ended compression. (*b*) Shear.

Table 9.1 Piezoelectric Accelerometer Characteristics

	Manufacturer			
	Endevco	Endevco	PCB	Kistler
Model	2222C	2292	302A	808A
Crystal material	Piezite P-8	Piezite P-8	Quartz	Quartz
Charge sensitivity (pC/g)	1.4	0.14	—	1.0
Voltage sensitivity (mV/g)	1.8	0.4	10	—
Capacitance (pF)	470	80	—	90
Internal time constant (s)	—	—	0.5	—
Frequency response (Hz)	20 to 8000	50 to 20,000	1 to 5000	0 to 7000
Mounted resonance frequency (Hz)	40,000	125,000	45,000	40,000
Transverse sensitivity	$\leq \pm 5\%$	$\leq \pm 5\%$	$\leq \pm 5\%$	$\leq \pm 5\%$
Range (g)	0 to 1000	0 to 20,000	0 to 500	0 to 10,000
Size (in.)	0.25 dia × 0.135	0.31 Hex × 0.3	$\frac{1}{2}$ Hex × 1.25	$\frac{1}{2}$ Hex × 0.90
Weight (grams)	0.5	1.3	23	20
Vibration (maximum g)	1000	1000	2000	—
Shock (maximum g)	10,000	20,000	5000	—
Temperature range (°F)	−100 to +350	−65 to +250	−100 to +250	−195 to +260
Application	General purpose	Shock	General purpose	General purpose

is used in many smaller transducers. Table 9.1 contains typical specifications for four common piezoelectric accelerometers.

B. Angular Seismic Accelerometers

It is difficult to design angular seismic accelerometers with both very high sensitivity for the desired angular acceleration and very low sensitivity for linear acceleration. However, recent developments in micromachining and fabrication of synthetic piezoelectric materials have resulted in a new transducer that allows both linear and angular acceleration to be measured simultaneously.

The design features of this accelerometer, shown in Fig. 9.8a, include a base, a center post, and two cantilever beams fabricated from a piezoelectric ceramic. The beams, labeled 1 and 2, are supported by the center post, and the entire assembly is enclosed in a case. The x and y axes describe the orientation of the piezobeams relative to the transducer base. Linear motion is sensed in the y direction and angular motion by α_z. The symbols $A, B, C,$ and D indicate locations of the lead-wire connections. The entire surface on the top and bottom of each piezobeam is an electrode.

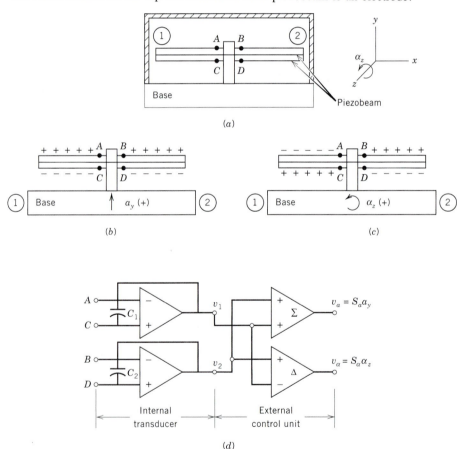

Figure 9.8 Schematic of a combined linear–angular accelerometer. (*a*) Dual piezobeam construction with connection points A, B, C, and D. (*b*) Charge generated for positive linear acceleration. (*c*) Charge generated for positive angular acceleration about the z axis. (*d*) Circuit used to extract linear and angular acceleration from the charge generated.

The charge condition for a positive linear acceleration is illustrated in Fig. 9.8b. With proper polarity in the piezobeam, the entire top surface of each beam becomes positively charged while the entire bottom surface is negatively charged because of the acceleration-induced stresses in the beams. Similarly, when the beam accelerates downward, the polarity of the electrical charge is reversed.

The charge condition for a positive angular acceleration is shown in Fig. 9.8c. With this inertia loading, the top of piezobeam 1 is negative while the bottom is positive and the top of piezobeam 2 is positive while the bottom is negative.

The electrical connections from points A, B, C, and D are used as input for the dual voltage follower–amplifier circuit shown in Fig. 9.8d. The output voltages from the voltage followers are

$$v_1 = \frac{q_{A/C}}{C_1} = \frac{k_1 a_y - k_2 \alpha_z}{C_1} = S_1 a_y - S_2 \alpha_z \tag{a}$$

where S_1 and S_2 are the voltage sensitivities for beam 1, and

$$v_2 = \frac{q_{B/D}}{C_2} = \frac{k_3 a_y + k_4 \alpha_z}{C_2} = S_3 a_y + S_4 \alpha_z \tag{b}$$

where S_3 and S_4 are the voltage sensitivities for beam 2. If voltages v_1 and v_2 are added and subtracted with sum and difference amplifiers, as shown in Fig. 9.8d, the result is

$$\begin{aligned}
v_a &= v_1 + v_2 = (S_1 + S_3)a_y + (S_4 - S_2)\alpha_z \cong S_a a_y \\
v_\alpha &= v_2 - v_1 = (S_3 - S_1)a_y + (S_2 + S_4)\alpha_z \cong S_\alpha \alpha_z
\end{aligned} \tag{9.18}$$

These equations indicate that v_a is proportional to the linear acceleration a_y if $S_4 = S_2$, and v_α is proportional to the angular acceleration α_z if $S_3 = S_1$. In theory, the two piezobeams can be made identical; however, it is impractical to adjust them mechanically. Thus, the power unit provided with the transducer contains carefully adjusted sensitivities to cancel the unwanted signals. Consequently, this accelerometer can only be used with its uniquely adjusted power unit. The voltage followers are located in the transducer, and the summing and difference amplifiers with gain adjustments are located in the power unit. A four-wire connection cable to provide two grounds and two positive supply voltages is required.

The piezobeam design exhibits a sensitivity of about 1000 mV/g (± 10-g range) for linear acceleration and either 0.5, 5, or 50 mV/ (rads/s^2) for angular acceleration. The usable frequency range for ± 5 percent magnitude error is 0.5 to 2000 Hz for an 8-kHz natural-frequency transducer. The major advantage of this transducer is that both linear and angular motion can be measured simultaneously. When three accelerometers are placed in a triaxial mounting, all six degrees of motion at a single location can be measured simultaneously. The major disadvantages are its limited range and the requirement that the power unit and transducer must be used together as a single unit.

9.5 PIEZOELECTRIC FORCE TRANSDUCERS

The seismic frame can also be used to support a piezoelectric crystal that senses the applied force $F(t)$, as shown in Fig. 9.1. If the base is attached to a fixed point, $m\ddot{x} = 0$ and Eq. 9.4 shows that $R(t) = F(t)$. The analysis for the ramp-

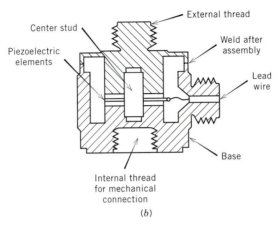

Figure 9.9 Cross section showing construction details of a piezoelectric force transducer.

hold excitation presented in Section 8.7 clearly shows the importance of having an extremely high natural frequency for a force transducer.

The cross section of a typical piezoelectric load cell is shown in Fig. 9.9. The seismic mass and base are separated by a two-piece piezoelectric sensor. The two-piece sensor permits the outer case to be the circuit ground while the interface at the center provides the circuit signal. The sensor is preloaded by torquing the seismic mass onto the central stud. Then, the outer case and seismic mass are welded together. The advantages of a piezoelectric load cell are its high output signal, high natural frequency, and large stiffness for its size. Unfortunately, the transducer's high natural frequency is degraded when the load cell is attached to another mass, because its effective seismic mass is so small. Typical characteristics of piezoelectric load cells are given in Table 9.2. See Section 10.10 for additional information.

9.6 PIEZOELECTRIC SENSOR CIRCUITS

Piezoelectric sensors are often employed in accelerometers and in force and pressure transducers to obtain very high natural frequencies, very low weight, and excellent sensitivity. Piezoelectric sensing elements are charge-generating devices that exhibit a high output impedance; therefore, special signal conditioning instruments, such as voltage followers and charge amplifiers, must be incorporated into the measurement circuit to convert the charge to an output voltage that can be measured and recorded. The operational and performance characteristics of these circuits are discussed in detail in this section.

9.6.1 Charge Sensitivity Model

The charge generated by a piezoelectric crystal was discussed in Section 5.7. Equation 5.14 indicates that the charge generated is directly proportional to the electrode area A and pressure p exerted on the crystal. This equation is recast for a given transducer to show that the charge generated is related to the measured quantity by a linear relationship of the form

$$q = S_q^* a \qquad \qquad (9.19)$$

Table 9.2 Piezoelectric Force Transducer Characteristics

	Manufacturer			
	Endevco	PCB	PCB	Kistler
Model	2103-500	208A03	200A05	912
Crystal material	Piezite P-1	Quartz	Quartz	Quartz
Charge sensitivity (pC/lb)	45	—	—	50
Voltage sensitivity (mV/lb)	150	10	1.0	—
Capacitance (pF)	200	—	—	58
Internal time constant (s)	—	2000	2000	—
Frequency response (Hz)	2 to 4000	0.1 to 14,000	0.1 to 14,000	To 14,000
Mounted resonant frequency (Hz)[a]	20,000	70,000	70,000	70,000
Range (lb)	0 to 500	±500	0 to −5000	−5000 to +100
Amplitude linearity	±3%	±1%	±1%	±1%
Size (in.)	$\frac{3}{4}$ Hex × 1.25	$\frac{5}{8}$ Hex × 0.625	0.65 Dia × 0.36	$\frac{5}{8}$ Hex × 0.500
Weight (grams)	57	25	14	17
Stiffness (lb/in.)	6×10^6	10×10^6	10×10^6	5×10^6
Vibration (maximum g)	—	2000	2000	—
Shock (maximum g)	—	10,000	10,000	10,000
Temperature range (°F)	−30 to +200	−100 to +250	−100 to +250	−240 to +150
Application	Axial force link	General purpose	Impact compression	Impact

[a]With no external mass attached.

where

q is the charge generated by the piezoelectric sensor (pC)

S_q^* is the charge sensitivity of the transducer (pC/g, pC/1b, pC/N, pC/psi, or pC/Pa)

a is the quantity being measured (acceleration, force, or pressure in g, lb, N, psi, or Pa)

This linear charge model is used in developing circuit sensitivities in all subsequent equations.

9.6.2 Voltage-Follower Circuit

A schematic diagram of a circuit containing a piezoelectric sensor and a voltage follower (unity-gain buffer amplifier) is shown in Fig. 9.10a. The piezoelectric sensor is denoted by the charge-generator symbol. The circuit has capacitances C_t of the transducer, C_c of the cable, C_s of the standardizing section in the amplifier end of the circuit, and C_b of the blocking capacitor used to protect the amplifier. The resistance R represents the combined effect of amplifier input impedance and load resistance in parallel with the amplifier. The equivalent circuit, shown in Fig. 9.10b, is obtained by combining the transducer, cable, and standardizing capacitances into a single capacitance C with the parallel capacitor law:

$$C = C_t + C_c + C_s \tag{a}$$

The differential equation that describes the behavior of the circuit in Fig. 9.10b is obtained as follows:

$$i = i_1 + i_2 = \dot{q} = S_q^* \dot{a} \tag{b}$$

where

$$i_1 = C\dot{v}_1 \quad \text{and} \quad i_2 = C_b(\dot{v}_1 - \dot{v}) = \frac{v_o}{R} \tag{c}$$

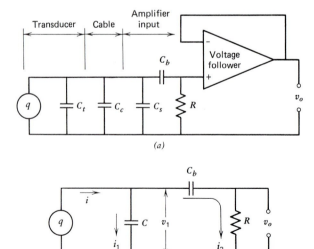

(a)

(b)

Figure 9.10 (a) Measurement circuit with a piezoelectric sensor and a voltage follower. (b) Equivalent circuit for analysis.

Substituting Eqs. c into Eq. b and simplifying yields

$$\dot{v}_o + \frac{v_o}{RC_{eq}} = \frac{S_q^*}{C}\dot{a} \qquad (9.20)$$

where

$$\frac{1}{C_{eq}} = \frac{1}{C} + \frac{1}{C_b} \qquad (d)$$

Typical values of C range from 300 pF to 10,000 pF, and values of C_b are usually on the order of 100,000 pF. For practical purposes, the equivalent capacitance C_{eq} of Eq. d is essentially the same as the combined source capacitance C given by Eq. a.

The frequency response function for this circuit is obtained by using $a = a_0 e^{j\omega t}$ for the quantity being measured and $v = v_o e^{j\omega t}$ for the response voltage. Substituting these equations into Eq. 9.20 gives the output voltage phasor as

$$v_o = \left(\frac{jRC_{eq}\omega}{1 + jRC_{eq}\omega}\right)\left(\frac{S_q^* a_0}{C}\right) \qquad (9.21)$$

The basic voltage sensitivity of the instrument is obtained from Eq. 9.21 as

$$S_v = \frac{v_o}{a_0} = \frac{S_q^*}{C} \qquad (9.22)$$

Equation 9.22 shows that voltage sensitivity of the measurement system of Fig. 9.10a depends on all source capacitances that contribute to C. The specific cable used during the measurement, environmental factors such as temperature and humidity, and dirt (including grease or sprays), which plays no role in a clean laboratory but may be significant in an industrial setting, can produce a change in the circuit voltage sensitivity.

The frequency response function H(ω) for the circuit in Fig 9.10a is obtained from the first term on the right-hand side of Eq. 9.21. Thus

$$H(\omega) = \frac{Cv_o}{S_q a_0} = \frac{jRC_{eq}\omega}{1 + jRC_{eq}\omega} = \frac{\omega\tau}{[1 + (\omega\tau)^2]^{1/2}}e^{j\phi} \qquad (9.23)$$

$$\tau = RC_{eq}$$

$$\phi = \frac{\pi}{2} - \tan^{-1}(\omega\tau)$$

where

  τ is the circuit time constant (s)
  ϕ is the phase angle of the output phasor relative to the input phasor (rad)

The low-frequency characteristics of this circuit are given in Fig. 9.11a, where $|H(\omega)|$ is shown as a function of $\omega\tau$. The nonlinearity of $|H(\omega)|$ with $\omega\tau$ is evident in Fig. 9.11a. The response curve of Fig. 9.11a is linearized by expressing $|H(\omega)|$ in

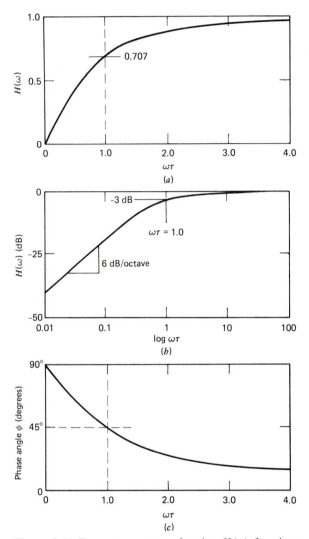

Figure 9.11 Frequency response function $H(\omega)$ for piezoelectric sensors at low frequencies. (*a*) Magnitude of $H(\omega)$ versus $\omega\tau$. (*b*) Magnitude of $H(\omega)$ in decibels versus $\log \omega\tau$. (*c*) Phase angle ϕ versus $\omega\tau$.

decibels and by plotting it as a function of $\log_{\omega\tau}$, as shown in Fig. 9.11*b*. This form of representation of the frequency response function is known as a *Bode diagram*. The phase angle, given as a function of $\omega\tau$, is shown in Fig. 9.11*c*. Since the attenuation is 30 percent and the phase shift is 45° when $\omega\tau = 1$, it is clear that a minimum value must be established for $\omega\tau$ that is much higher than unity. Signal attenuation and phase angle as a function of $\omega\tau$ are presented in Table 9.3.

It is evident from the values listed in Table 9.3 that $\omega\tau$ should be greater than 2π to limit the attenuation to 1 percent. This limit is equivalent to requiring that $f\tau > 1$ where f is the frequency of the input signal in Hz.

Table 9.3 Attenuation and Phase Angle as a function of $\omega\tau$

Percent Attenuation	$\omega\tau$	Phase Angle (deg)
1	7.02	8.1
2	4.93	11.5
5	3.04	18.2
10	2.06	25.8
20	1.33	36.9
30	1.00	45.0

9.6.3 Charge-Amplifier Circuit

The circuit diagram for a piezoelectric sensor and a charge amplifier is shown in Fig. 9.12. Two op-amps are connected in series to provide the required input impedance and gain. The first op-amp is the charge amplifier, which converts the charge q into the voltage v_2. It employs both capacitive C_f and resistive R_f feedback. The second op-amp is an inverting amplifier, which standardizes the system voltage sensitivity. It has a variable input resistance $R_1 = bR_{f1}(0 \leq b \leq 1.0)$ and a fixed feedback resistance R_{f1}. The circuit can be reduced to the form shown in Fig. 9.13 by combining the input (source) capacitances of the transducer C_t, the cable C_c, and the op-amp C_a into an effective source capacitance C. The unit calibration capacitance C_{cal} has no effect on circuit performance so long as *its input is not grounded*.

The differential equation for the circuit in Fig. 9.13 is obtained as follows:

$$i = i_1 + i_2 + i_3 = \dot{q} = S_q^* \dot{a}$$

$$= C\dot{v}_1 + \frac{v_1 - v_2}{R_f} + C_f(\dot{v}_1 - \dot{v}_2) = S_q^* \dot{a} \qquad \textbf{(9.24)}$$

Equation 9.24 can be expressed in terms of the output voltage v_o by observing that

$$v_2 = -G_1 v_1 \qquad \text{and} \qquad v_o = -G_{C2} v_2 = -\frac{1}{b} v_2$$

where

G_1 is the open-loop gain of the first op-amp
G_{C2} is the circuit gain of the second op-amp ($= 1/b$)
b is the potentiometer setting

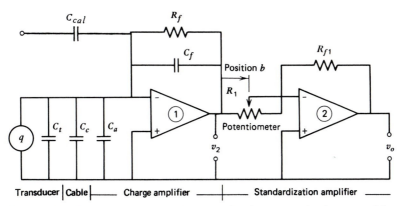

Figure 9.12 Measurement circuit with a piezoelectric sensor and a charge amplifier.

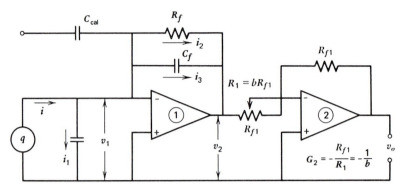

Figure 9.13 Equivalent circuit for analysis.

These equations combine to yield

$$\frac{1}{G_{C2}}\left[\frac{C}{G_1} + C_f\left(1 + \frac{1}{G_1}\right)\right]\dot{v}_o + \frac{1}{G_{C2}}\left(1 + \frac{1}{G_1}\right)\frac{v_o}{R_f} = S_q^*\dot{a} \tag{a}$$

Since G_1 is very large ($\geq 10^5$) for a typical op-amp, Eq. a reduces to

$$\dot{v}_o + \frac{v_o}{R_f C_{eq}} = \left(\frac{G_{C2}S_q^*}{C_{eq}}\right)\dot{a} \tag{9.25}$$

where

$$C_{eq} = \frac{C}{G_1} + C_f = C_f\left(1 + \frac{C}{C_f G_1}\right) \tag{9.26}$$

Equation 9.26 shows that the source capacitance has little effect on this measurement system because the term $C/C_f G_1$ will usually be very small when compared with unity. With reasonable limits on C so that $C/C_f G_1 \approx 0$, note that $C_{eq} = C_f$ and Eq. 9.25 becomes

$$\dot{v}_o + \frac{v_o}{R_f C_f} = \left(\frac{G_{C2}S_q^*}{C_f}\right)\dot{a} \tag{9.27}$$

A comparison of Eq. 9.27 with Eq. 9.20 shows that they are identical. This fact implies that the frequency response characteristics given in Fig. 9.11 are the same for both the voltage-follower and the charge-amplifier circuits.

The voltage sensitivity of the charge-amplifier circuit is given by

$$S_v = \frac{G_{C2}S_q^*}{C_f} = \frac{1}{b}\frac{S_q^*}{C_f} = \frac{S_q^{**}}{C_f} \tag{9.28}$$

This equation indicates that a charge amplifier has two parameters (b and C_f) that can be adjusted to control voltage sensitivity. The charge sensitivity S_q^* of a particular transducer can be standardized to an equivalent value of 1, 10, or 100 pC per unit. The standardized value is obtained by adjusting the potentiometer so that $S_q^*/b = S_q^{**}$ for the particular piezoelectric transducer in use.

Once the charge sensitivity is standardized, the instrument range is established by selecting the feedback capacitance. Typical instruments provide feedback capacitances ranging from 10 to 50,000 pF in a 1–2–5–10 sequence.

The time constant for the charge amplifier is given by Eq. 9.27 as $\tau = R_f C_f$. It is important to note that both R_f and C_f are contained in the charge amplifier. The time constant for the voltage follower is given by Eq. 9.23 as $\tau = RC_{eq}$. The important distinction for the voltage follower is that the equivalent capacitance C_{eq} is due to the combined capacitances of the transducer, the cable, and the amplifier. It is much more difficult to control C_{eq} in the voltage follower because the cable and transducer capacitances are external to the signal conditioning circuit.

Typical time constants that can be achieved with a number of piezoelectric transducers incorporated in a charge-amplifier circuit are presented in Table 9.4.

As indicated in Table 9.4, different time constants can be selected for each of the three ranges (short, medium, and long). For the high reciprocal sensitivities, with the charge amplifier adjusted to long, the time constant is 5,000,000 s (nearly 2 months). On the other hand, for very low reciprocal sensitivities, with the charge amplifier adjusted to short, the time constant is only 10 ms.

In comparing the relative merits of the charge-amplifier and voltage-follower circuits it is clear that the charge amplifier is preferred because of the following advantages:

1. The time constant is controlled by the feedback resistance and the feedback capacitance and is independent of the transducer and cable capacitance.
2. System performance is independent of transducer and cable capacitance so long as the source capacitance is less than the maximum allowed for a given amount of error (see Table 9.4).
3. Charge sensitivity can be standardized by using the position parameter b, which controls the gain of the standardization amplifier.
4. A wide range of voltage sensitivities and time constants are available by changing the feedback capacitance.

9.6.4 Built-in Voltage Followers

Microelectronic developments have progressed to the point where miniature voltage-follower amplifiers are incorporated in transducer housings. A P-channel MOSFET (Metal-Oxide-Semiconductor Field-Effect Transistor) unity-gain amplifier (voltage follower) is used for this application. The additional components required for this measurement circuit are a power supply, a cable to connect the transducer to the power supply, and a recording instrument, as shown in Fig. 9.14.

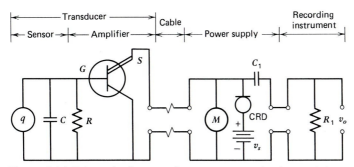

Figure 9.14 Schematic diagram of a measurement circuit containing a piezoelectric transducer with a built-in amplifier.

Table 9.4 Typical Sensitivities and Time Constants for a Piezoelectric Transducer in a Charge-Amplifier Circuit[a]

Reciprocal Sensitivity[b] (units of a/V) Range (pC/unit of a)			Feedback Capacitance C_f (pF)	Time Constant (s)			Maximum Source Capacitance for 0.5% Error (μF)
0.1 to 1.0	1.0 to 10	10 to 100		Short	Medium	Long	
50,000	5,000	500	50,000	50	50,000	5,000,000	0.1
20,000	2,000	200	20,000	20	20,000	2,000,000	0.1
10,000	1,000	100	10,000	10	10,000	1,000,000	0.1
5,000	500	50	5,000	5.0	5,000	500,000	0.1
2,000	200	20	2,000	2.0	2,000	200,000	0.1
1,000	100	10	1,000	1.0	1,000	100,000	0.1
500	50	5.0	500	0.5	500	50,000	0.05
200	20	2.0	200	0.2	200	20,000	0.02
100	10	1.0	100	0.1	100	10,000	0.01
50	5	0.5	50	0.05	50	5,000	0.005
20	2	0.2	20	0.02	20	2,000	0.002
10	1	0.1	10	0.01	10	1,000	0.001

[a] Adapted from the Kistler manual for a Model 504A charge amplifier.
[b] The reciprocal of the voltage sensitivity S_v given by Eq. 9.28.

Placing the voltage-follower circuit inside the transducer housing where it is adjacent to the piezoelectric crystal effectively removes the cable capacitance C_c from the charge-generating side of the circuit. A connecting cable is still required, but it is on the output side of the circuit, where its effect on the transmitted signal is small. The sensor capacitance C and the input resistance R are not significantly affected by environmental conditions, because these components are also protected inside the transducer housing. The resistance and capacitance are adjusted during assembly to give a nominal voltage sensitivity and a reasonable time constant τ.

The power supply shown in Fig. 9.14, which consists of a dc supply voltage v_s and a current-regulating diode (CRD), provides a nominal $+11$ V at the transistor (S) when there is no input signal. The capacitance C_1 shields the recording instrument from this dc voltage. The meter (M) monitors the transducer cable connection. If the meter shows zero, a short exists in the connection, but if the meter shows v_s, the circuit is open.

The equivalent circuit for this built-in voltage follower is illustrated in Fig. 9.15. Considering the input to the voltage follower,

$$\dot{v}_1 + \frac{v_1}{RC} = \left(\frac{S_q^*}{C}\right)\dot{a} = S_v\dot{a} \tag{9.29}$$

The MOSFET amplifier exhibits a unity gain; therefore, the output side yields

$$\dot{v}_o + \frac{v_o}{R_1C_1} = \dot{v}_2 = \dot{v}_1 \tag{9.30}$$

The steady-state frequency response function obtained from Eqs. 9.29 and 9.30 is

$$H(\omega) = \frac{Cv_o}{S_q^*a_0} = \left(\frac{jRC_1\omega}{1 + jRC_1\omega}\right)\left(\frac{jRC\omega}{1 + jRC\omega}\right) \tag{9.31}$$

The form of Eq. 9.31 indicates the presence of two time constants.

$$\tau = RC \qquad \text{and} \qquad \tau_1 = R_1C_1$$

The internal time constant $\tau = RC$ is fixed at the time of instrument assembly. Ideally, this value of τ should be very large; however, typical values range from 0.5 to 2000 s. The value of τ is limited because of constraints on R for amplifier current and on C for voltage sensitivity. The external time constant $\tau_1 = R_1C_1$ is controlled by the blocking capacitor C_1 (limited by the power supply) and by the input resistance R_1 of

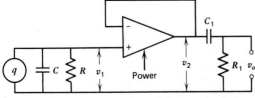

Figure 9.15 Equivalent voltage-follower circuit for analysis.

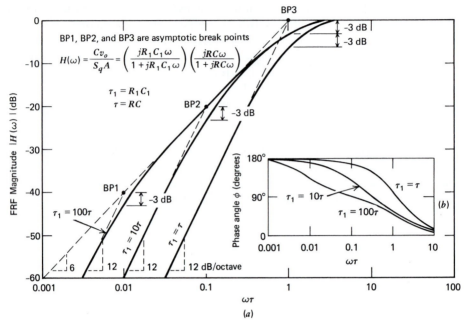

Figure 9.16 Frequency response function $H(\omega)$ for a piezoelectric transducer with a built-in amplifier. (*a*) Magnitude of $H(\omega)$ versus $\log \omega\tau$. (*b*) Phase angle ϕ versus $\log \omega\tau$.

the readout instrument (selected by the transducer user). Since these instruments have input resistances ranging from 0.01 MΩ to over 1 MΩ, the external time constant can easily become the controlling time constant. Magnitude and phase-angle response characteristics of this dual time-constant instrument are described by Eq. 9.30 and are shown in Fig. 9.16.

The Bode diagrams in Fig. 9.16 show the effect of the external time constant on instrument response for three time-constant ratios of 1, 10, and 100. When $\tau_1 = \tau$, the low-frequency response drops off at a rate of 12 dB per octave and the phase angle rapidly approaches 180 deg. At $\omega\tau = 1$ (break point BP3), the amplitude is attenuated 6 dB (down 50 percent) and a 90-deg phase shift occurs. When $\tau_1 > \tau$, signal attenuation occurs at even lower values of $\omega\tau$. To minimize error resulting from attenuation and phase shift, a low-frequency minimum on $\omega\tau$ must be maintained. The minimum values of $\omega\tau$ for errors of 2 and 5 percent for time-constant ratios τ_1/τ of 1, 10, and 100 are given in Table 9.5.

Errors can be limited to 2 percent when τ_1/τ is greater than 10 if $\omega\tau$ is greater than 5. Similarly, errors can be limited to 5 percent when τ_1/τ is greater than 10 if $\omega\tau$

Table 9.5 Minimum Values of $\omega\tau$ for Errors of 2 and 5 Percent[a]

τ_1/τ	2%	5%
1	7.00	4.36
10	4.93	3.06
100	4.90	3.04

[a] For the built-in voltage-follower circuit.

is greater than 3. The values of $\omega\tau$ listed in Table 9.5 for τ_1/τ equal to 100 are the same as those obtained for the single-time-constant systems described in Table 9.3.

The concept of placing a miniaturized voltage follower in the transducer housing yields the following advantages:

1. Voltage sensitivity S_v is fixed by the manufacturer. The user has no gains to adjust.
2. Cable length and cable capacitance have little effect on the output voltage.
3. High-level output voltage signals are obtained with a low level of cable noise.
4. Low-cost battery power supplies are adequate for field use.
5. The unit interfaces with most voltage recorders.

9.7   RESPONSE OF PIEZOELECTRIC CIRCUITS TO TRANSIENT SIGNALS

Piezoelectric sensors can alter transient signals since these signals almost always contain a dc component that is blocked by the sensor's ac-coupled response characteristics. The differential equation governing both voltage-follower and charge-amplifier circuits is of the same form:

$$\dot{v}_o + \frac{v_o}{RC} = S_v\dot{a} \tag{9.32}$$

The particular solution of Eq. 9.32 is

$$v_o = S_v e^{-t/RC} \int_0^t e^{\varphi/RC}\,\dot{a}(\varphi)\,d\varphi \tag{9.33}$$

where φ is a dummy variable of integration. This solution can be applied to a transient signal if $a(\varphi)$ is defined.

As an example, consider the case where $a(\varphi)$ is defined by the rectangular pulse illustrated in Fig. 9.17a. Although this pulse cannot be generated mechanically, it nevertheless provides a limiting case for judging the adequacy of the electronic circuit's low-frequency response.

The derivative $\dot{a}$, shown in Fig. 9.17b, is expressed as

$$\dot{a} = a_0\delta(\varphi) - a_0\delta(\varphi - t_1) \tag{a}$$

where $\delta(\varphi)$ is the Dirac delta function. This function has properties such that it is zero except when the argument is zero, and the area under the function is unity when integrated over the zero argument. Substituting Eq. a into Eq. 9.33 and integrating yields

$$v_o = S_v a_0 \left[u(t)e^{-t/RC} - u(t - t_1)e^{-(t-t_1)/RC} \right] \tag{9.34}$$

where $u(t - t_1)$ is the unit-step function obtained by integrating $\delta(\varphi)$. The output signal v_o, shown in Fig. 9.17c, has two significant characteristics. First, the output signal decays exponentially during the pulse duration and an error of $(1 - e^{-t/RC})$ develops with time. Second, an undershoot occurs at $t = t_1$, which is equal to the maximum error

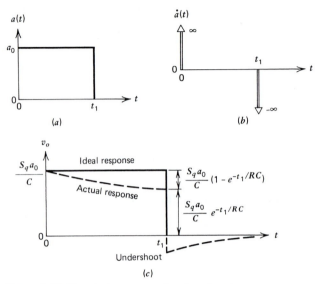

Figure 9.17 Electrical response of a piezoelectric sensor to a rectangular pulse (transient) input. (*a*) Rectangular input pulse. (*b*) Time derivative of the input pulse. (*c*) Output voltage.

associated with the rectangular pulse. The maximum error is estimated using a series expansion of the function $(1 - e^{-t_1/RC})$ to give

$$\mathscr{E}_{max} = \frac{t_1}{RC}\left[1 - \frac{1}{2}\frac{t_1}{RC} + \frac{1}{6}\left(\frac{t_1}{RC}\right)^2 - \cdots\right] \tag{b}$$

When t_1/RC is small compared with unity, Eq. b reduces to

$$\mathscr{E}_{max} = \frac{t_1}{RC} \qquad \text{or} \qquad \tau = RC = \frac{t_1}{\mathscr{E}_{max}} \tag{9.35}$$

The time constant $\tau = RC$ clearly controls the error that develops during the duration of the pulse. The error can be minimized by making τ/t_1 very large. The time constant necessary to limit the error in measuring a rectangular pulse is given in Table 9.6.

Triangular and half-sine pulse shapes are good examples of transient pulses. The minimum time constants required to limit specified measurement errors for these pulse shapes are also listed in Table 9.6. It is evident from this analysis that the rectangular pulse requires the largest time constant for a given level of error. The potential error occurring in a given measurement can be estimated by comparing the transient shape of the output signal with the pulses listed in Table 9.6. Also, undershoot with its exponential decay is clear evidence of an inadequate time constant.

Transducer systems utilizing the built-in voltage follower present a special problem because they exhibit two time constants. An analysis of this circuit (Fig. 9.14) with a rectangular pulse for $a(t)$ indicates that the two time constants can be combined into an effective time constant τ_e given by

$$\tau_e = \frac{\tau_1 \tau}{\tau_1 + \tau} \tag{9.36}$$

Table 9.6 Time-Constant Requirements to Limit Error

Pulse Shape		Time Constant		
		2% Error	5% Error	10% Error
Rectangular pulse		$50t_1$	$20t_1$	$10t_1$
Triangular pulse		$25t_1$	$10t_1$	$5t_1$
Half-sine pulse	A	$16t_1$	$6t_1$	$3t_1$
	B	$31t_1$	$12t_1$	$6t_1$

This effective time constant can be used with Table 9.6 to estimate the maximum errors associated with typical transient measurements by comparing the output pulse with the shapes listed.

A. Overview

A typical frequency response function for a piezoelectric transducer is shown in Fig. 9.18. At the low-frequency end of the frequency spectrum (below ω_1), the system

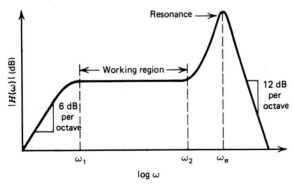

Figure 9.18 Typical frequency response function $H(\omega)$ for a piezoelectric transducer.

exhibits a rapid drop in amplitude (6 dB per octave). Also, periodic signals undergo serious phase distortion and transient signals exhibit exponential decay. Signal fidelity at this end of the frequency spectrum is controlled by the electrical RC time constant. Between ω_1 and ω_2, the system exhibits essentially zero phase shift and nearly constant output per unit input. Measurements should be limited to this range of the spectrum if possible. For frequencies higher than ω_2, mechanical resonance becomes important and causes amplitude magnification and phase distortion of signals with a periodic input. For transient signals a serious ringing occurs if the rise (or fall) time of the transient input is less than five times the natural period of the transducer.

In every instrumentation application, it is the engineer's responsibility to select instruments that will provide a measurement with a minimum of error regardless of input type or shape; therefore, it is important that all of the effects illustrated in Fig. 9.18 be understood and given careful consideration.

9.8   ACCELEROMETER CALIBRATION

The voltage sensitivity or calibration constant of a transducer is an important quantity. Transducer manufacturers provide calibration information, traceable to the National Institute of Standards and Technology (NIST), when the instrument is shipped to the customer. Users must recalibrate the instrument to ensure that the original calibration has not changed with use or abuse. Users generally employ comparison methods to recalibrate instruments. Calibration is a process where a known input is generated and instrument response is recorded so that a ratio of output–input can be established over the ranges of interest (frequency and magnitude).

A widely used calibration method (constant-acceleration method) requires only the simple act of rotating the accelerometer's sensitive axis in the earth's gravitational field. This produces a nominal change in acceleration of 2 g (-1 g to $+1$ g). The primary disadvantages of this simple method are limited range, local gravity variations, and absence of any check on frequency response characteristics. The method provides a quick means to ensure that an accelerometer is at least functional. Centrifuges are required to achieve higher levels of constant acceleration, but these facilities are costly and are not readily available in most laboratories.

A sinusoidal input, such as that produced by a vibration generator, provides calibration signals at different frequencies and different levels of acceleration. This sinusoidal input is widely used for both comparison and absolute calibration.

In the comparison method, two accelerometers (test and standard) are mounted as close as possible to one another on the moving head of a vibration table. Some calibration accelerometers are constructed so that the test accelerometer is mounted directly on the back of the calibration standard. This procedure is often referred to as the back-to-back calibration method. The standard accelerometer should be used only for calibration and never for testing. The calibration of the standard accelerometer must be directly traceable to NIST. Standard accelerometers, along with special charge amplifiers that are adjusted to give a constant sensitivity (usually 10.0 mV/g) over a broad range of frequencies, are available.

Linearity of the test accelerometer at a specified frequency is obtained from a graph of the output signal from the test accelerometer versus the output signal from the standard accelerometer as the level of acceleration is increased over a wide range. The sensitivity of the test accelerometer as a function of frequency is determined by varying the frequency of the vibration table and adjusting the acceleration level to maintain the signal from the standard accelerometer at a constant value. Accuracies of ±0.2 percent for the calibration constants of the test accelerometer are attainable with the comparison technique.

Several manufacturers have developed portable calibrating systems for field use. One such unit, shown in Fig. 9.19, uses a dual magnet system with two mechanically coupled moving coils. One coil operates as the driver and the other serves as a velocity sensor. The coils can be connected to external instrumentation for performing reciprocity calibration (Reference 18).

The local acceleration of gravity can be accurately measured by aligning the calibrator vertically, mounting the test accelerometer on the calibrator, and placing a small (less than 1 gm) nonmagnetic object on the accelerometer's free surface. At an acceleration level of ±1 g, the object will begin to "rattle." The onset of rattle can be detected accurately (±1 percent) in the accelerometer signal.

Absolute accelerometer calibration methods require precise measurement of frequency ω and displacement x, since peak acceleration magnitude is given by the equation

$$a_{max} = -\omega^2 x_{max} \qquad (9.37)$$

As an example of the magnitudes involved, a sinusoidal displacement of 0.001 in. at a frequency of 100 Hz gives a peak acceleration of 1.02 g. Thus, the peak-to-peak amplitudes to be accurately measured are extremely small. For frequencies below 100 to 200 Hz, reasonable results can be obtained with a microscope. For frequencies above 200 Hz, it is necessary to use proximity gages and interferometry (Reference 13) methods to measure the amplitudes with sufficient accuracy to employ Eq. 9.37 for absolute calibration of the test accelerometer.

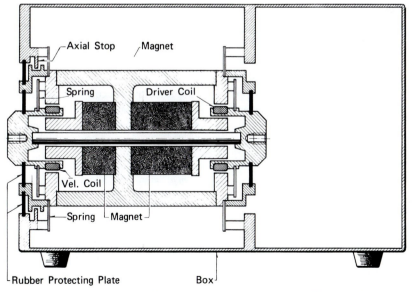

Figure 9.19 Construction details of a portable accelerometer calibrator. (Courtesy of Bruel and Kjaer.)

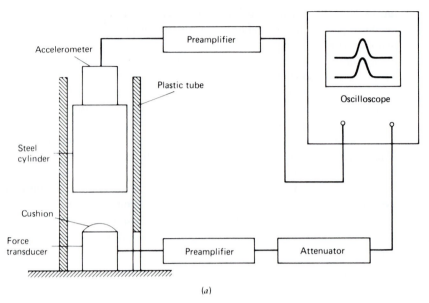

(a)

Figure 9.20 Schematic diagram of a gravimetric calibration system for accelerometers.

A simple impulse method of accelerometer calibration (Reference 18) is based on Newton's second law, which is written as

$$F = ma = mg\left(\frac{a}{g}\right) \tag{9.38}$$

A schematic diagram of the fixture used for impulse calibration is shown in Fig. 9.20. The system consists of a solid steel cylinder on which the test accelerometer is mounted, a plastic tube to guide the cylinder motion, a force transducer, and a rigid support base. The calibration is performed in two steps and requires measurement of three voltages.

First, the test mass (cylinder and accelerometer) is positioned on the force transducer and voltage v_{mg} is measured when the test mass is quickly removed (see Fig. 9.21a). Second, the test mass is dropped onto the force transducer while the transient impulse voltages v_f (force transducer) and v_a (accelerometer) are simultaneously measured (see Fig. 9.21b). Since the force and acceleration acting on these transducers are their output voltages divided by their sensitivities,

$$F_{mg} = mg = \frac{v_{mg}}{S_f} \qquad \text{Step one}$$

$$F = \frac{v_f}{S_f} \quad \text{and} \quad \frac{a}{g} = \frac{v_a}{S_a} \qquad \text{Step two} \tag{9.39}$$

where

S_f is the voltage sensitivity of the force transducer
S_a is the voltage sensitivity of the accelerometer

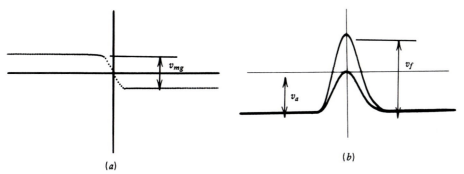

Figure 9.21 Typical voltage–time curves from a gravimetric accelerometer calibration. (*a*) Weight lifting. (*b*) Impacting. (Courtesy of PCB Piezotronics.)

Substitution of these equations into Eq. 9.38 gives S_a as

$$S_a = \left(\frac{v_a}{v_f}\right) v_{mg} \qquad (9.40)$$

Equation 9.40 shows that the unknown accelerometer sensitivity S_a is simply the impact voltage ratio (v_a/v_f) times the static voltage v_{mg}. Note that the force transducer sensitivity S_f does not enter into this accelerometer calibration technique.

In Fig. 9.20, the cushion material and the size of the mass control the duration time of the impact pulse and the drop height controls the amplitude. A typical calibration is performed over a range of amplitudes and time durations by using a combination of different drop heights and cushion materials. The preamplifiers shown in Fig. 9.20 can be any one of the three standard types (voltage follower, charge amplifier, or built-in voltage follower). The attenuator is a voltage divider used to vary force transducer signal amplitude. If the attenuator is adjusted to give $v_f = v_a$ during impact, the voltage v_{mg} obtained after the attenuator is set becomes a direct measure of accelerometer voltage sensitivity S_a. Digital oscilloscopes are extremely useful when using this calibration technique.

The major advantages of this calibration method are its portability and low cost. However, the method is dependent on local gravity, since the measurement of v_{mg} (step one) involves the weight (mg) of the impacting cylinder. Accuracies better than 1 percent are easily obtained.

9.9 DYNAMIC CALIBRATION OF FORCE TRANSDUCERS

Dynamic calibration can be performed by using the vibration exciter arrangement shown in Fig. 9.22. Known calibration masses m_c are attached in sequence to seismic mass m. The transducer base is then subjected to a sinusoidal oscillation, which is measured by the accelerometer. With this arrangement, the external force $F(t)$ applied to the force transducer by the calibration mass is

$$F(t) = m_c \ddot{y} = -m_c(\ddot{x} + \ddot{z}) \qquad (9.41)$$

The differential equation that describes the motion of the force transducer during calibration is obtained by substituting Eq. 9.41 into Eq. 9.3 to give

$$(m + m_c)\ddot{z} + C\dot{z} + kz = -(m + m_c)\ddot{x} \qquad (9.42)$$

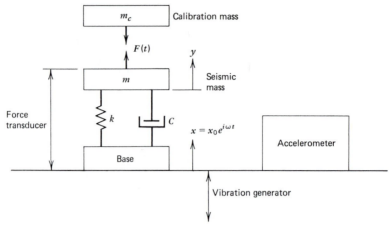

Figure 9.22 Calibration of a force transducer with a vibration generator.

The output signals v_f and v_a from the force transducer and the accelerometer, respectively, are

$$v_f = S_f(m + m_c)\ddot{x} \qquad \text{and} \qquad v_a = S_a\left(\frac{\ddot{x}}{g}\right) \qquad \textbf{(a)}$$

From these equations, it is obvious that the voltage ratio (v_f/v_a) is related to the sensitivity ratio (S_f/S_a) by

$$\frac{v_f}{v_a} = \frac{S_f}{S_a}(m + m_c)g = \frac{S_f}{S_a}(W + W_c) \qquad \textbf{(9.43)}$$

where

W is the weight of the seismic mass
W_c is the weight of the calibration mass

A graph of voltage ratio v_f/v_a as a function of weight W_c for a typical calibration is shown in Fig. 9.23. It is evident from Eq. 9.43 that the slope s is the sensitivity ratio S_f/S_a and that the horizontal axis intercept point is the seismic weight W. During calibration, several graphs of v_f/v_a should be obtained over a wide range

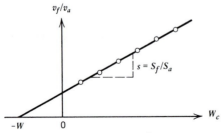

Figure 9.23 Voltage ratio v_f/v_a versus calibration weight W_c from a sinusoidal calibration of a force transducer.

of input amplitudes (to check linearity) and frequencies (to check for resonance effects) for each value of W_c. Once the slope s is established, the voltage sensitivity S_f is given by

$$S_f = sS_a \qquad (9.44)$$

The accelerometer sensitivity S_a is accurately known, since it comes from a standard accelerometer.

Calibration accuracies of ± 0.5 percent are possible with the sinusoidal method; however, care must be exercised to avoid the natural frequency of the force transducer system when adding calibration mass. The addition of the mass m_c decreases the natural frequency of the force transducer, and Eq. 9.43 is not valid near the resonance condition since the force voltage is too large.

9.9.1 Force Transducer Calibration by Impact

In structural testing, a half-sine input force is applied at a specified point by striking the structure with a hammer. Force transducers, installed on the hammerhead, measure these impact forces. A typical impact hammer with an impact head and an attached force transducer is shown in Fig. 9.24. The time duration of the impact force is controlled by the stiffness of the structure and impact head as well as by the mass of the hammer body. It is common practice to change the masses of the hammer body and impact head to achieve different impact time durations. It will be shown that these changes affect the calibration and performance of the force transducer.

A model (Reference 12) of the impact hammer–force transducer combination is shown in Fig. 9.25, where m_h is the mass of the impact head and m_b is the mass of the hammer body. The mass m_p of the pendulum and accelerometer is also used in the calibration. The differential equation for the force transducer is

$$m_e \ddot{z} + C\dot{z} + kz = -\left(\frac{1}{1+M}\right)F(t) \qquad (9.45)$$

where M is a mass ratio and m_e is an effective transducer mass given by

$$M = \frac{m_h}{m_b} \quad \text{and} \quad m_e = \frac{m_b m_h}{m_b + m_h} = \frac{m_h}{1+M} \qquad \text{(a)}$$

Figure 9.24 Impact hammer with attached force transducer and impact head. (Courtesy of PCB Piezotronics.)

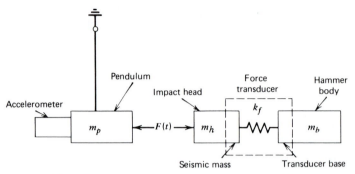

Figure 9.25 Calibration of an impact hammer with a pendulum system.

Equation 9.45 indicates that both the effective damping and the natural frequency of the transducer depend on m_e and M. The frequency response function obtained from Eq. 9.45 is

$$H(\omega) = \frac{1}{k - m_e\omega^2 + jC\omega} = \frac{1}{k(1 - r^2 + j2\,dr)} \tag{9.46}$$

Transducer output as a function of frequency is obtained from Eqs. 9.45 and 9.46 as

$$v_f(\omega) = S_z z = \left[\frac{S_z}{k}\left(\frac{1}{1 + M}\right)\right]\frac{F(\omega)}{(1 - r^2 + j2dr)} \tag{b}$$

The effective voltage sensitivity is contained in the first bracketed term of Eq. b. Thus,

$$S_f^* = \left[\frac{S_z}{k}\left(\frac{1}{1 + M}\right)\right] = S_f\left(\frac{1}{1 + M}\right) \tag{9.47}$$

where $S_f = S_z/k$ is the static sensitivity obtained from a static calibration of the transducer. It is evident that the effective sensitivity is changed by the user when either base (hammer body) or impact-tip masses are changed. It is possible to predict the new transducer sensitivity when one or both masses are changed, since S_f is not altered by changing system masses and m_b and m_h are known.

When a calibration is performed using the system shown in Fig. 9.25, the force $F(t)$ applied by the pendulum is

$$F(t) = m_p a = W_p\left(\frac{a}{g}\right) \tag{9.48}$$

where a is the acceleration of the pendulum during impact. The accelerometer output voltage during impact is

$$v_a = S_a\left(\frac{a}{g}\right) \tag{9.49}$$

Combining Eqs. b, 9.47, 9.48, and 9.49 gives

$$\frac{v_f}{v_a} = \frac{S_f^*}{S_a} W_p \tag{9.50}$$

The results of Eq. 9.50 are similar to those of Eq. 9.43 except that the seismic mass is not involved in Eq. 9.50. If the voltage ratio v_f/v_a is measured on, for example, a digital oscilloscope, the calibration constant S_f^* is

$$S_f^* = \left(\frac{v_f}{v_a}\right)\left(\frac{S_a}{W_p}\right) \tag{9.51}$$

Again, if standard accelerometers are used and care is exercised during the measurements, accurate calibration (better than ± 1 percent) can be achieved.

This calibration procedure for impact hammers is adequate for operating frequencies that are much lower than the natural frequency of the force transducer. At higher frequencies, stress-wave effects become important, and force measurements made with such hammer devices exhibit considerable error.

9.10   OVERALL SYSTEM CALIBRATION

Electronic measurement systems contain a number of components, such as power supplies, transducers, amplifiers, signal conditioning circuits, and recording instruments. It is possible to calibrate each component; however, this procedure is time-consuming and subject to the calibration errors associated with each component. A more precise and direct procedure establishes a single calibration constant for the complete system, which relates the recording-instrument reading to the quantity being measured.

System calibration involves determinations not only of voltage sensitivity, but also of the dynamic response characteristics of the system, such as rise time, overshoot, and time constant. An ideal method for system calibration does not require the transducer to be disconnected from either the measurement system or the structure to which it is mounted. This approach employs a known voltage input at the transducer end of the system. For piezoelectric transducers, a voltage generator is connected to the system through a calibration capacitor C_{cal}, as shown in Fig. 9.13. This charge-amplifier circuit is used for the discussion of system calibration that follows.

The calibration voltage v_{cal}^* needed to simulate a charge q_{cal}, is related to a_{cal} by Eq. 9.19. Thus,

$$v_{cal}^* = \frac{q_{cal}}{C_{cal}} = \frac{S_q^* a_{cal}}{C_{cal}} \tag{9.52}$$

Charge-amplifier voltage sensitivity is given by Eq. 9.28 as

$$S_v = \frac{S_q^*}{bC_f} \tag{9.53}$$

The charge-amplifier output voltage v_{cal} resulting from the calibration voltage v_{cal}^* is

$$v_{cal} = S_v a_{cal} = \frac{S_q^* a_{cal}}{bC_f} = \frac{C_{cal} v_{cal}^*}{bC_f} \tag{9.54}$$

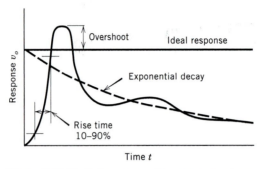

Figure 9.26 Typical response of a measurement system to a step input.

Equation 9.54 indicates that potentiometer position b can be adjusted to provide a standardized output voltage that includes any circuit-loading effects. The specified range is then established by selecting the feedback capacitance C_f.

The voltages used in calibration may be sinusoidal, periodic, or transient. Precision voltage sources are available that provide either a 0 to 10 V full-scale step pulse or a 0 to 10 V full-scale 100-Hz square wave. The voltage level is adjustable over the full range with a resolution of 0.02 percent of full scale and a linearity of ±0.25 percent of full scale. Precision calibration capacitors are available for use in calibrating charge amplifiers that do not have built-in calibration capacitors. It is important to avoid connecting a voltage source to the input of a charge amplifier without a series-connected calibration capacitance.

A step-input voltage is commonly used to test the total system because this signal thoroughly checks system fidelity. A typical response signal is shown in Fig. 9.26. This signal establishes the system rise time, overshoot, and exponential decay characteristics. All of these quantities are important because they imply measurement limitations. Unfortunately, the transducer's natural frequency is not excited by this calibration technique. Also, the natural frequency of the instrument system is difficult to interpret from the signal oscillations about the exponential decay trace.

Calibration of transducers and complete instrument systems must be performed periodically to ensure satisfactory performance. As the number of channels in a system increases, the need for accurate and efficient calibration procedures becomes more important. Calibration must be regarded as a vital step in the experimental process because the data collected are only as good as the calibration accuracy.

9.11 SOURCES OF ERROR WITH PIEZOELECTRIC TRANSDUCERS

Many factors affect the performance of piezoelectric transducers and the accuracies achieved when they are used in field or laboratory applications. These factors include transducer mass, sensitivity, cable noise, and mounting methods.

The mass of an accelerometer can affect the accuracy of the measurement by changing the dynamic response characteristics of a lightweight structure. The magnitude of the error can be estimated from

$$a_s = a_m(1 + m_a/m_s)$$

$$f_s = f_m \sqrt{(1 + m_a/m_s)}$$

(9.55)

where a and f are acceleration and frequency and subscripts s and m indicate structure and measured, respectively. To minimize the effect of the transducer on the process (in this case vibration) it is essential that the mass ratio m_a/m_s be minimized (5 percent or less).

The upper limit on the range (5000 to 10,000 g) of an accelerometer is controlled by the strength of the piezoelectric material. The lower limit is established by the noise imposed on the output signal by the amplifiers and the connecting cables. Usually, this noise level is equivalent to an acceleration signal of 0.0001 g. Accelerometers are also sensitive to motions perpendicular to the sensing axis. Maximum transverse sensitivity is less than ± 5 percent for most commercially available accelerometers.

The charge sensitivity S_q of a piezoelectric material depends on temperature, as shown in Fig. 9.27. These results indicate that corrections are needed when an accelerometer is used at temperatures other than the 20°C (68°F) reference temperature. Most general-purpose accelerometers function at temperatures up to 250°C (482°F). Depolarization of the piezoelectric material occurs at the Curie temperature.

The cable connecting the piezoelectric transducer to a charge amplifier can be a source of noise owing to three problems: ground loops, triboelectric effects, and electromagnetic interference.

Ground-loop currents develop in the cable shield when an electric potential exists between the transducer and the ground point on the charge amplifier. These loops are eliminated by electrically isolating the transducer from the structure and grounding the charge amplifier and test structure at the same point.

Triboelectric noise occurs in damaged cables when an electrical charge is generated by rubbing between one of the conductors and the adjoining insulating material (see Fig. 9.28a). The triboelectric charge often has the same frequency as the charge generated by the transducer because the cable vibrates with the structure; consequently, this phenomenon is difficult to detect from the test data. Before and after each experiment cables should be checked by shaking them to ensure that the output signal is zero when there is no input. Damage to cables is minimized by using proper attachment methods. A cable should be attached to and removed from a vibrating structure at a point of minimum motion, as illustrated in Fig. 9.28b. This practice minimizes flexing stresses in the cable that lead to cable failure and associated triboelectric effects.

Electromagnetic noise is generated in cables located near electrical equipment that produces large oscillating magnetic fields. Double-shielded cables or grounded conduits often can be used to reduce the magnitude of the noise. In cases where the lead

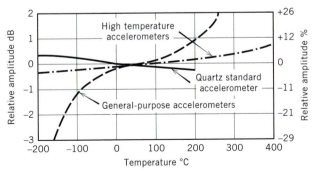

Figure 9.27 Typical charge sensitivity versus temperature characteristics for piezoelectric accelerometers.

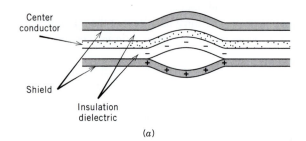

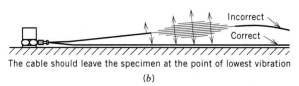

The cable should leave the specimen at the point of lowest vibration

(b)

Figure 9.28 (a) Generation of triboelectric noise by a damaged accelerometer cable. (b) Proper mounting of accelerometer cables to avoid cable whip.

wires cannot be relocated or satisfactorily shielded, special differential preamplifiers can be used to eliminate the noise by using the principle of common-mode rejection.

One of the most critical factors in acceleration measurements is the method used to attach the accelerometer to the structure, since improper mounting can compromise the usable frequency range of the accelerometer. The preferred mounting method is illustrated in Fig. 9.29a. The accelerometer is attached to the structure with a steel

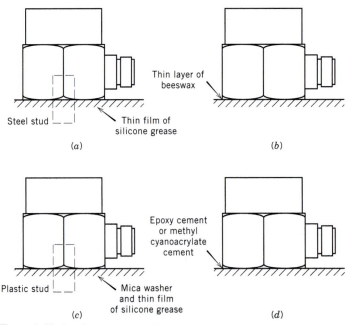

Figure 9.29 Accelerometer attachment methods. (a) Steel stud with a film of grease. (b) Bonding with beeswax. (c) Plastic stud with a mica washer and a film of grease. (d) Adhesive bonding.

stud. A thin layer of grease, applied before attachment, significantly improves contact between the accelerometer base and the structure by filling all of the interface voids. Optimum torque for a 10-32 NF steel stud is 1.76 N·m (15 lb·in.). The maximum temperature for this type of attachment is controlled by the accelerometer.

A second mounting method, shown in Fig. 9.29*b*, uses beeswax to bond the accelerometer to a structure. This mounting method gives nearly as effective a response as attaching the accelerometer with a stud; however, the wax is limited to a maximum temperature of 40°C (104°F). The strength of the bond, which depends on accelerometer mass and contact area, limits maximum accelerations to about 10 g. The advantage of this method is low cost, since no holes are drilled and tapped, which saves considerable time.

A thin film of silicone grease, a mica washer, and an electrically insulating plastic stud are used in the third method, illustrated in Fig. 9.29*c*, to prevent ground loops by isolating the accelerometer from the test item. Optimum torque for a 10-32 NF plastic stud is 1.76 N·m (15 lb·in.). The maximum temperature for this type of attachment is 250°C (482°F).

A fourth method of attachment involves bonding of the accelerometer to the structure and is illustrated in Fig. 9.29*d*. The advantages of this method are electrical isolation and elimination of the mounting holes. The frequency response of the transducer is not affected if the adhesive is strong and rigid; however, use of a low-modulus adhesive leads to erratic behavior. Epoxy adhesives are the most durable, though cyanoacrylate cement is easiest to apply as it polymerizes in minutes. The maximum temperature for this type of attachment is 100°C (212°F).

9.12 DISPLACEMENT MEASUREMENTS IN A FIXED REFERENCE FRAME

Displacement and velocity measurements relative to a fixed reference frame are considered in this section. The major difference between seismic and fixed-reference motion measurements is that in fixed reference measurement the base (see Fig. 9.1) does not move ($x = 0$) and Eq. 9.1 describes only the force required to move the sensing element. Since the spring in a fixed reference transducer has a very low spring rate and damping is small, this force is frequently due to inertia.

Measurements related to the extent of movement of an object are usually referred to as *displacement measurements*. Many of the sensor devices described in Chapter 5 can be used for displacement determinations when a fixed reference frame is available. Variable resistance (Section 5.4) and capacitance (Section 5.5) sensors are widely used for small static and dynamic displacement measurements (from a few microinches to small fractions of an inch). Differential transformers (Section 5.3) are used for larger displacement magnitudes (from fractions of an inch to several inches). Resistance potentiometers (Section 5.2) are used where less accuracy but greater range (small fractions of an inch to several feet) is required. In this section, applications of resistance potentiometers and photosensing transducers are described in detail, since they provide a convenient means for introducing several circuits that have been designed and perfected for displacement measurement.

9.12.1 Displacement Measurements with Resistance Potentiometers

Potentiometer devices, such as those shown in Fig. 5.2, are commonly used to measure linear and angular motions. A typical potentiometer circuit is shown in Fig. 9.30. This

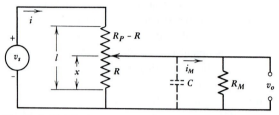

Figure 9.30 Displacement measuring circuit with a potentiometer as a sensor.

circuit consists of a potentiometer sensor with a resistance R_P, a recording instrument
with a resistance R_M, a power supply to provide a supply voltage v_s, and, in certain
instances, a capacitor C. The capacitor is used with wire-wound potentiometers to
smooth the output signal by momentarily maintaining the voltage as the wiper moves
from wire to wire along the helical resistance coil. In the following analysis of the
circuit, the effects of the capacitor are neglected.

The effects of the load imposed on the output signal by the voltage-measuring
instrument are obtained from an analysis of the circuit shown in Fig. 9.30. The
output voltage v_o can be expressed in terms of the various resistances as

$$v_o = i_M R_M = (i - i_M)R = v_i - i(R_P - R) \tag{9.56}$$

where

$R = (x/l)R_P$ is a resistance proportional to position x of the wiper on the potentiometer coil

R_M is the load resulting from the measuring instrument

Solving Eq. 9.56 for the output voltage v_o yields

$$v_o = \left[\frac{R_M/R_P}{(R_M/R_P) + (R/R_P) - (R/R_P)^2} \right] \left(\frac{R}{R_P} \right) v_s \tag{9.57}$$

Equation 9.57 indicates that the output voltage v_o of this circuit is a nonlinear function
of resistance R (and thus position x) unless resistance R_M of the measuring instrument
is large with respect to potentiometer resistance R_P. This nonlinear behavior can be
expressed in terms of a nonlinear factor η such that Eq. 9.57 becomes

$$v_o = (1 - \eta) \left(\frac{R}{R_P} \right) v_s \tag{9.58}$$

Values of the nonlinear term η as a function of resistance ratio R/R_P (wiper position)
for different values of R_M/R_P are shown in Fig. 9.31. Note that the resistance ratio
R_M/R_P must be at least 10 if significant deviations from linearity are to be avoided.

Figure 9.31 also indicates that the maximum deviation from linearity will occur at
a position x where $R/R_P = 0.5$ for all values of R_M. At such a position, the error $\mathcal{E}$
owing to nonlinear effects is obtained from Eq. 9.57 as

$$\mathcal{E} = \frac{v_o - v_o'}{v_o} = \frac{1}{1 + 4(R_M/R_P)} \tag{9.59}$$

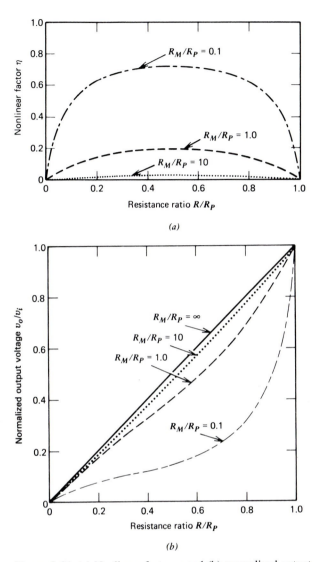

Figure 9.31 (a) Nonlinear factor η and (b) normalized output voltage v_o / v_i as a function of resistance ratio R/R_P (position) for different values of R_M / R_P.

where v_o and v'_o are open-circuit and indicated output voltages, respectively, from a measuring instrument when $R/R_P = 0.5$. For a measuring instrument with $R_M/R_P = 10$, the maximum error resulting from nonlinear effects is

$$\mathscr{E}_{max} = 0.0244 = 2.44 \text{ percent}$$

A second circuit employing a resistance potentiometer as a displacement sensor is shown in Fig. 9.32. This circuit utilizes only a potentiometer and a four-wire digital multimeter. The multimeter is set to provide a reading of the resistance placed between its two measuring terminals, identified as A and B in Fig. 9.32. The multimeter contains an internal constant-current generator that provides the current i_s, which

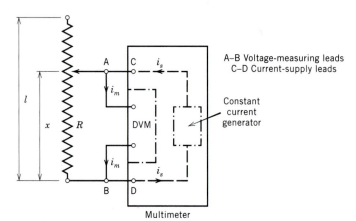

Figure 9.32 The use of a four-wire multimeter to measure R for a potentiometer sensor.

passes through the potentiometer. Note that the connection to the multimeter is made through terminals C and D. The voltage drop across the potentiometer

$$v_o = i_s R \tag{a}$$

is measured by the digital voltmeter contained in the multimeter. This reading of v_o is converted to a reading of R, by the logic circuits in the multimeter, and the resistance is displayed directly.

The use of four wires in this simple measurement is important because it reduces the error owing to contact resistance R_c at the terminals. The voltage drop at terminals C and D owing to R_c is

$$v_d = 2i_s R_c \tag{b}$$

The error v_d/v_o is reduced by connecting the voltmeter to the potentiometer at points A and B, which are at locations inside the loop defined by terminals C and D.

Of course, an error will still occur owing to the contact resistance at terminals A and B because

$$v_d = 2i_m R_c \tag{c}$$

The current i_m passing through the DVM is 100 to 1000 times less than the supply current i_s; therefore, the error owing to the voltage drop across the meter terminals is usually small enough to be neglected.

Calibration of a potentiometer-type displacement-measuring device can be accomplished by using a micrometer as the source of accurate displacements. The process of calibration can also be automated by comparing the output of the device to be calibrated with that from a previously calibrated instrument. Resolutions on the order of ±0.001 in. are possible. Sensitivities of 30 V/in. can be achieved by using potentiometers with a resistance of 2000 Ω/in. and the ability to dissipate a power of 0.5 W/in. of coil.

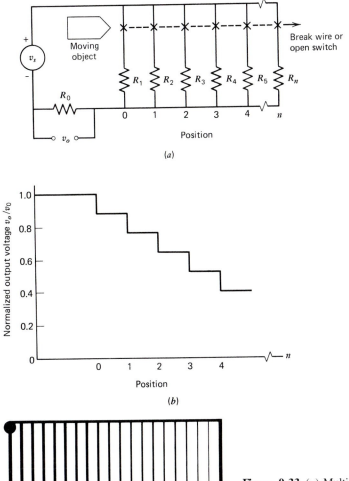

Figure 9.33 (*a*) Multiple-resistor circuit for displacement measurements. (*b*) Output voltage v_o from the multiple-resistor circuit. (*c*) Bonded parallel-resistance gage for sensing crack propagation in a fracture specimen.

9.12.2 Displacement Measurements with Multiple-Resistor Devices

Another type of variable resistance displacement-measuring device consists of a sequence of resistors in parallel, as shown in the circuit of Fig. 9.33*a*. With a fixed supply voltage v_s, the initial output voltage v_0 of the circuit is given by the expression

$$v_0 = \frac{R_0}{R_0 + R_e} v_s \tag{a}$$

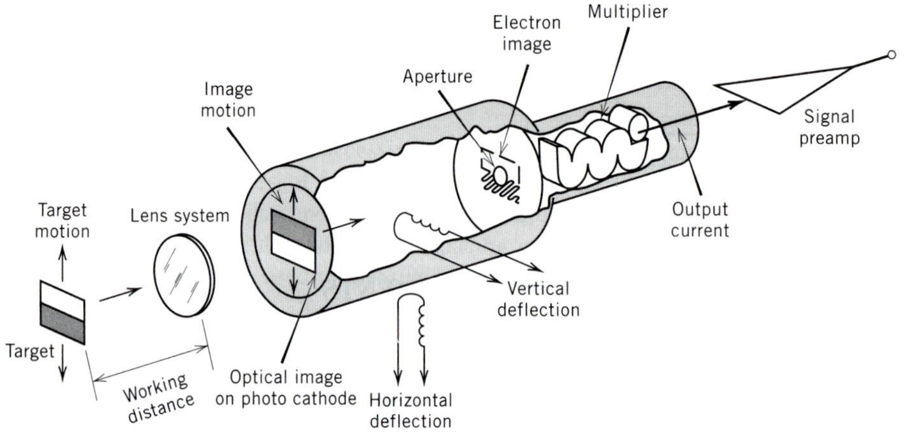

Figure 9.37 Schematic diagram of an optical-tracker sensing head. (Courtesy of Optron Corporation.)

The amplified signal can be used in two ways. First, a servo controller can be used to keep the discontinuity of the electron target image centered on the aperture. As the optical image moves, the servo controller changes the deflecting coil current so that the electron image remains stationary relative to the aperture. The current used in the coil is a measure of the target position. Second, the electron image can be scanned over the aperture by applying a sawtooth current to the deflection coil, which sweeps the electron image over the aperture. The position of the discontinuity is detected by the time required for the transition from a light to dark condition during a ramp period. Because the time for an intensity change varies as the object moves in the field of view, an output voltage proportional to this transition time relative to the ramp-time duration gives the position of the object. The sensitivity and range of motion that can be tracked are controlled by the lens systems employed and the working distance to the object.

The frequency response of optical trackers is from dc to 50 kHz. The biggest disadvantages in using these instruments are poorly defined discontinuities, nonuniform light sources, secondary targets, initial cost, and apparent change in position owing to variation of the angle of illumination with object motion. The major advantage is the ability to measure motion of objects without physical contact.

9.13.2 Video Camera Motion Analysis

The development of high-performance video cameras coupled with video recorders and digital computers has produced a new generation of full-field displacement-measurement systems. The major elements of these systems, shown in Fig. 9.38, consist of a video camera, a video recorder, a video monitor, a video processor, a computer with disk storage, a graphics workstation, and a printer/plotter unit.

Motion analysis with a video system consists of a number of basic steps. The first step involves recording of the images of interest. Numerous camera options are available that support a wide range of illumination conditions, fields of view, and picture frame rates ranging from 1 to 2000 frames per second. The data representing the image are usually stored on a videocassette by the video recorder. These data are available for editing, display, or video processing.

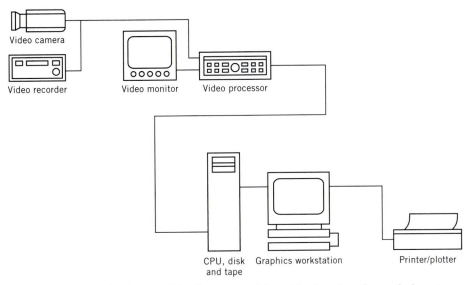

Figure 9.38 Schematic diagram of the elements used in a video-based motion-analysis system.

The second step is to reduce the vast amount of data available in a given frame to those data required for the motion analysis. The video processor is used to define the pixels[1] needed to describe the object being recorded. The video processor and monitor allow the operator to observe the moving object in outline form before the video image is digitized. This preview permits elimination of undesirable lighting and background effects from the image. Once adjustments are made, only data from selected points are digitized and stored in the computer memory. By using only selected pixels to define the motion of the object in the field of view, a smaller amount of data is stored and analyzed.

The third step is to analyze positions of points, lines, area centroids, and so on that are of interest. This analysis involves developing displacement–time histories for each feature on the object of interest. From these displacement–time histories, linear velocity and acceleration as well as angular position, velocity, and acceleration are determined. The computer performs the numerical computations in the differentiation processes and displays the results on the monitor. The printer/plotter is used to prepare hard copies for oral presentations and written reports.

Video technology is developing rapidly and is applicable to a wide variety of measurements in engineering and other fields. Although the method has limitations, it is versatile and is adaptable to numerical processing.

9.14   VELOCITY MEASUREMENTS

The principle of electromagnetic induction provides the basis for design of direct-reading, linear- and angular-velocity measuring transducers. The principle of operation for these two transducer types is shown in Fig. 9.39. For the linear-velocity measurements, a magnetic field associated with the velocity to be measured moves

---

[1]A picture is composed of a number of points called pixels. A video image composed of 500 rows by 500 columns has 250,000 picture elements (pixels) that describe the picture.

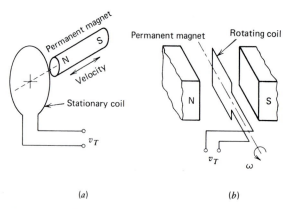

(a)   (b)

Figure 9.39 Schematic representations of transducers for
(a) linear-velocity measurements, and (b) angular-velocity
measurements.

with respect to a fixed conductor. For the angular-velocity measurements, a moving
conductor associated with the velocity to be measured moves with respect to a fixed
magnetic field. In either case, a voltage is generated that can be related to the desired
velocity. The basic equation relating voltage generated to velocity of a conductor in
a magnetic field can be expressed as

$$v_T = B\ell V \tag{9.60}$$

where

$\quad$ v_T is the voltage generated by the transducer
$\quad$ B is the component of the flux density (magnetic field strength) normal to the
$\quad\quad$ velocity
$\quad$ ℓ is the length of the conductor
$\quad$ V is the velocity

9.14.1 Linear-Velocity Measurements

A schematic representation of a self-generating linear-velocity transducer (LVT) is
shown in Fig. 9.40. The windings are installed in series opposition so that the induced
voltages add when the permanent magnet moves through the coil. Construction details
of a commercially available linear-velocity transducer are shown in Fig. 9.41.

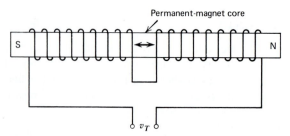

Figure 9.40 Schematic representation of a linear-velocity
transducer.

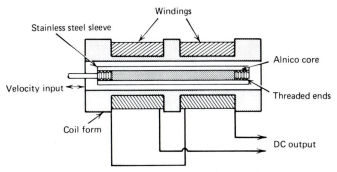

Figure 9.41 Cross section of a linear-velocity transducer.

The LVT is equivalent to a voltage generator connected in series with an inductance L_T and a resistance R_T, as shown in Fig. 9.42. The governing differential equation for this circuit with an LVT sensor and a recording instrument having an input resistance R_M is

$$L_T \frac{di}{dt} + (R_T + R_M)i = S_v V \tag{9.61}$$

where

S_v is the voltage sensitivity of the transducer [mV/(in./s)]
V is the time-dependent velocity being measured (in./s)
i is the time-dependent current flowing in the circuit

The frequency response function characteristics of an LVT are obtained by assuming a sinusoidal input velocity and output current in phasor form as done previously. Substitution of these phasors into Eq. 9.61 yields

$$i_0 = \frac{S_v V_0}{(R_M + R_T) + jL_T \omega} \tag{a}$$

where V_0 is the magnitude of the velocity and i_0 is the magnitude of the current. The current lags the velocity by a phase angle ϕ that is given by the expression

$$\tan \phi = \frac{L_T \omega}{R_M + R_T} \tag{b}$$

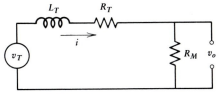

Figure 9.42 Linear-velocity transducer and circuit.

The output voltage v_o from the circuit shown in Fig. 9.42 is

$$v_o = iR_M = \frac{R_M S_v V_0}{(R_M + R_T) + jL_T\omega} e^{j\omega t} \tag{c}$$

The corresponding frequency response function for the circuit is

$$H(\omega) = \frac{v_o}{V} = \frac{v_o}{V_0 e^{j\omega t}} = \frac{R_M S_v}{(R_M + R_T) + jL_T\omega}$$

$$= \frac{R_M S_v e^{-j\phi}}{[(R_T + R_M)^2 + (L_T\omega)^2]^{1/2}} \tag{9.62}$$

Equation 9.62 also indicates that the output will be attenuated and the phase will be shifted relative to the input at the higher frequencies. The break frequency ω_c for a circuit controlled by a first-order differential equation occurs when the real and imaginary parts of the frequency response function are equal. This equality occurs when

$$\omega_c = \frac{R_M + R_T}{L_T} = \frac{R_M}{L_T}\left(1 + \frac{R_T}{R_M}\right) \tag{9.63}$$

Measurement error at the break frequency (-3 dB or 30 percent with a 45-deg phase shift) is much too large for most practical applications. A more realistic value of 2-percent error with a phase shift of 15.9 deg occurs when $\omega = 0.28 \, \omega_c$. Magnitude errors are less than 5 percent if use of the transducer is limited to frequencies below one-third of the break frequency.

Equations 9.62 and 9.63 clearly indicate that the sensitivity and frequency response of an LVT is dependent on the input resistance R_M of the recording instrument. The output voltage of an LVT sensor is relatively large [10 to 100 mV/(in./s)]; therefore, signal amplification is not usually required. Linearity within ± 1 percent can usually be achieved over the rated range of motion of the transducer. Common values for R_T are less than 10 Ω.

9.14.2 Angular-Velocity Measurements

Angular velocities can be measured by using either dc or ac generators. The angular velocity–terminal voltage relationship for dc motors and generators is given in Eq. 1.1. AC generators are used to measure average angular velocities. The number of cycles of voltage generated per revolution is an even integer depending on the number of poles. Therefore, the velocity readout can be performed with a simple frequency counter.

The measurement of instantaneous angular velocities presents a greater challenge. The dual-channel torsional converter shown in Fig. 9.43 provides a convenient circuit for measuring torsional vibrations in rotating machinery. This converter requires a pulsatile input signal from either an optical encoder (the preferred signal source) or an eddy-current probe that monitors a shaft-mounted gear. Gears often present problems since worn or bent gear teeth can generate spurious signal spacings that affect the

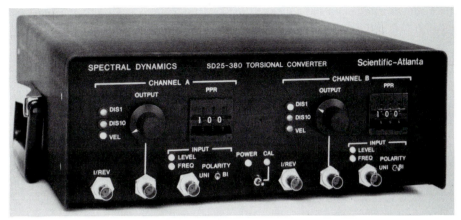

Figure 9.43 A dual-channel torsional converter. (Courtesy of Scientific-Atlanta, Inc.)

pulsatile signal. A detailed explanation of the internal circuitry of a torsional converter is beyond the scope of this book. A brief description of its use follows.

The torsional converter requires two inputs for each channel. The first input is the number of pulses per revolution (ppr) that the encoder or gear generates. The unit accepts values from 20 to 400 ppr in 1-ppr increments. The ppr value is set by thumb-wheel switches on the front panel. The second input is the pulsatile input voltage from the sensor. The unit accepts bipolar or unipolar voltage pulse trains in the range from ± 0.5 to ± 80 V over the frequency range from 50 pulses per second (pps) to 20,000 pps. These signals are processed internally to generate an output voltage signal with a sensitivity of 7.07 mV peak/deg peak (10 mV rms/deg peak to peak) over either of two frequency ranges.

1. Displacement switch position 1: 1 Hz to 1 kHz;
2. Displacement switch position 10: 10 Hz to 1 kHz.

An angular-velocity output signal is also available with a sensitivity of 0.05 mV peak/(deg/s) peak [0.0707 mV rms/(deg/s) peak to peak] over the frequency range 1 Hz to 1 kHz. The unit also has a 1/rev pulse that can be used to time other events with shaft position.

In addition, the unit has two sets of red and green LED lights that indicate unit performance in terms of input signal level and input frequency. A satisfactory input frequency is indicated by a green light, and an out-of-range input frequency is indicated by a red light. Similarly, a satisfactory signal level is indicated by a green light and an out-of-range signal level is indicated by a red light.

9.14.3 Laser-Doppler System

The Michelson interferometer is used in a noncontacting optical method for measuring velocities. The operating principle of the interferometric measurement method is illustrated in Fig. 9.44, which shows the laser beam divided into two beams, an internal reference beam and a signal beam. The signal beam is focused on the moving surface by a computer-controlled scanning mirror so that various points can be selected in a prescribed pattern. Back-reflected light from the signal beam is recombined with the reference beam. The signal-beam path length changes as the reflecting surface moves

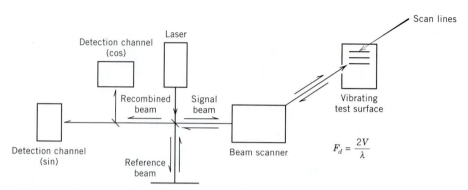

Figure 9.44 Schematic diagram illustrating the operating principle of a laser-Doppler velocity system.

so that the light intensity is modulated from bright to dark as the lights from the two beams reinforce or cancel one another. One complete cycle in intensity variation occurs with a surface displacement of half a wavelength, or $\lambda/2 = 314$ nm for a helium-neon laser. The reflected signal beam undergoes a Doppler shift with a Doppler frequency f_d given by

$$f_d = \frac{2V}{\lambda} \tag{9.64}$$

where V is the velocity of the moving surface. The direction of the velocity of the moving surface is resolved by using two detectors spaced one-quarter wavelength apart. One detector is called the cosine detector, and the other is called the sine detector. By using frequency modulation and detection schemes, an output voltage is available that gives the sign and magnitude of the surface velocity.

Typical instruments employ a 1-mW helium-neon laser and are effective with most surfaces up to a distance of 200 m. The instruments are designed with several velocity ranges with accuracies of ±3 percent of the full-scale range. Although accurate measurements can be made without contact with this instrument, its major disadvantage is its relatively high cost.

9.15   SUMMARY

A broad range of topics associated with motion measurement and dynamic force measurement were considered in this chapter. A single-degree-of-freedom vibration model, using relative motion between a seismic mass and a transducer base as a motion coordinate, describes the basic dynamic characteristics common to all seismic instruments. These instruments include seismic displacement, velocity, and acceleration transducers and dynamic force transducers. The steady-state frequency response function that describes the output from these transducers is that of a forced single-degree-of-freedom mechanical system. For transient signals, input-signal rise time must be greater than five times a transducer's natural period. A similar rule of thumb is required for measuring transient half-sine or triangular inputs, where pulse durations must be greater than five times the natural period of the transducer.

Motion measurement without a fixed reference requires a seismic transducer frame. Displacement and velocity transducers are effective above their natural frequency because the seismic mass essentially remains motionless while the base moves. For this reason, these instruments have soft springs and are large. Commonly used sensing elements for displacement transducers include linear variable differential transformers (LVDTs) and resistance strain gages mounted on soft elastic springs. Accelerometers, on the other hand, require stiff springs and a small seismic mass to provide high natural frequencies. Accelerometers are very lightly damped, so the usable frequency range is limited to 20 percent of their mounted natural frequency. The piezoelectric crystal is the most widely used sensing element in accelerometers.

A piezoelectric sensor is used to generate a charge proportional to acceleration, force, or pressure and is used with voltage-follower, charge-amplifier, and built-in voltage-follower interface devices. The charge amplifier is extremely versatile, but costly, whereas the built-in voltage follower is less expensive and easier to use. Voltage followers and charge amplifiers have a single RC time constant that controls the sensor's low-frequency response. The built-in voltage follower has dual time constants, one set by the manufacturer during assembly and the other by the user. Sinusoidal signals attenuate at 6 dB per octave for frequencies below the low-frequency cutoff. Also, small RC time constants produce signal undershoot when measuring transient signals.

Systems for measuring displacement and velocity with respect to a fixed reference point include resistance potentiometers, photoelectric displacement transducers, full-field optical displacement-measurement systems, and linear- and angular-velocity sensors.

REFERENCES

1. Bell, R. L.: "Development of 100,000 g Test Facility," *Shock and Vibration Bulletin* 40, Part 2, December 1969, pp. 205–214.
2. Broch, J. T.: "Mechanical Vibrations and Shock Measurements," available from Bruel and Kjaer Instruments, Inc., Marlborough, Mass., October 1980.
3. Bruel and Kjaer Instruments: "Piezoelectric Accelerometer and Vibration Preamplifier Handbook," available from Bruel and Kjaer Instruments, Inc., Marlborough, Mass., March 1978.
4. Bruel and Kjaer Instruments: *Technical Review,* a quarterly publication available from Bruel and Kjaer Instruments, Inc., Marlborough, Mass.
   a. "Vibration Testing of Components," no. 2, 1958.
   b. "Measurement and Description of Shock," no. 3, 1966.
   c. "Mechanical Failure Forecast by Vibration Analysis," no. 3, 1966.
   d. "Vibration Testing," no. 3, 1967.
   e. "Shock and Vibration Isolation of a Punch Press," no. 1, 1971.
   f. "Vibration Measurement by Laser Interferometer," no. 1, 1971.
   g. "A Portable Calibrator for Accelerometers," no. 1, 1971.
   h. "High Frequency Response of Force Transducers," no. 3, 1972.
   i. "Measurement of Low Level Vibrations in Buildings," no. 3, 1972.
   j. "On the Measurement of Frequency Response Functions," no. 4, 1975.
   k. "Fundamentals of Industrial Balancing Machines and Their Applications," no. 1, 1981.
   l. "Human Body Vibration Exposure and Its Measurement," no. 1, 1982.
   m. "Vibration Monitoring of Machines," no. 1, 1987.
   n. "Recent Developments in Accelerometer Design," no. 2, 1987.
   o. "Trends in Accelerometer Calibration," no. 2, 1987.

5. Change, N. D.: "General Guide to ICP Instrumentation," available from PCB Piezotronics, Inc., Depew, N.Y.

6. Crosswy, F. L., and H. T. Kalb: "Dynamic Force Measurement Techniques, Part I: Dynamic Compensation," *Instrum. Control Syst.,* February 1970, pp. 81–83.

7. Crosswy, F. L., and H. T. Kalb: "Dynamic Force Measurement Techniques, Part II: Experimental Verification," *Instrum. Control Syst.,* March 1970, pp. 117–121.

8. Dove, R. C., and P. H. Adams: *Experimental Stress Analysis and Motion Measurement,* Merrill, Columbus, Ohio, 1964.

9. Ewins, D. J.: *Modal Testing: Theory and Practice,* Research Studies Press Ltd., Letchworth, Hertfordshire, England, available from Bruel & Kjaer Instruments, Inc., Marlborough, Mass.

10. Graham, R. A., and R. P. Reed (eds.): "Selected Papers on Piezoelectricity and Impulsive Pressure Measurements," Sandia Laboratories Report, SAND78-1911, December 1978.

11. Graneek, M., and J. C. Evans: "A Pneumatic Calibrator of High Sensitivity," *Engineer,* July 13, 1951, p. 62.

12. Han, S. B., and K. G. McConnell: "Effect of Mass on Force Transducer Sensitivity," *Exp. Tech.,* vol. 10, no. 7, 1986, pp. 19–22.

13. Hariharan, P.: *Optical Interferometry,* Academic Press, Orlando, Fla. 1985.

14. Herceg, E. E.: "Handbook of Measurement and Control," available from Schaevitz Engineering, Pennsauken, N.J. 1976.

15. Jones, E., S. Edelman, and K. S. Sizemore: "Calibration of Vibration Pickups at Large Amplitudes," *J. Acoust. Soc. Am.,* vol. 33, no. 11, November 1961, pp. 1462–1466.

16. Jones, E., E. Lee, and S. Edelman: "Improved Transfer Standard for Vibration Pickups," *J. Acoust. Soc. Am.,* vol. 41, no. 2, February 1967, pp. 354–357.

17. Kistler, W. P.: "Precision Calibration of Accelerometers for Shock and Vibration," *Test Engineer,* May 1966, p. 16.

18. Lally, R. W.: "Gravimetric Calibration of Accelerometers," available from PCB Piezotronics, Inc., Depew, N.Y.

19. Luxmoore, A. R.: *Optical Transducers and Techniques in Engineering Measurement,* Applied Science, London, 1983.

20. Magrab, E. B., and D. S. Blomquist: *The Measurement of Time-Varying Phenomena: Fundamentals and Applications,* Wiley, New York, 1971.

21. Neubert, H. K. P.: *Instrument Transducers: An Introduction to Their Performance and Design,* Oxford Univ. Press (Clarendon) London/New York, 1975.

22. Otts, J. V.: "Force-Controlled Vibration Testing," Sandia Laboratories Technical Memorandum, SC-TM-65-31, February 1965.

23. Pennington, D.: "Piezoelectric Accelerometer Manual," Endevco Corp., Pasadena, Calif., 1965.

24. Perls, T. A., and C. W. Kissinger: "High Range Accelerometer Calibrations," Report No. 3299, National Bureau of Standards, June 1954.

25. Peterson, A. P. G., and E. E. Gross, Jr.: "Handbook of Noise Measurement," available from General Radio Company, Concord, Mass., 1972.

26. Schmidt, V. A., S. Edelman, E. R. Smith, and E. Jones: "Optical Calibration of Vibration Pickups at Small Amplitudes," *J. Acoust. Soc. Am.,* vol. 33, no. 6, June 1961, pp. 748–751.

27. Schmidt, V. A., S. Edelman, E. R. Smith, and E. T. Pierce: "Modulated Photoelectric Measurement of Vibration," *J. Acoust. Soc. Am.,* vol. 34, no. 4, April 1966, pp. 455–458.

28. Shock and Vibration Measurement Technology, notes from an applications-oriented short course available from Endevco, Inc., San Juan Capistrano, Calif.

29. Steel, W. H.: *Interferometry,* 2nd ed., Cambridge Univ. Press, London, New York, 1983.

30. Thomson, W. T.: *Theory of Vibration with Applications,* 3rd ed., Prentice–Hall, Englewood Cliffs, N.J., 1988.

EXERCISES

9.1 The expression $x = 8\cos(12t) + 6\sin(12t)$ is a harmonic function with a frequency of 12 rad/s. Show that x can be written as either $x = 10\cos(12t + \theta_1)$ or $x = 10\sin(12t + \theta_2)$. Determine the phase angles θ_1 and θ_2. Which angle is leading and which angle is lagging the reference phase? What is the reference phase for each expression of x?

9.2 A simple harmonic motion has an amplitude of 0.001 in. and a frequency of 100 Hz. Determine the maximum velocity (in./s) and the maximum acceleration (g) associated with the motion. If the amplitude of the motion doubles and the frequency remains the same, what is the effect on the maximum velocity and the maximum acceleration? If the frequency doubles and the amplitude of motion remains the same, what is the effect on the maximum velocity and the maximum acceleration?

9.3 Plot the real and imaginary parts of $H(\omega)$ (see Eq. 9.6) over the frequency range $0 < r < 2$ if $k = 1.0(10^6)$ lb/in. and $d = 0.02$. Compare these plots to Fig. 9.3. Explain the differences obtained.

9.4 A uniform rigid bar of mass m_b and length ℓ rotates in the vertical plane about the pin at O, as shown in Fig. E9.4. An accelerometer is mounted at point B at a distance b from pin O. Show that the tangential acceleration at point B is

$$A_B = -\left[\frac{3b}{2\ell}\right]g\sin(\omega_n t)$$

where $\omega_n = \sqrt{3g/2\ell}$ is the bar's pendulous natural frequency. Show that the acceleration indicated by the accelerometer is

$$A_{acc} = -\left(1 - \frac{3b}{2\ell}\right)g\sin(\omega_n t)$$

Why does the accelerometer indicate an incorrect acceleration? Hint: See Eq. 9.4 and evaluate the forces acting on the seismic mass m of the accelerometer.

9.5 Show that the mechanical response of an underdamped seismic instrument to a step input is given by Eq. 3.37. Sketch a typical response and compare your sketch with Fig. 9.4.

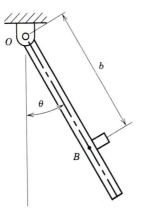

Figure E9.4

9.6 A velocity meter is being designed with a natural frequency of 5 Hz, damping of 5 percent, and a sensitivity of 8.3 mV/(in./s). The magnetic core weighs 0.20 lb and is mounted on soft springs. Determine the required spring constant for supporting the core. The velocity meter is mounted on a surface that is vibrating with a maximum velocity of 8.0 in./s. Determine the peak output voltage and phase angle if the frequency of vibration is (a) 8.0 Hz and (b) 20 Hz. Which measurement has the greatest error? Why? Would performance be improved by increasing the damping by a factor of 2? How could you increase the damping electrically?

9.7 An accelerometer is used to measure a periodic signal that can be expressed as

$$a(t) = a \sin(0.2\omega_n t) - 0.4a \sin(0.6\omega_n t)$$

where ω_n is the natural frequency of the accelerometer. The relative motion of the transducer can be expressed as

$$z(t) = b_i \sin(0.2\omega_n t - \phi_1) + b_3 \sin(0.6\omega_n t - \phi_3)$$

Determine the coefficients b_1 and b_3 and phase angles ϕ_1 and ϕ_3 when the transducer damping is (a) 5 percent and (b) 60 percent. Which damping condition gives the best modeling? Why?

9.8 Verify Eqs. 9.18 as the input–output relationship for the combined linear and angular accelerometer. How do both the natural frequency of the beam and damping affect each input–output relationship?

9.9 A piezobeam accelerometer is needed to measure a combined sinusoidal motion that is estimated to have a linear-velocity magnitude of 2.0 in./s and an angular-velocity magnitude of 6.0 rad/s at a frequency of 100 Hz. Should the model with an angular-acceleration sensitivity of 0.5, 5, or 50 mV/(rad/s^2) be selected?

9.10 An accelerometer with a charge sensitivity of 83 pC/g and a capacitance of 1000 pF is connected to a voltage follower with an input connector capacitance of 15 pF (in parallel with the cable capacitance), a 10,000-pF blocking capacitor, and a 100-MΩ resistance. A 10-ft-long cable with a capacitance of 312 pF connects the accelerometer and the voltage follower. Determine:
(a) the instrument's voltage sensitivity in mV/g
(b) the time-constant error between $\tau = RC$ and $\tau = RC_{eq}$
(c) the -3-dB low-cutoff frequency in Hz

9.11 A Kistler 808A accelerometer (see Table 9.1) is connected by 10 ft of 30 pF/ft transducer cable to a voltage follower, as shown in Fig. 9.10. Determine:
(a) the minimum blocking capacitance C_b if the error owing to C_b must be limited to 1 percent
(b) the gain required to obtain a voltage sensitivity of 10 mV/g
(c) the low-frequency time constant if $R = 1000$ MΩ

9.12 Typical characteristics for a charge-amplifier circuit are listed in Table 9.4. Estimate the minimum open-loop gain of the first op-amp when the feedback capacitance is (a) 10 pF and (b) 1000 pF. If the calibration capacitor $C_{cal} = 10{,}000$ pF in Fig. 9.13 is inadvertently grounded, how would this affect the performance of the charge amplifier?

9.13 The charge-amplifier sensitivity was set at 4.86 pC/g while measuring acceleration data. Later, it was discovered that the transducer's sensitivity is 8.46 pC/g. Determine the required correction factor to apply to the data.

9.14 An accelerometer with a charge sensitivity $S_q = 170$ pC/g and a capacitance $C_t = 10,000$ pF is used to measure low-level accelerations. The transducer cable is 30 ft long with a capacitance of 30 pF/ft. Assume that the op-amp has a minimum gain $G_1 > 20,000$. What minimum feedback capacitance can be used if source capacitance error is to be less than 0.5 percent? What is the unit sensitivity (g/V) if $C_f = 200$ pF and the transducer dial setting is (a) 0.085, (b) 0.170, and (c) 0.340? Which transducer setting gives the largest voltage signal for a given level of acceleration?

9.15 Verify the minimum values of $\omega\tau$ for 5-percent errors as given in Table 9.5 when $\tau_1/\tau = 1$, 10, and 100. What practical conclusion can be made about time-constant requirements?

9.16 A transient time history increases in a near linear fashion from 0 to the maximum value in 1.0 ms, remains at the maximum value for 3 ms, and then returns to zero in 0.50 ms. Based on this information, determine the minimum transducer natural frequency and time constant so that ringing and undershoot errors are bounded by 5 percent.

9.17 Show that the response of a dual time-constant system to a step input is

$$v = \frac{S_v A_0}{\tau - \tau_1} \left(\tau e^{-t/\tau_1} - \tau_1 e^{-t/\tau} \right)$$

Then show that the effective time constant given by Eq. 9.36 can be obtained by matching the initial slope of the actual response curve to that of an equivalent circuit described by

$$v = S_v A_0 e^{-t/\tau_e}$$

Compare results when (a) $\tau_1 = \tau$ and (b) $\tau_1 = 10\,\tau$.

9.18 The PCB 302A accelerometer listed in Table 9.1 is to be used to measure a half-sine shock loading with a time duration of 0.1 s. If the accelerometer is connected to a recording device with a 1.0-MΩ input resistance through a 1.0-μF coupling capacitor C_1 (see Figs. 9.14 and 9.15), estimate the peak reading error and the amount of undershoot.

9.19 A traveling microscope with a least count of 0.0001 in. is used to measure the peak-to-peak displacement during a sinusoidal calibration of an accelerometer. The microscope is focused on an object having a diameter of 0.0011 in., as shown in Fig. E9.19. For the readings shown, determine the acceleration (g) if the frequency of oscillation is
(a) 50 Hz   (c) 500 Hz
(b) 100 Hz   (d) 1000 Hz

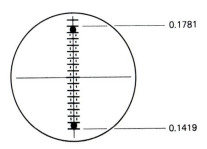

0.1781

0.1419

Figure E9.19

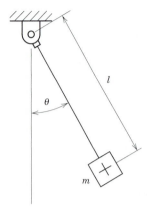

Figure E9.20

What is your estimate of the percent error in these measurements? Are any of these acceleration levels difficult to obtain if the moving mass weighs 0.35 lb and the vibration exciter can deliver a maximum force of 100 lb?

9.20 Several calibration methods require accurate knowledge of the local acceleration of gravity. A common method for obtaining this quantity utilizes a simple pendulum (see Fig. E9.20). If a pendulum having a length of 25 in. is used with a stopwatch that can be started and stopped consistently within 0.15 s, which variable, pendulum length or period of oscillation, must be measured most accurately? How can the 0.15-s uncertainty in the period measurement be overcome? What shape should the pendulum mass have so that its center of mass can most easily be located accurately? Would increasing the pendulum length or using multiple period measurements increase the accuracy of the measurement of g?

9.21 Data obtained from four gravimetric calibration tests are listed in the following table:

	Test Number			
	1	2	3	4
v_{mg} (mV)	46.6	46.5	46.7	46.5
v_f (mV)	19.48	78.1	302	1162
v_a (mV)	4.23	16.83	64.5	254.0

Determine the sensitivity S_a of the accelerometer. After the tests were conducted it was determined that the local acceleration of gravity was 31.30 ft/s^2 instead of the nominal 32.17 ft/s^2. What effect will this change have on the previously determined values of S_a?

9.22 The rms voltage ratio v_f/v_a versus calibration weight W_c curve from a typical sinusoidal calibration of a dynamic force gage is shown in Fig. 9.23. If a similar curve for a dynamic force gage being calibrated has a slope of 8.06 g/lb and intersects the vertical axis at a value of 0.403 V/V, determine:
(a) the sensitivity of the force gage if the sensitivity of the accelerometer is 6.20 mV/g
(b) the weight of the seismic mass

9.23 In a dynamic force gage calibration, the transducer sensitivity dial of the charge amplifier being used with the accelerometer was set at $b = 0.276$ and the transducer sensitivity dial of the charge amplifier being used with the force gage was set at $b = 1.00$. The charge sensitivity of the accelerometer is 2.76 pC/g. The feedback capacitor of the charge amplifier being used with the accelerometer was set at 100 pF and that of the force transducer was set at 2000 pF. The slope of the v_f/v_a versus W_c curve is 7.82 g/lb. Determine the charge sensitivity of the force gage.

9.24 Explain how Eq. 9.43 is changed, if at all, when the accelerometer in Fig. 9.22 is part of the calibration mass m_c so that the acceleration being measured is $\ddot{y}$ instead of $\ddot{x}$.

9.25 An impact hammer with attached force transducer is to be calibrated by impacting a mass suspended as a pendulum, as shown in Fig. 9.25. The sensitivity of the accelerometer is 7.30 mV/g. The peak voltage ratio v_f/v_a versus pendulum weight W_p curve has a slope of 7.19 g/lb. The hammerhead weighed 0.165 lb and the hammer body weighed 0.769 lb during the calibration tests. Show that the voltage sensitivity of the force gage as installed is 52.5 mV/lb. At a later date, the weight of the hammerhead was increased to 0.611 lb and the weight of the body was increased to 1.278 lb in order to obtain some desired impact-force characteristics. Show that the voltage sensitivity of the force gage changes to 43.2 mV/lb.

9.26 A pressure transducer having a charge sensitivity of 1.46 pC/psi is to be used with a charge amplifier to measure hydraulic pump pressures that range from 100 to 1000 psi. The charge amplifier has a calibration capacitor ($C_{cal} = 1000$ pF), as shown in Fig. 9.13. The required voltage sensitivity for this application is 5 mV/psi. Specify:
   (a) the transducer sensitivity setting b
   (b) the feedback capacitor setting C_f (see Table 9.4 for standard values)
   (c) the peak calibration charges to simulate 100 psi and 1000 psi
   (d) the peak calibration voltages required to simulate 100 psi and 1000 psi
   (e) the anticipated calibration output voltages corresponding to the two pressures
   How would this problem be changed if acceleration were specified in g for the measurement variable?

9.27 A PCB 200A05 (Table 9.2) force transducer is mounted on a lightweight hammer, as shown in Fig. 9.24. Using the impact-hammer calibration method shown in Fig. 9.25 and Eq. 9.47, devise a method to determine both the force transducer sensitivity S_f and the effective hammer masses (m_h and m_b) so that the mass ratio M can be calculated for various hammer-body and impact-tip mass combinations and new values for transducer sensitivity S_f^* can be predicted.

9.28 The "pop test" illustrated in Fig. E9.28 provides a convenient means for testing the overall performance of a pressure transducer. The pressure variation resulting from rupture of the diaphragm approaches a step input; therefore, typical step-input damped response occurs. During a specific test, the first three peaks were 9.01, 7.59, and 6.67 mV, while the first three valleys were 0, 1.778, and 2.92 mV for an initial chamber pressure of 3.0 psig. The final steady-state

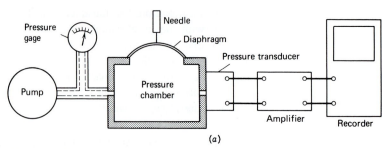

(a)

Figure E9.28

response was 5.0 mV. The three peaks occurred at 0.375, 1.125, and 1.875 ms, and the valleys occurred at 0, 0.750, and 1.50 ms. Determine
(a) the natural frequency of the transducer in Hz
(b) the damping in percent
(c) the transducer sensitivity in mV/psi

9.29 A charge amplifier has a 1000-pF calibration capacitance and is to be used with an accelerometer that has a charge sensitivity of 41.8 pC/g. A maximum acceleration of 80 g is to be measured. Using the charge-amplifier data in Table 9.4, select appropriate values for b, C_f, and v^*_{cal} if the recorder has a ± 10.0 V full-scale voltage range. What is the system's overall voltage sensitivity?

9.30 When a step-input voltage is applied to the calibration capacitor in a charge amplifier, the output voltage may look like that shown in Fig. 9.26. What terms in a typical measurement system contribute to the exponential decay? What is the source of the damped oscillation and rise time? Is the damped oscillation the result of the mechanical response of the transducer?

9.31 A 20-kΩ single-turn potentiometer has been incorporated into a displacement-measuring system as shown in Fig. E9.31. The potentiometer consists of 0.008-in. diameter wire wound onto a 1.25-in. diameter ring. The pulley diameters are 2.00 in. Determine:
(a) the minimum load resistance that can be used if nonlinearity error must be limited to 0.25 percent
(b) the smallest motion x that can be detected with this system

9.32 The potentiometer circuit shown in Fig. E9.32 is used to measure angular position θ. The capacitor C is used to reduce contact bounce, and the op-amp

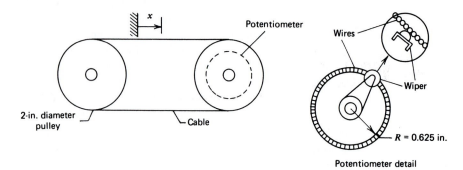

Figure E9.31

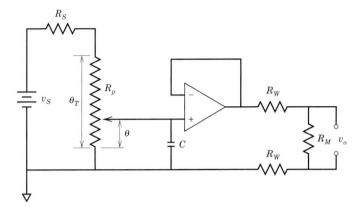

Figure E9.32

isolates the potentiometer from long lead wires and recording-instrument resistance loading. The potentiometer being used can rotate 320 deg, has a resistance $R_p = 4.0\,\text{k}\Omega$, and is capable of dissipating 0.02 W of power in most environments. Show that

(a)   the maximum voltage that can be applied to the potentiometer is 8.94 V
(b)   the value of series resistor R_s in order to protect the potentiometer if $V_s = 15.0$ V is 2.71 kΩ
(c)   the output sensitivity S is given by

$$S = \left(\frac{R_M}{R_M + 2R_W}\right)\left(\frac{R_p}{R_p + R_S}\right)\frac{V_S}{\theta_T}$$

where θ_T is the range of the potentiometer in degrees.

9.33   The circuit in Exercise 9.32 is to have a minimum sensitivity of 10.0 mV/deg when 400 ft of AWG 28 copper lead wire (66.2 Ω per 1000 ft) is being used. The op-amp is capable of driving an output circuit continuously at 20 mA, and a 10-V dc power supply is to be used.

(a)   What values of R_M and R_s would you select to achieve the desired sensitivity?
(b)   What electrical components would you add to the circuit to easily adjust the circuit's sensitivity?
(c)   How would you calibrate the system?

9.34   Assume that the current generated by the photoelectric sensor in Figs. 9.35 and 9.36 is given by $i = Ky$. Show that the governing differential equation for the sensor is

$$R_M C \dot{v}_o + v_o = R_M K y \qquad \text{for Fig. 9.35}$$

and

$$\frac{R_f C}{G}\dot{v}_o + \frac{1 + G}{G}v_o = -R_f K y \qquad \text{for Fig. 9.36}$$

where G is the op-amp's open-loop gain. Compare circuit behavior when G is large ($> 10,000$) and $R_M = R_f$. Why is the high-frequency performance improved?

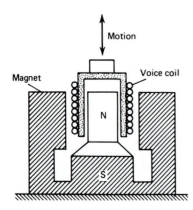

Motion

Magnet

Voice coil

N

S

Figure E9.35

9.35   The voice coil of a small speaker can be used as a velocity sensor. The coil, shown schematically in Fig. E9.35, has a diameter of 20 mm and has 100 turns of wire wrapped onto the nonmagnetic core. During calibration, the sensor exhibited a sensitivity of 60 mV/(m/s) over a range of frequencies. Estimate the magnetic field strength B in the gap occupied by the voice coil.

9.36   The magnetic flux density B (Wb/m^2) generated by an electric solenoid can be expressed as

$$B = 12.57(10^{-7})\frac{Ni}{\ell}$$

where N is the number of turns, i is the current in amperes, and ℓ is the length of the solenoid in meters. The ring magnet, shown in Fig. E9.36, is being designed to have a field strength of 0.10 Wb/m^2. The solenoid will consist of 1000 turns of wire wrapped uniformly over half of the 10-mm-diameter steel ring. Determine the required current i and estimate the voltage that would be induced in a single length of wire as it passed through the gap at a speed of 1.0 m/s.

9.37   A linear-velocity transducer (LVT) has an inductance of 16.5 mH, a resistance of 6.2 Ω, and a sensitivity of 30 mV/(in./s). Evaluate the performance of this sensor when it is used in conjunction with a recording instrument having an input resistance of (a) 100 Ω and (b) 1000 Ω if the measurement errors are limited to 5 percent in magnitude and 10 deg in phase.

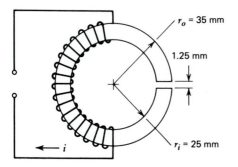

r_o = 35 mm

1.25 mm

r_i = 25 mm

i

Figure E9.36

9.38   A shaft has a 128-tooth gear mounted on it and is rotating at 2650 rpm. A magnetic proximity sensor gives a pulsatile signal as the gear rotates beneath it. What is the fundamental frequency of these pulses in Hz?

9.39   Determine the Doppler shift frequency when a helium-neon laser is being used to measure a 100 mm/s velocity.

9.40   The charge sensitivity, voltage sensitivity, and capacitance for an Endevco 2222C accelerometer transducer are listed in Table 9.1. Typical coaxial transducer cable has a nominal capacitance of 30 pF/ft. Show that approximately 123 in. of cable must have been connected to the transducer to give the voltage sensitivity listed in the table. Show that the 3-dB cutoff frequency is 2.0 Hz when the transducer is connected to a voltage follower with an input resistance of 100 MΩ. Show that the voltage sensitivity drops to 1.52 mV/g when the transducer is connected to the voltage follower with 15 ft of transducer cable. At what frequency (Hz) will the attenuation be 10 percent with this longer cable?

Chapter **10**

Analysis of Vibrating Systems

10.1 INTRODUCTION

Investigation of the vibrations associated with structures or machines usually involves one of the two situations illustrated in Fig. 10.1. The first type of vibration problem deals with a system that has one or more internal excitation sources S_P, as shown in Fig. 10.1a. The sources are usually characterized by measuring the structural response at some point Q on the external surface of the structure or machine with an accelerometer.

The second type of problem involves a known excitation force $F_P(t)$, which is applied at location P on the external surface of the system. A response $x_Q(t)$ is measured at some exterior point Q, as indicated in Fig. 10.1b. In both situations, the

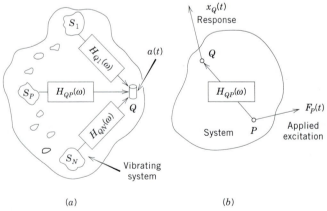

(a)	(b)

Figure 10.1 Two types of vibrating systems. (a) System with unknown internal excitation sources. (b) System with known excitation applied at an external point.

frequency response function $H_{QP}(\omega)$ for the structural system controls the magnitude and phase of the output signal recorded at point Q.

The output signal $a(t)$ in Fig. 10.1a results from the combined effects of the vibration sources S_P and their respective frequency response functions (FRFs) $H(\omega)$. Furthermore, the various FRFs depend on the design of the structure, the location of the source, and the location and mass of the transducer. The transducer's FRF also influences the output signal. If a transducer is moved, all of the FRFs change. At certain frequencies, vibration will disappear at some locations while it is maximum at other locations. Thus, transducer location significantly influences the output signal.

A typical response signal, recorded with a transducer, is illustrated in Fig. 10.2a. Such traces are extremely complex; however, when they are properly analyzed, they can provide much useful information relating to the vibrational characteristics of a system. Response signals $f(t)$ are typically analyzed in two domains: the time domain and the frequency domain.

In the time domain, two single-value descriptions of the complicated trace shown in Fig. 10.2a are the temporal mean and the temporal root mean square. The word *temporal* indicates that these single-value parameters are based on time averages.

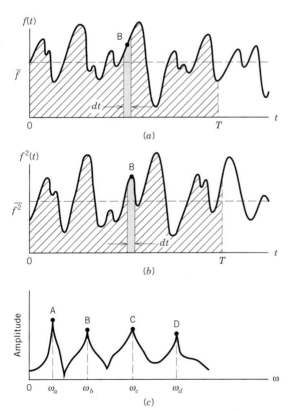

Figure 10.2 Methods of analyzing the signal $f(t)$. (a) Signal $f(t)$ and determination of the temporal mean. (b) Signal $f^2(t)$ and determination of the temporal mean square. (c) Frequency spectrum of the signal $f(t)$ showing significant frequency components.

10.1.1 Temporal Mean

The temporal mean $\overline{f}$, a statistical quantity (see discussion of mean value in Chapter 13), is defined as

$$\overline{f} = \lim_{T \to \infty} \frac{1}{T} \int_0^T f(t)\, dt \tag{10.1}$$

where T is the integration or averaging time. This time-averaging concept is illustrated in Fig. 10.2a, where $f(t)\, dt$ represents a differential area under the time-history curve at point B. Clearly, the rectangular area $\overline{f}T$ is equal to the area under the signal from 0 to T. Since fluctuations in $\overline{f}$ become smaller as time T increases, long averaging times are used to give accurate estimates of $\overline{f}$. The temporal mean $\overline{f}$ represents the signal's zero-frequency component, which is often referred to as the static or dc component.

10.1.2 Temporal Mean Square and Root Mean Square

Two other single-value statistical measures of a signal are the temporal mean square and the root mean square. The temporal mean square $\overline{f^2}$ is defined by

$$\overline{f^2} = \lim_{T \to \infty} \frac{1}{T} \int_0^T f^2(t)\, dt \tag{10.2}$$

which is a time average of the $f^2(t)$ trace, as shown in Fig. 10.2b.

The root mean square (A_{rms}) is the square root of the temporal mean square, so that

$$A_{\mathrm{rms}} = \sqrt{\overline{f^2}} \tag{10.3}$$

Unfortunately, $\overline{f}$, $\overline{f^2}$, and A_{rms} are not unique. Thus they provide little detail about the complicated motion producing the signal $f(t)$ unless a statistical probability density or other more restrictive description is included.

In the frequency domain, the data in the $f(t)$ signal are represented as a function of frequency ω. This representation, presented in Fig. 10.2c, gives the amplitudes of the frequency components as a function of the excitation frequency ω and is known as the frequency spectrum. The peaks at ω_a, ω_b, and so on can be related to the unbalance of rotating shafts, blade passage rates in a blower-type mechanism, gear mesh rates, ball-bearing noise, structural resonances, and the like. It is evident that a frequency spectrum is more useful than $\overline{f}$, $\overline{f^2}$, or A_{rms}, because it indicates discrete frequencies that are related to the operating characteristics of the system.

10.2 SINUSOIDAL SIGNAL ANALYSIS

Sinusoidal functions, which arise in solutions for underdamped second-order systems, were used previously to describe the response of a transducer and a recorder. The sinusoid is also an important function in signal analysis, where it is used to characterize signals from vibrating systems. Consider, for example, the function

$$f(t) = B \cos(\omega t + \phi) = B \cos \theta \tag{10.4}$$

where B is the amplitude and $\theta = \omega t + \phi$ is the combined argument. Equation 10.4 represents both sine and cosine functions. For example, $\phi = 0$ gives a cosine function, and $\phi = \pi/2$ gives a sine function.

The Euler formula

$$e^{\pm j\theta} = \cos\theta \pm j\sin\theta \tag{a}$$

relates sine and cosine functions to the complex exponential function. Thus,

$$\cos\theta = \frac{e^{j\theta} + e^{-j\theta}}{2} \qquad \sin\theta = \frac{e^{j\theta} - e^{-j\theta}}{2j} \tag{10.5}$$

Substituting these equations into Eq. 10.4 gives a real-valued function $f(t)$, which can be written as

$$f(t) = B\cos(\omega t + \phi) = \left(\frac{B}{2}e^{j\phi}\right)_{ccw} e^{j\omega t} + \left(\frac{B}{2}e^{-j\phi}\right)_{cw} e^{-j\omega t}$$

$$= ce^{j\omega t} + c^* e^{-j\omega t} \tag{10.6}$$

where c and c^* are two counter-rotating vectors in the complex plane, each with an amplitude $B/2$, as illustrated in Fig. 10.3. Counter-rotating vectors introduce the concept of negative frequency; that is, the vector representing $(+\omega t)$ rotates in a counterclockwise (ccw) direction (the $e^{j\theta}$ term), while the vector c^* representing $(-\omega t)$ rotates in a clockwise (cw) direction (the $e^{-j\theta}$ term). In Fig. 10.3, the real-axis projections at point P_1 of these two vectors add to give the cosine response (point P) with magnitude $B\cos(\omega t + \phi)$, whereas the imaginary-axis projections at points P_2 cancel. The reference line for measuring all angles is the real axis.

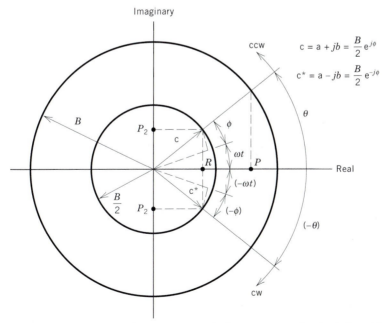

Figure 10.3 Counter-rotating vectors c and c^* generate a real-valued sinusoid $B\cos(\omega t + \phi)$.

From Eq. 10.6, it is clear that vectors c and c^* may be expressed as complex conjugates

$$c = a + jb = \sqrt{a^2 + b^2} \, e^{j\phi} = \frac{B}{2} e^{j\phi} \qquad \text{(ccw rotation)}$$

$$c^* = a - jb = \sqrt{a^2 + b^2} \, e^{-j\phi} = \frac{B}{2} e^{-j\phi} \qquad \text{(cw rotation)} \qquad \textbf{(10.7)}$$

where

$$a = \frac{B}{2} \cos \phi \qquad b = \frac{B}{2} \sin \phi \qquad \tan \phi = \frac{b}{a} \qquad \textbf{(10.8)}$$

A sinusoidal signal with a constant offset D can be expressed as

$$f(t) = D + B \cos (\omega t + \phi) \qquad \textbf{(10.9)}$$

The temporal mean $\overline{f}$ of this signal is obtained from Eqs. 10.9 and 10.1 as

$$\overline{f} = D + B \left[\frac{\sin (\omega T + \phi) - \sin \phi}{\omega T} \right] \cong D \qquad \textbf{(10.10)}$$

This result indicates that $\overline{f} = D$, because the bracketed term vanishes with increasing time T.

The temporal root mean square A_{rms} is obtained from Eqs. 10.9 and 10.2 as

$$A_{rms}^2 = D^2 + \frac{B^2}{2} + 2DB \left[\frac{\sin (\omega T + \phi) - \sin \phi}{\omega T} \right] + \frac{B^2}{2} \left[\frac{\sin (2\omega T + 2\phi) - \sin 2\phi}{2\omega T} \right]$$

which reduces to

$$A_{rms}^2 \cong \frac{B^2}{2} + D^2 = B_{rms}^2 + D^2 \qquad \textbf{(10.11a)}$$

$$A_{rms} = \sqrt{B_{rms}^2 + D^2} \qquad \textbf{(10.11b)}$$

for either large values of ωT or ωT = multiples of 2π. This result indicates that the sinusoidal term fluctuates at frequencies of ω and 2ω, with an amplitude that decreases with time T. Equation 10.11a illustrates two important points. First, if $B = 0$,

$$A_{rms} = D \qquad \textbf{(10.12a)}$$

Second, if $D = 0$ (the case for an ac-coupled instrument),

$$A_{rms} = B_{rms} = \frac{B}{\sqrt{2}} = 0.707 \, B = \sqrt{2c \, c^*} \qquad \textbf{(10.12b)}$$

where B_{rms} is the sinusoidal rms amplitude. Note that the vectors c and c^* are also related to B_{rms} and that the B_{rms} term fluctuates with frequency 2ω.

The rms amplitude B_{rms} and the peak amplitude B are related by Eq. 10.12b *for sinusoidal functions only*. Under no circumstances can the A_{rms} amplitude for an arbitrary signal be converted to an equivalent peak amplitude by using Eq. 10.12b. An examination of Eq. 10.11 makes this point clear, because the mean square includes the square of the mean.

This analysis implies that an ac-coupled voltmeter measures only the B_{rms} signal value because ac-coupling eliminates D. Consequently, two voltmeters are needed for analysis of a signal: an ac-coupled meter to measure B_{rms}, and a dc meter to measure D. This combination of two voltmeters provides both B_{rms} and D, which are needed to describe the signal. The overall rms amplitude A_{rms} can be computed from Eq. 10.11.

Several techniques are used to present the frequency spectra of the sinusoidal signal given by Eq. 10.9. These presentation techniques, shown in Fig. 10.4, include single-sided $(+\omega)$ and double-sided $(\pm\omega)$ frequency spectra for $\bar{f}$, $\overline{f^2}$, and A_{rms}.

In each case, mean value D occurs at zero frequency and is the same in both single-sided and double-sided representations. The sinusoidal amplitudes $(B, B/2, B^2/2, B^2/4, B/(2\sqrt{2}), $ or $B/\sqrt{2})$ occur at frequencies of $(\pm\omega)$ or $(+\omega)$. The magnitude depends on whether a single-sided or a double-sided presentation technique is used.

10.3   CHARACTERISTICS OF SIGNALS

Three types of signals, periodic, transient, and random, are encountered in vibration analysis. Periodic signals are repetitive with a period T; transient signals vary over a relatively short time period and are zero for all other times; random signals contain all frequencies with random phase. Characteristics of these three types of signals are described in this section.

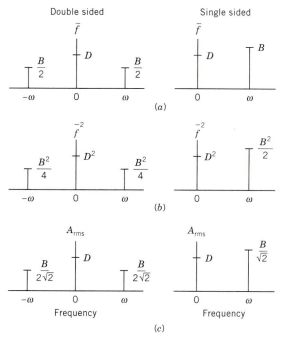

Figure 10.4 Three different frequency spectra for the function $D + B\cos(\omega t + \phi)$: (a) amplitude, (b) mean square, and (c) root mean square.

10.3.1 Periodic Signals

A periodic signal is a common dynamic response for vibrating structures. This signal repeats itself every T seconds; that is, $f(t + T) = f(t)$. The fundamental frequencies ω_0 and f_0 are related to the period T by

$$\omega_0 = 2\pi \left(\frac{1}{T}\right) = 2\pi f_0 \tag{a}$$

A. Fourier Series

Any periodic and continuous real-valued function $f(t)$ may be written in terms of a Fourier series, which is given in terms of an exponential summation by the equation

$$f(t) = \sum_{p=-\infty}^{+\infty} c_p e^{j p \omega_0 t} \tag{10.13a}$$

where the complex Fourier coefficient c_p is given by

$$c_p = \frac{1}{T} \int_t^{t+T} f(t) e^{-j p \omega_0 t} \, dt \tag{10.13b}$$

The use of Eq. 10.13a provides a means for relating a function $f(t)$ in the time domain to a function $f(\omega)$ in the frequency domain.

The complex Fourier coefficients c_p are obtained by integrating over one period T, with the integration starting at any time t. Coefficients c_p are determined to fit $f(t)$ with the function $(e^{\pm j p \omega_0 t})$ over one fundamental period T. These coefficients occur in complex conjugate pairs as previously defined by Eqs. 10.7 and 10.8. Thus

$$c_p = a_p + j b_p$$
$$c_{-p} = a_p - j b_p = c_p^* \tag{10.14}$$

From Eqs. 2.37 and 2.38, the magnitude and phase of c_p are

$$|c_p| = \sqrt{a_p^2 + b_p^2} = \sqrt{c_p c_p^*} \quad \text{(magnitude)}$$

$$\tan \phi_p = \frac{b_p}{a_p} \quad \text{(phase)} \tag{10.15}$$

The complex conjugate phase is the negative of ϕ_p.

The complex form of the Fourier series defined by Eq. 10.13a has other useful properties. First, since c_p^* is the complex conjugate of c_p, only one-half of the coefficients (those for positive values of p) have to be determined. Second, only real coefficients a_p result when $f(t)$ is an even function ($b_p = 0$) and $f(-t) = f(t)$. Third, only imaginary coefficients b_p result when $f(t)$ is an odd function ($a_p = 0$) with $f(t) = -f(t)$. Knowledge of these properties often simplifies a theoretical analysis. In analysis of experimental signals, however, the instant when $t = 0$ is usually arbitrarily assigned and the coefficients c_p are complex quantities with both real and imaginary parts.

B. Mean

From Eq. 10.13b, it is clear that the temporal mean $\overline{f} = c_0$. It is possible to use a single period T and obtain a good estimate of $\overline{f}$, since the integrals of all harmonics go to zero over the period T.

C. Mean Square

The mean square of the signal is obtained by substituting Eq. 10.13 into Eq. 10.2. Integration gives

$$A_{rms}^2 = \frac{1}{T} \int_0^T [f(t)]^2 \, dt = c_0^2 + 2\sum_{p=1}^{\infty} |c_p|^2 \qquad (10.16)$$

which is known as Parseval's formula. Substituting Eq. 10.7 into Eq. 10.16 gives

$$A_{rms}^2 = c_0^2 + \sum_{p=1}^{\infty} \frac{1}{2} B_p^2 = c_0^2 + \sum_{p=1}^{\infty} B_{rms_p}^2 \qquad (10.17)$$

The importance of these results is that the amplitude squared of each frequency component contributes to A_{rms}^2. It is also evident from Eq. 10.17 that many different periodic signals will give the same A_{rms} value, so that there is no unique relationship between a multiple-component periodic signal and A_{rms}.

10.3.2 Transient Signals

A transient signal shows pulselike variations over a relatively short period of time T_d and is essentially zero or constant for all other times. A typical transient signal is illustrated in Fig. 10.5a.

To show the difference between periodic and transient behavior, consider the rectangular pulse of amplitude A, duration T_d, and repeat period T, as shown in Fig. 10.5b. Let

$$T = \beta T_d \qquad \textbf{(a)}$$

with a fundamental frequency

$$\omega_0 = \frac{2\pi}{T} = \frac{2\pi}{\beta T_d} \qquad \textbf{(b)}$$

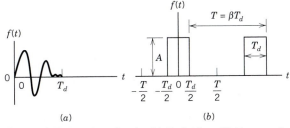

(a)   (b)

Figure 10.5 Transient signals. (a) Definition. (b) Rectangular pulses of duration T_d and period T.

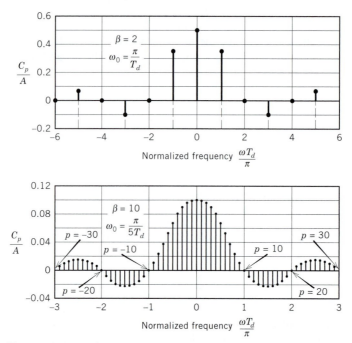

Figure 10.6 Fourier coefficients used to describe a rectangular pulse where $T = \beta T_d$ with $\beta = 2$ and 10.

These rectangular pulses become isolated in time with increasing values of β, and essentially become a single pulse of duration T_d when β is very large.

Combining Eqs. (a) and (b) with Eq. 10.13b gives only real-valued coefficients c_p as

$$c_p = \frac{AT_d}{T}\left[\frac{\sin\left(p\pi/\beta\right)}{p\pi/\beta}\right] = \frac{A}{\beta}\,\text{sinc}\,\left(p\pi/\beta\right) = \frac{A}{\beta}\,\text{sinc}\,x_p \qquad \textbf{(10.18)}$$

where $x_p = p\pi/\beta$. The results for c_p from Eq. 10.18 are presented in Fig. 10.6 for $\beta = 2$ and 10. Examination of this figure shows that the magnitudes c_p/A are inversely proportional to β, the frequency components increase with β and are much closer together, and the zeros occur when $(p/\beta) = k$, with $k = 1, 2, 3$, and so on.

It is clear that representing transient signals with a Fourier series requires extremely large β and that a large number of coefficients are involved. Thus, the Fourier series is not a suitable method for analysis of transient signals.

A. Fourier Transform

The Fourier transform is obtained from Eq. 10.13b by defining a new Fourier coefficient as $C(\omega) = c_p T$ and by taking the limit as T is allowed to go to infinity. The result is

$$C(\omega) = \int_{-\infty}^{+\infty} f(t)e^{-j\omega t}\,dt \qquad \textbf{(10.19a)}$$

and

$$f(t) = \frac{1}{2\pi} \int_{-\infty}^{+\infty} C(\omega)e^{j\omega t}\,d\omega \qquad \textbf{(10.19b)}$$

These relations, which are used in transient signal analysis, indicate that all frequencies are present in the representation. Each frequency component makes an infinitesimal amplitude contribution to $f(t)$, which is

$$\text{amplitude} = \frac{1}{2\pi}\, C(\omega)\, d\omega \qquad \textbf{(a)}$$

Equations 10.19 indicate that $C(\omega)$ is a spectral density. The spectral density is expressed in terms of units of $f(t)$ per hertz, such as lb/Hz when $f(t)$ has force units. Thus, a subtle difference exists between a frequency spectrum and a spectral density. A frequency spectrum is a display of discrete sinusoidal amplitudes as a function of discrete frequencies (i.e., c_p versus $p\omega_0$). A spectral density is a continuous display of amplitude density as a function of frequencies [i.e., $C(\omega)$ versus ω].

To show the differences between a frequency spectrum and spectral density, consider the limit of the rectangular pulse shown in Fig. 10.5*b* as T goes to infinity. Integration of Eq. 10.19a over the interval $-T_d/2$ to $T_d/2$ gives

$$C(\omega) = AT_d \frac{\sin(\omega T_d/2)}{(\omega T_d/2)} = AT_d \, \text{sinc}\left(\frac{\omega T_d}{2}\right) \qquad \textbf{(10.20)}$$

Comparing Eqs. 10.18 and 10.20 shows the differences. The amplitude of continuous function $C(\omega)$ is AT_d, compared with AT_d/T for discrete function c_p. The argument of the sinc function is $(\omega T_d/2)$ if ω is continuous and $(p\pi/\beta)$ if $\omega = p\omega_0$ is discrete. The differences are small when $\beta \geq 10$, as indicated by the continuous and discrete forms of representation shown in Fig. 10.7. A discrete frequency spectrum can be converted to a spectral density by multiplying the discrete frequency spectrum by T. However, β must be greater than 10 in order to use the Fourier series method to simulate the Fourier transform method in transient signal analysis. The reasons for this requirement become more understandable with the discussion of digital frequency analyzers in Sections 10.7 and 10.8.

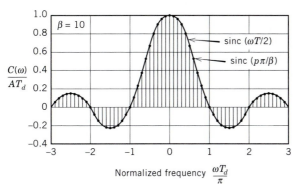

Figure 10.7 Comparison of spectral density $C(\omega)$ and $C_p T_0 = C(p\omega_0)$ for a rectangular pulse of amplitude A and duration T_d.

10.3.3 Random Signals

A random signal contains all frequencies with random phase. Furthermore, a random signal may be stationary with A_{rms} constant, or nonstationary with A_{rms} time dependent. To properly define a stationary random process, it is necessary to determine the statistical probability density, the mean, and the mean square of the signal. In the following description of random processes, coverage is limited to stationary processes that can be represented with a Gaussian probability density. With this limitation, auto-correlations and cross-correlations can be used to develop the required relationships.

A. Autocorrelation

The temporal autocorrelation function $\phi_{11}(\tau)$ is defined by the relation

$$\phi_{11}(\tau) = \lim_{T \to \infty} \frac{1}{T} \int_0^T f_1(t) f_1(t + \tau) \, dt \tag{10.21}$$

where $f_1(t)$ is a function of the type illustrated in Fig. 10.8a. The function $\phi_{11}(\tau)$ indicates how well the value $f_1(t)$ at point A and the time-shifted value $f_1(t + \tau)$ at point B correlate with one another over a time interval T. The correlation is positive for a given τ when the product $f_1(t) f_1(t + \tau)$ is positive, as shown in Fig. 10.8b. The time-average value of the area under the curve is the autocorrelation function ϕ_{11}. This function is plotted as a single point C in Fig. 10.8c. Repeating this time-averaging

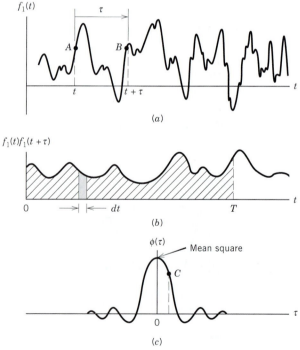

Figure 10.8 An autocorrelation calculation. (a) Definition of $f_1(t)$. (b) Positive correlation at time shift τ. (c) Typical distribution of an autocorrelation function.

process for many values of τ gives the complete distribution of the autocorrelation function, as shown in Fig. 10.8c.

The autocorrelation is the mean square of the signal when $\tau = 0$. Note that Eq. 10.21 reduces to

$$\phi_{11}(0) = \lim_{T \to \infty} \frac{1}{T} \int_0^T f_1^2(t) \, dt \tag{10.22}$$

This result is identical to the definition of the mean square $\overline{f^2}$ given by Eq. 10.2.

B. Mean-Square Spectral Density

For a random process, the autocorrelation function decreases and goes to zero as τ increases, because the events become uncorrelated when they are separated in time. This behavior indicates that $\phi_{11}(\tau)$ is a real-valued even function in the τ time domain. Clearly $\phi_{11}(\tau)$ can be represented with a Fourier transform as:

$$\phi_{11}(\tau) = \frac{1}{2\pi} \int_{-\infty}^{+\infty} S_{11}(\omega) e^{j\omega\tau} \, d\omega \tag{10.23a}$$

and

$$S_{11}(\omega) = \int_{-\infty}^{+\infty} \phi_{11}(\tau) e^{-j\omega\tau} \, d\tau \tag{10.23b}$$

The function $S_{11}(\omega)$ has only real values because $\phi_{11}(\tau)$ is a real even function. It represents a continuous double-sided frequency spectrum called the mean-square spectral density. When $\tau = 0$, Eq. 10.23a reduces to

$$\phi_{11}(0) = \frac{1}{2\pi} \int_{-\infty}^{+\infty} S_{11}(\omega) \, d\omega \tag{10.24}$$

This result shows that the mean square $\overline{f_1^2}$ is the area under the mean-square spectral density curve. Equation 10.24 is the random equivalent of Parseval's formula (which applies to periodic functions).

C. Cross-correlation

The temporal cross-correlation function $\phi_{12}(\tau)$ between two signals $f_1(t)$ and $f_2(t)$ is defined by the expression

$$\phi_{12}(\tau) = \lim_{T \to \infty} \frac{1}{T} \int_0^T f_1(t) \, f_2(t + \tau) \, dt \tag{10.25}$$

This relation measures the correlation of $f_2(t + \tau)$ with $f_1(t)$. The temporal cross-correlation function $\phi_{12}(\tau)$ is a real-valued function, but it is generally not an even function. A typical cross-correlation is shown in Fig. 10.9, where $\phi_{12}(\tau)$ exhibits a peak value at τ_0. In fact, cross-correlations often have several peaks at different times τ_0, which depend on the relationship between $f_1(t)$ and $f_2(t)$. Usually, time τ_0 is associated with physically significant phenomena.

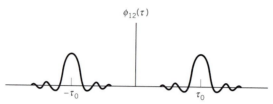

Figure 10.9 Cross-correlation indicating a significant time shift τ_0 between signals f_1 and f_2.

D. Cross-spectral Density

As was the case with autocorrelation and mean-square spectral density, the cross-correlation function $\phi_{12}(\tau)$ can be represented by a Fourier transform with its cross-spectral density as

$$\phi_{12}(\tau) = \frac{1}{2\pi} \int_{-\infty}^{+\infty} S_{12}(\omega)\, e^{j\omega\tau}\, d\omega \qquad \textbf{(10.26a)}$$

for the cross-correlation and

$$S_{12}(\omega) = \int_{-\infty}^{+\infty} \phi_{12}(\tau)\, e^{-j\omega\tau}\, d\tau \qquad \textbf{(10.26b)}$$

for the cross-spectral density.

The equations and concepts developed here are used later in developing and understanding digital frequency analyzer concepts.

10.4 LUMPED MASS-SPRING VIBRATION MODELS

In Chapters 2, 3, 7, and 8, simple vibration models were introduced that incorporated a single mass-spring-damper system. Solutions obtained were in terms of the system's FRF $|H(\omega)|$, which described changes in amplitude of the vibratory response with frequency ω. The solutions also gave angle $\phi(\omega)$, which showed the phase-shift dependence on frequency ω. A more complex two-degree-of-freedom vibration model is introduced in this section. Since this model is treated in detail in standard textbooks on vibrations, this section focuses on interpretation of the solutions to enhance understanding of dynamic behavior in vibration analysis.

The two-degree-of-freedom model, shown in Fig. 10.10, consists of masses (m_1 and m_2), springs (k_1, k_2, and k_3), and dampers (c_1, c_2, and c_3). External excitation forces $F_1(t)$ and $F_2(t)$ are applied to the masses. Coordinates x_1 and x_2 describe mass positions relative to their fixed static equilibrium positions (SEP on Fig. 10.10). The resulting differential equations of motion for this model are

$$(m_1)\ddot{x}_1 + (c_1 + c_2)\dot{x}_1 + (-c_2)\dot{x}_2 + (k_1 + k_2)x_1 + (-k_2)x_2 = F_1(t)$$

$$(m_2)\ddot{x}_2 + (-c_2)\dot{x}_1 + (c_2 + c_3)\dot{x}_2 + (-k_2)x_1 + (k_2 + k_3)x_2 = F_2(t) \qquad \textbf{(10.27)}$$

These equations can be expressed in matrix form as

$$[m]\{\ddot{x}\} + [c]\{\dot{x}\} + [k]\{x\} = \{F\} \qquad \textbf{(10.28)}$$

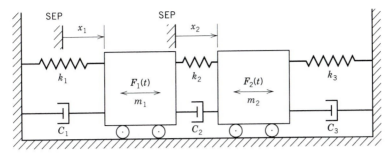

Figure 10.10 Two-degree-of-freedom model showing system parameters, coordinates, and exciting forces.

where $\{\ddot{x}\}$, $\{\dot{x}\}$, $\{x\}$, and $\{F\}$ are column matrices and

$$[m] = \begin{bmatrix} m_{11} & m_{12} \\ m_{21} & m_{22} \end{bmatrix} = \begin{bmatrix} m_1 & 0 \\ 0 & m_2 \end{bmatrix}$$

$$[c] = \begin{bmatrix} c_{11} & c_{12} \\ c_{21} & c_{22} \end{bmatrix} = \begin{bmatrix} (c_1 + c_2) & (-c_2) \\ (-c_2) & (c_2 + c_3) \end{bmatrix} \qquad \textbf{(10.29)}$$

$$[k] = \begin{bmatrix} k_{11} & k_{12} \\ k_{21} & k_{22} \end{bmatrix} = \begin{bmatrix} (k_1 + k_2) & (-k_2) \\ (-k_2) & (k_2 + k_3) \end{bmatrix}$$

Both free undamped vibration and steady-state forced vibration responses are of interest. An efficient method to determine these responses is to assume that the excitation and the response have the form $F_i(t) = F_i e^{j\omega t}$ and $x_i(t) = A_i e^{j\omega t}$. Then, Eqs. 10.28 lead to

$$D_{11}A_1 + D_{12}A_2 = F_1$$

$$D_{21}A_1 + D_{22}A_2 = F_2 \qquad \textbf{(10.30)}$$

where the coefficients D_{mn} that represent the dynamic stiffness of the system are given by

$$D_{mn} = \left(k_{mn} - m_{mn}\omega^2\right) + j c_{mn}\omega \qquad \textbf{(10.31)}$$

For example,

$$D_{11} = \left(k_{11} - m_{11}\omega^2\right) + j c_{11}\omega$$
$$= \left[(k_1 + k_2) - \omega^2 m_1 + j (c_1 + c_2)\omega\right]$$

These relations show that the dynamic stiffness depends on the masses, the dampers, and the frequency in addition to the spring rates.

10.4.1 Undamped Natural Frequency and Mode Shape

If the damping and excitation force terms are zero in Eq. 10.30, it can be shown that the amplitudes A_1 and A_2 are nonzero when the determinant Δ of the dynamic stiffness vanishes. Accordingly,

$$\Delta = D_{11}D_{22} - D_{12}D_{21} = m_1 m_2 (\omega_1^2 - \omega^2)(\omega_2^2 - \omega^2) = 0 \qquad \textbf{(10.32)}$$

This relation is known as the characteristic frequency equation for a two-degree-of-freedom system. The roots ω_1 and ω_2 give the natural frequencies. The values of A_1 and A_2 in Eq. 10.30 must satisfy a ratio for each of the two natural frequencies so that

$$\frac{A_2}{A_1} = -\frac{D_{11}(\omega_i)}{D_{12}(\omega_i)} = -\frac{D_{21}(\omega_i)}{D_{22}(\omega_i)} \qquad \text{for } i = 1 \text{ and } 2 \tag{10.33}$$

Amplitude ratios for ω_1 and ω_2 describe the two mode shapes. The response of the undamped free vibration is given by writing x_1 and x_2 as

$$\begin{Bmatrix} x_1 \\ x_2 \end{Bmatrix} = B_1 \begin{Bmatrix} u_{11} \\ u_{21} \end{Bmatrix} e^{j\omega_1 t} + B_2 \begin{Bmatrix} u_{12} \\ u_{22} \end{Bmatrix} e^{j\omega_2 t} \tag{10.34}$$

First Mode   Second Mode

where vectors $\{u\}$ are called modal vectors and define the mode shape for the first and second natural frequencies, respectively. The constants B_1 and B_2 are the modal amplitudes of motion that depend on the initial conditions x_1, $\dot{x}_1$, x_2, and $\dot{x}_2$ at $t = 0$.

10.4.2 Forced Vibration Response (Direct Solution)

The forced vibration response is obtained from Eq. 10.30 by solving for A_1 and A_2 to obtain

$$x_1 = \left(\frac{D_{22}}{\Delta}\right) F_1 e^{j\omega t} + \left(\frac{-D_{12}}{\Delta}\right) F_2 e^{j\omega t}$$

$$x_2 = \left(\frac{-D_{21}}{\Delta}\right) F_1 e^{j\omega t} + \left(\frac{D_{11}}{\Delta}\right) F_2 e^{j\omega t} \tag{10.35}$$

where the system determinant Δ is defined in Eq. 10.32 with damping included. The determinant Δ is a minimum when $\omega = \omega_1$ or ω_2, which causes a resonant response. Equation 10.35 may be written in terms of the frequency response functions $H_{mn}(\omega)$ as

$$x_1 = H_{11}(\omega) F_1 e^{j\omega t} + H_{12}(\omega) F_2 e^{j\omega t}$$

$$x_2 = H_{21}(\omega) F_1 e^{j\omega t} + H_{22}(\omega) F_2 e^{j\omega t} \tag{10.36}$$

10.4.3 Forced Vibration Response (Modal Solution)

In the modal solution, it is assumed that a forced vibration response is obtained by summing the modal responses so that the relationship for steady-state motion can be expressed as:

$$\{x(t)\} = [u]\{q(t)\} \tag{10.37}$$

where $q(t)$ are generalized coordinates that need to be determined. Substituting Eq. 10.37 into Eq. 10.28 and using matrix algebra leads to the equations for the m natural modes.

$$m_m \ddot{q}_m + c_m \dot{q}_m + k_m q_m = Q_m \qquad (10.38)$$

where the mth excitation force Q_m is given by

$$Q_m = \underbrace{u_{1m} F_1 + u_{2m} F_2}_{2 \text{ DOF}} = \underbrace{\sum_{k=1}^{N} u_{km} F_k}_{N \text{ DOF}} \qquad (10.39)$$

Two facts are important in interpreting Eq. 10.39. First, force F_k cannot excite an mth mode if the modal parameter $u_{km} = 0$. Second, Eq. 10.39 is valid for both static and dynamic loads, since static loads are dynamic loads with $\omega = 0$. This second observation is important when static loads are applied at one or more locations of a structure and then released suddenly to set the structure into vibration. It is clear that applying a single load or a combination of loads so that $Q_m = 0$ does not excite the mth mode and the mth mode is absent from any vibration data collected from the structure.

Equation 10.38 is a second-order differential equation with constant coefficients that exhibits a steady-state solution given by

$$q_m = \frac{Q_m}{D_m} e^{j\omega t} = \frac{u_{1m} F_1 + u_{2m} F_2}{D_m} e^{j\omega t} \qquad (10.40)$$

The modal dynamic stiffness D_m is given by

$$D_m(\omega) = k_m - m_m \omega^2 + j c_m \omega = k_m \left[\left(1 - r_m^2 \right) + j 2 d_m r_m \right] \qquad (10.41)$$

where

r_m is the mth dimensionless modal frequency ratio (ω / ω_m)
d_m is the mth dimensionless modal damping ratio

It is clear from Eq. 10.41 that the mth modal natural frequency is

$$\omega_m = \sqrt{\frac{k_m}{m_m}} \qquad (10.42)$$

and the mth modal damping ratio is

$$d_m = \frac{c_m}{2\sqrt{m_m k_m}} \qquad (10.43)$$

Substituting Eq. 10.40 into Eq. 10.37 gives steady-state modal responses

$$x_1 = \left[u_{11} \left(\frac{u_{11}}{D_1} \right) + u_{12} \left(\frac{u_{12}}{D_2} \right) \right] F_1 e^{j\omega t} + \left[u_{11} \left(\frac{u_{21}}{D_1} \right) + u_{12} \left(\frac{u_{22}}{D_2} \right) \right] F_2 e^{j\omega t}$$

$$(10.44)$$

$$x_2 = \left[u_{21} \left(\frac{u_{11}}{D_1} \right) + u_{22} \left(\frac{u_{12}}{D_2} \right) \right] F_1 e^{j\omega t} + \left[u_{21} \left(\frac{u_{21}}{D_1} \right) + u_{22} \left(\frac{u_{22}}{D_2} \right) \right] F_2 e^{j\omega t}$$

The bracketed terms in Eq. 10.44 represent the frequency response functions $H_{ij}(\omega)$. These functions may be expressed using either the direct or the modal solutions for x_k.

$$H_{11}(\omega) = \left(\frac{D_{22}}{\Delta}\right) = \left[u_{11}\left(\frac{u_{11}}{D_1}\right) + u_{12}\left(\frac{u_{12}}{D_2}\right)\right]$$

$$H_{12}(\omega) = \left(\frac{-D_{12}}{\Delta}\right) = \left[u_{11}\left(\frac{u_{21}}{D_1}\right) + u_{12}\left(\frac{u_{22}}{D_2}\right)\right]$$

$$H_{21}(\omega) = \left(\frac{-D_{21}}{\Delta}\right) = \left[u_{21}\left(\frac{u_{11}}{D_1}\right) + u_{22}\left(\frac{u_{12}}{D_2}\right)\right] \tag{10.45}$$

$$H_{22}(\omega) = \left(\frac{D_{11}}{\Delta}\right) = \left[u_{21}\left(\frac{u_{21}}{D_1}\right) + u_{22}\left(\frac{u_{22}}{D_2}\right)\right]$$

To measure mode shapes, it is apparent from Eqs. 10.45 that a single excitation source is preferred. Otherwise, the response is dependent on several sets of modal parameters instead of just one set. If multiple excitation is employed, cross-coupling occurs and the responses are much more difficult to interpret.

Consider a study with excitation applied at a single point ($F_2 = 0$). The structural response with $\omega = \omega_1$ is to be analyzed. In this case, Eqs. 10.44 and 10.45 show the relation between $H_{11}(\omega)$ and $H_{21}(\omega)$ and the modal parameters u_{11} and u_{21}. Note that u_{11}/D_1 is common to all terms multiplying F_1 at frequency ω_1. This fact suggests that applying an excitation force at location 1, while moving an accelerometer from location 1 to location 2, will give a signal that reveals the first mode shape. The modal parameter u_{i1} varies from location to location and describes the first modal vector. A similar result occurs when $\omega = \omega_2$, and the same procedure gives the modal parameters u_{i2}. Similarly if $F_1 = 0$ with F_2 the single point excitation, it is clear that the same modal vectors are obtained.

In general, it can be shown that

$$H_{mn}(\omega) = \sum_{k=1}^{N} \frac{u_{mk}u_{nk}}{D_k} = \underset{\underset{\text{Mode}}{\text{First}}}{\frac{u_{m1}u_{n1}}{D_1}} + \underset{\underset{\text{Mode}}{\text{Second}}}{\frac{u_{m2}u_{n2}}{D_2}} + \cdots \tag{10.46}$$

where the ratio (u_{nk}/D_k) refers to excitation force F_n at location n and reflects its excitation efficiency in the kth mode. The modal parameter u_{mk} refers to the response at location m. Equation 10.46 represents the basis for experimental modal analysis. It shows that the maximum response at any natural frequency is contaminated to some degree by adjacent modal responses. However, if the natural frequencies differ by a sufficient amount and the damping is small, the adjacent modal responses do not produce large errors in the analysis of the vibration response.

10.5 CONTINUOUS VIBRATION MODELS

Modal analysis is a powerful method for interpreting vibration behavior, because each natural frequency has a corresponding mode shape. These mode shapes are summed to provide information characterizing structures subjected to both transient and steady-state vibration. In this section, continuous vibration models are introduced to show differences in behavior when compared with lumped mass-spring models and to provide guidance for conducting vibration experiments.

10.5.1 Fundamental Equation of Motion

The uniform beam, shown in Fig. 10.11, is used to represent a continuous vibrating structural element. The beam deflection $y(x, t)$ is governed by the equation

$$m\frac{\partial^2 y}{\partial t^2} + c\frac{\partial y}{\partial t} + \frac{\partial^2}{\partial x^2}\left(EI\frac{\partial^2 y}{\partial x^2}\right) = F(x, t) \qquad (10.47)$$

where

m and c are the mass and damping per unit length
EI is the bending stiffness (modulus times inertia)

The excitation forces, also expressed per unit length, are assumed to be separable, so that

$$F(x, t) = P(x)f(t) \qquad (10.48)$$

Natural modes of vibration of continuous systems are described by a mode shape function $V_k(x)$ that is equivalent to the modal vector $\{u\}$ of lumped mass-spring systems. This function must satisfy Eq. 10.47 when $c = 0$ and $F(x, t) = 0$ as well as the boundary conditions for each natural frequency ω_k. Free undamped vibration is expressed by summing the modes to give

$$y(x, t) = \sum_{k=1}^{\infty} V_k(x)\,[A_k \cos(\omega_k t) + B_k \sin(\omega_k t)] \qquad (10.49)$$

where A_k and B_k depend on the initial conditions.

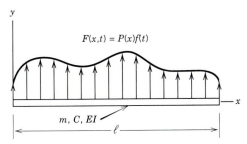

Figure 10.11 A beam used to model a continuous system.

The mode shapes of continuous systems are orthogonal and

$$\int_0^\ell m V_k(x) V_r(x) \, dx = \begin{cases} 0 & \text{for } r \neq k \\ m_k & \text{for } r = k \end{cases} \tag{10.50a}$$

where m_k is the kth mode modal mass and

$$\int_0^\ell V_r(x) \frac{\partial^2}{\partial x^2} \left[EI \frac{\partial^2 V_k(x)}{\partial x^2} \right] dx = \begin{cases} 0 & \text{for } r \neq k \\ k_k & \text{for } r = k \end{cases} \tag{10.50b}$$

where k_k is the kth mode modal stiffness.

10.5.2 Steady-State Modal Solution

To develop a steady-state modal solution, the excitation force is separated into spatial $P(x)$ and temporal $f(t)$ functions, as indicated in Eq. 10.48, and $y(x, t)$ is taken as

$$y(x, t) = \sum_{k=1}^\infty V_k(x) \, q_k(t) \tag{10.51}$$

where q_k is a generalized coordinate that remains to be determined. Then, substituting Eqs. 10.48 and 10.51 into Eq. 10.47, integrating, assuming that damping is proportional to mass, and using Eq. 10.50,

$$m_k \ddot{q}_k + c_k \dot{q}_k + k_k q_k = Q_k f(t) \qquad \text{for } k = 1, 2, 3, \cdots \tag{10.52}$$

where Q_k is the modal excitation force. Because Eq. 10.52 is identical to Eq. 10.39, the lumped mass-spring model and the continuous system give the same controlling equation. The natural frequencies and modal damping ratios for the continuous system are given by Eqs. 10.42 and 10.43, respectively, if the continuous definitions of m_k, c_k, and k_k are employed.

The modal excitation force Q_k is

$$Q_k = \int_0^\ell P(x) \, V_k(x) \, dx \tag{10.53}$$

This relation is an integral equivalent to Eq. 10.39 for the lumped mass-spring model with its discrete description of Q_m. The modal excitation force shows the effectiveness of $P(x)$ in exciting the natural frequency ω_k.

The steady-state solution to Eq. 10.52 when $f(t) = e^{j\omega t}$ is given by Eq. 10.40 as

$$q_k(t) = \left(\frac{1}{D_k} \right) Q_k \, e^{j\omega t} = H_k(\omega) \, Q_k e^{j\omega t} \tag{10.54}$$

where D_k is the modal dynamic stiffness defined by Eq. 10.41 and $H_k(\omega) = 1/D_k$ is the modal frequency response function. The modal frequency response function is the same as that for a single-degree-of-freedom system.

The displacement $y(b, t)$ is obtained by combining Eqs. 10.51 and 10.54 to give

$$y(b, t) = \sum_{k=1}^{\infty} V_k(b)\, q_k(t) = \sum_{k=1}^{\infty} V_k(b)\, Q_k\, H_k(\omega)\, e^{j\omega t} \tag{10.55}$$

Since Eq. 10.55 is valid for each frequency ω, this relation can be rewritten in the frequency domain as

$$Y(\omega) = \sum_{k=1}^{\infty} V_k(b)\, Q_k\, H_k(\omega)\, F(\omega) = H_b(\omega)\, F(\omega) \tag{10.56}$$

where $f(t)$ has a frequency spectrum of $F(\omega)$ and $y(t)$ has a frequency spectrum of $Y(\omega)$.

Note that the value of $H_b(\omega)$ changes with the measurement location. If either $V_k(b)$ or Q_k is zero, then the kth mode of vibration is absent from the measured data. Also, it is clear that $H_b(\omega)$ changes with the excitation force distribution, because Q_k depends on the load distribution $P(x)$ (see Eq. 10.53). A given mode shape is determined by changing the location of the accelerometer (by varying b). The dominant term in Eq. 10.56 is $V_k(b)H_k(\omega)$ when ω approaches ω_k. This procedure gives the mode shape function $V_k(x)$ if the input is constant as the measurement locations are varied.

A concentrated excitation load allows a point-to-point frequency response function to be developed. The solution of Eq. 10.52 for a concentrated load P_a applied at $x = a$ gives

$$Y(\omega) = \sum_{k=1}^{\infty} V_k(b)\, V_k(a)\, H_k(\omega)\, P_a F(\omega) = H_{ba}(\omega)\, P_a\, F(\omega) \tag{10.57}$$

where $H_{ba}(\omega)$ is the output at $x = b$ due to a concentrated excitation force at $x = a$. $H_{ba}(\omega)$ is also the sum of all modal response functions, modified by corresponding mode shapes, evaluated at a and b. The product of the modal response functions $V_k(a)V_k(b)$ plays the same role as modal parameters $u_{mk}u_{nk}$ do in Eq. 10.46. The advantage of using concentrated excitation forces is that it leads to an input–output relationship for specific points. With distributed loads, only the output points b are explicitly defined.

Equation 10.57 also indicates the procedure for obtaining mode shapes from $Y(\omega)$. First, the peak points in a graph of Y versus ω correspond to resonances; that is, $H_k(\omega)$ peaks at ω_k. Second, each peak value is related to its corresponding mode shape $V_k(\omega)$ evaluated at both the point of excitation ($x = a$) and the point of motion measurement ($x = b$). Third, the kth mode shape can be estimated by either fixing the excitation point a and varying the measurement location b or vice versa. A major difficulty occurs when either a measurement point or an excitation point coincides with a node point (no motion) for a given natural mode. If this is the situation, that mode is absent in the experimental data. Thus, several different force input and sensor output locations must be used to avoid missing a given mode of vibration. It is easy to miss significant information when testing an unknown structure. Selecting appropriate accelerometers as well as locating them during an experiment requires considerable skill.

10.6 THE LINEAR INPUT–OUTPUT MODEL

In spite of the mathematical complexities required to describe the motion of systems that vibrate with many degrees of freedom, the motion itself is governed by relatively simple linear differential equations. The solutions are complex because both the input and output signals contain many frequency components. The solutions are important because the frequency response functions $H(\omega)$ contain information regarding the resonance frequencies and the amplitudes at the different resonances. In this section, relationships between a linear system, illustrated in Fig. 10.12, and the concepts introduced in Sections 10.2 and 10.5 are discussed.

In a linear system, the input excitation is $f(t)$ and the output is $y(t)$. If a sinusoidal excitation of

$$f(t) = F_0 e^{j\omega t} \tag{a}$$

is applied to a linear system, the response is sinusoidal at the same frequency such that

$$y(t) = Y_0 e^{j\omega t} = H(\omega) F_0 e^{j\omega t} \tag{b}$$

where Y_0, the output magnitude, is

$$Y_0 = H(\omega) F_0 \tag{c}$$

Periodic signals are composed of discrete frequency components described by a discrete frequency spectrum, and transient signals are composed of all frequency components described by a continuous frequency spectrum. Equation c is valid for all frequencies. For periodic signals it is written as

$$Y_p = H(p\omega_0) F_p \tag{10.58a}$$

and for transient signals with a continuous spectrum it is written as

$$Y(\omega) = H(\omega) F(\omega) \tag{10.58b}$$

These relations indicate that it is possible to determine the functional form of $H(\omega)$ by measuring input and output frequency spectra. These measurements are made at discrete frequencies ($p\omega_0$) or continuously (all ω's) provided that the magnitude and phase are retained.

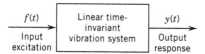

Figure 10.12 Block diagram of a general linear system.

The frequency spectrum of the input signal is determined from Eqs. 10.13b and 10.19a as

$$F_p = \frac{1}{T} \int_t^{t+T} f(t) \, e^{-jp\omega_0 t} \, dt$$

(10.59)

$$F(\omega) = \int_{-\infty}^{+\infty} f(t) \, e^{-j\omega t} \, dt$$

The output signal is determined by substituting Eqs. 10.13a and 10.19 b into Eq. 10.58 to obtain

$$y(t) = \sum_{p=-\infty}^{\infty} Y_p \, e^{jp\omega_0 t} = \sum_{p=-\infty}^{\infty} H(p\omega_0) \, F_p \, e^{jp\omega_0 t}$$

or **(10.60)**

$$y(t) = \frac{1}{2\pi} \int_{-\infty}^{+\infty} Y(\omega) \, e^{-j\omega t} \, d\omega = \frac{1}{2\pi} \int_{-\infty}^{+\infty} H(\omega) \, F(\omega) \, e^{-j\omega t} \, d\omega$$

Equations 10.60 indicate that $y(t)$ can be obtained directly for an arbitrary input function $f(t)$. The process amounts to multiplication of frequency components in the frequency domain [either $y_p = H(p\omega_0) F_p$ or $Y(\omega) = H(\omega) F(\omega)$] and then superimposing these frequency components.

10.6.1 Impulse Response

Consider the unit impulse function shown in Fig. 10.13*a*, which is defined by

$$f(t) = \delta(t - \tau)$$ **(10.61)**

where $\delta(t - \tau)$ is the Dirac delta function. This function is zero for all time except $t = \tau$, where it is equal to infinity. The function with an ordinate of infinity and a duration of zero is represented by a rectangular area of width Δt and height $1/\Delta t$ in a limiting process as Δt goes to zero. The word *impulse* is commonly used to describe a class of excitation signals with a very short duration input regardless of units involved. Only if $f(t)$ is a force is the result an impulse as defined in dynamics.

The system response $y(t)$, owing to $\delta(t-\tau)$, is called the impulse response function $h(t - \tau)$, as shown in Fig. 10.13*b*. The quantity $\eta = t - \tau$ is a relative time that is measured relative to a time shift τ. Clearly, $h(\eta) = 0$ for $\eta < 0$.

The impulse response function $h(\eta)$ is used to obtain $y(t)$ for an arbitrary input $f(t)$, as illustrated in Fig. 10.13*c*. The impulse $[f(\tau)d\tau]$ at time τ replaces the Dirac delta impulse shown in Fig. 10.13*a*. The corresponding response at time t (where $t > \tau$) resulting from this impulse becomes

$$[f(\tau) \, d\tau] \, h(t - \tau)$$ **(a)**

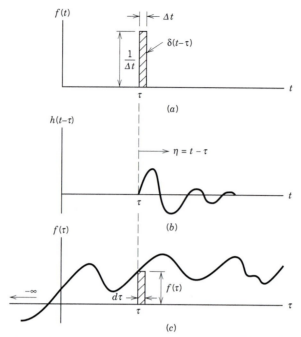

Figure 10.13 The convolution concept. (*a*) A Dirac delta function. (*b*) An impulse-response function. (*c*) An arbitrary input/time history.

Because the system is linear, these responses are superimposed in the time domain at time t by integrating with respect to τ to obtain

$$y(t) = \int_{-\infty}^{+\infty} f(\tau)\, h(t - \tau)\, d\tau \qquad (10.62)$$

In vibration analysis, this relation is often called the convolution integral. The upper integration limit can be either t or ∞ because $h(t - \tau) = 0$ when $\tau > t$. Equation 10.62 is a general input–output relationship that gives $y(t)$, as did Eq. 10.60. However, the processes are completely different. One involves multiplication in the frequency domain before integration over frequency using Fourier transforms, whereas the other involves integration in the time domain using the impulse response function.

The frequency response function $H(\omega)$ and the impulse response function $h(t - \tau)$ are also related to one another. Consider the case when a Dirac delta function is applied at $\tau = 0$. Then, Eq. 10.62 shows that $y(t) = h(t)$, and Eq. 10.59 shows that $F(\omega) = 1$ for all frequencies, so that Eq. 10.60 becomes

$$h(t) = \frac{1}{2\pi} \int_{-\infty}^{+\infty} H(\omega)\, e^{j\omega t}\, d\omega$$

$$H(\omega) = \int_{-\infty}^{+\infty} h(t)\, e^{-j\omega t}\, dt \qquad (10.63)$$

It is evident that the impulse response function and the frequency response function form a Fourier transform pair.

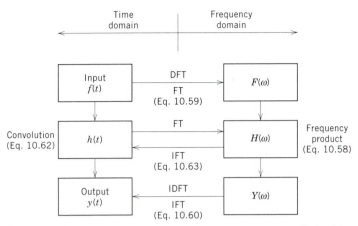

Figure 10.14 Illustration of linear system input–output relationships showing Fourier transform as the bridge between convolution and frequency multiplication processes.

The results of this section are summarized in the block diagram shown in Fig. 10.14. The left-hand blocks represent the time-domain processes, and the right-hand blocks represent the frequency-domain processes. The Fourier transforms provide the conversion between the time- and frequency-domain descriptions.

The left-hand blocks represent the time-domain convolution process to obtain $y(t)$ from the input time history $f(t)$ as given by Eq. 10.62. The right-hand blocks represent frequency-domain multiplication described by Eq. 10.58. The process of transferring from the frequency domain to the time domain is either the inverse discrete Fourier transform (DFT) for periodic signals or the inverse Fourier transform (FT) for transient signals, described by Eqs. 10.60.

10.6.2 Random Input–Output Relationships

For random signals, the output mean-square spectral density is related to the input mean-square spectral density by

$$S_{yy}(\omega) = |H(\omega)|^2 \, S_{ff}(\omega)$$

$$G_{yy}(\omega) = |H(\omega)|^2 \, G_{ff}(\omega)$$

(10.64)

where

$S_{yy}(\omega)$ and $S_{ff}(\omega)$ are the output and input double-sided mean-square spectral densities, respectively

$G_{yy}(\omega)$ and $G_{ff}(\omega)$ are the output and input single-sided mean-square spectral densities, respectively

Similarly, the cross-spectral density is related to the input mean-square spectral density by

$$S_{fy}(\omega) = H(\omega) \, S_{ff}(\omega)$$

$$G_{fy}(\omega) = H(\omega) \, G_{ff}(\omega)$$

(10.65)

where

  $S_{fy}(\omega)$ is the output dual-sided cross-spectral density
  $G_{fy}(\omega)$ is the output single-sided cross-spectral density

Phase information is not preserved in Eq. 10.64, but both magnitude and phase information are available in Eq. 10.65. These equations are useful in performing frequency analysis of random signals.

10.7   BASICS OF A DIGITAL FREQUENCY ANALYZER

A digital frequency analyzer is an instrument used to measure frequency spectra. Since digital analyzers have many advantages, they are rapidly replacing analog frequency analyzers. In this section, the basic principles and operating characteristics of this powerful instrument are reviewed. It is shown that aliasing, window functions, sample functions, and leakage play important roles in frequency analysis. The fast Fourier transform algorithm, developed by Cooley and Tukey (Reference 6), is used to determine a discrete Fourier frequency spectrum from a digitized input signal $f(t)$.

10.7.1 Time Sampling Process

The digital conversion process involves sampling an analog signal $x(t)$ at a specified sampling frequency f_s over a specified period of time T with an analog-to-digital converter. This sampling process is shown in Fig. 10.15, where $w(t)$ is the window function of duration T that defines the portion of the signal $x(t)$ that is to be analyzed,

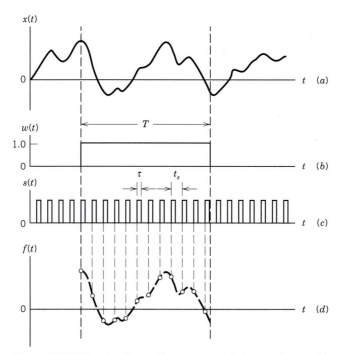

Figure 10.15 The signal-sampling process. (*a*) Analog signal $x(t)$. (*b*) Window function $w(t)$. (*c*) Sample function $s(t)$. (*d*) Digitized function $f(t)$.

and $s(t)$ is the sample function consisting of rectangular pulses of duration τ spaced at a sampling period t_s to generate a sample frequency $f_s = 1/t_s$. The sampling time τ is small compared with the sampling period t_s. Finally, $f(t)$ is the digitized version of $x(t)$ that is analyzed. It is clear that, if f_s is too small, $f(t)$ will not contain sufficient points to properly define $x(t)$.

The sample period t_s and analysis period T are related by

$$T = (N - 1)\, t_s \tag{10.66}$$

where N is the total number of samples. Note that there are $N - 1$ intervals in N samples. In order to use the fast Fourier transform (FFT) algorithm N must be a power of 2. The most common values are 2^{10} (1024) and 2^{11} (2048). The output function $f(t)$ is given by

$$f(t) = w(t)\, s(t)\, x(t) \tag{10.67}$$

where $f(t)$ represents $x(t)$ by a sequence of discrete numbers that are equally spaced in time.

This approach raises a number of questions. Can a frequency analysis of $f(t)$ give a good estimate of $x(t)$? Are $w(t)$ and $s(t)$ important in the FFT process? It will be shown that $w(t)$ introduces digital filtering, and $s(t)$ is related to the aliasing problem. Both digital filtering and aliasing are a consequence of convolution in the frequency domain.

10.7.2 Convolution

Equation 10.67 indicates that a Fourier transform of $f(t)$ is equivalent to taking a Fourier transform of the product of several time functions. As an example, consider the product of two time functions such that

$$f_1(t) = f_2(t)\, f_3(t) \tag{10.68}$$

where each time function $f_i(t)$ has its own Fourier spectrum $F_i(\omega)$ as defined by Eq. 10.19a. To show that these frequency spectra relate to each other, substitute $f_1(t)$ into Eq. 10.19a to give

$$F_1(\omega) = \int_{-\infty}^{+\infty} f_1(t) e^{-j\omega t}\, dt = \int_{-\infty}^{+\infty} f_2(t) f_3(t) e^{-j\omega t}\, dt \tag{a}$$

Substitution of appropriate expressions for $f_2(t)$ and $f_3(t)$, along with some careful manipulation and interpretation, gives

$$F_1(\omega) = \frac{1}{2\pi} \int_{-\infty}^{+\infty} F_2(\nu) F_3(\omega - \nu)\, d\nu \tag{10.69}$$

where ν is a dummy integration variable. Equation 10.69 is known as frequency-domain convolution. It was previously noted that frequency-domain multiplication, as in Eq. 10.58, is equivalent to time-domain convolution, as in Eq. 10.62. Here, a corollary is found where time-domain multiplication leads to frequency-domain convolution. The results of Eq. 10.69 show that for each frequency ω, integration of $F_2(\nu) F_3(\omega - \nu)$ over the range of ν gives a unique value for $F_1(\omega)$.

A. Digital Filtering

Digital filtering that occurs in the FFT process is explained with the convolution integral. Consider the situation illustrated in Fig. 10.16 where $F_3(\omega - \nu)$ has unity magnitude over a frequency bandwidth from $-\pi B$ to πB and a magnitude of zero elsewhere. Let $\omega = \omega_1$, which defines points A and C in Fig. 10.16b and points D and E in Fig. 10.16c. For point A, $\nu_A = \omega_1 + \pi B$, which gives point A' in Fig. 10.16a. Similarly, point C in Fig. 10.16b gives point C' in Fig. 10.16a. Also, if $\omega = -\omega_1$, then points A and C correspond to points A'' and C'' at $-\omega_1$ in Fig. 10.16a.

Integration of Eq. 10.69 for this example gives

$$F_1(\omega_1) = \frac{1}{2\pi} \int_{\omega_1-\pi B}^{\omega_1+\pi B} F_2(\nu) \, d\nu = \overline{F}_2(\omega_1) \, B \qquad \textbf{(b)}$$

This result indicates that $F_1(\omega_1)$ is $(\frac{1}{2\pi})$ times the area under the $F_2(\nu)$ curve at $\omega = \omega_1$ in Fig. 10.16a. The quantity $F_1(\omega_1)$, which plots as the single point D in Fig. 10.16c, is a frequency-averaged value of $\overline{F}_2(\omega_1)$ multiplied by B. A similar result occurs at $\omega = -\omega_1$ and gives point E.

This example indicates that $F_3(\omega - \nu)$ serves as a digital filter in examining the frequency content of $F_2(\nu)$ with changing values of ω. Thus, ω is the filter center frequency and $F_1(\omega)$ is proportional to the frequency-averaged value of $F_2(\nu)$. As B

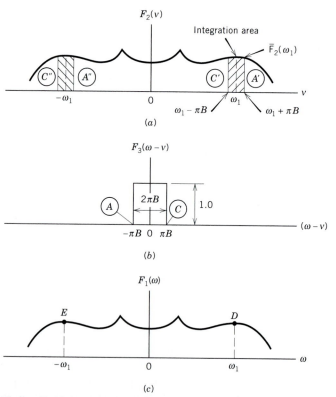

Figure 10.16 Illustration of the frequency convolution integral.
(a) Function $F_2(\nu)$. (b) Function $F_3(\omega - \nu)$. (c) Function $F_1(\omega)$.

becomes smaller, $F_1(\omega)$ becomes closer to the actual value of $F_2(\nu)$ at each point. This observation suggests that it may be possible to multiply an input signal by an appropriate time function in the time domain to obtain an output signal that is automatically filtered in the frequency domain. In other words, $F_3(\omega - \nu)$ is the digital filter characteristic that is due to time function $f_3(t)$. The window function $w(t)$ serves this purpose in digital frequency analyzers.

B. The Aliasing Process

The effects of the window function are ignored in this discussion, and attention is focused on the relationship

$$f(t) = s(t)x(t) \tag{c}$$

because it is of the same form as Eq. 10.68. Since the sample function $s(t)$ is periodic with a frequency f_s, the discrete Fourier transform generates a discrete frequency spectrum given by

$$S(p\omega_s) = \frac{2}{\beta}\left[\frac{\sin(p\pi/\beta)}{(p\pi/\beta)}\right] = \frac{2}{\beta} \text{ sinc }(p\pi/\beta) \tag{d}$$

where $p = 0, 1, 2, \ldots$ and $\beta = t_s/\tau$ is very large. The Fourier transform of $s(t)$ includes all time and all discrete pulses so that

$$S(\omega) = \int_{-\infty}^{+\infty} s(t)e^{-j\omega t}\,dt$$

$$= \lim_{N\to\infty} 2\tau N \text{ sinc }(p\omega_s\tau/2) = \delta(p\omega_s - \omega) \tag{10.70}$$

where N is the number of pulses. Note that the sample function $S(\omega)$ has a frequency spectrum represented with a series of Dirac delta functions spaced at $p\omega_s$, as shown in Fig. 10.17. $S(\omega)$ acts as the filter function used in analyzing the frequency spectrum $X(\nu)$.

The output frequency spectrum $F(\omega)$, shown in Fig. 10.17c, is due to the convolution integral of the input-signal frequency spectrum $X(\nu)$ with the sample-function frequency spectrum $S(\omega - \nu)$. Since $S(\omega - \nu)$ is represented with Dirac delta functions, the convolution integration of Eq. 10.69 for the center Dirac delta function, where $\nu = 0$, is

$$F(\omega) = \frac{1}{2\pi}\int_0^0 \delta(\omega - \nu)X(\nu)\,d\nu$$

$$= \frac{X(\omega)}{2\pi}\int_{-0}^{+0} \delta(\omega - \nu)\,d\nu = \frac{X(\omega)}{2\pi} \tag{10.71}$$

Note that the integral of δ over zero is one. This result shows that the output spectrum is the same as the input spectrum except for a scale factor of 2π. Thus, the solid portion of the curve for $F(\omega)$ is due to the convolution of $S(\omega - \nu)$ and $X(\nu)$ for those values of ω that make $(\omega - \nu)$ zero.

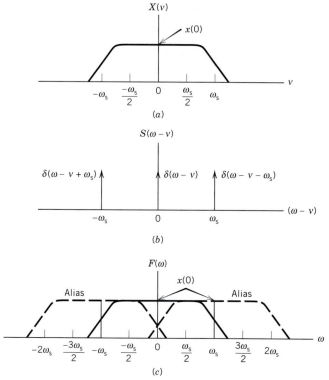

Figure 10.17 Aliasing process. (*a*) Original frequency spectrum. (*b*) Sample-function frequency spectrum. (*c*) Combined frequency spectrum.

Considering the Dirac delta functions in Eq. 10.71 of the form $\delta(\omega - v - \omega_s)$ gives the dashed curve on the right-hand side of Fig. 10.17c that is centered on $(+\omega_s)$. Similarly, changing $\delta(\omega - v)$ to $\delta(\omega - v + \omega_s)$ generates the dashed curve centered on $(-\omega_s)$. Thus, it is clear that $F(\omega)$ contains $F(v)$ centered at zero and many other values of $F(v)$ centered on frequencies of $\pm\omega_s$, $\pm2\omega_s$, and so on. It is important to have only the zero-centered representation of $F(\omega)$ in Fig. 10.17. However, the aliases centered at $\omega = \pm\omega_s$ (dashed lines) distort this information, because their frequency spectrum extends into the region of interest. Figure 10.17c shows that the alias curves can distort the true frequency spectrum near zero as well as in the $\omega_s/2$ region.

This overlapping of frequency spectra is eliminated by limiting the input signal $x(t)$ to frequencies less than $\omega_s/2$ (the Nyquist frequency). The limit is imposed with a low-pass filter with a sharp cutoff, as shown in Fig. 4.30. It is common practice for the input signal to be attenuated 40 dB at the Nyquist frequency so that the reflected error is attenuated more than 80 dB in the useful region of the analyzer (0 to ω_b). If the filter attenuation rate is 120 dB per octave, the usable frequency range is about 0.4 ω_s. Digital frequency analyzers are equipped with antialiasing filters to eliminate the alias problem. These filters must have break frequencies ω_b that can be varied to match the different sample rates selected in employing the digital frequency analyzer.

10.7.3 Filter Leakage

The discrete Fourier transform (DFT) is used to determine the frequency components of that portion of the input signal that is contained within the window period T. In employing the DFT method, the sample is assumed to be periodic. Because this assumption is generally not true, it leads to the filter-leakage phenomenon. Filter leakage is due to discontinuities that occur between real and sampled signals. These discontinuities occur at the window ends, as indicated by the two sinusoids shown in Fig. 10.18. Sinusoid A has exactly two complete cycles within the window of period T, so that this time function is a perfect fit and will repeat itself in the next sample without discontinuities in either magnitude or slope at $t = T$. Sinusoid B has 2.25 cycles within the window period; consequently, this signal is not periodic when referenced to the window period T. The slope or magnitude discontinuities owing to mismatch between the window period and the true signal period produce extraneous frequency components in the frequency analysis.

To illustrate the effects of filter leakage, consider the simple sinusoidal signal

$$x(t) = \sin (N\omega_0 t) \tag{a}$$

where N is the number of cycles within the analysis window of duration T. Recall that frequency $\omega_0 = 2\pi/T$ is the fundamental frequency of the analyzer. A discrete Fourier transform of Eq. a gives

$$c_p = \sqrt{2\left(\frac{\text{sinc}^2 \ [(p + N)\pi]}{(1 - N/p)} + \frac{\text{sinc}^2 \ [(p - N)\pi]}{(1 + N/p)}\right)} \tag{b}$$

for both positive and negative values of p.

The positive frequency components $p > 0$ from Eq. b are shown in Fig. 10.19 for values of $N = 10$, 10.25, and 10.5. In Fig. 10.19a, there is only one frequency

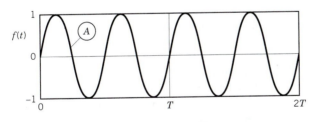

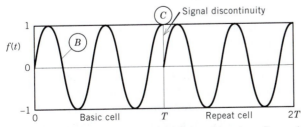

Figure 10.18 Plot of two sinusoidal time histories. Curve A fits perfectly within analysis window period T while curve B has a magnitude and slope discontinuity at its ends.

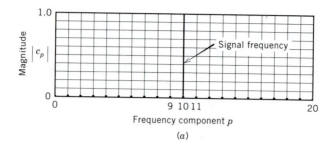

(a)

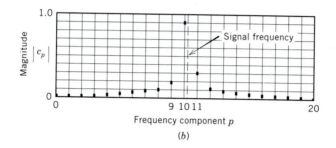

(b)

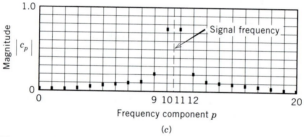

(c)

Figure 10.19 Discrete Fourier frequency components of a sinusoid for N cycles: (a) $N = 10.00$ cycles, (b) $N = 10.25$ cycles, and (c) $N = 10.50$ cycles. All points in (b) and (c) represent filter leakage.

component at $p = 10$, as the others all vanish. In this case, the signal frequency and the 10th analyzer frequency coincide exactly. The function $x(t)$ repeats in a continuous fashion within the window period T as does curve A in Fig. 10.18. This is the correct frequency analysis, since the true magnitude of the signal is recovered at the correct frequency.

When $N = 10.25$, there are 10.25 complete cycles in window period T, so that slope and magnitude discontinuities occur at $t = T$. The corresponding frequency spectrum is shown in Fig. 10.19b. When $N = 10.5$, there is only a slope discontinuity at $t = T$ and the corresponding frequency spectrum is shown in Fig. 10.19c. It is clear that signal information is now spread over a number of extraneous frequency components because of these signal discontinuities. This filter leakage results from analyzing a signal with a frequency ω by using discrete terms at frequencies of $p\omega_0$ where $p\omega_0 \neq \omega$.

It is evident that serious errors can occur in the measurement of the magnitudes of the frequency components. In Fig. 10.19c, the peak values are 0.637, an error of

36.3 percent when compared with a peak value of unity. This is the maximum error that can occur in analyzing a signal with a single frequency component. This result suggests that for analyzing periodic functions either additional data processing or a wider digital filter with a flat top and low side lobes is required to obtain an improved estimate of the amplitude.

10.7.4 Block Diagram

A block diagram of the operation of a digital frequency analyzer is shown in Fig. 10.20. The input signal $x(t)$ is attenuated and is modified by a tunable electronic antialiasing filter. This filtered signal, $y(t)$, is mixed with low-level random noise before analog-to-digital (A/D) conversion occurs in order to suppress the ADC noise in the low-level random noise. The ADC output is multiplied by a window function and stored in RAM. Two memories are often used so that one can be loading while the contents of the other are being processed. The FFT program calculates the discrete Fourier transform of the signal and stores the results in a memory contained in the spectrum averager. These averaged frequency components are often processed by a log converter for display in a dB format on an analog plotter. The digital results may be transferred to a computer for processing or to a digital plotter for hard copy.

10.8   USING A DIGITAL FREQUENCY ANALYZER

Four major topics are covered in this section. The first describes use of discrete Fourier transform frequency components to give periodic spectra, a transient Fourier transform spectral density, and a mean-square spectral density. The second topic addresses the ideal filter and introduces figures of merit with which to rate filters. These figures of merit are applied to four commonly employed digital filters. The third topic describes frequency component averaging to remove amplitude uncertainty when measuring a random signal. The fourth covers the application of these four filters to the analysis of periodic, transient, and random signals.

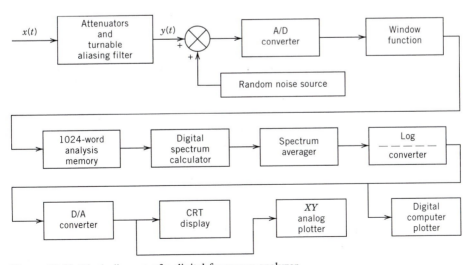

Figure 10.20 Block diagram of a digital frequency analyzer.

10.8.1 Relationships for Frequency Analyzers

The fast Fourier transform calculates discrete Fourier transform frequency components that are equivalent to those given by Eq. 10.13b. The discrete version of Eq. 10.13b is obtained by noting that the kth data point $f(t_k) = f(k)$ occurs at time $t_k = kt_s = kT/(N - 1)$, that $\omega_0 T = 2\pi$, that the integration on T is replaced by a summation on k, and that $dt = t_s$. Then Eq. 10.13b becomes

$$c_p = \left(\frac{1}{N - 1}\right) \sum_{k=0}^{N-1} f(k) \; e^{[-j(2\pi/N-1)pk]}$$

$$= \left(\frac{1}{N - 1}\right) \sum_{k=0}^{N-1} w(k) \; y(k) \; e^{[-j(2\pi/N-1)pk]} \tag{10.72}$$

where $w(k)$ is the time-domain window function and $y(k)$ is a digitized form of the filtered analog signal $y(t)$.

Equations 10.14 and 10.15 indicate that c_p has real and imaginary parts a_p and b_p, which contain both magnitude and phase information. Frequency analyzers display a single-sided frequency spectrum that corresponds to $p = 0, 1, 2, \ldots, N_c$, where N_c is the maximum number of frequency components. Consequently, the magnitude and phase of these $N_c + 1$ frequency components are of interest. The pth harmonic's single-sided magnitude is related to $|c_p|$ by

$$B_p = 2|c_p| = 2\sqrt{c_p c_p^*} \tag{10.73}$$

which has a root mean square amplitude given by

$$B_{\mathrm{rms}_p} = 0.707 \, B_p = \sqrt{2} \, |c_p| = \sqrt{2 c_p c_p^*} \tag{10.74}$$

The mean value has a special relationship given by

$$B_0 = c_0 \tag{10.75}$$

The phase angle ϕ_p given by Eq. 10.15 is the same for either Eq. 10.72 or Eq. 10.73. It is clear from these equations that either a_p and b_p or the magnitude $|c_p|$ and the phase angle ϕ_p must be stored in the memory of the analyzer. The required information is then determined by simple scaling, which can be used to produce different spectral displays. These frequency components (also called either spectral lines or lines) are spaced at intervals Δf on the output from the analyzer where

$$\Delta f = \frac{\omega_0}{2\pi} = \frac{1}{T} \tag{10.76}$$

The pth frequency component is then given by

$$f_p = p \, \Delta f \tag{10.77}$$

A. Periodic Signals

The frequency spectrum for periodic signals is displayed as magnitudes of the discrete Fourier frequency components. Equations 10.73 and 10.75 are used to display these component amplitudes as a peak display. However, if it is necessary to display the component amplitudes on a rms display, then Eqs. 10.74 and 10.75 are used. The rms of the overall signal is found by using Eq. 10.17. The mean value c_0 is used in both display situations and in the overall rms calculation. The mean value is independent of time and is treated differently from the amplitude of the frequency components.

B. Transient Signals

The Fourier transform used in dealing with transient signals generates a continuous frequency spectrum $F(\omega)$. Equation 10.20 indicates that the discrete Fourier transform generates a discrete frequency spectrum that is equal to the continuous spectrum when the discrete spectrum is multiplied by the window period T. Thus, the required display for $F(p\omega_0)$ is obtained by multiplying c_p by T so that

$$F(p\omega_0) = F(p\Delta f) = c_p T = |c_p| T\, e^{j\phi_p} = D_p e^{j\phi_p} \tag{10.78}$$

for $p = 0, 1, 2, \ldots, N_c$, where $D_p = |c_p| T$ is the magnitude of the continuous frequency spectrum at each discrete component.

C. Random Signals

Random signals are characterized by the dual-sided mean-square spectral density function $S(\omega)$ (see Eq. 10.23b) defined over the region $-\infty < \omega < \infty$. The single-sided mean-square spectral density (MSSD) is defined over the region $0 < \omega < \infty$ by

$$G(\omega) = \begin{cases} S(0) & \text{for } \omega = 0 \\ 2S(\omega) & \text{for } \omega > 0 \end{cases} \tag{10.79}$$

The quantity $G(\omega)$ represents the mean square per hertz and is commonly called the power spectral density (PSD). $G(\omega)$ is estimated by:

$$G(p\omega_0) = G(p\Delta f) = \left(B_{\mathrm{rms}_p}\right)^2 T$$

$$= 2|c_p|^2 T = 2c_p c_p^* T = D_p \qquad \text{for } p = 1, 2, \cdots, N_c \tag{10.80}$$

$$G(0) = c_0^2 T \qquad \text{for } p = 0$$

where D_p is the discrete magnitude displayed as the frequency spectrum. It is evident from the form of Eq. 10.80 that all phase information is lost in the mean-square spectral density because $c_p c_p^*$ represents only magnitude information.

Further interpretation of the MSSD is obtained by writing the autocorrelation function as given by Eq. 10.24 in terms of $G(\omega)$. This gives

$$\phi(0) = \frac{1}{2\pi} \int_0^\infty G(\omega)\, d\omega \tag{a}$$

The continuous integral can be written as a summation of the discrete MSSD frequency components of $G(p\omega_0)$ by noting that $d\omega = 2\pi\Delta f = 2\pi/T$. Substituting Eq. 10.80 into Eq. a, converting to a summation notation, and using all N_c terms gives

$$\phi(0) = B_0^2 + \sum_{p=1}^{N_c} \left(B_{\mathrm{rms}_p}\right)^2 = c_0^2 + \sum_{p=1}^{N_c} 2|c_p|^2 \tag{10.81}$$

This result, which is the same as Eqs. 10.16 and 10.17, is used to calculate the overall rms value of a random signal.

In many analyzers, the selection of units for the frequency-spectrum display instructs the on-board computer to use appropriate conversion factors when calculating the display. The foregoing equations indicate that all displays are related to the spectrum magnitude $|c_p|$.

10.8.2 Filter Characteristics

An ideal filter acts like a Dirac delta function, but there is no practical way to implement such a filter in the time domain. A reasonable alternative to this ideal filter transmits a rectangular pulse with a gain of unity within a passband of B Hz and with a gain of zero elsewhere. Such a filter transmits all frequencies within the passband and blocks all frequency components outside the passband, as shown in Fig. 10.21a. The transmission characteristics of a practical filter, shown in Fig. 10.21b, are described by four parameters in the frequency domain: center frequency, bandwidth, ripple, and selectivity.

The center frequency is defined in terms of lower and upper frequency limits, as shown in Fig. 10.21a. For constant bandwidth analyzers, the center frequency is given by

$$f_0 = \frac{f_u + f_l}{2} \tag{10.82}$$

where f takes on values of $(p\Delta f)$ in a digital analyzer.

The filter's bandwidth is a measure of its ability to separate frequency components and indicates the instrument's resolution. Two measures, the effective bandwidth and the 3-dB bandwidth, are commonly employed. The effective bandwidth is related to the convolution integral of Eq. 10.69. In this case, the input signal is assumed to be white noise with a MSSD of unity. Integrating the square of the transmitted amplitude gives the cross-hatched area shown in Fig. 10.21b. This area is replaced by an equivalent area for an ideal rectangular filter with unity height and bandwidth B. The 3-dB bandwidth is the difference in hertz between the half power points (filter amplitude of 0.707 or filter squared amplitude of 0.5). In most practical filters, there is little difference between the two measures of bandwidth. The 3-dB bandwidth is usually employed, since it is the easier of the two to measure.

Ripple is related to the oscillations in signals transmitted over the frequency limits, as shown in Fig. 10.21b. It characterizes amplitude uncertainty at a given frequency.

Selectivity indicates the ability of a filter to separate adjacent frequency components with widely varying amplitudes. Shape factor (SF) is an important parameter used to

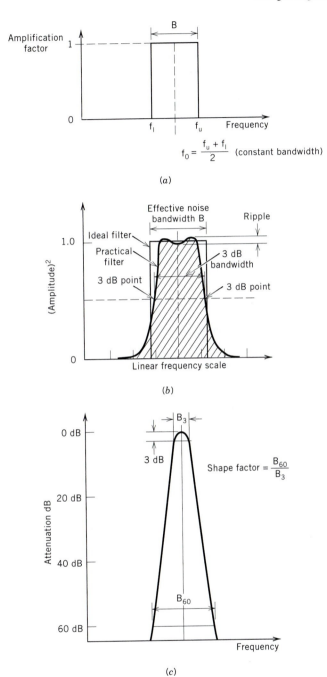

$$f_0 = \frac{f_u + f_l}{2} \quad \text{(constant bandwidth)}$$

(a)

(b)

$$\text{Shape factor} = \frac{B_{60}}{B_3}$$

(c)

Figure 10.21 Filter characteristics. (a) An ideal filter. (b) A practical filter. (c) A shape factor. (Courtesy of Bruel and Kjaer Instruments, Inc.)

Table 10.1 Window Function Coefficients

Function	a_0	a_1	a_2	a_3	a_4
Rectangular	1	0	0	0	0
Hanning	1	1	0	0	0
Kaiser–Bessel	1	1.298	0.244	0.003	0
Flat-top	1	1.933	1.286	0.388	0.032

describe selectivity. Shape factor is defined in terms of filter bandwidth corresponding to attenuations of 60 dB and 3 dB, as shown in Fig. 10.21c.

$$SF = \frac{B_{60}}{B_3} \tag{10.83}$$

A shape factor of unity is desirable but unrealistic. This definition of SF, Eq. 10.83, covers a dynamic range of 1000 to 1.

10.8.3 Four Common Window Functions

Four commonly employed window functions available in digital frequency analyzers are rectangular, Hanning, Kaiser–Bessel, and flat-top. These window functions are defined by the series function

$$w(t) = a_0 - a_1 \cos(\omega_0 t) + a_2 \cos(2\omega_0 t)$$

$$- a_3 \cos(3\omega_0 t) + a_4 \cos(4\omega_0 t)$$

$$\text{for } 0 < t < T \tag{10.84}$$

$$w(t) = 0 \quad \text{for } t > T$$

where ω_0 is the fundamental frequency of the analyzer. The coefficients a_i, given in Table 10.1, are scaled so that the area under each window function is unity. Graphic displays of the functions are shown in Fig. 10.22. These results show that the rect-

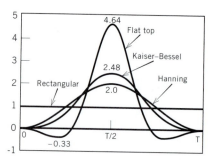

Figure 10.22 Characteristics of rectangular, Hanning, Kaiser–Bessel, and flat-top window functions in the time domain. (Courtesy of Bruel and Kjaer Instruments, Inc.)

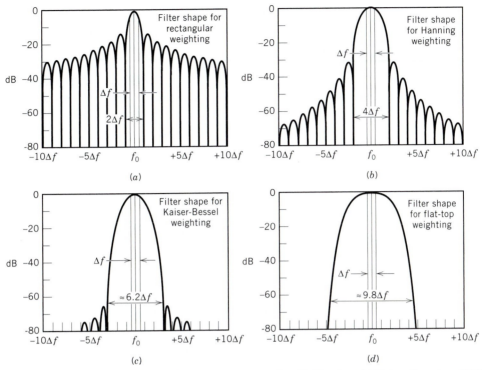

Figure 10.23 Digital filter characteristics for (*a*) rectangular, (*b*) Hanning, (*c*) Kaiser–Bessel, and (*d*) flat-top window functions. (Courtesy of Bruel and Kjaer Instruments, Inc.)

angular window transmits all time data with the same weight. The other three window functions emphasize the window center and go to zero on both ends to reduce errors owing to filter leakage at the window ends.

Comparisons of several frequency-domain characteristics of these four window functions are given in Table 10.2.

Bandwidth (either noise or 3 dB) increases with complexity of the window functions while ripple decreases. The last four columns in Table 10.2 are used to describe the selectivity of the window function in separating closely spaced frequency components with widely varying amplitudes. These selectivity factors are compared in Fig. 10.23, where filter characteristics of each window function are given. Selection of the best window function depends on the type of signal being analyzed.

Table 10.2 Characteristics of Four Window Functions

Window	Noise Band-width (Δf)	3-dB Band-width (Δf)	Ripple (dB)	Highest Side Lobe (dB)	Side Lobe Falloff Rate (dB/Decade)	60-dB Bandwidth (Δf)	Shape Factor
Rectangular	1.0	0.89	3.92	−13.3	20	665	750
Hanning	1.5	1.44	1.42	−31.5	60	13.3	9.2
Kaiser–Bessel	1.80	1.71	1.02	−66.6	20	6.1	3.6
Flat-top	3.77	3.72	0.01	−93.6	0	9.1	2.5

10.8.4 Uncertainty in the Magnitude of Spectral Lines

There is uncertainty in the determination of the magnitude of any spectral line. The amount of uncertainty depends on the type of signal being analyzed. For a stationary Gaussian random signal, a statistical analysis of the magnitude of a given line shows that its variation is also a stationary Gaussian random process. The mean value and standard deviation of the magnitude describe the uncertainty for a given line. For Gaussian white noise (bandwidth limited) with a zero mean value and record duration T_r, the normalized mean-square error is given by

$$\epsilon^2 = \frac{1}{BT_r} \qquad \textbf{(a)}$$

provided that the BT_r product (bandwidth times averaging time) is greater than 5. The same expression holds for mean-square spectral-density lines when B is replaced by an equivalent filter bandwidth B_e. The total record length T_r is related to the individual analyzer sample length T by

$$T_r = N_r T \qquad \textbf{(b)}$$

where N_r is the number of statistically independent window periods used to make up the spectral line average. The BT product can be written as

$$B_e T_r = N_r (B_e T) = N_r \qquad \textbf{(c)}$$

since $B_e T \cong 1$ for most digital frequency analyzers. The corresponding relative standard deviation ϵ_r of the line amplitude comes from combining Eqs. a and c and by noting that $\epsilon_r = \epsilon/2$. Thus,

$$\epsilon_r = \frac{1}{2\sqrt{B_e T_r}} = \frac{1}{2\sqrt{N_r}} = \frac{\alpha_p}{\overline{D}_p} \qquad \textbf{(10.85)}$$

where $\overline{D}_p$ is the average mean square value and α_p is the standard deviation of the pth line.

This result shows that the relative error, indicated by ϵ_r, can be controlled by increasing N_r. For example, for $\epsilon_r < 0.05$, $N_r = 100$. Since line magnitude is a Gaussian process, there is a 68.3 percent probability that the reading is within $\pm\epsilon_r$, a 95.4 percent probability that the reading is within $\pm 2\epsilon_r$, and a 99.7 percent probability that the reading is within $\pm 3\epsilon_r$. Note that $\epsilon_r = 0.05$ corresponds to about ± 0.45 dB and $3\epsilon_r = 0.15$ corresponds to about ± 1.3 dB.

Linear averaging is often used in determining the average spectral magnitude. The standard average formula for the pth spectral line is given by

$$|\overline{D_p}|^2 = \frac{1}{N_r} \sum_{k=1}^{N_r} |D_p(k)|^2 \qquad \textbf{(d)}$$

Implementation of Eq. d may be difficult because the average $\overline{D}_p$ is lost if a user decides to stop gathering data before all N_r records are analyzed. This deficiency is overcome by using an averaging method given by

$$|\overline{D_p(k)}|^2 = \frac{1}{k}|D_p(k)|^2 + \frac{k-1}{k}|\overline{D_p(k-1)}|^2 \qquad \textbf{(10.86)}$$

for $1 \leq k \leq N_r$. This procedure gives a correctly scaled rms result for $\overline{D}_p(k)$ for any value of k within the specified range. Averaging is mandatory when working with random signals. Usually, 75 to 200 window samples are required to achieve acceptable results in analysis of random signals. Ideally, when working with periodic signals averaging is not required. However, periodic and transient signals contain random noise that contributes to the measured frequency spectrum. This noise comes from instrument electronics and from small fluctuations in the periodic signal. If the signal-to-noise ratio is high, then the number of window samples required to reduce error owing to noise is small. The lower the signal-to-noise ratio, the greater the number of samples that must be averaged in order to ensure that the random noise does not significantly bias the frequency spectrum.

10.8.5 Summary of Window Use

The four different window functions provided in commercial spectrum analyzers have different advantages depending on the signal being processed.

A. Periodic Signals

The rectangular window is poorly suited for use in general periodic analysis. However, it is excellent for use with special periodic signals such as pseudorandom signals and order-tracking experiments. Pseudorandom signals are periodic signals with many frequency components in random phase. In tracking experiments, the sample frequency is a multiple of a machine event, such as 100 times rpm. In these cases, analyzer and measured frequencies match, so that leakage is avoided. The Hanning window is a good general-purpose window function for use with periodic signals. The Kaiser–Bessel window is best suited for cases requiring greater selectivity, such as separation of two closely spaced frequency components with widely different amplitudes. The flat-top window is excellent for determining periodic amplitudes, but is poor in frequency accuracy owing to its flat top. The flat-top window is fully effective in determining spectral amplitudes only if the frequency components are separated by $6 \Delta f$ or more.

B. Transient Signals

The rectangular window is generally required for transient analysis, particularly when measuring Fourier transform frequency spectra. The Hanning window is used in special cases when a transient signal persists for a long time compared with the window duration T or if repetitive impulses are employed to obtain repetitive input signals. The Kaiser–Bessel and flat-top windows are not recommended for this type of signal.

C. Random Signals

The rectangular window is poorly suited for use with random signals. This window can be used only with pseudorandom signals when there is no filter leakage owing to mismatch of frequency components. The Hanning window is recommended as a general-purpose random-analysis window. The Kaiser–Bessel window gives good frequency resolution but is not recommended for general-purpose analysis of random signals because it is slower to implement than the Hanning window. The flat-top window is not recommended for random-signal analysis because of its very wide center lobe.

10.9 ACCELEROMETER CROSS-AXIS SENSITIVITY

Accelerometers are sensitive to motion in a plane that is perpendicular to their primary sensing direction. Cross-axis sensitivity is primarily due to manufacturing imperfections. Even though maximum transverse sensitivity is less than 5 percent, Han and McConnell (Reference 11) have shown that cross-axis sensitivity can seriously degrade vibration measurements. McConnell and Han (Reference 15) have modeled accelerometer cross-axis characteristics and applied them to triaxial accelerometers. The purpose of acceleration measurements is to obtain correct linear acceleration components of structural motion at a given point from which vibration analysis is performed.

10.9.1 Single Accelerometer Cross-Axis Coupling Model

Consider an accelerometer oriented such that its primary sensing axis is the z axis and its cross-axis (perpendicular) plane contains the x and y axes, as shown in Fig. 10.24. The accelerometer's sensitivity is represented by vector $\overline{\mathbf{S}}_0$, which is described by

$$\mathbf{S}_0 = S_0 \sin \phi \cos \theta \, \mathbf{i} + S_0 \sin \phi \sin \theta \, \mathbf{j} + S_0 \cos \phi \, \mathbf{k} \tag{a}$$

where

ϕ is the angle of the sensitivity vector with respect to the z axis
θ is the angle in the xy plane that locates the vertical plane containing vector S_0 and the z axis
$\mathbf{i}$, $\mathbf{j}$, and $\mathbf{k}$ are unit vectors in the x, y, and z directions, respectively

The acceleration that is to be measured can be described as

$$\mathbf{a} = a_x \mathbf{i} + a_y \mathbf{j} + a_z \mathbf{k} \tag{b}$$

The accelerometer's output voltage is directly proportional to a dot product of the acceleration and sensitivity vectors. Thus,

$$v_z = S_0 \sin \phi \cos \theta \, a_x + S_0 \sin \phi \sin \theta \, a_y + S_0 \cos \phi \, a_z$$

$$= S_{zx} a_x + S_{zy} a_y + S_{zz} a_z \tag{10.87}$$

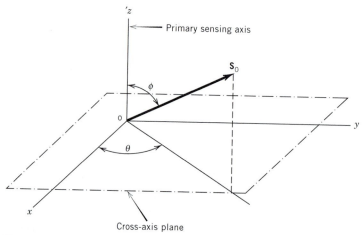

Figure 10.24 Coordinate axes and voltage-sensitivity vector $\mathbf{S}_0$.

where

$$S_{zx} = S_0 \sin\phi \cos\theta \qquad \text{(10.88a)}$$

$$S_{zy} = S_0 \sin\phi \sin\theta \qquad \text{(10.88b)}$$

are the cross-axis voltage sensitivities, and

$$S_{zz} = S_0 \cos\phi \qquad \text{(10.88c)}$$

is the primary voltage sensitivity. Note that angle ϕ is small so that $\cos\phi \cong 1$ and $\sin\phi \cong \phi$. These equations show that output voltage is independent of the coordinate axes used, since this voltage depends on a dot product the value of which is independent of the coordinates employed.

10.9.2 Triaxial Accelerometer Model

A triaxial accelerometer has three orthogonal axes, called x, y, and z, that correspond to the respective primary sensing axes. The output voltage for each primary sensing axis is given by equations similar to Eqs. 10.87 and 10.88. The acceleration vector and its components are described by Eq. b for all accelerometers in the triaxial assembly. Thus, the output voltage for each accelerometer becomes

$$S_{xx}\, a_x + S_{xy}\, a_y + S_{xz}\, a_z = v_x$$

$$S_{yx}\, a_x + S_{yy}\, a_y + S_{yz}\, a_z = v_y \qquad \text{(10.89)}$$

$$S_{zx}\, a_x + S_{zy}\, a_y + S_{zz}\, a_z = v_z$$

The voltage sensitivity S_{ip} assigned to each cross axis is arbitrary in the sense that the x-, y-, and z-coordinate directions are determined during transducer assembly. However, since the output voltage is independent of the coordinate axes selected, owing to dot product characteristics, calibration is sufficient to establish a suitable set of voltage sensitivities for a given triaxial accelerometer.

10.9.3 Correcting Acceleration Voltage Readings

True acceleration readings are obtained from Eqs. 10.89 by dividing each equation by its primary-axis voltage sensitivity S_{pp}. This gives equations that relate the actual acceleration components a_i to the apparent (uncorrected) acceleration components b_p. Thus, Eqs. 10.89 become

$$\epsilon_{xx} a_x + \epsilon_{xy} a_y + \epsilon_{xz} a_z = b_x$$

$$\epsilon_{yx} a_x + \epsilon_{yy} a_y + \epsilon_{yz} a_z = b_y \qquad \text{(10.90)}$$

$$\epsilon_{zx} a_x + \epsilon_{zy} a_y + \epsilon_{zz} a_z = b_z$$

where

$$\epsilon_{pi} = \frac{S_{pi}}{S_{pp}} \tag{10.91}$$

is the cross-axis sensitivity for that direction expressed relative to the primary-axis sensitivity and

$$b_p = \frac{v_p}{S_{pp}} \tag{c}$$

is the apparent acceleration reported by the pth direction transducer. It is observed from Eq. 10.91 that $\epsilon_{xx} = \epsilon_{yy} = \epsilon_{zz} = 1$. Equation 10.90 can be written in terms of ϵ_{pi} in matrix notation as

$$[\epsilon]\{a\} = \{b\} \tag{10.92}$$

Equations 10.90 and 10.92 indicate that apparent acceleration b_p is modified by the cross-axis terms $\epsilon_{pi} a_i$ and is not directly equal to a_p even though $\epsilon_{pp} = 1$. Only if the corresponding ϵ_{pi} terms or the acceleration terms are zero does $b_p = a_p$.

The inverse of Eq. 10.92 is then written as

$$\{a\} = [C]\{b\} \tag{10.93}$$

where matrix element C_{pi} is a multiplier of apparent acceleration b_i for direction p. Table 10.3 shows typical ranges of C_{pi} values when ϵ_{pi} is bounded by ± 5 and ± 10 percent.

The reference values used in Table 10.3 are unity for the C_{pp} terms and $C_{pi} = -\epsilon_{pi}$ for the cross-axis terms. It is clear that 5 percent cross-axis error causes a maximum of 1 percent error in the C_{pp} terms while C_{pi} is in error by about 5.8 percent. The 10 percent cross-axis error causes about 5 percent error in C_{pp}. The cross-axis C_{pi} terms are in error by about 13 percent for cross-axis error of ± 10 percent. Thus, approximations of $C_{pp} \approx 1$ and $C_{pi} \approx -\epsilon_{pi}$ are acceptable when cross-axis error is bounded by ± 5 percent.

With the approximations $C_{pp} \approx 1$ and $C_{pi} \approx -\epsilon_{pi}$, Eq. 10.93 can be written as

$$a_x = b_x - \epsilon_{xy} b_y - \epsilon_{xz} b_z$$

$$a_y = -\epsilon_{yx} b_x + b_y - \epsilon_{yz} b_z \tag{10.94}$$

$$a_z = -\epsilon_{zx} b_x - \epsilon_{zy} b_y + b_z$$

Table 10.3 Coefficient C_{pi} and Errors from Nominal Values

ϵ_{pi}	C_{pp} Range of Values[a]	C_{pi} Range of Values[b]	C_{pi} Nominal Value	C_{pi} Max. Error %
0.05	0.990 to 1.010	0.0478 to 0.052	0.05	5.8
0.10	0.951 to 1.052	0.0874 to 0.113	0.10	$\cong 13$

[a] Range of actual values to be compared with nominal value of unity.
[b] Range of actual values to be compared with nominal C_{pi} value.

Equations 10.94 are linear, convenient to apply, and acceptable for cross-axis coupling bounded by ±5 percent.

10.9.4 Application to Modal Analysis Signals

Experimental FRFs control the quality of experimental modal analysis. A steel free-free beam with a cross section 25.4 by 28.6 mm and a length of 2.337 m has a triaxial accelerometer mounted on its left end, as shown in Fig. 10.25. This beam is used as a source of known structural vibration for demonstrating that the approximate correction matrix $[C]$ is able to remove cross-axis contamination from experimental data. Two impact-force input directions, Y and S, are used. The Y input excites only Y-direction vibration. The S input causes nearly equal motion in both the Y and Z directions. The Y-direction acceleration FRF resulting from Y-direction excitation is shown in Fig. 10.26a. In this case, both raw data and corrected data plot the same, since little or no motion is induced in the X and Z directions. Thus no corrections are required for this case.

Impact in the S direction causes significant motion in both the Y and Z directions. The Y-direction acceleration FRF is shown in Fig. 10.26b. In this case, there are significantly different responses in the raw data at points A and B, which are not obtained in Fig. 10.26a for the Y-input case. This response at point A corresponds to motion induced in the Z direction and is measured by the Y-direction accelerometer because of its cross-axis sensitivity. The compensated Y-direction data are also shown in Fig. 10.26b. Thus, compensation calculations based on Eqs. 10.94 are very effective in removing cross-axis signal contamination. A more detailed discussion of how to use cross-axis sensitivity corrections in modal analysis is given in Reference 10.

10.9.5 Cross-Axis Resonance

Experience shows that the cross-axis natural frequencies are less (about half in some cases) than the primary sensing axis natural frequency. This fact implies that Eqs.

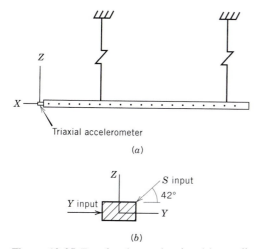

(a)

(b)

Figure 10.25 Free-free beam showing (a) coordinate directions and transducer location, and (b) impact directions Y and S.

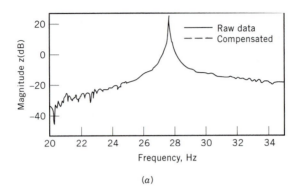

(a)

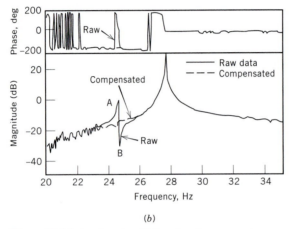

(b)

Figure 10.26 Acceleration Y-direction frequency response functions. (a) Y-direction input. (b) S-direction input.

10.94 are frequency dependent as well, since the ϵ_{pi} terms change with frequency. Thus, cross-axis calibration must be performed over a broad range of frequencies, a task that is difficult because the input motion must be pure linear in order to achieve accurate calibrations. Further research is needed to provide understanding of the significance of cross-axis sensitivity in different measurement applications.

10.10 FORCE TRANSDUCER–STRUCTURE INTERACTION

Often a force transducer is significantly altered when it is attached to a structure to measure dynamic forces. There are two common application environments. First, a load cell may be attached to a hammer in order to measure impulsive forces, as shown in Fig. 10.27a. Second, a force transducer may be attached to a structure on one end and to the armature of a vibration exciter on the other end, as shown in Fig. 10.27b. The use of force transducers in each of these environments is explored in this section.

In the piezoelectric force transducer shown in Fig. 9.9, the seismic mass and the base mass are nearly equal and are separated by the piezoelectric sensing element. In other designs, the seismic mass may represent only 15 to 20 percent of the total mass of the load cell. In this case it is important to connect the seismic-mass end to the structure under test.

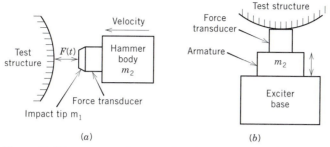

(a) (b)

Figure 10.27 Force transducer. (*a*) Impulse hammer. (*b*) Force transducer attached to a structure and to the moving head of a vibration exciter.

10.10.1 General Two-Degree-of-Freedom Force Transducer Model

In order to study how a load cell is influenced by its environment, a more general dynamic model is useful. Consider a force transducer that consists of two masses m_1 (seismic end) and m_2 (base end), a spring k and damper c (primarily owing to the piezoelectric sensor), and two excitation forces $F_1(t)$ and $F_2(t)$, as shown in Fig. 10.28*a*. In Fig. 10.28*a*, $F_1(t)$ acts on the seismic-mass end of the transducer, which is in contact with the structure under test, and is the external force to be sensed by the sensing element, and $F_2(t)$ is the external force acting on the base mass m_2. A comparison of Fig. 10.28*a* with Fig. 10.10 shows that $F_1(t) = -F_1(t)$, $k_2 = k$, $c_2 = c$, $k_1 = k_3 = 0$, and $c_1 = c_3 = 0$. Then, Eqs. 10.27 reduce to

$$(m_1)\ddot{x}_1 + (c)\dot{x}_1 + (-c)\dot{x}_2 + (k)x_1 + (-k)x_2 = -F_1(t)$$
$$(m_2)\ddot{x}_2 + (-c)\dot{x}_1 + (c)\dot{x}_2 + (-k)x_1 + (k)x_2 = -F_2(t)$$

(10.95)

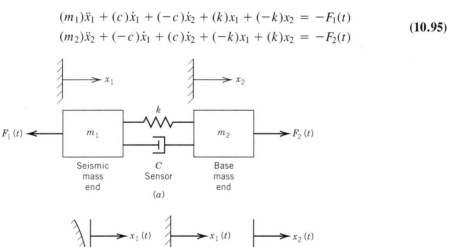

(a)

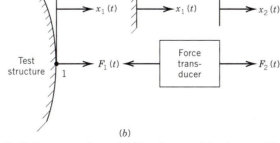

(b)

Figure 10.28 Force transducer. (*a*) Two-degree-of-freedom model. (*b*) Attached to a test structure.

The steady-state responses to sinusoidal excitations of $F_1(t) = F_1 e^{j\omega t}$ and $F_2(t) = F_2 e^{j\omega t}$ are $x_1 = A_1 e^{j\omega t}$ and $x_2 = A_2 e^{j\omega t}$. Let $S = k + jc\omega$ be the transducer's apparent stiffness. Then, Eqs. 10.95 reduce to the following algebraic equations:

$$(S - m_1\omega^2)A_1 - SA_2 = -F_1$$

$$-SA_1 + (S - m_2\omega^2)A_2 = F_2$$

(a)

The determinant of the A_1 and A_2 coefficients in Eq. a is given by

$$\Delta(\omega) = -[(m_1 + m_2)S - m_1 m_2\,\omega^2]\omega^2$$

$$= -k(m_1 + m_2)(1 - r^2 + j2dr)\omega^2$$

(b)

The natural frequency occurs when the real part of Eq. b is zero. This requirement gives two values, the first of which is zero, which implies that the transducer is free to move as a rigid body (called a semidefinite system). The other natural frequency ω_n and corresponding damping ratio d are defined by

$$\omega_n = \sqrt{k/m_e} \qquad m_e = m_1 m_2/(m_1 + m_2)$$

$$d = c/(2\sqrt{km_e}) \qquad r = \omega/\omega_n$$

(10.96)

where m_e is the effective mass and r is the dimensionless frequency ratio. Equations 10.96 show that the effective mass is not the seismic mass, but rather a combination of masses, and that the transducer's natural frequency and damping change when either mass changes. Application of Kramer's rule to Eqs. a gives the steady-state responses

$$A_1 = \frac{SF_2 - (S - m_2\omega^2)F_1}{\Delta(\omega)}$$

and

(c)

$$A_2 = \frac{(S - m_1\omega^2)F_2 - SF_1}{\Delta(\omega)}$$

The piezoelectric-sensor output voltage can be written as

$$v_f = H_f^e(\omega)\frac{S_z}{C_f}(A_2 - A_1)$$

(d)

where S_z is the displacement charge sensitivity (C per unit of displacement) and $H_f^e(\omega)$ is the transducer's electrical low-frequency characteristic (as indicated by the superscript e), which can be expressed as

$$H_f^e(\omega) = \frac{jR_f C_f\omega}{1 + jR_f C_f\omega}$$

(e)

where $R_f C_f$ is the force-transducer time constant. Substituting A_1 and A_2 from Eq. c into Eq. d gives

$$v_f = S_f H_f(\omega)\left[\frac{m_2 F_2}{(m_1 + m_2)} + \frac{m_1 F_2}{(m_1 + m_2)}\right]$$

(10.97)

where $S_f = S_z/(C_f k) = S_q/C_f$ is the voltage sensitivity (mV per unit of force) and $S_q = S_z/k$ is the transducer charge sensitivity (C per unit of force). The transducer frequency response function $H_f(\omega)$ is given by

$$H_f(\omega) = \left(\frac{j R_f C_f \omega}{1 + j R_f C_f \omega} \right) \left(\frac{1}{1 - r^2 + j2dr} \right) \qquad (10.98)$$

electrical   mechanical

where r is defined in Eqs. b and 10.96 and is referenced to the installed transducer natural frequency. The FRF described in Eq. 10.98 is the typical electrical and single-degree-of-freedom mechanical response predicted in Chapter 7. However, the effective inertia m_e in this model is application dependent. Before Eq. 10.97 can be used to predict force F_1, a relationship is needed between F_1 and F_2. Two cases are considered here, the impact hammer shown in Fig. 10.27a and the vibration exciter shown in Fig. 10.27b.

A. Impact Hammer

The impact hammer in Fig. 10.27a is characterized by force F_2 being zero, since the dominant input force comes from hammer inertia. In this case, mass m_1 represents the combined seismic and impact-tip masses and m_2 represents the combined transducer-base and hammer-body masses. Thus, when F_2 is zero, Eq. 10.97 becomes

$$v_f = \left[\frac{m_2 S_f}{(m_1 + m_2)} \right] H_f(\omega) \, F_1 = S_f^* H_f(\omega) \, F_1 \qquad (10.99)$$

where S_f^* is the effective voltage sensitivity (mV per unit of force), which depends on masses m_1 and m_2 and is identical in form to Eq. 9.47. It is common practice to change these masses to achieve different contact times since the impact time depends on the interface spring constant, the hammer mass, and the characteristics of the test structure. It is also clear from Eq. 10.96 that the transducer's natural frequency and damping are significantly altered by changing either tip mass m_1 or hammer-body mass m_2. These equations describe impact-hammer behavior quite accurately over a broad range of mass ratios so long as there are no glancing blows and the person swinging the hammer has a hand mass that is small compared with hammer mass. Hand mass can be a problem when dealing with very light hammers, since F_2 is no longer zero during impact and some of the hand's inertia is transferred to m_2, which can change the effective voltage sensitivity S_f^*.

B. Vibration Exciter Arrangement

A force transducer that is attached to a test structure on one end and to a vibration exciter on the other end is shown in Fig. 10.27b. For this case, mass m_1 represents the transducer's seismic mass that is in contact with the structure plus the mass of any connection device, such as bolts or mounting blocks. If an accelerometer is attached to the test structure at the same spot as the force transducer, then the accelerometer mass must be included in m_1. Mass m_2 includes the transducer's base mass as well as the vibration exciter's armature mass.

Force F_1 represents the force transmitted directly to the test structure at the transducer-structure interface, and force F_2 is the force applied to the vibration exciter's armature. The required expression between forces F_1 and F_2 is obtained as follows. The amplitude of structural motion at point 1 in Fig. 10.28b is related to the structure's driving-point displacement FRF $H_{11}(\omega)$ and excitation force F_1. Thus,

$$A_1 = H_{11}(\omega)\, F_1 \tag{f}$$

Amplitude A_1 is also given by Eq. c. Equating Eqs. c and f and solving for F_2 in terms of F_1 gives

$$F_2 = \left[\frac{\Delta(\omega)H_{11}(\omega) + S - m_2\omega^2}{S} \right] F_1 \tag{10.100}$$

Equation 10.100 shows how force F_1 is modified for feedback to the exciter. The test structure $[H_{11}(\omega)]$, force-transducer characteristics $[\Delta(\omega)]$, and exciter's inertia term $[m_2]$ affect this input–output relationship. Thus, it is seen that the useful force available to excite the structure depends on the dynamic characteristics of each element in the testing system.

Substitution of Eq. 10.100 into Eq. 10.97 gives

$$v_f = S_f H_f(\omega)\left[1 - m_1 H_{11}(\omega)\, \omega^2\right] F_1 \tag{10.101}$$

where $S_f = S_q/C_F = S_z/(C_f k)$ is the voltage sensitivity (mV per unit of force) of the load cell. The transducer FRF is given by

$$H_f(\omega) = \left(\frac{jR_f C_f \omega}{1 + jR_f C_f \omega} \right)\left(\frac{1}{1 + jc\omega/k} \right) \tag{10.102}$$

$$\underset{\text{electrical}}{} \qquad \underset{\text{mechanical}}{}$$

which has no transducer resonance when compared with Eq. 10.98. The mechanical part of Eq. 10.102 contributes to signal attenuation and phase shift. The significance of this term can be obtained when the mechanical term is written as

$$1 + \frac{jc\omega}{k} = 1 + \frac{j2d_{bt}\omega}{\omega_{bt}} \tag{g}$$

where

$$\omega_{bt} = \sqrt{k/m_s} \text{ is the bare transducer's natural frequency}$$
$$d_{bt} = c/2\sqrt{km_s} \text{ is the bare transducer's damping ratio}$$
$$m_s \text{ is the bare transducer's seismic mass}$$

The word *bare* implies that the end of the transducer that is attached to the structure is a free surface whereas the other end is attached to a large mass. The bare transducer's natural frequency is the one specified by the manufacturer. Consider, for example, that $f_{bt} = 20$ kHz and $d_{bt} \cong 0.01$. Then the imaginary part corresponding to a 20-kHz signal in Eq. g is 0.02 and causes a 1.14-deg phase shift. Thus, this term is usually unimportant for frequencies up to the bare transducer's natural frequency so long as the transducer has light damping of a few percent or less.

An interesting fact about Eq. 10.101 is that $\Delta(\omega)$ cancels out and leaves only the mechanical transducer characteristic given in Eq. 10.102. This result means that there is no transducer mechanical resonance to contend with in terms of the generated voltage signal. Transducer resonance affects the input–output force ratio in Eq. 10.100 as well as the ratio of motion A_1/A_2 in Eqs. c. It is clear from Eq. 10.101 that for the structure under test, the transducer's seismic mass plus the connection mass and possibly an accelerometer mass (m_1) can cause significant measurement errors. This error is application dependent, since both $H_{11}(\omega)$ and m_1 usually change from application to application.

In order to obtain some feel for how $H_{11}(\omega)$ affects force measurement, assume that a test structure is represented by a single-degree-of-freedom system described by

$$m_s \ddot{x}_1 + c_s \dot{x}_1 + k_s x_1 = F_1(t) \tag{h}$$

where m_s, c_s, and k_s are the structural mass, damping, and stiffness. Substitution of the corresponding structural driving point $H_{11}(\omega)$ into Eq. 10.101 gives an error ratio multiplier of

$$\frac{\hat{v}_f}{v_f} = ER = \left[\frac{k_s - m_s\omega^2 + jc_s\omega}{k_s - (m_1 + m_s)\,\omega^2 + jc_s\omega} \right]$$

$$= \left[\frac{1 - r^2 + j2d_sr}{1 - (1 + M)r^2 + j2d_sr} \right] \tag{10.103}$$

where v_f is the true voltage signal, and M is the mass ratio m_1/m_s. A plot of ER for $M = 0.01$ and $d_s = 0.01$ is shown in Fig. 10.29 around structural resonance. This plot shows that force is first overestimated (27 percent) at point A and then underestimated (22 percent) at point B at the most critical frequencies near structural

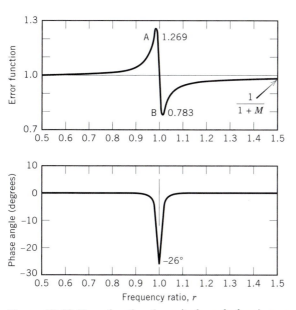

Figure 10.29 Error function (magnitude and phase) as a function of frequency for $M = 0.01$ and structural damping of 0.01.

Table 10.4 Maximum and Minimum Error Function Values

Mass Ratio m	Damping d	Error Function			Phase (deg)
		Maximum	Minimum	Resonance	
1	0.01	20.9	0.02	0.02	−174
1	0.05	7.07	0.10	0.10	−152
0.1	0.10	1.25	0.765	0.894	−26.9
0.1	0.05	1.56	0.610	0.707	−51.4
0.1	0.01	4.89	0.196	0.196	−135
0.01	0.05	1.05	0.949	0.995	−5.1
0.01	0.01	1.27	0.783	1.013	−26.0
0.001	0.05	1.005	0.995	1.001	0
0.001	0.01	1.025	0.975	1.001	−2.3

resonance. Equation 10.103 and Fig. 10.29 show that significant errors occur only at frequencies near the structure's resonance that lead to shifting of structural resonance frequency as well as peak FRF values. Also note that significant phase shift occurs in this resonance frequency range. Equation 10.103 shows that measurement error approaches a value of $1/(1 + M)$ for frequencies well above structural resonance. Table 10.4 gives maximum, minimum, and resonance values of ER as well as phase at structural resonance. This table indicates that transducer seismic mass, connector mass, and possibly accelerometer mass in contact with the structure (m_1) must be very small to avoid serious force-measurement errors when measuring forced response of lightweight and lightly damped structures.

10.11 SUMMARY

Investigations of the vibrations associated with structures or machines usually involve one of the following two situations: a system that has one or more internal excitation sources and the structural response at some point on the surface is measured with an accelerometer, or a system where a known excitation force is applied at one point on the surface and the response is measured at some other point. In both of these situations, the frequency response function $H(\omega)$ for the structural system controls the magnitude and phase of the output signal. Response signals are typically analyzed in two domains: the time domain and the frequency domain.

In the time domain, two single-valued parameters, the temporal mean and the temporal root mean square, are used to describe a signal. The temporal mean $\overline{f}$ is a time average of the $f(t)$ signal. It represents the signal's zero frequency component (the dc component) and is defined as

$$\overline{f} = \lim_{T \to \infty} \frac{1}{T} \int_0^T f(t) \, dt \tag{10.1}$$

The temporal mean square $\overline{f^2}$ is a time average of $f^2(t)$. Thus,

$$\overline{f^2} = \lim_{T \to \infty} \frac{1}{T} \int_0^T f^2(t) \, dt \tag{10.2}$$

The root mean square (A_{rms}) is the square root of the temporal mean square:

$$A_{rms} = \sqrt{\overline{f^2}} \qquad (10.3)$$

The quantities $\overline{f}$, $\overline{f^2}$, and A_{rms} provide little detail about the complicated motion producing the signal.

In the frequency domain, the data in the signal is represented as a function of frequency ω. This representation (known as a frequency spectrum) gives amplitudes of the frequency components as a function of the frequency ω. A frequency spectrum is more useful than $\overline{f}$, $\overline{f^2}$, or A_{rms} because it indicates discrete frequencies related to the operating characteristics of the system.

A digital frequency analyzer is an instrument used to measure frequency spectra. Processing of an analog signal by a digital frequency analyzer involves multiplication of the analog signal by both a window function w(t) with a duration T and a sample function $s(t)$ that selects the points to be digitized. The digital sequence of equally spaced numbers that is generated represents the original analog signal. The discrete frequency components are then processed internally to provide discrete periodic frequency components, transient continuous spectrum frequency components, and random mean square spectral density (power spectral density) frequency components.

An ideal filter acts like a Dirac delta function, but no practical way exists to implement such a filter in the time domain. An alternative to this ideal filter transmits a rectangular pulse with a gain of unity within a passband and with a gain of zero elsewhere. Rectangular, Hanning, Kaiser–Bessel, and flat-top filters are commonly available in digital frequency analyzers. The rectangular filter is excellent for use with transient signals when measuring Fourier transform frequency spectra. The Hanning window is a good general purpose filter for both periodic signals and random signals. It is also used for transient signals that persist for a long time compared with the window duration T or if repetitive impulses are used to obtain repetitive input signals. The Kaiser-Bessel window is used for cases involving two closely spaced frequency components with widely different amplitudes. The flat-top window is excellent for determining periodic amplitudes but exhibits poor frequency accuracy due to its "flat top."

In the continuous model for multi-degree-of-freedom motion each natural frequency has a unique mode shape, and modal loading is represented as an integral of the product of the spatial distribution of the excitation force and a mode shape. This integral product shows that some modes of vibration will never be excited by certain excitation load distributions. For example, point forces that act at a node point for a particular mode of vibration can not excite that vibration mode.

Finally, a force transducer in a dynamic environment, where the base mass accelerates, is considered. Analysis of a typical application of a force transducer between a vibration exciter armature and a structure under test indicates that the measured voltage is given by the equation

$$v_f = S_f H_f(\omega)\left[1 - m_1 H_{11}(\omega)\,\omega^2\right] F_1 \qquad (10.101)$$

where $H_{11}(\omega)$ is the structure's driving point receptance (x_1/F_1) and m_1 is the seismic mass that includes all attachment masses between the load cell's sensor and the structure under test. $H_f(\omega)$ is the measurement system's FRF that shows no mechanical resonance. The term $m_1 H_{11}(\omega)\,\omega^2$ causes significant measurement errors in the region of the test structure's natural frequency, the very frequency range where the most accurate measurements are required.

REFERENCES

1. Bendat, J. S., and A. G. Piersol: *Engineering Applications of Correlation and Spectral Analysis,* Wiley, New York, 1980.
2. Bendat, J. S., and A. G. Piersol: *Random Data: Analysis and Measurement Procedures,* 2nd ed., Wiley, New York, 1986.
3. Broch, J. T.: "Non-Linear Systems and Random Vibration: Selected Reprints from Technical Review," available from Bruel and Kjaer Instruments, Inc., Marlborough, Mass., January 1972.
4. Broch, J. T.: "Mechanical Vibration and Shock Measurement," available from Bruel and Kjaer Instruments, Inc., Marlborough, Mass., October 1980.
5. Bruel and Kjaer Instruments: *Technical Review,* a quarterly publication available from Bruel and Kjaer Instruments, Inc., Marlborough, Mass.
   a. "Measurement and Description of Shock," no. 3, 1966.
   b. "Vibration Testing," no. 3, 1967.
   c. "High Frequency Response of Force Transducers," no. 3, 1972.
   d. "On the Measurement of Frequency Response Functions," no. 4, 1975.
   e. "Digital Filters and FFT Technique," no. 1, 1978.
   f. "Discrete Fourier Transform and FFT Analyzers," No. 1, 1979.
   g. "Zoom-FFT," No. 2, 1980.
   h. "Cepstrum Analysis," no. 3, 1981.
   i. "System Analysis and Time Delay Spectrometry," Part 1, no. 1, 1983.
   j. "System Analysis and Time Delay Spectrometry," Part II, no. 2, 1983.
   k. "Dual Channel FFT Analysis," Part I, no. 1, 1984.
   l. "Dual Channel FFT Analysis," Part II, no. 2, 1984.
   m. "The Hilbert Transform," no. 3, 1984.
   n. "Vibration Monitoring of Machines," no. 1, 1987.
   o. "Windows to FFT Analysis," Part I, no. 3, 1987.
   p. "Windows to FFT Analysis," Part II, no. 4, 1987.
6. Cooley, J. W., and J. W. Tukey: "An Algorithm for the Machine Calculation of Complex Fourier Series," *Mathematics of Computation,* vol. 19, 1965, pp. 297–301.
7. Crandall, S. H., and W. D. Mark: *Random Vibration in Mechanical Systems,* Academic Press, New York, 1963.
8. Doebelin, E. O.: *System Modeling and Response,* Wiley, New York, 1980.
9. Ewins, D. J.: *Modal Testing: Theory and Practice,* Research Studies Press Ltd., Letchworth, Hertfordshire, England, 1984, available from Bruel & Kjaer Instruments, Inc., Marlborough, Mass.
10. Han, S.: "Effects of Transducer Cross-Axis Sensitivity on Modal Analysis," Ph.D. dissertation, Iowa State University, Ames, Iowa, 1988.
11. Han, S., and K. G. McConnell: "Effect of Mass on Force Transducer Sensitivity," *Exp. Tech.,* vol. 10, no. 7, 1986, pp. 19–22.
12. Han, S., and K. G. McConnell: "The Effects of Transducer Cross-Axis Sensitivity in Modal Analysis," *Proceedings 7th International Modal Analysis Conference,* Las Vegas, Nev., January 1989, vol. 1, pp. 505–511.
13. Korn, G. A.: *Random Process Simulation and Measurement,* McGraw–Hill, New York, 1966.
14. McConnell, K. G.: "Errors in Using Force Transducers," *Proceedings 8th International Modal Analysis Conference,* Kissimmee, Fla., 1990, vol. 2, pp. 884–890.

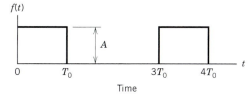

f(t)

A

0   T_0 $3T_0$ $4T_0$ *t*

Time

Figure E10.1

15. McConnell, K. G., and S. Han: "A Theoretical Basis for Cross-Axis Corrections in Tri-Axial Accelerometers," *Proceedings 9th International Modal Analysis Conference,* Florence, Italy, April 1991, vol. 1, pp. 171–175.
16. Pandit, S. M.: *Modal and Spectrum Analysis: Data Dependent Systems in State Space,* Wiley, New York, 1991.
17. Randall, R. B.: "Frequency Analysis," available from Bruel and Kjaer Instruments, Inc., Marlborough, Mass., 1987.
18. Rao, S. S.: *Mechanical Vibrations,* 2nd ed., Addison–Wesley, Reading, Mass., 1990.
19. Stroud, K. A.: *Fourier Series and Harmonic Analysis,* Stanley Thornes (Publishers) Ltd., Cheltenham, U.K., 1986.
20. Thomson, W. T.: *Theory of Vibration with Applications,* 3rd ed., Prentice–Hall, Englewood Cliffs, N.J., 1988.

EXERCISES

10.1   Determine the temporal mean and root mean square values for the periodic signals shown in Fig. E10.1.

10.2   Determine the temporal mean and root mean square values for the periodic signal shown in Fig. E10.2.

10.3   Verify the relationships given in Eqs. 10.6, 10.7, and 10.8.

10.4   Verify Eqs. 10.10 and 10.11a. Plot the $B^2/2$ error term as a function of ωT over the range $0 < \omega T < 20\pi$ when $\phi = 0$. What conclusions can you draw from this plot?

10.5   Determine the complex Fourier series coefficients for the signal shown in Fig. E10.1. Plot the magnitude and phase of these components for the first 20 terms. Compare this plot with Fig. 10.6.

10.6   Determine the complex Fourier series coefficients for the signal in Fig. E10.2. Plot the magnitude for the first 20 terms and compare this plot with Fig. 10.6.

10.7   Compare the temporal mean and root mean square values for the periodic signal in Fig. E10.1 as determined by Eqs. 10.1 and 10.2 with those from Eqs. 10.16 and 10.17.

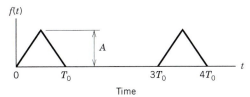

f(t)

A

0   T_0 $3T_0$ $4T_0$ *t*

Time

Figure E10.2

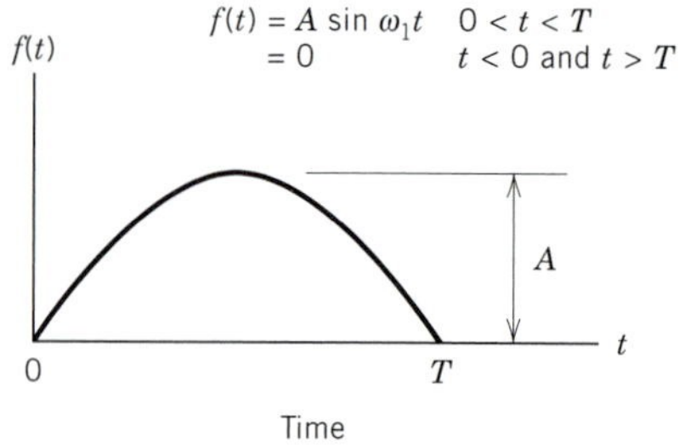

$f(t) = A \sin \omega_1 t \quad 0 < t < T$

$\quad = 0 \qquad\qquad t < 0 \text{ and } t > T$

$f(t)$

A

0 $\qquad\qquad T$ $\quad t$

Time

Figure E10.9

10.8   Repeat Exercise 10.7 for the periodic signal shown in Fig. E10.2.

10.9   Determine the frequency spectrum for the half-sine pulse shown in Fig. E10.9 by using a Fourier series approach ($\beta > 10$) and the Fourier transform approach in Eq. 10.19a. Compare the results. What must be done to compare these frequency spectra?

10.10   The mean-square spectral density for a random signal is a constant S_0 from $-\omega_0$ to ω_0 as shown in Fig. E10.10. Determine the corresponding root mean square (see Eq. 10.24) and autocorrelation function (Eq. 10.23a). Plot the autocorrelation function.

10.11   Determine the natural frequencies and mode shapes for a system where $k_1 = k$, $k_2 = 3k$, $k_3 = 0$, $m_1 = 2m$, and $m_2 = 3m$.

10.12   Verify the meaning of $H_{mn}(\omega)$ in Eqs. 10.36 and 10.45.

In the following problems the mode shapes for a simply supported beam (see Fig. E10.13a) are given by

$$V_k(x) = \sin \frac{k \pi x}{\ell}$$

10.13   If the excitation force is a uniformly distributed value of P_0 as shown in Fig. E10.13b, show that the modal excitation force Q_k (see Eq. 10.53) is given by

$$Q_k = \frac{\ell P_0}{k \pi}[1 - \cos(k\pi)] = \begin{cases} \dfrac{2\ell p_0}{k \pi} & k \text{ is odd} \\ 0 & k \text{ is even} \end{cases}$$

What is the implication of using this type of vibration excitation force for this simple system?

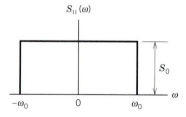

$S_{11}(\omega)$

S_0

ω

$-\omega_0 \qquad 0 \qquad \omega_0$

Figure E10.10

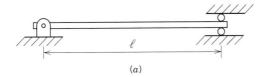

(a)

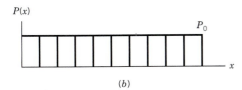

(b)

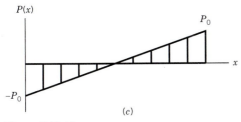

(c)

Figure E10.13

10.14   If an accelerometer is attached to the middle of the beam in Fig. E10.13a and the uniformly distributed excitation force in Fig. E10.13b is used (see problem 10.13), which natural frequencies should be observed?

10.15   If the triangular shaped excitation load shown in Fig. E10.13c is applied to the beam, show that the modal excitation force Q_k [see Eq. (10.53)] is given by

$$Q_k = \frac{-P_0\ell}{k\pi}(1 + \cos k\pi) = \begin{cases} -\dfrac{2P_0\ell}{k\pi} & k \text{ is even} \\ 0 & k \text{ is odd} \end{cases}$$

What is the implication of using this type of excitation force distribution?

10.16   When a point excitation force P_0 is applied at $x = x_0$, the distributed load $P(x)$ is given by $P(x) = P_0\delta(x - x_0)$. Show that the corresponding modal excitation force becomes

$$Q_k = P_0 \sin \frac{k\pi x_0}{\ell}$$

What natural frequencies are not excited if $x_0 = \ell/3$?

10.17   If an accelerometer is mounted at $x_t = \ell/4$ and a point excitation load is applied at $x_0 = \ell/3$ (see problem 10.16), what natural frequencies are observed in the accelerometer data?

10.18   A half-sine impulse force (see Fig. E10.9) with $A = 100$ lb and $T = 5$ ms is applied to a single-degree-of-freedom system ($k = 500$ lb/in., natural frequency $= 50$ Hz, and 5-percent critical damping). Using the frequency-domain input–output processes (see Fig. 10.14 and Eqs. 10.58, 10.59, and 10.60), calculate the single-degree-of-freedom response to this impulse. *Hint:* Solve this problem using a computer program to calculate the required Fourier transforms, FRF, and output time history.

10.19   For a single-degree-of-freedom undamped mechanical system

$$h(t) = \frac{1}{m\omega_n} \sin(\omega_n t)$$

and

$$H(\omega) = \frac{1}{k - m\omega^2} = \frac{1}{m(\omega_n^2 - \omega^2)}$$

Verify that Eqs. 10.63 are valid for this system when $h(t)$ is substituted into the $H(\omega)$ expression and integrated. Do not try to substitute $H(\omega)$ into the $h(t)$ expression, as this integral is difficult to evaluate without the use of special integral tables.

10.20   Verify that Eq. 10.69 can be obtained from Eq. 10.68 when $f_2(t)$ and $f_3(t)$ are replaced by their corresponding Fourier transforms.

10.21   Explain why antialiasing filters must have high attenuation rates (120 dB per octave) in order to have high break frequencies ($\omega_b = 0.4\ \omega_s$). Recall that most digital analyzers employ 14-bit or higher A/D converters.

10.22   Filter leakage causes the amplitude of a periodic frequency component to be distributed over many adjacent frequency components, as shown in Figs. 10.19*b* and *c*. For the cases shown, use Parseval's formula (Eqs. 10.16 or 10.17) to obtain the component's amplitude by summing terms 7 through 14 from Eq. b in section 10.7.3. Show that the effective rms amplitude is within 2.5 percent of the original value of unity.

10.23   Show that Eqs. 10.73 and 10.74 can be obtained from Eqs. 10.7, 10.8, and 10.14. Why are Eqs. 10.73 and 10.74 important in understanding and using a digital frequency analyzer?

10.24   How much is the relative error reduced when the number of independent data windows is doubled from 75 to 150 in measuring the pth frequency component from a random process?

10.25   When analyzing two periodic frequency components that have large amplitude differences (on the order of 40 dB) and are separated by two spectral lines, which window function or functions should be used to obtain the best magnitude estimate and best frequency resolution?

10.26   Why is it inappropriate to use a Hanning or flat-top window function when analyzing a single-event transient signal.

10.27   Develop Eq. 10.86 from first principles.

10.28   If the maximum accelerometer cross-axis sensitivity is ±5 percent, what is the value of angle ϕ in degrees?

10.29   Starting with Eq. 10.87 for the single-axis cross-axis sensitivity model, develop Eqs. 10.89 and 10.90.

10.30   When each accelerometer has a ±5 percent acceleration cross-axis sensitivity, verify that C_{pp} in Table 10.3 ranges from 0.990 to 1.010.

10.31   Show that Eqs. 10.94 are a reasonable inversion of Eqs. 10.90 in view of the results in Table 10.3.

10.32   Assume that the operator's hand has an effective mass equal to the mass of the hammer. How much will the sensitivity of the impact hammer be changed?

10.33   Verify that Eq. 10.103 can be obtained from Eq. 10.101 and the solution to Eq. h of Section 10.10.1.

Chapter 11

Temperature Measurements

11.1 INTRODUCTION

Temperature, unlike other quantities such as length, time, or mass, is an abstract quantity that must be defined in terms of the behavior of materials as the temperature changes. Some examples of material behavior that have been used in the measurement of temperature include change in volume of a liquid, change in length of a bar, change in electrical resistance of a wire, change in pressure of a gas at constant volume, and change in color of a lamp filament.

Several temperature scales have been developed over time to provide a suitable reference for the level of thermodynamic activity associated with temperature changes. Gabriel D. Fahrenheit in 1715 introduced the Fahrenheit scale, with 180 divisions (degrees) between the freezing point (32°F) and the boiling point (212°F) of water. Anders Celsius in 1742 divided the same interval, between the freezing and boiling points of water, into 100 divisions, and Linnaeus later set the zero value of this scale at the freezing point of water. This scale was initially known as centigrade (reflecting the 100 divisions), but in 1948 the name was changed to Celsius, in honor of Anders Celsius. Two other temperature scales are used to describe absolute temperatures, where the zero value is set equal to the thermodynamic minimum. These scales are the Kelvin and Rankine scales, defined by

$$\theta_K = T_C + 273.15 \tag{11.1a}$$

$$\theta_R = T_F + 459.67 \tag{11.1b}$$

where

θ_K and θ_R are absolute temperature in Kelvin and Rankine, respectively
T_C and T_F are temperatures in Celsius and Fahrenheit, respectively

Using absolute temperatures, the ideal gas law can be written as

$$pv = R\theta \tag{11.2}$$

413

where

p is the absolute pressure
v is the specific volume
R is the universal gas constant
θ is the absolute temperature

Temperature is related to the kinetic energy of the molecules at a localized region in a body; however, this kinetic energy cannot be measured directly and the temperature inferred. To circumvent this difficulty, the International Temperature Scale has been defined in terms of the behavior of a number of materials at thermodynamic fixed points.

The International Temperature Scale is based on 17 fixed points, which cover the temperature range from $-270.15°C$ to $1084.62°C$. Most of these 17 points correspond to an equilibrium state during a phase transformation of a particular material (see Table 11.1). The fixed points associated with either melting or freezing of a material are determined at a pressure of one standard atmosphere (1 atm).

Between selected fixed points, the temperature is defined by the response of specified sensors with empirical equations to provide for the interpolation of the temperature. Several different definitions are provided in the International Temperature Scale of 1990 for very low temperatures (approaching absolute zero). At these temperatures, a helium gas thermometer is employed to measure pressure and the temperature is determined from the relationship between temperature and pressure. Between the triple point of e-H_2 (13.8033 K) and the freezing point of Ag (961.78°C), the temperature is defined with a platinum resistance thermometer. Platinum resistance thermometers are calibrated at specified sets of fixed points with carefully defined interpolation equations.

Table 11.1 Fixed Points on the International Temperature Scale (1990 Definitions)

Fixed Point No.	Material	State	Temperature
1	He	Vapor	-270.15 to -268.15
2	e-H_2[a]	Triple Point[b]	-259.3467
3	e-H_2	Vapor	≈ -256.16
4	e-H_2	Vapor	≈ -252.85
5	Ne	Triple Point	-248.5939
6	O_2	Triple Point	-218.7916
7	Ar	Triple Point	-189.3442
8	Hg	Triple Point	-38.8344
9	H_2O	Triple Point	0.01
10	Ga	Melting	27.7646
11	In	Freezing	156.5985
12	Sn	Freezing	231.928
13	Zn	Freezing	419.527
14	Al	Freezing	660.323
15	Ag	Freezing	961.78
16	Au	Freezing	1064.18
17	Cu	Freezing	1084.62

[a] e-H_2: hydrogen at the equilibrium concentration of orthomolecular and paramolecular forms.
[b] Triple point: Temperature at which the solid, liquid, and vapor phases are in equilibrium.

Above the freezing point of Ag the temperature is defined using optical pyrometers to measure radiation and Planck's law to relate this radiation to the temperature. A more detailed description of the International Temperature Scale of 1990 is provided in Reference 21.

11.2 EXPANSION METHODS FOR MEASURING TEMPERATURE

When materials are subjected to temperature changes ($\Delta T = T - T_0$), they expand or contract according to

$$\Delta \ell = \alpha \ell_0 \Delta T \tag{11.3}$$

where

$\Delta \ell$ is the change in length
ℓ_0 is the length at the reference temperature T_0
α is the temperature coefficient of expansion

The temperature coefficient of expansion α is too small for most materials [α is of the order of $20(10^{-6})/°C$] to permit an easy measurement of $\Delta \ell$. Thus Eq. 11.3 cannot be used to determine the temperature T directly from length-change $\Delta \ell$ measurements.

Expansion methods are employed to measure temperature, but the approach is less direct than that indicated by Eq. 11.3. The liquid-in-glass thermometer consists of a glass capillary tube with a bulb containing a volume of liquid (usually Hg). When the temperature changes, the liquid volume expands much more than the glass capillary and bulb because of the difference in α between the fluid and the glass. The differential change in volume causes the liquid to extend in the capillary tube. A scale etched on the glass is used to convert the extension of the fluid in the capillary tube to the temperature of the thermometer.

The liquid-in-glass thermometer is an accurate device ($\pm 0.2°C$ to $2°C$), depending on the design of the thermometer and the procedures used to measure the temperature. However, it has two disadvantages. First, an operator is needed to read the output, so the thermometer cannot be used in a closed-loop control system. Second, the time required to reach the equilibrium temperature is excessive for many practical applications. Thus, the common thermometer used by a physician to measure a patient's temperature has been replaced in many instances with a resistance detector that has a digital readout.

A second temperature-measurement technique based on differential expansion is the bimetallic-strip thermometer. The principle of operation of the bimetallic-strip thermometer is illustrated in Fig. 11.1. Strips of two different metals are welded together to form a laminated-beam structure. The beam is straight when its temperature T is the same as the welding temperature T_w. However, when the temperature of the beam changes, the two metals expand (or contract) by different amounts because the expansion coefficients of the two metals are different ($\alpha_1 \neq \alpha_2$). This differential expansion between the top and bottom layers causes the beam to deform into a segment of a circular arc. The radius of curvature ρ of the circular arc is

$$\rho = \frac{\left[3(1 + r_h)^2 + (1 + r_h r_E)\left(r_E^2 - \frac{1}{r_h r_E}\right)\right] h}{6(\alpha_1 - \alpha_2)(1 + r_h)^2 \Delta T} \tag{11.4}$$

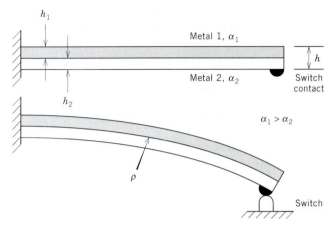

Figure 11.1 Action of a bimetallic-strip thermometer. (*a*) Straight beam when $T = T_w$. (*b*) Curved beam when $T > T_w$.

where

$r_h = h_2/h_1$ the thickness ratio
$r_E = E_2/E_1$ the modulus ratio

The sensitivity of the bimetallic-strip thermometer is maximized by making the radius of curvature ρ small for a given temperature change ΔT. This is accomplished by reducing the thickness h and maximizing $(\alpha_1 - \alpha_2)$. Two common materials used in constructing the strips are stainless steel, with $\alpha_1 = 16(10^{-6})/°C$, and Invar, with $\alpha_2 = 0.02(10^{-6})/°C$.

The bimetallic strip is very useful in constructing thermostats used in controlling temperatures. In this application, the bimetallic strip carries one of the contacts for a switch, as shown in Fig. 11.1*b*. When $(T - T_w)$ is sufficient, the beam element rotates and closes the switch, which activates either a heating or cooling system so as to bring the temperature T in close correspondence with T_w.

In many instances the throw of a straight beam element is not sufficient to reliably activate switches for small ΔTs. To enhance the sensitivity of the switch to temperature, the metallic difference strips are often formed into much larger spiral or helical structures where the increased length of the bimetallic element increases the distance moved by the active switch contact.

11.3   RESISTANCE THERMOMETERS

Resistance thermometers consist of a sensor element that exhibits a change in resistance with a change in temperature, a signal conditioning circuit that converts the resistance change to an output voltage, and appropriate instrumentation to record and display the output voltage. Two different types of sensors are normally employed: resistance temperature detectors (RTDs) and thermistors.

Resistance temperature detectors are simple resistive elements formed of such materials as platinum, nickel, or a nickel-copper alloy known commercially as Balco.

---

[1]Balco is a trade name for a product of the W.B. Driver Co.

These materials exhibit a positive coefficient of resistivity and are used in RTDs because they are stable and provide a reproducible response to temperature over long periods of time.

Thermistors are fabricated from semiconducting materials, such as oxides of manganese, nickel, or cobalt. These semiconducting materials, which are formed into the shape of a small bead by sintering, exhibit a high negative coefficient of resistivity. In some special applications requiring very high accuracy, doped silicon or germanium is used as the thermistor material.

The equations governing the response of RTDs and thermistors to a temperature change and the circuits used to condition their outputs are different; therefore, they are treated separately in the following subsections.

11.3.1   Resistance Temperature Detectors (RTDs)

A typical RTD consists of a wire coil sensor with a framework for support, a sheath for protection, a linearizing circuit, a Wheatstone bridge, and a voltage display instrument. The sensor is a resistive element that exhibits a resistance–temperature relationship given by the expression

$$R = R_0(1 + \gamma_1 T + \gamma_2 T^2 + \cdots \gamma_n T^n) \qquad (11.5)$$

where

$\gamma_1, \gamma_2, \cdots, \gamma_n$ are temperature coefficients of resistivity

R_0 is the resistance of the sensor at a reference temperature T_0. The reference temperature is usually specified as $T_0 = 0°C$.

The number of terms retained in Eq. 11.5 for any application depends on the material used in the sensor, the range of temperature, and the accuracy required in the measurement. Resistance–temperature curves for platinum, nickel, and copper, which illustrate typical nonlinearities in resistance R with temperature T for each of these materials, are shown in Fig. 11.2. For a limited range of temperature, the linear form of Eq. 11.5

$$\frac{\Delta R}{R_0} = \gamma_1(T - T_0) \qquad (11.6)$$

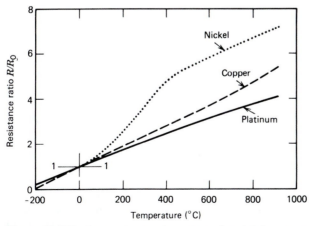

Figure 11.2 Resistance–temperature curves for nickel, copper, and platinum.

is often used to relate resistance change to temperature change. When the error owing to neglecting nonlinear terms becomes excessive, either linearizing circuits can be used to compensate for the nonlinearities or additional terms can be retained from Eq. 11.5 to relate the measured ΔR to the unknown temperature T. Retaining the temperature coefficients γ_1 and γ_2 from Eq. 11.5 yields the second-order relationship

$$\frac{\Delta R}{R_0} = \gamma_1(T - T_0) + \gamma_2(T - T_0)^2 \tag{11.7}$$

Equation 11.7 is more cumbersome to employ, but it provides more accurate results over a wider temperature range.

Sensing elements are available in a wide variety of forms. The different RTD sensors shown in Fig. 11.3 illustrate the wide range of commercially available products. One widely used sensor consists of a high-purity (99.99 percent) platinum wire wound about a ceramic core and hermetically sealed in a ceramic capsule. Platinum is the superior material for precision thermometry. It resists contamination and corrosion, and its mechanical and electrical properties are stable over long periods of time. The platinum wire coils are stress relieved after winding, immobilized against strain, and artificially aged during fabrication to provide for long-term stability. Drift is usually less than 0.1°C when such a sensor is used at its upper temperature limit.

The sensing element is usually protected by a sheath fabricated from stainless steel, glass, or a ceramic. Such sheaths are pressure tight to protect the sensing element from the corrosive effects of both moisture and the process medium. Lead wires from the sensor exit from the sheath through a specially designed seal. The method of sealing between the sheath and the lead wires depends on the upper temperature limit of the sensor. Epoxy cements are used for the low-temperature range (< 260°C), and glass and ceramic cements are used for the high-temperature range (> 260°C). Since the temperature at the sheath exit is usually much lower than the process temperature

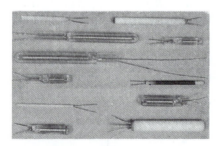

(a)

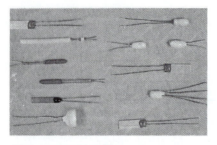

(b)

Figure 11.3 Selection of wire-wound and thin-film resistance temperature detectors. (a) Wire wound. (b) Thin film. (Courtesy of Omega Engineering, Inc.)

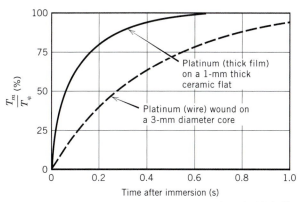

Figure 11.4 Response time for wire-wound and thick-film platinum RTDs immersed in a fluid with T_* at $t = 0$.

being monitored, lead wires insulated with Teflon or impregnated fiberglass are often suitable for use with process temperatures as high as 750°C (1380°F).

The sensors shown in Fig. 11.3a are immersion-type transducers that are inserted in the medium to measure fluid temperatures. The response time in this application is relatively long (a time of 1 to 5 s is required to approach 100 percent response), as shown in Fig. 11.4. This relatively long response time for immersion thermometers is not usually a serious concern, since the rate of change of liquid temperatures in most processes is small.

Platinum RTDs are also constructed using either thick- or thin-film technologies. With both of these approaches, a film of platinum is placed on a thin, flat, ceramic substrate and encapsulated with a glass or ceramic coating. Both the thick- and thin-film methods of fabrication permit the resistance (typically 100 Ω) of the sensor to be developed on a substrate with significantly smaller volume and mass. As a result, the response time of film-type RTDs is reduced appreciably, as indicated in Fig. 11.4.

A distinction must be made between the sensing element and the probe. Previous paragraphs have described wire-wound and film-type sensing elements (Fig. 11.3). The probe (see Fig. 11.5) is an assembly consisting of the sensing element, a sheath,

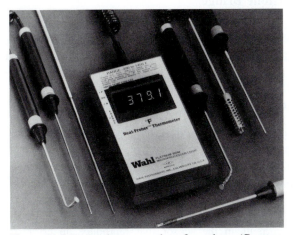

Figure 11.5 Probe elements and configurations. (Courtesy of Wahl Instruments, Inc.)

a seal, lead wires, and a connector. The sheath is usually a closed-end tube that protects the sensing element from the corrosive effects of the medium being measured. The sheath also protects the junction between the rugged lead wires and the fragile wires from the sensing element.

The probe configuration is designed to accommodate specific applications. Examples of probes for measuring temperatures in liquids, in gases (at rest and flowing), and on surfaces are illustrated in Fig. 11.5. The probes protect the sensing element; however, they add a large mass to the sensor and markedly degrade the response time for the sensor.

When response time must be minimized, the probe assembly is eliminated and the sensor is exposed directly to the medium. Rapid temperature changes are common on surfaces; therefore, modifications have been made in film-type RTDs used for surface measurements to improve the response time. These modified RTD surface sensors incorporate a photoetched grid fabricated from Balco foil and resemble strain gages (see Fig. 11.6a). These foil sensors are available on either polyimide or glass-fiber-reinforced epoxy carriers. Similar wire-grid sensors are available with either Teflon or phenolic-glass carriers or as free filaments. The sensors with carriers are bonded to the surface with an adhesive suitable for the temperature range to be encountered. The free filaments are normally mounted by flame spraying. The response time of a thin-film sensor compares favorably with a small thermocouple; therefore, the measurement of rapidly changing surface temperatures is possible.

An example of a foil-type, dual-grid resistance temperature detector is shown in Fig. 11.6b. The construction detail shows two thin-foil sensing elements connected in series and laminated in a glass-fiber-reinforced epoxy-resin matrix. One of the two sensing elements is fabricated from nickel and the other from Manganin. These two materials were selected because they exhibit equal but opposite nonlinearities in their resistance–temperature characteristics over a significant temperature range. By connecting the nickel and Manganin in series, the nonlinear effects cancel and the composite sensor provides a linear response with temperature over the range $-269°$ to $24°C$ ($-452°$ to $75°F$). The bondable RTD is fabricated with integral printed-circuit terminals to provide for easy attachment of the lead wires.

11.3.2 RTDs and the Wheatstone Bridge

The output from a resistance temperature detector is a resistance change $\Delta R/R$ that can be conveniently monitored with a Wheatstone bridge, as illustrated schematically in Fig. 11.7. The RTD is installed in one arm of the bridge, a decade resistance box is placed in an adjacent arm, and two matched precision resistors are inserted in the remaining arms to complete the bridge. Careful consideration must be given to the lead wires, since any resistance change $\Delta R/R$ in the lead wires will produce an error in the readout. With the three-lead-wire arrangement shown in Fig. 11.7, any temperature-induced resistance change in the lead wires is canceled. The three-lead-wire arrangement was discussed in Section 7.7.1. The Wheatstone bridge, shown in Fig. 11.7, can be balanced by adjusting the decade resistance box. In the null position, the reading on the box is exactly equal to the resistance of the RTD. The temperature is then determined from a table of resistance versus temperature for the specific RTD being used. Care must be exercised in powering the bridge to avoid excessive currents in the sensor that reproduce an error owing to self-heating.

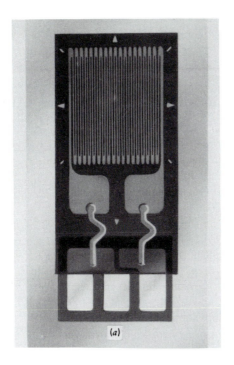

(a)

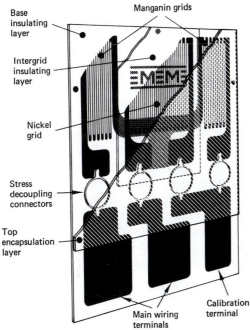

(b)

Figure 11.6 (*a*) Resistance temperature detector for surface temperature measurements. (*b*) Cryogenic linear temperature sensor. (Courtesy of Micro-Measurements Division, Measurements Group, Inc., USA.)

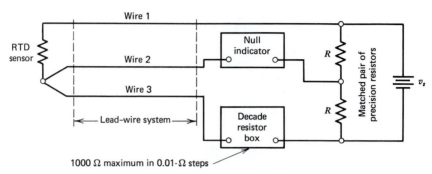

Figure 11.7 Wheatstone bridge circuit with lead-wire compensation and manual reading of the output from a resistance temperature detector.

Self-heating of the sensor occurs because of the power that must be dissipated by the sensor. This power p_T is given by the expression

$$p_T = i^2 R_T \qquad (11.8)$$

For example, the power p_T dissipated by an RTD placed in a Wheatstone bridge with equal resistances R_T in each of the four arms and supplied with a voltage v_s is

$$p_T = \frac{v_s^2}{4R_T} \qquad (11.9)$$

The increase in temperature from self-heating ΔT_{sh} required to dissipate p_T is

$$\Delta T_{sh} = F_{sh}\, p_T \qquad (11.10)$$

where F_{sh} is a self-heating factor (°C/mW).

The magnitude of F_{sh} is provided by the manufacturer of the sensing element. However, it is evident that F_{sh} is related to heat transfer from the sensor to the surroundings. Clearly, F_{sh} will decrease with an increase either in the area of the sensor or in the convection heat-transfer coefficient. A typical value for F_{sh} is 0.5°C/mW for a wire-wound platinum RTD in still air. If this sensor ($R_T = 100\ \Omega$) is placed in an equal-arm bridge with a 1 V dc supply voltage, then $p_T = 2.5$ mW and $\Delta T_{sh} = 1.25$°C. Thus, an error of 1.25°C will be made in the measurement of temperature unless a correction is made to compensate for the self-heating effect.

Another circuit that can be employed for automatic readout is the constant-current potentiometer circuit, shown in Fig. 11.8. The output voltage $v_o = iR_T$ from this circuit can be monitored with a digital voltmeter. If a constant current of 1 mA is supplied to the sensor, the output of the digital voltmeter converts easily to resistance ($R_T = v_o/i$). The temperature is then determined from a resistance–temperature relation for the sensor. Errors owing to resistance changes in the lead wires are also eliminated by using the four-lead-wire system shown in this figure.

The circuits shown in Figs. 11.7 and 11.8 provide simple and accurate methods for measuring the sensor resistance. However, since the sensor resistance R_T is a nonlinear function of temperature T, data processing requires a nonlinear temperature relation. The Callendar–Van Dusen equation is often used.

$$\frac{\Delta R}{R_0} = \alpha \left[T - \delta \left(\frac{T}{100} - 1 \right)\left(\frac{T}{100} \right) - \beta \left(\frac{T}{100} - 1 \right)\left(\frac{T^3}{100} \right) \right] \qquad (11.11)$$

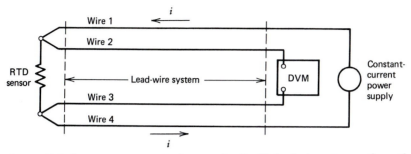

Figure 11.8 Constant-current potentiometer circuit with lead-wire compensation and automatic reading of the output from an RTD sensor.

where

R_0 is the resistance at $T = 0°C$

α is the temperature coefficient of resistivity $[0.00392 \; \Omega/(\Omega \cdot °C)]$

δ is the second-order temperature coefficient (1.49)

β is a higher order temperature coefficient (0 for $T > 0$ and 0.11 for $T < 0$)

The values of α, δ, and β are typical for platinum RTDs. More exact values for these coefficients are obtained by calibrating the sensor at several fixed points (see Table 11.1) located within the range of temperatures of interest.

Unfortunately, the conversion of $\Delta R/R_0$ to temperature is time-consuming, which prevents direct display of the temperature on the DVM.

A. Common Errors

Four common errors encountered when RTDs are used to measure temperature result from lead-wire effects, stability, self-heating, and sensitivity of the RTD to strain. Lead-wire errors can be minimized by making the lead wires as short as possible. The total resistance of the leads should always be less than 1 percent of the sensor resistance. The effect of lead-wire resistance is to increase the apparent resistance of the sensor and thus cause a zero shift (offset) and a reduction in sensitivity. The error resulting from temperature-induced resistance changes in the lead wires can be eliminated by using the three- or four-lead-wire systems described in Figs. 11.7, 11.8, and 7.11*b*.

Stability of the sensors is usually assured by aging the elements during the manufacturing process. Stability may become a source of error when the upper temperature limit of a sensor is exceeded either by design or by accident. Whenever the upper temperature limit of a sensor is exceeded, new temperature measurements should be repeated until stable and reproducible readings are obtained. Stability can also be affected by the polymeric carrier used with bondable RTDs. These carriers have a finite life and lose their strength at temperatures in excess of 120°C (250°F).

Self-heating errors are produced when excitation voltages or currents are used in the signal conditioning circuits. Usually there is no reason for large excitation signals, since an RTD is a high-output sensor [a typical output is about 1 mV/(V·°C) for a platinum RTD]. Self-heating errors can be minimized by limiting the power dissipation in the RTD to less than 2 mW. In those applications where small temperature changes are to be measured and very high sensitivity is required, sensors with large surface areas should be employed. Sensors with large surface areas can dissipate larger amounts of heat; therefore, higher excitation voltages can be used without introducing self-heating errors.

Bonded RTD sensors resemble strain gages and, in fact, they respond to strain. Fortunately, the strain sensitivity of the sensor is small in comparison with the temperature sensitivity. A bonded RTD with a nickel sensor exhibits an apparent temperature change of 1.7°C (3°F) when subjected to an axial tensile strain of 1000 μm/m along the filaments of the gage grid. The magnitude of the strain effect is such that it can be neglected in most applications.

11.3.3 Thermistors

Thermistors are temperature-sensitive resistors fabricated from semiconducting materials, such as oxides of nickel, cobalt, or manganese and sulfides of iron, aluminum, or copper. Thermistors with improved stability are obtained when oxide systems of manganese-nickel, manganese-nickel-cobalt, or manganese-nickel-iron are used. Conduction is controlled by the concentration of oxygen in the oxide semiconductors. An excess or deficiency of oxygen from exact stoichiometric requirements results in lattice imperfections known as Schottky defects and Frankel defects. *N-type oxide semiconductors* are produced when the metal oxides are compounded with a deficiency of oxygen that results in excess ionized metal atoms in the lattice (Frankel defects). *P-type oxide semiconductors* are produced when there is an excess of oxygen that results in a deficiency of ionized metal atoms in the lattice (Schottky defects).

Semiconducting oxides, unlike metals, exhibit a decrease in resistance with an increase in temperature. The resistance–temperature relationship for a thermistor can be expressed as

$$\ln(R/R_0) = \beta(1/\theta - 1/\theta_0)$$

or

$$R = R_0 e^{\beta(1/\theta - 1/\theta_0)} \tag{11.12}$$

where
  R is the resistance of the thermistor at temperature θ
  R_0 is the resistance of the thermistor at reference temperature θ_0
  β is a material constant that ranges from 3000 to 5000 K
  θ and θ_0 are absolute temperatures, K

The sensitivity S of a thermistor is obtained from Eq. 11.12 as

$$S = \frac{\Delta R/R}{\Delta T} = -\frac{\beta}{\theta^2} \tag{11.13}$$

For $\beta = 4000$ K and $\theta = 298$ K, the sensitivity S equals $-0.045/$K, which is more than an order of magnitude higher than the sensitivity of a platinum resistance thermometer ($S = +0.0035/$K). The very high sensitivity of thermistors results in a large output signal and good accuracy and resolution in temperature measurements. For example, a typical thermistor with $R_0 = 2000$ Ω and $S = -0.04/$K exhibits a response $\Delta R/\Delta T = 80 \ \Omega/\mathrm{K}$. This very large resistance change can be converted to a voltage with a simple two-wire potentiometric circuit. The voltage change associated with a temperature change as small as 0.0005 K can be easily and accurately monitored.

Equation 11.12 indicates that the resistance R of a thermistor decreases exponentially with an increase in temperature. Typical response curves for a family of thermistors are shown in Fig. 11.9. Since the output from a thermistor is nonlinear, precise

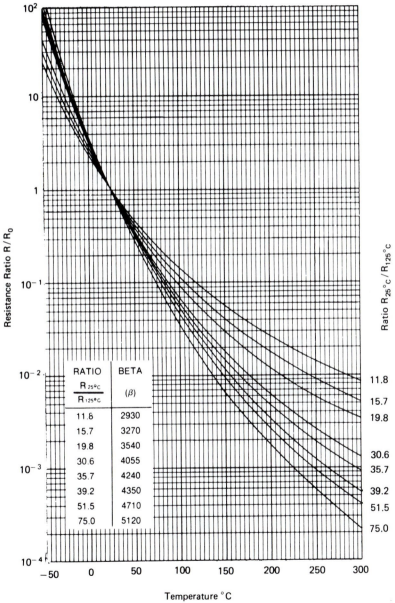

Figure 11.9 Resistance as a function of temperature for different thermistors. (Courtesy of Thermometrics, Inc.)

determinations of temperature must be made by measuring the resistance R and using a calibration table similar to the one presented in Appendix A (Table A.1). Linearity of the output can be improved by using modified potentiometer circuits; however, these circuits reduce the sensitivity and output of the thermistor and are not recommended.

Thermistors are produced by mixing two or more semiconducting oxide powders with a binder to form a slurry. Small drops (beads) of the slurry are formed over the lead wires, dried, and fired in a sintering furnace. During sintering, the metallic oxides shrink onto the lead wires and form an excellent electrical connection. The beads are then hermetically sealed by encapsulating the beads with glass. The glass coating improves stability of the thermistor by eliminating water absorption into the metallic oxide. Thermistor beads, such as those shown in Fig. 11.10, are available in diameters that range from 0.005 to 0.060 in. (0.125 to 1.5 mm). Thermistors are also produced in the form of disks, wafers, flakes, rods, and washers to provide sensors of the size and shape required for a wide variety of applications.

A large variety of thermistors are commercially available with resistances (at the reference temperature T_0) that vary from a few ohms to several megohms. When a thermistor is selected for a particular application, the minimum resistance at high temperature must be sufficient to avoid overloading of the readout device. Similarly, the maximum resistance at low temperature must not be so high that noise becomes a serious problem. Thermistors with a resistance $R_0 = 3000 \ \Omega$, which varies from a low of about 2000 Ω to a high of 5000 Ω over the temperature range, are commonly employed.

Thermistors can be used to measure temperatures from a few degrees above absolute zero to about 315°C (600°F). They can be used at higher temperatures; however, stability begins to decrease significantly above this limit. The range of a thermistor is usually limited to about 100°C (180°F), particularly if it is part of an instrumentation system with a readout device that has been compensated to provide nearly linear output. The accuracy of thermistors depends on the techniques employed to mea-

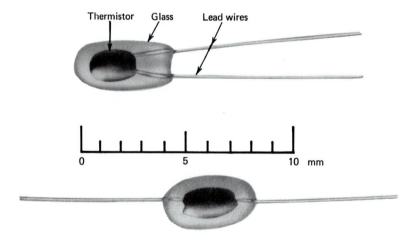

b

Figure 11.10 Bead-type thermistors. (Courtesy of Thermometrics, Inc.)

sure $\Delta R/R$ and to calibrate the sensor. With proper techniques and glass-encapsulated thermistors, temperatures of 125°C can be measured with an accuracy of 0.01°C. Long-term drift data indicate that stabilities better than 0.003°C/yr, when cycled between 20° and 125°C, can be achieved.

The accuracy of the measurement of temperature with a thermistor depends on the instrumentation system employed and the method used to account for the nonlinear response. Since the change in resistance is so large ($\Delta R/\Delta T = 80 \ \Omega/K$), a common multimeter (4 or $4\frac{1}{2}$ digits) can be employed to measure R within $\pm 1 \ \Omega$ as indicated in Fig. 11.11. No bridge or potentiometer circuits are required. If the readings of resistance are processed in a data-acquisition system with a computing microprocessor, the temperature can be approximated very closely by using the Steinhart–Hart relation

$$\frac{1}{\theta} = A + B \ln R_T + C \, (\ln R_T)^3 \tag{11.14}$$

where

θ is the absolute temperature in K

A, B, and C are coefficients determined from calibration curves similar to those shown in Fig. 11.9

When the data points are selected to span a range near the center of the operating range of the transistor (about 100°C), Eq. 11.14 is extremely accurate.

When thermistors are used to measure temperature, errors resulting from lead-wire effects are usually small enough to be neglected even for relatively long lead wires. The sensitivity of a thermistor is high; therefore, the change in resistance ΔR_T resulting from a temperature change is much greater than the small change in resistance of the lead wires resulting from the temperature variation. Also, the resistance of the thermistor is large relative to the resistance of the lead wires ($R_T/R_L \approx 1000$); consequently, any reduction in sensitivity of the sensor because of lead-wire resistance is negligible.

Errors may occur as a result of self-heating, since the power ($p = i^2 R_T$) dissipated in the thermistor will heat it above its ambient temperature. Recommended practice limits the current flow through the thermistor to a value such that the temperature rise resulting from the $i^2 R_T$ power dissipation is smaller than the precision to which the temperature is to be measured. A typical thermistor with $R_T = 5000 \ \Omega$ is capable of dissipating 1 mW with an increase in temperature of 1°C. This corresponds to a self-heating factor $F_{sh} = 1°C/mW$. Thus, if the temperature is to be determined with an accuracy of 0.5°C, the power to be dissipated should be limited to less than 0.5 mW. This limitation establishes a maximum value for the current i at

$$i = \sqrt{p/R_T} = \sqrt{0.0005/5000} = 316 \ \mu A$$

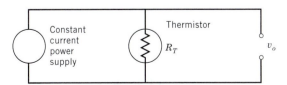

Figure 11.11 Constant-current potentiometer circuit within a digital multimeter used to measure R_T directly.

In this example, it would be prudent to limit the current i to approximately 100 μA. Adequate response can be obtained even at these low currents because the sensitivity of a thermistor is so very high. Precise measurements of Δv_o can be made easily with a digital millivoltmeter.

11.4   THERMOCOUPLES

A thermocouple is a simple temperature sensor that consists of two dissimilar materials in thermal contact. The thermal contact, called a junction, may be made by twisting wires together or by welding, soldering, or brazing two materials together. The junctions may also be formed by pressing the two materials together with sufficient pressure. An example of a single thermocouple junction is shown in Fig. 11.12a.

The operation of a thermocouple is based on a combination of thermoelectric effects that produce a small open-circuit voltage when two thermocouple junctions are maintained at different temperatures. The classic diagram of the dual-junction thermocouple circuit is shown in Fig. 11.12b, where junctions J_1 and J_2 are maintained at temperatures T_1 and T_2, respectively. The thermoelectric voltage v_o is a nonlinear function of temperature that can be represented by an empirical equation having the form

$$v_o = C_1(T_1 - T_2) + C_2(T_1^2 - T_2^2) \tag{11.15}$$

where

C_1 and C_2 are thermoelectric constants that depend on the materials used to form the junctions

T_1 and T_2 are junction temperatures

The generation of the open-circuit voltage indicated by Eq. 11.15 is due to the Seebeck effect (Reference 22), which is produced by diffusion of electrons across the interface between the two materials. The electric potential of the material accepting electrons becomes negative at the interface zone, whereas the potential of the material providing the electrons becomes positive. Thus, an electric field is established by the flow of electrons across the interface. When this electric field becomes sufficient to balance the diffusion forces, a state of equilibrium with respect to electron migration is established. Because the magnitude of the diffusion force is controlled by the temperature of the thermocouple junction, the electric potential developed at the junction provides a measure of the temperature.

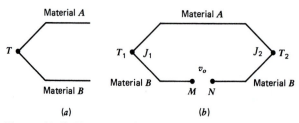

Figure 11.12 Thermocouple sensor and circuit for measuring the temperature difference $T_1 - T_2$. (a) Single junction. (b) Dual junction.

In addition to the Seebeck effect, two other basic thermoelectric effects occur in a thermocouple circuit. The Peltier effect (Reference 19) and the Thompson effect (Reference 23) are named for the scientists who first observed and explained these thermoelectric phenomena.

The Peltier effect occurs when a current flows in the thermocouple circuit. The presence of the current i in the thermocouple circuit produces the well-known self-heating effect, where the Joule heat transfer is $q = i^2R$. However, the Peltier heat transfer is in addition to the Joule heating effect. The Peltier heat transfer is given by

$$q_P = \pi_{AB} i \tag{11.16}$$

where

q_P is the heat transfer in watts (W)
π_{AB} is the Peltier coefficient for the A to B couple

It should be noted that $\pi_{AB} = -\pi_{BA}$ and the Peltier coefficient depends on the direction of current flow through the junction. This fact implies that heat will transfer from the junction to the environment at junction J_1 and from the environment to the junction at junction J_2. This dual-junction heat transfer, illustrated in Fig. 11.13, is the basis of a Peltier refrigerator, which is a cooling device without moving parts.

The Thompson effect is another thermoelectric interaction that affects the behavior of a thermocouple circuit. This effect involves the generation or absorption of heat q_T whenever a temperature gradient and a current exist in a conductor. The Thompson effect, illustrated in Fig. 11.14, results in a quantity of heat q_T being transferred, which is given by

$$q_T = \sigma i (T_1 - T_2) \tag{11.17}$$

where σ is the Thompson coefficient that depends on the conductor material.

Both the Peltier and Thompson effects produce voltages that contribute to the output of a thermocouple circuit and affect the accuracy of the measurement of temperature. Both effects can be minimized by severely limiting the current i that flows through the thermocouple circuit (Fig. 11.12b) during the measurement of v_o.

The thermocouple circuit of Fig. 11.12b is used to sense an unknown temperature T_1, while junction 2 is maintained at a known reference temperature T_2. Since the reference temperature T_2 is known, it is possible to determine the unknown temperature T_1 by measuring the voltage v_o. It is clear from Eq. 11.15 that the response of a thermocouple is a nonlinear function of the temperature. Also, experience has shown that Eq. 11.15 is not a sufficiently accurate representation of the voltage–temperature rela-

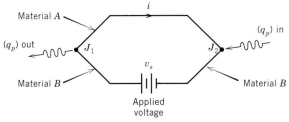

Figure 11.13 Heat transfer in and out of thermoelectric junctions owing to the Peltier effect.

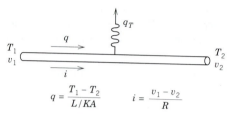

$$q = \frac{T_1 - T_2}{L/KA} \qquad i = \frac{v_1 - v_2}{R}$$

Figure 11.14 Heat transfer from a homogeneous conductor owing to current flow through a temperature gradient.

tionship to be used when precise measurements of temperature are required. Accurate conversion of the output voltage v_o to $(T_1 - T_2)$ is achieved either by using calibration (lookup) tables or by using a higher order polynomial instead of Eq. 11.15. Examples of lookup tables for Chromel-Alumel (Table A.2), Chromel-constantan (Table A.3), copper-constantan (Table A.4), and iron-constantan (Table A.5) thermocouples are presented in Appendix A. It is important to note that the reference temperature is $T_2 = 0°C$ (32°F) in these tables.

The higher order polynomials used for temperature determinations are of the form

$$T_1 - T_2 = a_0 + a_1 v_o + a_2 v_o^2 + \cdots + a_n v_o^n \qquad (11.18)$$

where $a_0, a_1, \cdots, a_n$ are coefficients specified for each pair of thermocouple materials, and $T_1 - T_2$ is the difference in junction temperature in °C. The polynomial coefficients for six different types of thermocouples are given in Appendix A (Table A.6).

11.4.1 Principles of Thermocouple Behavior

The practical use of thermocouples is based on the following six operating principles, which are illustrated in Fig. 11.15.

1. A thermocouple circuit must contain at least two dissimilar materials and at least two junctions (Fig. 11.15a).
2. The output voltage v_o of a thermocouple circuit depends only on the difference between junction temperatures $(T_1 - T_2)$ and is independent of the temperatures elsewhere in the circuit if no current flows in the circuit (Fig. 11.15b).
3. If a third metal C is inserted into either leg (A or B) of a thermocouple circuit, the output voltage v_o is not affected, provided that the two new junctions (A/C and C/A) are maintained at the same temperature, for example, $T_i = T_j = T_3$ (Fig. 11.15c).
4. The insertion of an intermediate metal C into junction 1 does not affect the output voltage v_o, provided that the two junctions formed by insertion of the intermediate metal (A/C and C/B) are maintained at the same temperature T_1 (Fig. 11.15d).
5. A thermocouple circuit with temperatures T_1 and T_2 produces an output voltage $(v_o)_{1-2} = f(T_1 - T_2)$, and one exposed to temperatures T_2 and T_3 produces an output voltage $(v_o)_{2-3} = f(T_2 - T_3)$. If the same circuit is exposed to temperatures T_1 and T_3, the output voltage $(v_o)_{1-3} = f(T_1 - T_3) = (v_o)_{1-2} + (v_o)_{2-3}$ (Fig. 11.15e).

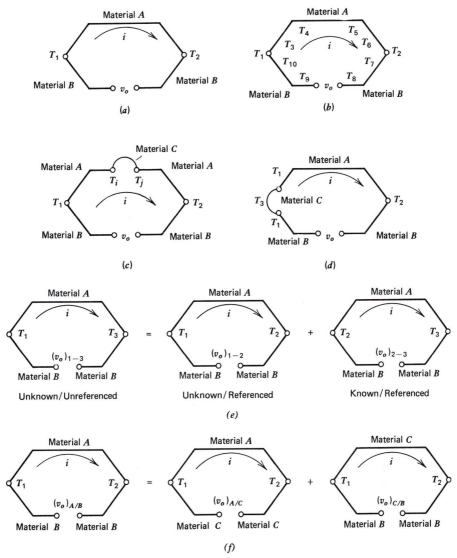

Figure 11.15 Typical situations encountered during use of thermocouples. (*a*) Basic thermocouple circuit. (*b*) Output depends on ($T_1 - T_2$) only. (*c*) Intermediate metal in circuit. (*d*) Intermediate metal in junction. (*e*) Voltage addition from identical thermocouples at different temperatures. (*f*) Voltage addition from different thermocouples at identical temperatures.

6. A thermocouple circuit fabricated from materials A and C generates an output voltage $(v_o)_{A/C}$ when exposed to temperatures T_1 and T_2, and a similar circuit fabricated from materials C and B generates an output voltage $(v_o)_{C/B}$. Furthermore, a thermocouple fabricated from materials A and B generates an output voltage $(v_o)_{A/B} = (v_o)_{A/C} + (v_o)_{C/B}$ (Fig. 11.15*f*).

The six principles of thermoelectric behavior are important because they provide the basis for the design, circuitry, and application of thermocouples to temperature measurements.

The first principle formalizes the experimental observation that a thermocouple circuit must be fabricated with two different materials so that two junctions are formed. The output voltage v_o has been observed to be a nonlinear function of the difference in temperature $(T_1 - T_2)$ at these two junctions. For clockwise current flow, as illustrated in Fig. 11.15a, the output voltage v_o can be expressed as

$$v_o = e_{B/A}T_1 + e_{A/B}T_2 \qquad \textbf{(a)}$$

where

$e_{B/A}$ is the junction potential per unit temperature at a junction as the current flows from material B to material A

$e_{A/B}$ is the junction potential per unit temperature at a junction as the current flows from material A to material B

Since $e_{B/A} = -e_{A/B}$, Eq. a can be written in its more familiar form as

$$v_o = e_{B/A}(T_1 - T_2) \qquad \textbf{(b)}$$

Experiments indicate that the relationship between v_o and the temperature difference $(T_1 - T_2)$, as expressed by Eq. b, depends on the two materials used to fabricate the thermocouple and is nonlinear. Since Eq. b is nonlinear, the junction potential $e_{B/A}$ is not a constant with respect to temperature.

Thermocouple calibration tables, such as Tables A.2, A.3, A.4, and A.5 in Appendix A, are used to relate temperature difference $(T_1 - T_2)$ to a measured output voltage v_o. Since an unknown temperature T_1 is being measured, the reference temperature T_2 must be known. The calibration information presented in Tables A.2, A.3, A.4, and A.5 is based on a reference temperature $T_2 = 0°C$ (32°F). If the reference temperature T_2 is not 0°C, but rather some other known value, such as 100°C, it is still possible to determine T_1, but the procedure involves application of the fifth principle of thermoelectric behavior.

The second principle indicates that the voltage output v_o from a thermocouple circuit is not influenced by the temperature distribution along the conductors except at points where connections are made to form junctions (see Fig. 11.15b). This principle provides assurance that the output voltage v_o of the thermocouple circuit is independent of the length of the lead wires and the temperature distribution along their length. The output voltage v_o is determined only by the junction temperatures.

The third principle deals with insertion of an intermediate conductor (such as copper lead wires or a voltage-measuring instrument) into one of the legs of a thermocouple circuit (see Fig. 11.15c). The effect of inserting material C into the A-B-type thermocouple can be determined by writing the equation for the output voltage v_o as

$$v_o = e_{B/A}T_1 + e_{A/C}T_i + e_{C/A}T_j + e_{A/B}T_2 \qquad \textbf{(c)}$$

Since $e_{B/A} = -e_{A/B}$ and $e_{A/C} = -e_{C/A}$, Eq. c can be written as

$$v_o = e_{B/A}(T_1 - T_2) + e_{A/C}(T_i - T_j) \qquad \textbf{(d)}$$

Note, however, that temperature gradients along the length of the lead wires result in heat transfer and affect the junction temperature. Equation d indicates that the effect of the A/C junctions is eliminated if $T_i = T_j$. A similar analysis with the third metal

C inserted in leg B of the thermocouple shows that the effect of B/C junctions is eliminated if $T_i = T_j$. The third principle verifies that insertion of a third material C into the circuit will have no effect on the output voltage v_o, provided that the junctions formed in either leg A or leg B are maintained at the same temperature $T_i = T_j = T_3$.

The fourth principle deals with insertion of an intermediate metal into a junction during fabrication or use of a thermocouple. Such a situation occurs when junctions are formed by twisting the two thermocouple materials A and B together and soldering or brazing the connection with an intermediate metal C (see Fig. 11.15d). The influence of the presence of the intermediate metal in the junction can be evaluated by considering the expression for output voltage v_o, which can be written as

$$v_o = e_{B/C}T_1 + e_{C/A}T_1 + e_{A/B}T_2 \tag{e}$$

Since $e_{C/A} = e_{C/B} + e_{B/A}$, Eq. e reduces to

$$v_o = e_{B/A}(T_1 - T_2) \tag{f}$$

Equation f verifies that the output voltage v_o is not affected by the presence of a third material C inserted during fabrication of the thermocouple if the two junctions B/C and C/A are at the same temperature.

The fifth principle deals with the relationship between output voltage v_o and the reference junction temperature. As mentioned previously, Tables A.2, A.3, A.4, and A.5 are based on a reference temperature of $0°C$ ($32°F$). In some instances, it may be more convenient to use a different reference temperature (for example, boiling water at $100°C$). The effect of this different reference temperature can be accounted for by using the equivalent thermocouple system, illustrated in Fig. 11.15e. The output from the equivalent system, which incorporates two thermocouple circuits, is

$$(v_o)_{1-3} = f(T_1 - T_3) = (v_o)_{1-2} + (v_o)_{2-3} \tag{11.19}$$

Use of Eq. 11.19 for the case of an arbitrary reference temperature T_3 can be illustrated by considering the example of a copper-constantan thermocouple exposed to an unknown temperature T_1. Assume that the arbitrary reference temperature T_3 is maintained at $100°C$ and that an output voltage $(v_o)_{1-3} = 8.388$ mV is recorded under these conditions. The voltage $(v_o)_{2-3}$ of Eq. 11.19 can be determined from Table A.4, because it is known that $T_2 = 0°C$ and $T_3 = 100°C$. Thus, $(v_o)_{2-3} = -(v_o)_{3-2} = -4.277$ mV. Solving Eq. 11.19 for $(v_o)_{1-2}$ yields

$$(v_o)_{1-2} = (v_o)_{1-3} - (v_o)_{2-3}$$
$$= 8.388 - (-4.277) = 12.665 \text{ mV}$$

Table A.4 indicates that an output voltage $(v_o)_{1-2} = 12.665$ mV would be produced by a temperature $T_1 = 261.7°C$. The same procedure can be used to correct for any known reference temperature.

The sixth principle illustrates the use of voltage addition (superposition) to analyze thermocouple circuits fabricated from different materials, as shown in Fig. 11.15f. The output voltage for the equivalent circuit is

$$(v_o)_{A/B} = (v_o)_{A/C} + (v_o)_{C/B}$$

or

$$(v_o)_{A/B} = (v_o)_{A/C} - (v_o)_{B/C} \tag{11.20}$$

By employing this principle, calibration tables can be developed for any pair of materials if the calibrations for the individual materials are paired with a standard thermocouple material, such as platinum. For example, materials A and B, when paired with the standard material C, would provide $(v_o)_{A/C}$ and $(v_o)_{C/B} = -(v_o)_{B/C}$. The calibration for a junction formed by using materials A and B could then be determined by using Eq. 11.20. Use of this principle eliminates the need to calibrate all possible combinations of materials [for n thermoelectric materials, $n(n-1)$ calibrations are necessary]. Instead, by calibrating all n materials against the standard reference material, platinum, only $(n-1)$ calibrations are required.

11.4.2 Thermoelectric Materials

The thermoelectric effect occurs whenever a thermocouple circuit is fabricated from any two dissimilar metals; therefore, a large number of materials are suitable for use in thermocouples. In most instances, materials are selected to:

1. Provide long-term stability at the upper temperature levels.
2. Ensure compatibility with available instrumentation.
3. Minimize cost.
4. Maximize sensitivity over the range of operation.

The sensitivities of several materials in combination with platinum are presented in Table 11.2. The results from Table 11.2 can be used to determine the sensitivity S at $0°C$ ($32°F$) of a thermocouple fabricated from any two materials listed in the table. For instance, the sensitivity of a Chromel-Alumel thermocouple is determined from Eq. 11.20 as

$$S_{\text{Chromel/Alumel}} = 25.8 - (-13.6) = 39.4 \ \mu V/°C$$

Table 11.2 Thermoelectric Sensitivity S of Several Materials in Combination with Platinum at $0°C$ ($32°F$)

Material	Sensitivity S		Material	Sensitivity S	
	$\mu V/°C$	$\mu V/°F$		$\mu V/°C$	$\mu V/°F$
Bismuth	-72	-40	Copper	$+6.5$	$+3.6$
Constantan	-35	-19.4	Gold	$+6.5$	$+3.6$
Nickel	-15	-8.3	Tungsten	$+7.5$	$+4.2$
Alumel	-13.6	-7.6	Nicrosil	$+15.4$	$+8.6$
Nisil	-10.7	-5.9	Iron	$+18.5$	$+10.3$
Platinum	0	0	Chromel	$+25.8$	$+14.3$
Mercury	$+0.6$	$+0.3$	Germanium	$+300$	$+167$
Carbon	$+3$	$+1.7$	Silicon	$+440$	$+244$
Aluminum	$+3.5$	$+1.9$	Tellurium	$+500$	$+278$
Lead	$+4$	$+2.2$	Selenium	$+900$	$+500$
Silver	$+6.5$	$+3.6$			

Table 11.3[a] Sensitivity as a Function of Temperature for Seven Types of
Thermocouples (mV/°C)

Temperature (°C)	E	J	K	N	R	S	T
−200	25.1	21.9	15.3	9.9	—	—	15.7
−100	45.2	41.1	30.5	20.9	—	—	28.4
0	58.7	50.4	39.5	26.1	5.3	5.4	38.7
100	67.5	54.3	41.4	29.7	7.5	7.3	46.8
200	74.0	55.5	40.0	33.0	8.8	8.5	53.1
300	77.9	55.4	41.4	35.4	9.7	9.1	58.1
400	80.0	55.1	42.2	37.0	10.4	9.6	61.8
500	80.9	56.0	42.6	—	10.9	9.9	—
600	80.7	58.5	42.5	—	11.3	10.2	—
700	79.8	62.2	41.9	—	11.8	10.5	—
800	78.4	—	41.0	—	12.3	10.9	—
900	76.7	—	40.0	—	12.8	11.2	—
1000	74.9	—	38.9	—	13.2	11.5	—

[a]From NBS Monographs 125 (1974) and 161 (1978).

It is important to recall that the sensitivity S of a thermocouple is not constant; the output voltage v_o is a nonlinear function of the difference in junction temperatures $(T_1 − T_2)$. Sensitivity S as a function of temperature for the seven most frequently used material pairs is listed in Table 11.3.

The letters E, J, K, N, R, S, and T are designated by the ANSI standards (Reference 1). The material pairs used in these thermocouple junctions are defined in Table 11.4.

The voltage output v_o as a function of temperature for several popular types of thermocouples is shown in Fig. 11.16. This graphic display shows that E-type (Chromel-constantan) thermocouples generate the largest output voltage at a given temperature; unfortunately, they have an upper temperature limit of only 1000°C (1832°F). The upper limit of the temperature range is increased (but the sensitivity is decreased) to 1260°C (2300°F) with K-type (Chromel-Alumel) thermocouples; to 1538°C (2800°F) with S-type (platinum 10 percent rhodium-platinum) thermocouples; and to 2800°C (5072°F) with G-type (tungsten-tungsten 26 percent rhenium) thermocouples. The operating temperature ranges, together with the span of output voltages, for most of the popular types of thermocouples are listed in Table 11.5.

Long-term thermal stability is an important property of a thermocouple installation if temperature is to be monitored over very long periods of time. A relatively new

Table 11.4 Materials Employed in the Standard Thermocouples

Type	Positive Material	Negative Material
E	Chromel	Constantan
J	Iron	Constantan
K	Chromel	Alumel
N	Nicrosil	Nisil
R	Platinum 13% Rhodium	Platinum
S	Platinum 10% Rhodium	Platinum
T	Copper	Constantan

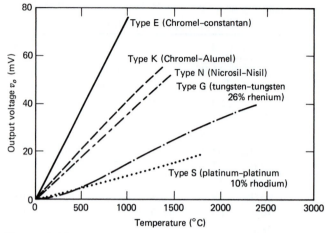

Figure 11.16 Output voltage v_o versus temperature T_1 with a reference temperature $T_2 = 0°C$ for several types of thermocouples.

thermocouple has been introduced, type N, with Nicrosil against Nisil, which exhibits a very high thermoelectric stability. Thermal instabilities in several of the other standard thermocouples, such as types E, J, K, and T, occur after 100 to 1000 h exposure to temperature.

The most important negative effect of thermal instabilities is the gradual and cumulative drift in the output voltage v_o during long exposure of the thermocouple to elevated temperatures. This effect is largely due to compositional changes caused by oxidation (particularly internal oxidation). Both the Nisil and Nicrosil alloys have been formulated to eliminate internal oxidation and substantially reduce external oxidation.

Table 11.5 Operating Range and Voltage Span for Several Types of Thermocouples

Type of Thermocouple	Temperature Range		Voltage Span (mV)
	°C	°F	
Copper-constantan	−185 to 400	−300 to 750	−5.284 to 20.805
Iron-constantan	−185 to 870	−300 to 1600	−7.52 to 50.05
Chromel-Alumel	−185 to 1260	−300 to 2300	−5.51 to 51.05
Chromel-constantan	0 to 980	32 to 1800	0 to 75.12
Nicrosil-Nisil	−270 to 1300	−450 to 2372	−4.345 to 47.502
Platinum 10% rhodium-platinum	0 to 1535	32 to 2800	0 to 15.979
Platinum 13% rhodium-platinum	0 to 1590	32 to 2900	0 to 18.636
Platinum 30% rhodium-platinum	38 to 1800	100 to 3270	0.007 to 13.499
Platinel 1813–Platinel 1503	0 to 1300	32 to 2372	0 to 51.1
Iridium-60% rhodium 40% iridium	1400 to 1830	2552 to 3326	7.30 to 9.55
Tungsten 3% rhenium-tungsten 25% rhenium	10 to 2200	50 to 4000	0.064 to 29.47
Tungsten-tungsten 26% rhenium	16 to 2800	60 to 5072	0.042 to 43.25
Tungsten 5% rhenium-tungsten 26% rhenium	0 to 2760	32 to 5000	0 to 38.45

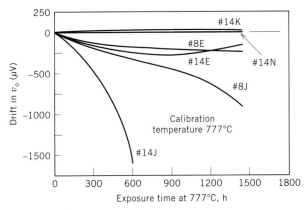

Figure 11.17 Drift in output voltage v_o from different types of thermocouples as a function of exposure time at a constant aging temperature of 777°C. (After Burley.)

The long-term drift in the output from N-, E-, J-, and K-type thermocouples is shown as a function of time at a constant aging temperature of 777°C in Fig. 11.17. The thermal drift of the J thermocouple fabricated from AWG No. 14 wire is excessive after only 100 to 200 h. Increasing the wire size to AWG No. 8 improves the time stability of the J thermocouple, but it remains inadequate for extended applications. The type E thermocouple, although superior to the J thermocouple, also shows excessive voltage drift. Only the K and N thermocouples exhibit the stability required for measurement of temperatures up to 777°C for at least 1500 h.

The stability with exposure time of the K- and N-type thermocouples at higher temperatures is shown in Fig. 11.18. It is evident that the performance of the K-type thermocouple degrades rapidly at temperatures exceeding 1050°C. However, the drift in voltage with exposure time for the N-type thermocouple is less than 100 μV (about 2°C) after 1000 h at temperatures up to 1200°C.

11.4.3 Reference Junction Temperature

Since a thermocouple circuit responds to a temperature difference $(T_1 - T_2)$, it is essential that the reference junction be maintained at a constant and accurately known temperature T_2. Four common methods are used to maintain the reference temperature.

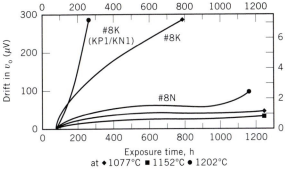

Figure 11.18 Drift in output voltage v_o for K and N thermocouples as a function of exposure time at different aging temperatures. (After Burley.)

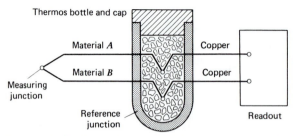

Figure 11.19 The ice-bath method for maintaining a reference temperature at 0°C (32°F).

The simplest technique utilizes an ice and water bath, as illustrated in Fig. 11.19. The reference junction is immersed in a mixture of ice and water in a thermos, which is capped to prevent heat loss and temperature gradients. Water (sufficient only to fill the voids) must be removed, and ice must be replaced periodically to maintain a constant reference temperature. Such an ice bath can maintain the water temperature (and thus the reference temperature) to within 0.1°C (0.2°F) of the freezing point of water.

A very high quality reference-temperature source employs thermoelectric refrigeration (Peltier cooling effect). Thermocouple wells in this unit contain distilled, deionized water maintained at precisely 0°C (32°F). The outer walls of the wells are cooled by the thermoelectric refrigeration elements until the water in the wells begins to freeze. The increase in volume of the water as it begins to freeze on the walls of the wells expands a bellows, which contacts a microswitch and deactivates the refrigeration elements. The cyclic freezing and thawing of the ice on the walls of the wells accurately maintains the temperature of the wells at 0°C (32°F). This automatic precise control of temperature can be maintained over extended periods of time.

The electrical bridge method, illustrated in Fig. 11.20, is usually used with potentiometric strip-chart recording devices to provide automatic compensation for reference junctions that are free to follow ambient temperature conditions. This method incorporates a Wheatstone bridge with a resistance temperature detector (RTD) as the active element into the thermocouple circuit. The RTD and the reference junctions of the thermocouple are mounted on a reference block that is free to follow the ambi-

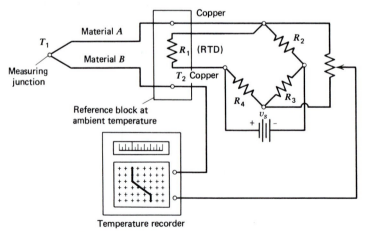

Temperature recorder

Figure 11.20 The electrical bridge method of compensation for changes in the reference temperature.

ent temperature. As the ambient temperature of the reference block varies, the RTD changes resistance. The bridge is designed to produce an output voltage that is equal but opposite to the voltage developed in the thermocouple circuit as a result of the changes in temperature T_2 from 0°C (32°F). Thus, the electrical bridge method automatically compensates for changing ambient conditions. This method is widely used with potentiometric recording devices used to display one or more temperatures over long periods of time when it is not practical to maintain the simple ice bath.

A different type of reference temperature control is obtained by using an oven that maintains a fixed temperature higher than any expected ambient temperature. The system is practical because heating is easier than cooling; however, the thermoelectric voltage–temperature tables must be corrected for the higher reference junction temperature.

A popular method, illustrated in Fig. 11.21, which eliminates the need for reference junction temperature corrections, employs two ovens at different temperatures to simulate a reference temperature of 0°C (32°F). In Fig. 11.21, each of the two junctions (Chromel-Alumel) in the first oven produces a voltage of 2.66 mV at an oven temperature of 65.5°C (150°F). This total voltage of 5.32 mV is canceled by the double junction of Alumel-copper and copper-Chromel in the second oven at a temperature of 130°C (266°F). The net effect of the four junctions in the two ovens is to produce the thermoelectric equivalent of a single reference junction at a temperature of 0°C (32°F).

11.4.4 Fabrication and Installation Procedures

Proper installation of a thermocouple may involve fabrication of the junction, selection of the lead wires (diameter and insulation), and placement of the thermocouple on the surface of the component or in the fluid at the point where the temperature is to be measured. The recommended procedure for forming a thermocouple junction, illustrated in Fig. 11.22, consists of butting the two wires together and fastening by welding, brazing, or soldering to form a small bead of material around the junction. The diameter of the wire used in the fabrication of the thermocouple depends on the dynamic response required of the thermocouple and the degree of hostility of the environment in which the thermocouple must operate. When temperature fluctuations are rapid, the diameter must be small and any protective sheathing must be eliminated to minimize thermal lag. Wire diameters as small as 0.0125 mm (0.0005 in.) are routinely employed when response time becomes an important factor in the temperature measurement. On the other hand, thermocouples are often required to operate over long periods of time at high temperatures in reducing or oxidizing atmospheres. In these applications, heavy-gage wires are used with relatively large-diameter junctions so that part of the junction can be sacrificed to extend the period of stable operation.

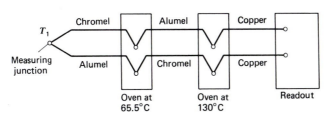

Figure 11.21 Double-oven method for reference junction control.

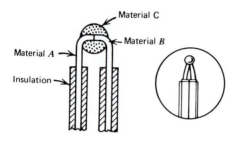

Figure 11.22 Fabrication details for a thermocouple junction.

Thermocouples are widely used to measure fluid temperatures in tanks, pipes, boilers, reactors, and so on. The thermocouple is usually protected from the fluid by metal wells or by mounting in a metal probe insulated with swaged and compacted ceramic powders, as indicated in Fig. 11.23. The exposed thermocouple junction (Fig. 11.23a) is used to measure the temperature of noncorrosive gases where rapid response of the sensor is necessary. The ungrounded junction (Fig. 11.23b) is used in corrosive fluids where the temperature is changing slowly with time. The thermocouple is totally insulated from the probe tube with ceramic powder (usually MgO). The grounded junction (Fig. 11.23c) is used in corrosive fluids and gases where moderate response time is required.

Surface installations are usually made by welding or brazing the thermocouple to the surface. When possible, the thermocouple is embedded in a shallow, small-diameter hole prior to welding. Care should be exercised in minimizing the weld material and in maintaining the geometry of the surface. In some cases, the thermocouples are adhesively bonded to the surface. Special filler epoxies are usually employed as the adhesive, because they can be cured in a few hours at room temperature. Ceramic powders (usually Al_2O_3) are used as fillers for the epoxy to markedly improve the thermal coefficient of conductivity $k = 8$ (Btu·in.)/(h·ft^2·°F). Some examples of surface attachment methods are shown in Fig. 11.24.

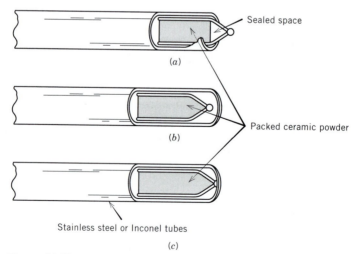

Figure 11.23 Thermocouple probe assemblies with different junction configurations. (a) Exposed junction. (b) Ungrounded junction. (c) Grounded junction. (Courtesy of Omega Engineering, Inc.)

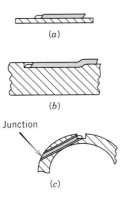

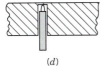

Junction

(d)

Figure 11.24 Examples of surface attachment techniques for thermocouples. Junction mounted (a) directly on the surface, (b) in a groove, (c) in a chordal hole in the tube wall, and (d) from the rear surface.

A. Lead Wires

The material used to provide insulation for the lead wires is determined by the maximum temperature to which the thermocouple will be subjected. Types of insulation and their temperature limits are listed in Table 11.6.

For higher temperature applications, thermocouple wire is available with a ceramic insulation swaged into a metal sheath. For extremely high temperature applications (2315°C or 4200°F), ceramic (Beryllia) tubes are often used to insulate the wires.

In some installations, it is necessary to separate the measuring and the reference junction by an appreciable distance. In these instances, extension wires are inserted between the measuring junction and the reference junction. The extension wires are fabricated from the same materials as the thermocouple junctions and, hence, they will exhibit approximately the same thermoelectric properties. The primary advantage of using extension wires is improvement in the properties of the wire. For example,

Table 11.6 Characteristics of Thermocouple-Wire Insulation

Material	Abrasion Resistance	Flexibility	Temperature (°C)	
			Max.	Min.
Polyvinyl chloride	Good	Excellent	105	−40
Polyethylene	Good	Excellent	75	−75
Nylon	Excellent	Good	150	−55
Teflon-FEP	Excellent	Good	200	−200
Teflon-PFA	Excellent	Good	260	−267
Silicon rubber	Fair	Excellent	200	−75
Nextel braid	Fair	Good	1204	−17
Glass braid	Poor	Good	482	−75
Refrasil braid	Poor	Good	871	−75
Kapton	Excellent	Good	316	−267

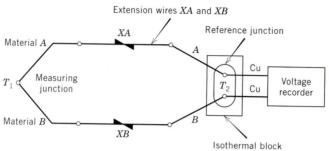

Figure 11.25 Schematic illustration of the use of extension wires and copper lead wires in a thermocouple circuit.

stranded wire of smaller diameter with PVC insulation can be used for the extension wires to provide a lower cost sensing system that is easier to install. The changes in diameter and type of insulation are possible because the extension wire is removed from the temperature T_1 and is at or near room temperature.

Ordinary copper lead wire is used to connect the voltage recording instrument to the sensing system, as shown in Fig. 11.25. A convenient technique that is often employed involves the use of an isothermal block, which provides the reference temperature and eliminates the effect of the voltage produced by the Cu/A and B/Cu junction.

11.4.5 Recording Instruments for Thermocouples

In the previous subsections it was shown that thermocouples are voltage generators; however, the voltage generated by a thermocouple is small (see Table 11.3); therefore, small voltage losses owing to current flow in the circuit ($\Delta v_o = iR$) can produce significant errors in the temperature measurements. In addition, when a temperature gradient or a current flow exists in the circuit, both Peltier and Thompson effects act to produce parasitic heating, which can also produce significant errors in the temperature measurements. Thus, the output voltage v_o from a thermocouple circuit must be measured and recorded with an instrument that has a high input impedance so that current flow in the circuit is minimized. The most common instruments used to measure and record the output voltage v_o from a thermocouple circuit are the digital voltmeter (DVM), the strip-chart recorder, and the oscilloscope. Each of these high-input-impedance instruments exhibits $R_M = 10^6 \ \Omega$ or higher.

The digital voltmeter is ideal for static and quasi-static measurements, where it can be used in the manual mode of operation or as the voltage-measuring component of a data-logging system for automatic recording at a rate of approximately 48 points per second. A digital voltmeter with an input impedance of 10 MΩ will limit current flow in the thermocouple circuit to 10^{-10} A with a thermocouple voltage of 1 mV. Under such conditions, the iR losses and junction heating and cooling owing to the Peltier and Thompson effects are negligible. A commercial DVM that has been adapted to provide a readout directly in terms of degrees (°F or °C) for several different types of thermocouples is shown in Fig. 11.26. This meter has a resolution of 0.1° with a $4\frac{1}{2}$-digit digital display. A microprocessor is incorporated in the DVM circuit to linearize the analog voltage output. Linearization is accomplished by measuring the analog voltage output with an integrating-type DVM (without display). The digital code

Figure 11.26 A digital thermometer that provides readout directly in terms of degrees for several types of thermocouples. (Courtesy of Soltec Corp.)

corresponding to this voltage is used as input to the microprocessor to determine the temperature by using the polynomial expression given in Eq. 11.18. The digital output from the microprocessor is displayed to give a direct reading of the temperature. With instruments that incorporate microprocessors, the lookup tables given in Appendix A are not required.

These direct-reading instruments eliminate the need for the reference junction by using the method of cold junction compensation, in which the thermoelectric voltages resulting from the reference junction are subtracted from the signal from the measuring junction in accordance with Eq. 11.15. This is accomplished by incorporating an isothermal block in the instrument and measuring its temperature with a thermistor or an RTD, as indicated in Fig. 11.20b. In digital instruments with microprocessors, subtraction and compensation are performed by digital processing of the signals from the measuring junction and the thermistor on the isothermal block.

The strip-chart recorder, with its servomotor-driven, null-balance potentiometric circuit (see Section 3.4.1), is an ideal instrument for quasi-static temperature measurements over long periods of time. Such strip-chart recorders are usually equipped with a bridge-compensation device that provides the reference junction required for thermocouple operation. In these instances, the scale of the recorder is calibrated to read temperature directly for a particular type of thermocouple.

High-frequency variations in temperature can be recorded with an oscilloscope. The use of an oscilloscope is straightforward, because its input impedance and sensitivity are usually well matched to the thermocouple circuit.

The oscillograph is not recommended for use with thermocouples, because the galvanometers in the oscillograph have a low input impedance. Any oscillograph used for temperature measurements with thermocouples must be equipped with a high-impedance input amplifier to limit current flow in the thermocouple circuit; otherwise, significant errors will occur as a result of iR losses and Thompson and Peltier effects.

Similarly, use of analog dc voltmeters without preamplifiers is not recommended. Although analog meters equipped with preamplifiers with an input impedance of 10^6 Ω are commercially available, the difference in cost between these meters and digital voltmeters is modest; therefore, cost is not usually a factor, and use of these analog meters should be avoided.

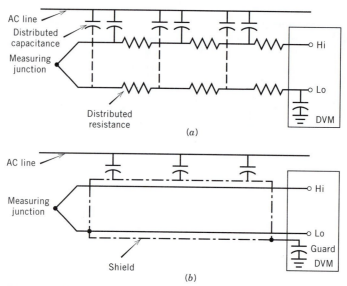

Figure 11.27 Shielding technique to suppress noise in a thermocouple circuit. (*a*) Capacitive coupling to generate a noise voltage. (*b*) Shielding on the wire and a guard on the DVM.

11.4.6 Noise Suppression in Thermocouple Circuits

Since the voltage signals from thermocouple circuits are usually low (1 to 10 mV), the noise voltages imposed on the signal must be minimized to maintain the accuracy of the measurement. The noise voltages are generated by capacitive coupling of the thermocouple extension wires with nearby power lines that run parallel to the extension wires. This capacitive coupling is illustrated in Fig. 11.27*a*.

Two techniques are effective in minimizing the noise signal. The first technique, shown in Fig. 11.27*b*, involves shielding the thermocouple wires so that the noise voltage is generated in the shield and not in the thermocouple leads. The shield is connected to the guard[2] on the DVM. The noise voltage is eliminated by capacitively coupling the guard to the system ground. Properly shielded extension wires and guarded DVM circuits reduce the electrical noise appreciably.

Low-pass filters (see Section 6.9) can also be employed to eliminate the noise voltages. The frequency of noise generated by capacitively coupled power lines is 60 Hz. In most temperature measurements, the frequency associated with temperature fluctuations is much lower than 60 Hz. For this reason, low-pass filters, which pass signals with $f < 10$ Hz and severely attenuate signals with $f \geq 60$ Hz, are very effective in reducing the magnitude of the noise voltages.

11.5   INTEGRATED-CIRCUIT TEMPERATURE SENSORS

The integrated-circuit temperature sensor is a semiconductor device that provides an output current i proportional to absolute temperature θ_A when an input voltage v_s

---

[2]The guard is a metal box that surrounds the circuitry in the DVM. The guard acts to shield the enclosed circuits from capacitively coupled noise voltages.

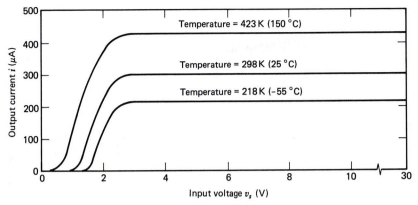

Figure 11.28 Input-voltage–current–temperature characteristics of a two-terminal integrated-circuit temperature transducer.

(between 4 and 30 V) is applied across its two terminals. This type of temperature sensor is a high-impedance, constant-current regulator over the temperature range from $-55°$ to $150°C$ $(-70°$ to $300°F)$. The sensor exhibits a nominal current sensitivity S_i of 1 $\mu A/K$. The current sensitivity is controlled by an internal resistance that is laser trimmed during production to give an output of 298.2 μA at a temperature of 298.2 K (26°C). Input-voltage–current–temperature characteristics of a typical sensor are shown in Fig. 11.28.

The integrated-circuit temperature sensor is ideally suited for remote temperature measurements since it acts like a constant-current source and, as a result, lead-wire resistance R_L has no effect on the output voltage of the sensor circuit at the recording instrument. A well-insulated pair of twisted lead wires can be used for distances of several hundred feet. Also, many of the problems associated with the use of thermocouples or RTD devices, such as small output signals, the need for precision amplifiers and linearization circuitry, cold junction compensation, and thermoelectric effects at connections, are not encountered with this sensor.

The output voltage v_o from the temperature-sensor circuit is controlled by a series resistance R_s, as shown in Fig. 11.29. Since the temperature sensor serves as a current source, this output voltage can be expressed as

$$v_o = iR_s = S_i\theta_A R_s = S_T\theta_A \tag{11.21}$$

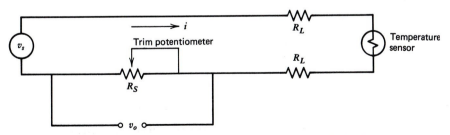

Figure 11.29 Integrated-circuit temperature sensor in a circuit with lead-wire resistance and a series output resistance, with a trim potentiometer for standardizing sensitivity.

where

S_i is the current sensitivity of the sensor
R_s is the series resistance across which the output voltage is measured
θ_A is the absolute temperature
i is the current output at absolute temperature θ_A
S_T is the voltage sensitivity of the circuit

The output resistance R_s often contains a trim potentiometer, as shown in Fig. 11.29, which is used to standardize the output voltage to a value such as 1 mV/K or 10 mV/K. This trim adjustment also permits the calibration error at a given temperature to be adjusted so as to improve accuracy over a given range of temperatures.

Unfortunately, the integrated-circuit temperature sensor is limited to a range of temperatures from −55°C to 150°C. It is an excellent sensor over this range.

11.6   DYNAMIC RESPONSE OF TEMPERATURE SENSORS

Previous sections have introduced several temperature sensors, such as bimetallic strips, RTDs, thermistors, and thermocouples. These sensors are all intended to measure temperature over a relatively small region of a much larger body. They have different operating principles but exhibit several common characteristics, which include dynamic response and sources of error. They are also calibrated using similar techniques. All these sensors are considered in the following descriptions of dynamic response, sources of error, and calibration.

Temperature sensors are classified as first-order systems, since their dynamic response is controlled by a first-order differential equation that describes the rate of heat transfer between the sensor and the surrounding medium. Consider the sensor at a time t in a medium with a temperature T_m, as shown in Fig. 11.30. A heat balance leads to the transient differential equation

$$q = hA(T_m - T) = mc \frac{dT}{dt} \tag{11.22}$$

where

q is the rate of heat transfer to the sensor by convection
h is the convection heat-transfer coefficient

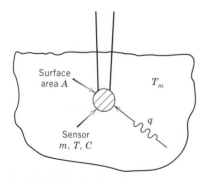

Figure 11.30 Model of temperature sensor leading to Eq. 11.22.

A is the surface area of the sensor through which heat passes
m is the mass of the sensor
c is the specific heat capacity of the sensor

Equation 11.22 can also be expressed as

$$\frac{dT}{dt} + \frac{hA}{mc}T = \frac{hA}{mc}T_m \tag{11.23}$$

Solving Eq. 11.23 for the homogeneous part yields

$$T = C_1 e^{-t/\beta} \tag{11.24}$$

where

C_1 is a constant of integration
β is the time constant for the sensor

$$\beta = \frac{mc}{hA} \tag{11.25}$$

A complete solution of Eq. 11.23 requires specification of the temperature T_m as a function of time and the initial conditions. Two examples provide valuable insight into the dynamic behavior of temperature sensors: the response to step-function inputs and the response to ramp-function inputs.

Consider first the response of a temperature sensor to a step-function input (sensor is suddenly immersed in a fluid medium maintained at constant temperature T_m). In this example, the particular solution of Eq. 11.23 is $T = T_m$; therefore, the general solution is

$$T = C_1 e^{-t/\beta} + T_m \tag{11.26}$$

For the initial condition $T(0) = 0$, the integration constant $C_1 = -T_m$ in Eq. 11.26; thus, the final expression for temperature T as a function of time t for the step-function input is

$$\frac{T}{T_m} = (1 - e^{-t/\beta}) \tag{11.27}$$

Results of this relationship, shown in Fig. 11.31, indicate that a temperature sensor requires considerable time before it begins to approach the temperature of the surrounding medium T_m. The temperature of the sensor is within 5 percent of T_m at $t = 3\beta$ and within 2 percent of T_m at $t = 3.91\beta$. The response time can be shortened by reducing the time constant β. Smaller values of β are obtained by designing a sensor with a small mass, a large surface area, and a low specific heat capacity (see Table A.7 for the heat capacities of selected thermocouple alloys). An example of a rapid-response thermocouple is shown in Fig. 11.32. This type of thermocouple is fabricated from a thin sheet of foil having a thickness of approximately 0.012 mm (0.0005 in.). The foil elements are mounted on a thin polymeric carrier to facilitate bonding to a surface. The time constant β for the foil thermocouple with a large surface area and a small mass ranges from 2 to 5 ms, depending primarily

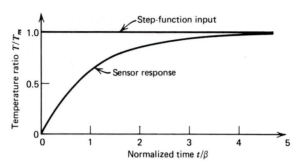

Figure 11.31 Response of a temperature sensor to a step-function input.

on the convective heat-transfer coefficient h that exists between the sensor and the medium.

A second example that shows dynamic behavior involves the response of a temperature sensor to a ramp-function input, such as the one illustrated in Fig. 11.33. The sensor and the surrounding medium are initially at the same temperature; thereafter, the temperature of the medium increases linearly with time so that

$$T_m = bt \tag{a}$$

Solving Eq. 11.23 for the particular solution gives

$$T = b(t - \beta) \tag{b}$$

where b is the slope of the temperature–time ramp function, as illustrated in Fig. 11.33. The general solution of Eq. 11.23 for the ramp-function input is

$$T = C_1 e^{-t/\beta} + b(t - \beta) \tag{11.28}$$

For the initial condition $T(O) = T_m(0) = 0$, the integration constant in Eq. 11.28 is $C_1 = b\beta$; therefore, the response of a temperature sensor to a ramp-function input can be expressed as

$$\begin{aligned} T &= b\beta e^{-t/\beta} + b(t - \beta) \\ &= bt - b\beta(1 - e^{-t/\beta}) \end{aligned} \tag{11.29}$$

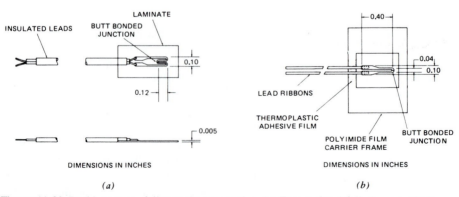

Figure 11.32 Rapid-response foil-type thermocouples. (*a*) Encapsulated foil element. (*b*) Free-filament foil element. (Courtesy of Omega Engineering, Inc.)

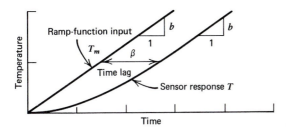

Figure 11.33 Response of a temperature sensor to a ramp-function input.

The results of Eq. 11.29, shown in Fig. 11.33, indicate that the initial response of the sensor is sluggish; however, after a short initial interval, the sensor tracks the rise in temperature of the medium surrounding the sensor with the correct slope but with a time lag equal to β. In Eq. 11.29 the exponential term is important during the initial response and the linear term dominates the long-term response, since the exponential term decreases rapidly with time and becomes negligible when $t > 3\beta$. Since the lag time β can be determined from a simple experiment, accurate temperature measurements can be made for times greater than 3β by correcting for the time lag. Sensors with small time constants should be used to reduce the time lag and the transient period so that the errors in the dynamic response are small enough to be acceptable.

11.7   SOURCES OF ERROR IN TEMPERATURE MEASUREMENTS

Many different errors arise in measuring temperature with such sensors as thermocouples, RTDs, and the like. Many of these sources of error, such as loading by the voltage recorder, precision of the readout, effects of noise, and dynamic response, have been described in previous sections. In this section, a source of error unique to temperature sensors, namely, insertion errors, will be discussed. Insertion errors are the result of heating or cooling of the junction that changes the junction temperature T from the medium temperature T_m. Insertion errors are classified as conduction errors, recovery errors, and radiation errors. Conduction and recovery errors are described in this section. Radiation errors are discussed in Section 11.9.

To illustrate the error owing to conduction of heat, consider measurement of the surface temperature at a point on a massive solid body, as shown in Fig. 11.34. Assume that the sensor is well bonded to the surface so that the thermal contact resistance can be neglected. The lead wires from the sensor pass through a fluid (usually air) with a

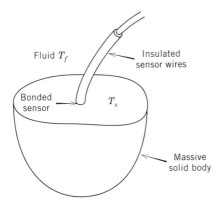

Figure 11.34 Temperature measurement of a massive solid with a surface-bonded sensor.

temperature T_f. Assume that the fluid temperature T_f is lower than that of the solid, and observe that the sensor and lead wires conduct heat away from the solid. The heat transfer (per unit time and area) by the sensor and its lead wires is usually much greater than the convective heat loss from the surface to the fluid (air). As a result, temperature gradients occur in the solid that cause heat to flow toward the sensor. Thus, the temperature of the sensor T is depressed below that of the surface and the sensor provides a low reading of T_s.

Hennecke and Sparrow (Reference 12) have determined the errors owing to conduction of the sensor and the lead wires. The results of this analysis are formulated neatly into three dimensionless groups:

$$\frac{T_s - T}{T_s - T_f} \tag{11.30a}$$

$$\frac{\sqrt{(kA)_e/R}\ \tanh\left[L/\sqrt{(kA)_e R}\right]}{\pi r_1 k_s} \tag{11.30b}$$

$$\frac{h_s r_1}{k_s} \tag{11.30c}$$

where

T, T_s, T_f are the sensor, surface, and fluid temperatures, respectively
$\quad\quad k_s$ is the coefficient of thermal conductivity of the solid
$\quad\quad h_s$ is the convective heat transfer between the solid and fluid
$\quad\quad R$ is the radial thermal resistance of the lead wires given by

$$R = \frac{\frac{1}{2\pi r_i h} + \ln_e(r_i/r_w)}{2\pi k_i} \tag{11.30d}$$

$\quad\quad r_i$ is the outer radius of the insulated lead wire
$\quad\quad r_w$ is the radius of the conductor in the lead wire
$r_1 = \sqrt{2}r_w$ is an equivalent radius of the lead wires
$\quad\quad L$ is the length of the lead wires
$\quad\quad (kA)_e$ is an effective conductivity-area product defined by

$$(kA)_e = k_w A_w + k_i A_i \tag{11.30e}$$

A_w, A_i are the cross-sectional areas of the wire and insulation
k_w, k_i are the coefficients of thermal conductivity of the wire and insulation

The first dimensionless number, defined by Eq. 11.30a, gives the difference between T and T_s and is used for the ordinate of the graphs shown in Fig. 11.35. The second dimensionless number, defined by Eq. 11.30b, is used for the abscissa of the graphs shown in Fig. 11.35. The Biot number, defined by Eq. 11.30c, is used as a parameter in the graphs of Fig. 11.35.

The magnitude of the abscissa number is an indication of the thermal resistance of the sensor compared with that of the solid body. For a fixed Biot number, the error is accentuated when the thermal resistance of the sensor lead wires is low compared with

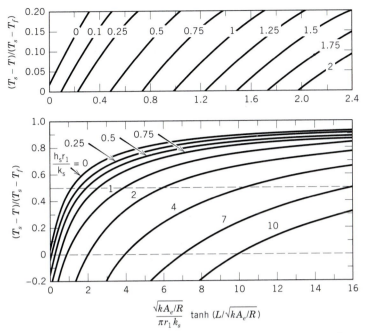

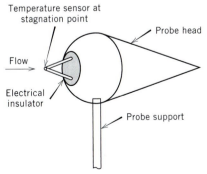

Figure 11.35 Nondimensional graph giving measurement errors for a surface-mounted thermocouple (See Reference 12).

that of the solid body. Note also that temperature errors are minimized by increasing the Biot number.

A second error occurs when temperature sensors are inserted in a gas that is moving at a high velocity. This measurement produces a recovery error resulting from stagnation of the flow when the temperature probe is inserted in the flow field. In this application, the total temperature T_t is sensed by an adiabatic or stagnation probe, as illustrated in Fig. 11.36. The total temperature is given by

$$T_t = T + \frac{V^2}{2Jgc_p} = T + T_v \qquad (11.31)$$

Figure 11.36 Schematic drawing of an adiabatic temperature probe.

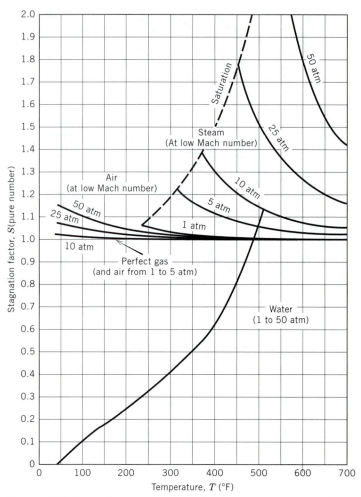

Figure 11.37 Stagnation recovery factor for air, water, and steam. (Reference 6).

where

T is the static or free-stream temperature
V is the velocity of the gas
J is the mechanical equivalent of heat
c_p is the specific heat of the gas at constant pressure

The term $V^2/2Jgc_p$ in Eq. 11.31 is defined as the dynamic temperature. In real fluids, the stagnation recovery factor is introduced to account for differences between the total temperature and the stagnation temperature T_{st} observed in experiments with stagnation probes. Accordingly, Eq. 11.31 is modified to read

$$T_{st} = T + ST_v \tag{11.32}$$

where S is the stagnation recovery factor, which is given for air, water, and steam in Fig. 11.37. Thus, it is possible to establish the free-stream temperature T by using

Eq. 11.32 after employing stagnation probes to measure T_{st}, determining T_v from velocity measurements, and establishing S from Fig. 11.37.

11.8 CALIBRATION METHODS

Calibration of temperature sensors is usually accomplished by using the freezing-point (or boiling-point) method, the melting-wire method, or the comparison method.

The *freezing-point method* is the easiest and most frequently employed calibration technique. With this approach, the temperature sensor is immersed in a melt of pure metal that has been heated in a furnace to a temperature above the melting point. The temperature of the melt is slowly reduced while a temperature–time record, similar to the one shown in Fig. 11.38, is made. As the metal changes state from liquid to solid, the temperature remains constant at the freezing-point temperature T_F and provides an accurate reference temperature for calibration. The particular metal selected for the bath is determined by the temperature required for the calibration. Usually, a sensor should be calibrated at three points within its temperature range (preferably the minimum, the midpoint, and the maximum). The freezing points of a number of metals are listed in Table 11.1. These data indicate that the freezing-point approach can be used to provide calibration temperatures from a low of 232°C (450°F) with tin to a high of 1084.6°C (1984.3°F) with copper. The metals must be pure, since small quantities of an impurity can significantly affect the freezing point and thus affect the calibration. Metals used as freezing-point standards are commercially available for temperatures ranging from 125°F to 600°F in 25°F increments. These standards are accurate to ±1°F.

The lower range of the temperature scale is often calibrated by using the *boiling phenomenon*. The temperature sensor is immersed in a liquid bath, and heat is added slowly until the fluid begins to boil and a stable calibration temperature is achieved. Atmospheric pressure must be considered in ascertaining the boiling point of any liquid, since pressure variations can significantly affect the calibration results. For example, reducing the atmospheric pressure from 29.922 to 26.531 in. of Hg results in a decrease in the saturation (boiling) temperature of water from 212°F to 206°F.

The *melting-wire method* of calibration is used with thermocouples. With this approach, the hot junction of the thermocouple is made by connecting the two dissimilar wires with a pure third metal, such as silver or tin. As the hot junction is heated, the output voltage v_o drops to zero. The output voltage v_o immediately prior to the

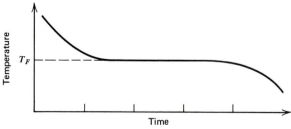

Figure 11.38 Typical temperature–time curve for a metal during solidification.

voltage drop is associated with the melting point (calibration temperature T_M) of the specific material used for the joint.

The *comparison method* utilizes two temperature sensors: one of unknown quality and one of reference or standard quality. Both sensors are immersed in a liquid bath that is temperature cycled over the range of interest. The response of the standard sensor gives the temperature of the bath at any time that can be used as the calibration temperature for the unknown sensor. The standard temperature sensor must be calibrated periodically to ensure its accuracy, and it should not be used for any purpose other than calibration.

11.9 RADIATION METHODS (PYROMETRY)

As the temperature of a body increases, it becomes increasingly difficult to measure the temperature with resistance temperature detectors, thermistors, or thermocouples. The problems associated with measurement of high temperatures by means of these conventional methods (lack of stability, breakdown of insulation, etc.) provided motivation for initial developments in *pyrometry* (inferring temperature from a measurement of the radiation emitted by the body).

By employing the principles of radiation, methods have been developed to measure surface temperatures without contacting the body. These noncontact methods have eliminated the problems of stability and insulation failure that plague the sensor methods of measurement of very high temperatures.

Two different radiation methods are widely employed. The first, which is described as optical pyrometry, compares the brightness of light radiating from a body with a known standard. The second method uses a photon detector to measure the photon flux density that varies with the temperature of the surface. Instruments are commercially available that permit temperature to be measured at a point or over an entire field by using photon detectors with suitable optical arrangements and electronic systems.

11.9.1 Principles of Radiation

The electromagnetic waves and particles emitted from the surface of a body are referred to as radiation. This radiation is often described in terms of photons that propagate from each emission point to another (receiving) surface. At the receiving surface, the photons are absorbed, reflected, or transmitted.

The intensity (power) of the radiation E_b from a black (ideal) surface is related to the absolute temperature θ as

$$E_b = \sigma\theta^4 \tag{11.33}$$

where

σ is the Stefan–Boltzmann constant [$5.67\,(10^{-8})\text{W}/(m^2 \cdot K^4)$]
E_b is the power radiated (W/m^2)

The radiation emitted from a heated surface has many different wavelengths. The electromagnetic spectrum, as shown in Fig. 11.39, covers a range of wavelengths; however, thermal radiation is concerned with the light spectrum, which exhibits wavelengths from about 300 nm to 20 μm.

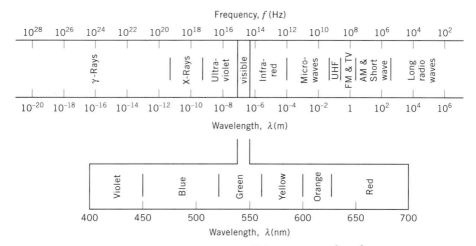

Figure 11.39 Electromagnetic spectrum with the light spectrum enlarged.

Max Planck developed a relationship that describes the radiation power E_λ in terms of the absolute temperature θ and the wavelength λ of the radiation.

$$E_\lambda = \frac{2\pi c^2 h}{\lambda^5 (e^{hc/k\lambda\theta} - 1)} = \frac{C_1}{\lambda^5 (e^{C_2/\lambda\theta} - 1)} \qquad (11.34)$$

where

h is Planck's constant $[6.626\,(10^{-34})\,(\text{J·s})]$
c is the velocity of light $[299.8(10^6)(\text{m/s})]$
k is Boltzmann's constant $[1.381(10^{-23})(\text{J/K})]$
C_1 is a constant $[2\pi c^2 h = 3.75(10^{-16})\ \text{W} \cdot m^2]$
C_2 is a constant $[hc/k = 1.44(10^{-2})\ \text{m·K}]$

The spectral radiation intensity E_λ is the amount of power emitted by radiation of wavelength λ from a flat surface at temperature θ into a hemisphere. It is evident from Eq. 11.34 that the spectral radiation intensity E_λ depends on both wavelength λ and temperature θ. A plot of E_λ versus λ for several different temperatures is shown in Fig. 11.40. Note that E_λ exhibits a maximum at a specific wavelength, which depends on temperature. Observe that the wavelength associated with the peak E_λ increases as the temperature decreases. The wavelength λ_p associated with the peak in E_λ can be expressed as

$$\lambda_p = \frac{2898(10^{-6})}{\theta} \qquad (11.35)$$

The area under each of the curves in Fig. 11.40 is the total power E_t emitted at the particular temperature θ. Thus

$$E_t = \int_\lambda E_\lambda \, d\lambda = 5.67(10^{-8})\theta^4\ \text{W/m}^2 \qquad (11.36)$$

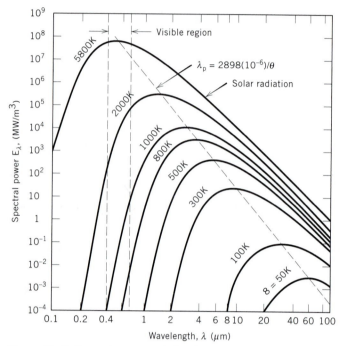

Figure 11.40 Spectral power emission from a black surface at different temperatures.

Equation 11.36 is the Stefan–Boltzmann law with the emissivity ε equal to unity ($\varepsilon = 1$). From the previous discussion it is evident that

1. The total power E_t increases as a function of θ^4.
2. The maximum value of spectral radiation intensity E_λ occurs at shorter wavelengths as the temperature increases.

Both of these physical principles are used as the basis for a measurement of temperature.

11.9.2 The Optical Pyrometer

The optical pyrometer, illustrated schematically in Fig. 11.41a, is used to measure temperature over the range from 700°C to 4000°C (1300°F to 7200°F). The radiant energy emitted by the body is collected with an objective lens and focused onto a calibrated pyrometer lamp. An absorption filter is inserted in the optical system between the objective lens and the pyrometer lamp when the temperature of the body exceeds 1300°C (2370°F). The radiant energy from both the hot body and the filament of the pyrometer lamp is then passed through a red filter with a sharp cutoff below $\lambda = 0.63$ μm. The light transmitted through this filter is collected by an objective lens and focused for viewing with an ocular lens. The image observed through the eyepiece of the pyrometer is that of the lamp filament superimposed on a background intensity owing to the hot body. The current to the filament of the pyrometer lamp is adjusted until the brightness of the filament matches that of the background. Under a matched condition, the filament disappears (hence, the commonly used name, disappearing-filament optical pyrometer), as illustrated in Fig. 11.41b. The current required to

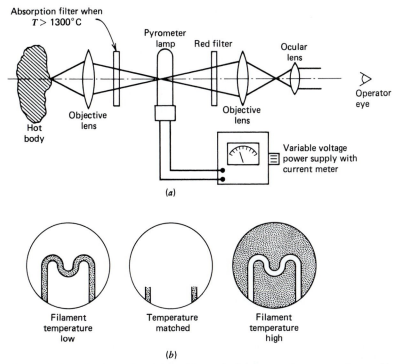

Figure 11.41 Schematic illustration of the optical system and pyrometer lamp with filament brightness adjustment in an optical pyrometer. (a) The optical system. (b) Filament brightness adjustment.

produce the brightness match is measured and used to establish the temperature of the hot body. Pyrometers are calibrated by visually comparing the brightness of the tungsten filament with a blackbody source of known temperature ($\varepsilon = 1$).

When the brightness of the background and the filament are matched, it is evident that E_λ for these two objects is the same. Thus, Eq. 11.34 gives

$$\frac{\varepsilon}{e^{C_2/\lambda_r\theta} - 1} = \frac{1}{e^{C_2/\lambda_r\theta_f} - 1} \tag{11.37}$$

where

> λ_r is the wavelength of the red filter (0.63 μm)
> ε is the emissivity of the surface of the hot body at $\lambda = 0.63$ μm
> θ_f is the temperature of the filament
> θ is the unknown surface temperature

When $\theta < 4000°$C (7200°F), the term $e^{C_2/\lambda_r\theta} \gg 1$ and Eq. 11.37 reduces to

$$\theta = \frac{1}{\lambda_r(\ln \varepsilon)/C_2 + 1/\theta_f} \tag{11.38}$$

It is obvious from Eq. 11.38 that $\theta = \theta_f$ only when $\varepsilon = 1$. If $\varepsilon \neq 1$, then $\theta \neq \theta_f$, and Eq. 11.38 must be used to determine the temperature θ from the temperature θ_f indicated by the pyrometer. The emissivities of a number of materials (oxidation-free surface) are listed in Table 11.7.

Table 11.7^{a,b} Emissivity ε of Engineering Materials at $\lambda = 0.65$ μm

Material	Solid	Liquid	Material	Solid	Liquid
Beryllium	0.61	0.61	Tantalum	0.49	—
Carbon	0.80–0.93	—	Thorium	0.36	0.40
Chromium	0.34	0.39	Titanium	0.63	0.65
Cobalt	0.36	0.37	Tungsten	0.43	—
Columbium	0.37	0.40	Uranium	0.54	0.34
Copper	0.10	0.15	Vanadium	0.35	0.32
Iron	0.35	0.37	Zirconium	0.32	0.30
Manganese	0.59	0.59	Steel	0.35	0.37
Molybdenum	0.37	0.40	Cast Iron	0.37	0.40
Nickel	0.36	0.37	Constantan	0.35	—
Platinum	0.30	0.38	Monel	0.37	—
Rhodium	0.24	0.30	90 Ni–10 Cr	0.35	—
Silver	0.07	0.07	80 Ni–20 Cr	0.35	—
			60 Ni–24 Fe–16 Cr	0.36	—

^{a}From ASME Performance Test Codes PTC 19.3, 1974.
^{b}A more complete table is given in Vol. 27 Supplement, Omega Engineering, 1991.

If the emissivity of a surface is not known precisely, then an error will occur when Eq. 11.38 is used to determine the temperature θ. The change in temperature as a function of change in emissivity is obtained from Eq. 11.38 as

$$\frac{d\theta}{\theta} = -\frac{\lambda\theta}{C_2}\frac{d\varepsilon}{\varepsilon} \tag{11.39}$$

Since $\lambda\theta/C_2 < 0.1$ for $\theta < 2000°$C ($3630°$F), errors in temperature measurements are mitigated considerably with respect to errors in emissivity. For example, at a temperature of 1500 K, a 20 percent error in emissivity produces only a 1.3 percent error in temperature.

For relatively low to intermediate temperatures, a portion of the surface can be coated with either a black paint or a black ceramic layer to provide an emissivity ε approaching one. For very high temperatures, a hole can be drilled in the body, with a depth-to-diameter ratio of six or more. This hole acts as a blackbody with $\varepsilon \approx 1$, and the temperature measured by focusing the optical pyrometer on the hole represents the correct temperature of the object.

The disappearing-filament optical pyrometer is an accurate instrument, and if the emissivity of the hot body is accurately known, the error in a temperature measurement is usually less than 1 percent.

11.9.3 Infrared Pyrometers

In many applications, regardless of the temperature, the measurement must be made without contacting the body. The optical pyrometer described previously is effective for temperatures above 700°C (1290°F), where a significant amount of radiant power is emitted in the visible-light region of the spectrum. At lower temperatures, the radiation emissions are concentrated in the infrared regions and are not visible to the human eye.

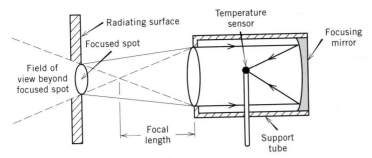

Figure 11.42 Schematic illustration of a radiometer.

Infrared pyrometers employ the infrared portion of the spectrum by using a thermal detector to measure the temperature of the surface of the body emitting the infrared waves. A schematic illustration of a thermal detector, sometimes called a radiometer, is shown in Fig. 11.42. The lens collects the infrared radiation emitted from the area included in the focused spot and collimates the radiation as indicated. The radiation is reflected from the end mirror and focused on a temperature sensor. Thermocouples or thermistors are employed as temperature sensors. The equilibrium temperature of the sensor is a direct measurement of the magnitude of the radiation absorbed by the sensor. The magnitude of the radiation gives the temperature of the emitting surface as indicated by Eq. 11.36.

Target size and distance from the lens to the object are critical in the operation of infrared pyrometers. The field of view of an infrared pyrometer depends on the focal length and diameter of the collecting lens. The optical system of the instrument collects all of the radiation from the objects in the field of view, and the reading represents an average of these temperatures. To show the importance of this averaging effect on the accuracy of an infrared pyrometer, consider the objects arranged in the field of view of the pyrometer in Fig. 11.43. In this illustration, object A covers the entire field of view and the reading represents the average surface temperature of object A. However, if object A is removed from the field, then object B and the wall are both included in the field of view. The indicated temperature will be between the temperature of the wall and that of object B, and will depend on the relative area of each object in the circular field of view.

Most infrared pyrometers have a fixed-focal-length collecting lens that defines the field of view. This field of view is usually expressed in terms of a d/D ratio, where d is the distance from the lens to the object and D is the diameter of the field at the position d. Note that the diameter of the field is equal to the diameter of the

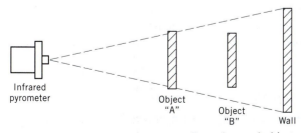

Figure 11.43 Example showing the effect of several objects in the field of view of an infrared pyrometer.

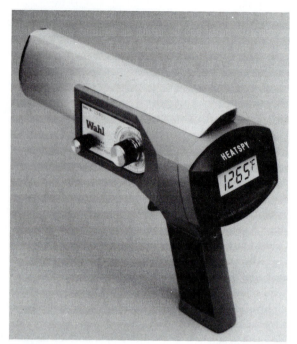

Figure 11.44 Commercial infrared pyrometer. (Courtesy of Wahl Instruments, Inc.)

collecting lens when d is twice the focal length of the lens. General-purpose infrared pyrometers use lenses with focal lengths between 500 and 1500 mm, although close-focus instruments use lenses with focal lengths between 10 and 100 mm, and long-range instruments use lenses with 10-m focal lengths. It is even possible to employ fiber optics to transmit the radiation from the source to the sensor.

Emissivity affects the reading from an infrared pyrometer just as it affects the reading from an optical pyrometer. When the emissivity is less than one, the radiation power actually emitted from the surface of the body is less than expected and the instrument gives a reading lower than the true surface temperature. The manufacturers of infrared pyrometers accommodate the emissivity error by installing an emissivity compensator on the instrument. The emissivity compensator is a calibrated gain adjustment that increases the amplification of the sensor signal to compensate for the power lost owing to an emissivity less than one. This gain adjustment can also be used to correct for transmission losses that occur when viewing the object through glass or plastic portholes, smoke, dust, or vapors.

A commercial infrared pyrometer is shown in Fig. 11.44. This instrument covers the temperature range from $-20°$ to $1000°$C or 0 to $2000°$F. Emissivity is adjustable from 0.2 to 1.0. The temperature, which is continuously updated 2.5 times per second, is displayed on a 4-digit LCD. Accuracies are specified as ± 0.3 percent of the full-scale reading.

11.9.4 Photon Detector Temperature Instruments

A second approach that uses radiation to measure temperature employs a photon detector. The instruments equipped with photon detectors differ from those with temperature

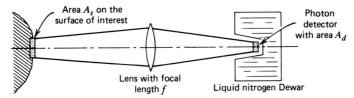

Figure 11.45 Schematic illustration of a temperature-measuring instrument with a photon detector sensor.

detectors in two ways. First, the response time of the photon detector is several orders of magnitude faster than that of the thermal detector. This advantage is used to develop instruments capable of scanning a field and producing images depicting the temperature distributions over an area of a surface. Second, the photon detector must be maintained at a very low temperature during operation, and it is necessary to have a source of liquid nitrogen.

A *photon detector* is a sensor that responds by generating a voltage that is proportional to the photon flux density ϕ impinging on the sensor. A schematic diagram of a photon detector system for measuring temperature is shown in Fig. 11.45. The photons emitted from a small area A_s of a surface (not necessarily hot) are collected by a lens and are focused on a photon detector of area A_d. The photon flux density ϕ at the detector, when the optical system is focused, can be expressed as

$$\phi = \frac{kD^2\varepsilon}{4f^2}g(T) \tag{11.40}$$

where

    k is the transmission coefficient of the lens and filter
    D is the diameter of the lens
    f is the focal length of the lens
  $g(T)$ is a known function of the temperature of the surface
    ε is the emissivity of the surface

The output voltage v_o from the detector, as a result of the flux density ϕ, is

$$v_o = k_t\frac{D^2}{4f^2}\varepsilon g(T) \tag{11.41}$$

where k_t is the system sensitivity, which includes the transmission coefficient of the lens, the amplifier voltage gain, and the detector sensitivity. The system sensitivity k_t is essentially a constant; however, a zoom lens is employed in a typical instrument to provide for different fields of view where the solid angle may range from 3.5 to 40°. The term $\varepsilon g(T)$ depends only on the temperature of the surface and its emissivity. A typical response curve, shown in Fig. 11.46, indicates that the output voltage varies as a function of the cube of the temperature. Thus Eq. 11.41 can be simplified to

$$v_o = K\varepsilon T^3 \tag{11.42}$$

where K is a calibration constant for the instrument. In practice, K is determined by calibrating the instrument with a blackbody source ($\varepsilon = 1$) over an appropriate

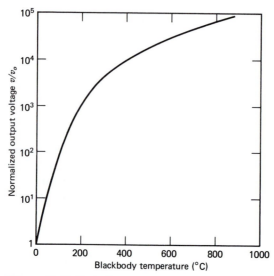

Figure 11.46 Typical response curve for an indium-antimonide photon detector.

range of temperatures. When the instrument is used for temperature measurements, the emissivity ε of the surface must be considered, since it may differ significantly from one, as shown in Table 11.7. Any correction required is easily made by substituting the correct value of the emissivity into Eq. 11.42 and solving for the required temperature T. Thus

$$T = \left(\frac{v_o}{k\varepsilon}\right)^{1/3} \tag{11.43}$$

Errors in temperature owing to inaccuracies in emissivity are mitigated by one-third, since differentiation of Eq. 11.43 gives $dT/T = -(\frac{1}{3})(d\varepsilon/\varepsilon)$.

Many different commercial instruments employ the photon detector; therefore, it is difficult to list specifications that cover the full range of products. Typical specifications for scanners such as that shown in Fig. 11.47 indicate that they are used to measure temperatures in the range from $-20°C$ to $1600°C$ ($0°F$ to $2900°F$) with a sensitivity of $0.1°C$ ($0.2°F$) at $30°C$ ($86°F$).

Figure 11.47 Infrared imaging and analysis system. (Courtesy of AGEMA Infrared Systems.)

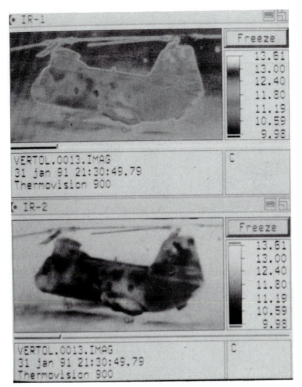

Figure 11.48 Example of a temperature distribution recorded with an infrared imaging system. (Courtesy of AGEMA Infrared Systems.)

A recent innovation with this type of instrument permits determination of temperature distributions over extended regions of a body. This improved capability is accomplished by inserting two mechanically driven cylindrical lenses into the optical path. As the two lenses are oscillated, a region of the surface of the body is scanned. At any instant, a relatively small target area is determined. To reduce the scanning time, several photon detectors (four to six) are incorporated in the instrument. Since the entire surface of the body is scanned in a short period of time, a full-field photograph of the temperature distribution representing an x–y array of the many small target areas can be obtained. A single frame typically contains 28,000 individual temperature measurements (280 lines with 100 elements per line). The voltage output can be displayed on a TV monitor in either gray scale or color. Photon-detector-type instruments can complete a scan of a field in about 40 ms. If a video recorder is used to store the images, the system can be used to study full-field dynamic temperature distributions. An example of a full-field temperature distribution is shown in Fig. 11.48.

11.10   SUMMARY

Temperature is an abstract quantity and as such must be defined in terms of the behavior of materials as the temperature changes. This is accomplished by defining the temperature associated with phase transformations in several different materials over the temperature range from $-259°C$ to $1084°C$.

The different sensors available for temperature measurement include resistance temperature detectors (RTDs), thermistors, expansion thermometers, integrated-circuit sensors, thermocouples, and pyrometers. Each type of sensor or instrument has advantages and disadvantages; selection of the proper sensor for a particular application is usually based on considerations of temperature range, accuracy requirements, environment, dynamic response requirements, and available instrumentation.

The most frequently used temperature-measuring sensor is the thermocouple, since it is a low-cost transducer that is easy to fabricate and install. The signal output is relatively low and must be measured and recorded with an instrument having a high input impedance so that current flow in the circuit is minimized; otherwise, significant errors can be introduced. The nonlinear output of the thermocouple is a disadvantage; however, modern instruments used to record the output voltage incorporate a microprocessor to linearize the output and give readout in terms of temperature directly. The range of temperatures that can be measured with thermocouples is very large, from $-185°C$ to $2800°C$.

Resistance-based temperature sensors (RTDs and thermistors) are usually employed when a high sensitivity is required. Because of the high-voltage output, higher accuracies can be achieved. The RTD-type sensor is available in coil form for fluid temperature measurements and in a bondable grid form for surface temperature measurements. These sensors are easy to install, and the instrumentation used to monitor the output signal is inexpensive and easy to employ. Thermistors are used in many commercial temperature recorders because their high-voltage output permits a reduction in complexity and cost of the readout system. The range of thermistors is limited, and the output is extremely nonlinear.

The advantages and disadvantages of the four most popular point sensors for measuring temperature are given in Fig. 11.49.

Bimetallic thermometers are used primarily in control applications where long-term stability and low cost are important considerations. They are often used to activate switches in on–off (bang-bang) temperature-control situations where precise control is not required.

Pyrometers are used primarily to monitor the extremely high temperatures associated with metallurgical processes. The disappearing-filament optical pyrometer has been a reliable instrument in many industrial applications for several decades. In more recent years, with the development of thermal and photon detectors, radiation methods of temperature measurement have been extended into the lower temperature range. A significant advantage of instruments that use either thermal or photon detectors as sensors is their ability to measure temperature without contacting the specimen in applications involving thin films, paper, or moving bodies.

An advantage of photon detector instruments, with the rapid-response time of the detector, is their ability to scan the field of view and to establish temperature distributions over extended areas. The disadvantages of photon-detector-based instruments are their relatively high cost and the need for liquid nitrogen to cool the photon detector.

Temperature sensors are first-order systems that respond to a step change in temperature in a manner that is described by the equation

$$T = T_m(1 - e^{-t/\beta})$$ **(11.27)**

Errors resulting from the time required for heat transfer can be minimized by reducing the time constant β for the sensor.

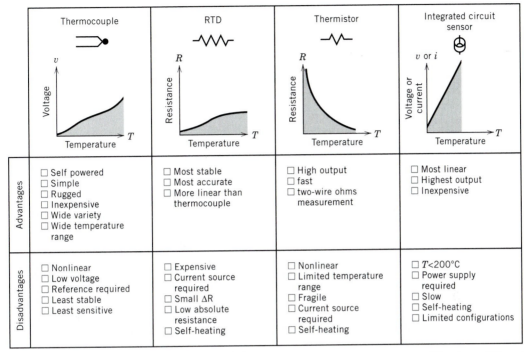

Figure 11.49 Advantages and disadvantages of the most common temperature sensors. (Courtesy of Omega Engineering, Inc.)

REFERENCES

1. American National Standards Institute (ANSI): Standard MC 96.1–1975, Instrument Society of America, 1976.
2. American Society of Mechanical Engineers: *Temperature Measurement,* Supplement to ASME Performance Test Codes, PTC 19.3, 1974.
   (a) Chapter 4, "Resistance Thermometers."
   (b) Chapter 5, "Liquid-in-Glass Thermometers."
   (c) Chapter 7, "Optical Pyrometers."
3. American Society for Testing and Materials: *Manual on the Use of Thermocouples in Temperature Measurement,* ASTM STP 470A, March 1974.
4. Baker, H. D., E. A. Ryder, and N. H. Baker: *Temperature Measurement in Engineering,* vols. 1 and 2, Wiley, New York, 1953, 1961.
5. Becker, J. A., C. B. Green, and G. L. Pearson: "Properties and Uses of Thermistors," *Trans. Am. Inst. Electr. Eng.,* vol. 65, November 1946, pp. 711–725.
6. Benedict, R. P.: *Fundamentals of Temperature, Pressure, and Flow Measurement,* 3rd ed., Wiley, New York, 1984.
7. Benedict, R. P., and R. J. Russo: "Calibration and Application Techniques for Platinum Resistance Thermometers," *Trans. ASME, Journal of Basic Engineering,* June 1972, pp. 381–386.
8. Cook, N. H., and E. Rabinowicz: *Physical Measurement and Analysis,* Addison–Wesley, Reading, Mass., 1963.
9. Curtis, D. J., and G. J. Thomas: "Long Term Stability and Performance of Platinum Resistance Thermometers for Use to 1063°C," *Metrologia,* vol. 4, no. 4, October 1968, pp. 184–190.

10. Finch, D. I.: "General Principles of Thermoelectric Thermometry," *Temperature*, vol. 3, part 2, Reinhold, New York, 1962.

11. Harrison, T. R.: *Radiation Pyrometry and Its Underlying Principles of Radiant Heat Transfer*, Wiley, New York, 1960.

12. Hennecke, D. K., and E. M. Sparrow: "Local Heat Sink on a Convectively Cooled Surface—Application to Temperature Measurement Error," *Int. J. Heat Mass Transfer*, vol. 13, 1970, pp. 287–304.

13. Kinzie, P. A.: *Thermocouple Temperature Measurement*, Wiley, New York, 1973.

14. Magison, E. C.: *Temperature Measurement in Industry*, Instrument Society of America, Research Triangle Park, N.C., 1990.

15. McGee, T. D.: *Principles and Methods of Temperature Measurement*, Wiley, New York, 1988.

16. National Bureau of Standards: "Liquid-in-Glass Thermometry," *NBS Monograph 150*, 1975.

17. Nutter, G. D.: "Radiation Thermometry," *Mechanical Engineering*, part 1, June 1972, p. 16, part 2, July 1972, p. 12.

18. Omega Engineering: *Temperature Measurement Handbook*, Omega Engineering, Stamford, Conn., 1981.

19. Peltier, J. C. A.: "Investigation of the Heat Developed by Electric Currents in Homogeneous Materials and at the Junction of Two Different Conductors," *Ann. Chim. Phys.*, vol. 56 (2nd ser.) 1834, p. 371.

20. Powell, R. L., W. J. Hall, C. H. Hyink, Jr., L. L. Sparks, G. W. Burns, M. G. Scroger, and H. H. Plumb: "Thermocouple Reference Tables Based on IPTS-68," *NBS Monograph 125*, March 1974.

21. Preston-Thomas, H.: "The International Temperature Scale of 1990 (ITS-90)," *Metrologia*, vol. 27, 1990, pp. 3–10.

22. Seebeck, T. J.: *Evidence of the Thermal Current of the Combination Bi-Cu by its Action on Magnetic Needle*, Royal Academy of Science, Berlin, 1822–1823, p. 265.

23. Thompson, W.: "On the Thermal Effects of Electric Currents in Unequal Heated Conductors," *Proceedings of the Royal Society*, vol. 7, May 1854.

EXERCISES

11.1 Why are the fixed points on the International Temperature Scale important?

11.2 Describe how you would calibrate a temperature sensor using the triple point of water and the freezing point of tin.

11.3 How is a temperature sensor calibrated if it is to operate over the temperature range from 1200°C to 1600°C?

11.4 A bimetallic strip is fabricated from stainless steel and Invar with thicknesses of 0.5 mm and 1.0 mm, respectively. Determine the radius of curvature of the strip if it undergoes a temperature change of

(a) 120°C   (c) 230°C

(b) 70°F   (d) −80°F

The mechanical and thermal properties of stainless steel and Invar are as follows:

Stainless steel: $E = 28(10^6)$ psi or 193 GPa

$\alpha = 9.6(10^{-6})/°F$ or $17.3(10^{-6})/°C$

Invar: $E = 21(10^6)$ psi or 145 GPa

$\alpha = 0.6(10^{-6})/°F$ or $1.1(10^{-6})/°C$

11.5   From the results shown in Fig. 11.2, determine the temperature coefficients of resistivity γ_1, γ_2, and γ_3 in Eq. 11.5 for an RTD fabricated from platinum for the temperature range from $-200°C$ to $1000°C$.

11.6   From the results shown in Fig. 11.2, determine the temperature coefficients of resistivity γ_1, γ_2, and γ_3 in Eq. 11.5 for an RTD fabricated from copper for the temperature range from $-200°C$ to $1000°C$.

11.7   From the results shown in Fig. 11.2, determine the temperature coefficients of resistivity γ_1, γ_2, and γ_3 in Eq. 11.5 for an RTD fabricated from nickel for the temperature range from $0°C$ to $1000°C$.

11.8   Repeat Exercise 11.5 for a temperature range from $0°C$ to $700°C$.

11.9   Repeat Exercise 11.6 for a temperature range from $0°C$ to $500°C$.

11.10   Repeat Exercise 11.7 for a temperature range from $0°C$ to $400°C$.

11.11   From the results shown in Fig. 11.2, determine the temperature coefficients of resistivity γ_1 and γ_2 in Eq. 11.7 for an RTD fabricated from platinum for the temperature range from $100°C$ to $400°C$.

11.12   An RTD fabricated from platinum exhibits a temperature coefficient of resistivity $\gamma_1 = 0.003902/°C$. Assume that γ_2 is negligible. If the resistance of the sensor is $100\ \Omega$ at $0°C$, find the resistance at

    (a) $-240°C$ (c) $90°C$ (e) $600°C$

    (b) $-120°C$ (d) $260°C$ (f) $900°C$

11.13   Show that lead-wire effects are completely eliminated by using the three-wire system illustrated in Fig. 11.7 to connect an RTD into a Wheatstone bridge.

11.14   Show that lead-wire effects are completely eliminated by using the four-wire system illustrated in Fig. 11.8 to connect an RTD into a constant-current potentiometer circuit.

11.15   Using the data for temperature as a function of time shown in Fig. 11.4, determine the response time constant for

    (a)   a platinum thick-film RTD

    (b)   a platinum wire-wound RTD

11.16   A platinum RTD with a resistance of $100\ \Omega$ at $0°C$ is used in the constant-current potentiometer circuit shown in Fig. 11.8. If the current i equals 5 mA, determine the output voltage v_o at the following temperatures:

    (a) $-240°C$ (c) $90°C$ (e) $600°C$

    (b) $-120°C$ (d) $260°C$ (f) $900°C$

11.17   Determine the error resulting from self-heating of an RTD if its self-heating factor is $F_{sh} = 0.35°C/mW$. The sensor ($R_T = 100\ \Omega$) is placed in an equal-arm bridge with a supply voltage of 5 V.

11.18   Using the Callendar–Van Dusen equation prepare a graph showing the non-linearities in $\Delta R/R_0$ with temperature. Cover the range from

    (a) $-200°C$ to $200°C$ (c) $400°C$ to $900°C$

    (b) $0°C$ to $600°C$ (d) $-200°C$ to $800°C$

11.19   Describe four common errors encountered in measuring temperature with an RTD. Indicate procedures that can be taken to minimize each of these errors.

11.20   Verify Eq. 11.13.

11.21   If $\beta = 4350$ K and $R_0 = 3000\ \Omega$ at $T_0 = 298$ K, determine the resistance of a thermistor at

    (a) $-80°C$ (c) $0°C$ (e) $75°C$

    (b) $-40°C$ (d) $50°C$ (f) $150°C$

11.22   Write an engineering brief describing the relative merits of the RTD and thermistor as temperature sensors. Also indicate the disadvantages of each sensor.

11.23   Recommend a sensor, either an RTD or a thermistor, to be used to control a process where the temperature is

(a)   400°C   (b)   90°C   (c)   −200°C

11.24   The thermistor described in Exercise 11.21 is connected in a constant-current potentiometer circuit (see Fig. 11.11). If the current is 10 mA, prepare a graph showing the output voltage v_o as the temperature increases from −50° to 300°C. Use $R_T = 100\ \Omega$ at $T_0 = 25°C$.

11.25   Determine the constants A, B, and C in Eq. 11.14 by using the data given in Fig. 11.9 for a thermistor with

(a) $\beta = 3270$ (b) $\beta = 4240$ (c) $\beta = 4710$

11.26   If the self-heating error is to be limited to 0.5°C for a thermistor with $R_T = 5000\ \Omega$, determine the maximum current that can be used with the constant-current potentiometer circuit shown in Fig. 11.11. Let F_{sh} be

(a)   0.5°C/mW   (b)   1.0°C/mW   (c)   2.0°C/mW

11.27   List the primary advantages and disadvantages of
(a)   liquid-in-glass thermometers
(b)   bimetallic thermometers
(c)   pressure thermometers

11.28   Prepare a graph showing $T_1 - T_2$ as a function of output voltage using Eq. 11.18 and the coefficients given in Table A.5 for
(a)   Chromel-Alumel (type K)
(b)   Chromel-constantan (type E)
(c)   copper-constantan (type T)

11.29   Compare the results of Exercise 11.28 with the results listed in Tables A.2, A.3, and A.4.

11.30   A DVM is being used to measure the output voltage v_o from a copper-constantan thermocouple, as shown in Fig. E11.30.
(a)   Determine the output voltage v_o indicated by the DVM.
(b)   If the DVM reading changes to 2.078 mV, what is the new temperature T_1?
(c)   Do temperatures T_2 or T_3 influence the measurement? Why?

11.31   A DVM is being used to measure the output voltage v_o from a copper-constantan thermocouple, as shown in Fig. E11.31. Determine the output voltage v_o indicated by the DVM.

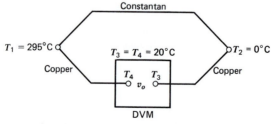

Figure E11.30

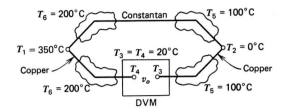

Figure E11.31

11.32   A DVM is being used to measure the output voltage v_o from a Chromel-Alumel thermocouple, as shown in Fig. E11.32.
   (a)   Determine the output voltage v_o indicated by the DVM.
   (b)   If the DVM reading changes to 20.470 mV, what is the new temperature T_1?
   (c)   Does the copper-Alumel junction at the DVM influence the reading on either (a) or (b)?

11.33   A DVM is being used to measure the output voltage v_o from an iron-constantan thermocouple, as shown in Fig. E11.33.
   (a)   Determine the temperature T_1 associated with a DVM reading of 14.123 mV.
   (b)   Does the separation at junction 1 influence the measurement of T_1? List any assumptions made in reaching your answer.
   (c)   How far can the junctions be separated before errors will develop? Explain.

11.34   A DVM is being used to measure the output voltage v_o from an iron-constantan thermocouple, as shown in Fig. E11.34.
   (a)   Determine the output voltage v_o indicated by the DVM.
   (b)   If the DVM reading changes to 21.333 mV, what is the new temperature T_1?

11.35   A DVM having an input impedance of 10 MΩ is being used to measure the output voltage v_o of the iron-constantan thermocouple shown in Fig. E11.35. The thermocouple is fabricated from AWG No. 20 wire having a resistance of 0.357 Ω per double foot (1 ft of iron plus 1 ft of constantan). The distance between junctions 1 and 2 is 40 ft. Determine:
   (a)   the iR drop resulting from the long lead wires
   (b)   the output voltage v_o indicated by the DVM
   (c)   the temperature associated with the output voltage indicated by the DVM

11.36   A Chromel-constantan thermocouple is accidentally grounded at both the active and reference junctions, as shown in Fig. E11.36. If the resistance of the thermocouple is 3 Ω and the resistance of the ground loop is 0.2 Ω, determine the error introduced into the measurement of temperature T_1.

11.37   The extension wires of an iron-constantan thermocouple were improperly wired to produce the situation illustrated in Fig. E11.37. Determine the error introduced into the measurement of temperature T_1.

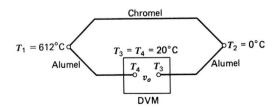

Figure E11.32

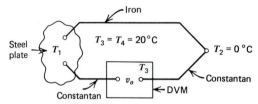

Figure E11.33

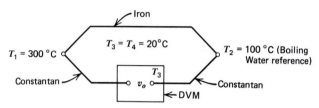

Figure E11.34

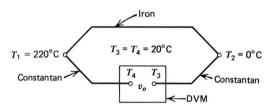

Figure E11.35

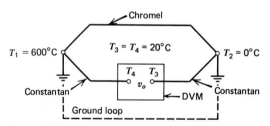

Figure E11.36

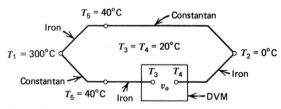

Figure E11.37

11.38 Write an engineering brief describing the seven different material combinations employed in standard thermocouples.

11.39 A thermocouple installation made with AWG No. 14 wire is to be used to measure a temperature of 777°C. Determine the error resulting from drift after an exposure of 500 h if the thermocouples are type

(a) J   (c) K
(b) E   (d) N

11.40 Stability improvement occurs with larger diameter wire. At an exposure of 1100 h, indicate the error in a measurement of a temperature of 1202°C with an N-type thermocouple fabricated from AWG No. 8 wire.

11.41 Write an engineering brief describing three methods for controlling reference junction temperature.

11.42 Write an engineering brief describing the modern instruments used to measure the voltage v_o from a thermocouple.

11.43 Continue the engineering brief of Exercise 11.42. Describe cold junction compensation methods and linearization techniques.

11.44 Write a set of instructions to be used by laboratory technicians to minimize noise in thermocouple circuits.

11.45 Write an engineering brief comparing an integrated-circuit temperature sensor with

(a) a thermocouple   (b) a thermistor   (c) an RTD

11.46 Verify Eqs. 11.24 and 11.25.

11.47 Verify Eqs. 11.28 and 11.29.

11.48 Develop an expression for the response of a temperature sensor to the truncated ramp type of input function shown in Fig. E11.48 for

(a) $0 < t \le t_0$ (b) $t > t_0$

11.49 Outline the procedure you would follow to determine the lag time associated with thermocouple response to a ramp-function type of input.

11.50 Determine the error in measuring surface temperature T_s owing to conduction of the sensor and lead wires if $T_f = 0.9T_s$ and

$$\frac{\sqrt{(kA)_e/R}\,\tanh[L/\sqrt{(kA)_e/R}]}{\pi r_1 k_s} = 8$$

$$\frac{h_s r_1}{k_s} = 1$$

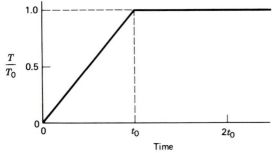

Figure E11.48

11.51   Repeat Exercise 11.50 for the following values of $h_s r_1 / k_s$:

(a)   2   (b)   4   (c)   10

11.52   Evaluate the dynamic temperature $T_v = V^2 / 2Jgc_p$ for air as V varies from 0 to 1000 mph.

11.53   Write an engineering brief to be used by a laboratory technician in calibrating thermocouples for use at temperatures ranging from 125° to 600°F.

11.54   Graph the results of Planck's law (Eq. 11.34) for temperatures of 100°C, 200°C, 500°C, 1000°C, and 2000°C.

11.55   Use the results of Exercise 11.54 to verify Eq. 11.35.

11.56   Use the results of Exercise 11.54 for $T = 1000$°C to verify Eq. 11.36.

11.57   Use Eq. 11.38 to prepare a graph showing the relationship between θ and θ_f if the emissivity ε is

(a)   0.1   (c)   0.3   (e)   0.6
(b)   0.2   (d)   0.4   (f)   0.8

11.58   Verify Eq. 11.39.

11.59   Use Eq. 11.39 to prepare a graph of error dT/T versus temperature T over the range from 1000°C to 4000°C for an optical pyrometer if the emissivity ε of the surface is in error by $d\varepsilon/\varepsilon$ equal to

(a)   0.05   (c)   0.20
(b)   0.10   (d)   0.50

11.60   Write an engineering brief for a laboratory technician that describes the operation of an infrared pyrometer.

11.61   Determine k in Eq. 11.43 by using the response curve for an indium antimonide photon detector shown in Fig. 11.46.

11.62   Write an engineering brief comparing the optical pyrometer with an instrument employing a photon detector.

Chapter 12

Fluid Flow Measurements

12.1 INTRODUCTION

Fluid flow measurements, expressed in terms of either volume flow rate or mass flow rate, are used in many applications, such as industrial process control, city water systems, petroleum or natural-gas pipeline systems, and irrigation systems. The fluid involved in the measurement may be a liquid, a gas, or a mixture of the two (mixed-phase flow). The flow can be confined or closed (as in a pipe or conduit), semiconfined (as in a river or open channel), or unconfined (as in the wake behind a jet). In each situation, several methods of flow measurement can be used to determine the required flow rates. Several of the more common measurement techniques are discussed in this chapter.

The concept of mass flow rate can be visualized by considering confined flow in a circular pipe, as show in Fig. 12.1. The local mass flow rate dm/dt through area dA surrounding point P (see Fig. 12.1a) can be expressed as

$$\frac{dm}{dt} = \rho V \, dA$$

where

ρ is the density of the fluid at point P
V is the velocity of the fluid at point P in a direction normal to area dA

The total mass flow rate $\dot{m}$ through the cross section of pipe containing point P is

$$\dot{m} = \int \frac{dm}{dt} = \int_A \rho V \, dA \qquad \textbf{(12.1a)}$$

Equation 12.1a is valid for any plane area, and both fluid density and fluid velocity can vary over the cross section. When the fluid is either a liquid or a gas, with small changes in pressure, the density is treated as a constant; therefore, it can be factored out of the second integral in Eq. 12.1 to give

$$\dot{m} = \rho \int_A V \, dA = \rho Q \qquad \textbf{(12.1b)}$$

where $Q = \int V \, dA$ is the volume flow rate. $\qquad \textbf{(12.1c)}$

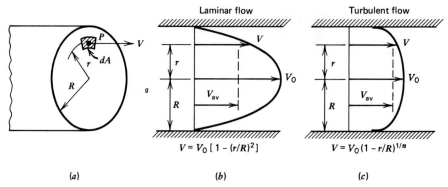

Figure 12.1 Mass flow rate in a closed conduit. (*a*) General concept, (*b*) velocity profile, and (*c*) velocity profile.

For laminar flow, a parabolic velocity profile exists, as shown in Fig. 12.1*b*, which can be expressed as

$$V = V_0 \left(1 - \frac{r^2}{R^2}\right) \tag{12.2}$$

where

V_0 is the centerline velocity
R is the inside radius of the pipe
r is a position parameter

When the density of the fluid is a constant, Eq. 12.1b can be used to define an average velocity V_{av}. Thus

$$\dot{m} = \rho \int_A V \, dA = \rho V_{av} A \tag{12.3}$$

For laminar flow in a circular pipe, Eq. 12.3 gives

$$\dot{m} = \rho V_0 \int_0^{2\pi} \int_0^R \left(1 - \frac{r^2}{R^2}\right) r \, dr \, d\theta$$

$$= \rho \pi R^2 \frac{V_0}{2} = \rho A V_{av} \tag{12.4}$$

where $V_{av} = V_0/2$.

In fully developed turbulent flow in a smooth pipe, as shown in Fig. 12.1*c*, the velocity profile is of the form

$$V = V_0 \left(1 - \frac{r}{R}\right)^{1/n} \tag{12.5}$$

where the exponent n depends on Reynolds number Re. For a circular pipe, Re is

$$Re = \frac{\rho V_{av} D}{\mu}$$

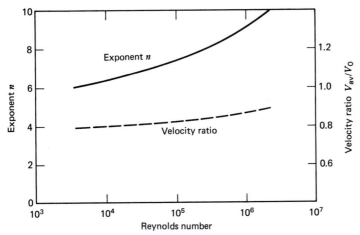

Figure 12.2 Exponent n and velocity ratio V_{av}/V_0 as a function of Reynolds number.

where μ is the absolute viscosity of the fluid and D is the diameter of the pipe. For the case of the circular pipe it can be shown that the average velocity V_{av} is related to the centerline velocity V_0 by the expression

$$V_{av} = \frac{2n^2}{(n + 1)(2n + 1)} V_0 \tag{12.6}$$

Values of the exponent n as a function of Reynolds number Re are shown in Fig. 12.2 together with the corresponding velocity ratios V_{av}/V_0. From the laminar-flow results of Eq. 12.4 and the turbulent-flow results of Eq. 12.6, it is evident that mass flow rates can be established from velocity measurements at a point only if the velocity profile is accurately known.

An approximation for the integral of Eq. 12.3, which can be applied for any flow measurement and is widely used, is the finite sum representation

$$\dot{m} = \sum_{i=1}^{n} \rho_i A_i V_i = \rho V_{av} A \tag{12.7}$$

As an example, consider the open-channel-flow problem illustrated in Fig. 12.3. The mass flow rate through each area A_i is summed to obtain the total mass flow rate.

In the following sections, different flow-measurement methods are described that determine either velocity measurements at a point or average velocity over a cross section. Usually, instruments inserted into the flow (insertion-type transducers) provide velocity information at a point. Other devices (or obstructions), such as orifices, nozzles, and weirs, alter the basic flow in such a way that changes in pressure can be related to the average flow rate.

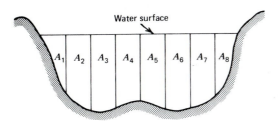

Figure 12.3 Measurement of mass flow rate in an open channel.

12.2 FLOW VELOCITY (INSERTION-TYPE TRANSDUCERS)

In this section, transducers designed to measure velocity at a point in the flow field are discussed. The types of transducers considered are pitot tubes, hot-wire and hot-film anemometers, drag-force transducers, turbine meters, and vortex-shedding devices.

12.2.1 Pitot Tube (Incompressible Flow)

The use of a pitot tube to measure velocity at a point in a fluid is illustrated schematically in Fig. 12.4. The velocity of the fluid at point O is measured by inserting the open-ended pitot tube just downstream from the point. As the fluid particles move from point O to point S (the stagnation point at the center of the pitot-tube opening), their velocity decreases to zero as a result of the increased pressure at point S. A second vertical section of pipe, called a piezometer tube, above point O measures the static pressure, and the pitot tube measures the total pressure at the stagnation point S. The dynamic pressure is the difference between the total pressure and the static pressure. For the streamline from point O to point S, the Bernoulli equation gives

$$\frac{p_o}{\gamma} + \frac{V_o^2}{2g} = \frac{p_s}{\gamma} + \frac{V_s^2}{2g} \tag{12.8}$$

where

V_o and V_s are velocities at points O and S
g is the local acceleration owing to gravity
γ is the weight per unit volume of the fluid
p_o and p_s are the static and stagnation pressures, respectively

Since $V_s = 0$, the dynamic pressure $p_d = (p_s - p_o)$ is

$$p_d = p_s - p_o = \frac{\gamma V_o^2}{2g} = \gamma h \tag{12.9}$$

where h is the dynamic head defined in Fig. 12.4. The velocity at point O, as obtained from Eq. 12.9, is

$$V_o = \sqrt{2g\left(\frac{p_s - p_o}{\gamma}\right)} = \sqrt{2gh} \tag{12.10}$$

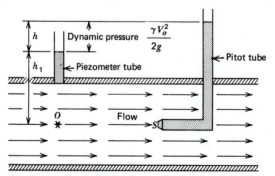

Figure 12.4 Velocity measurement with a pitot tube.

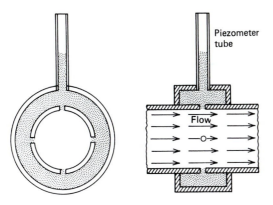

Figure 12.5 Piezometer ring for making static-pressure measurements.

Velocity measurements made with a pitot tube require accurate static-pressure measurements. Slight geometric errors in the pressure tap, such as a rounded corner or a machining burr, can lead to significant errors in the static-pressure measurement. To minimize such errors, static-pressure measurements are often made with a piezometer ring, as illustrated in Fig. 12.5. The use of multiple holes around the periphery of the tube minimizes the static-pressure errors.

Pitot-static tubes are compact, efficient, velocity-measuring instruments that combine static-pressure measurements and stagnation-pressure measurements into a single unit, as illustrated schematically in Fig. 12.6. The static pressure recorded by a pitot-static tube is usually lower than the true static pressure because of the increase in velocity of the fluid near the tube. However, this difference between the indicated and true static pressures can be accounted for by employing a calibration coefficient C_I in Eq. 12.10. Thus

$$V_o = C_I \sqrt{2g\left(\frac{p_s - p_o'}{\gamma}\right)} \tag{12.11}$$

where

C_I is an experimentally determined calibration constant for the tube
p_o' is the indicated static pressure

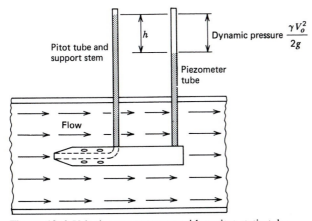

Figure 12.6 Velocity measurement with a pitot-static tube.

In the pitot-static tube designed by Prandtl, the static-pressure tap is located at the point where the drop in static pressure owing to the increase in velocity of the fluid near the tube is exactly equal to the increase in static pressure owing to fluid stagnation along the leading edge of the support stem. Thus, for the Prandtl pitot-static tube, the instrument coefficient C_I is unity. For other pitot-static tubes, the instrument coefficient C_I must be determined by calibration.

The stagnation pressure p_s is easy to measure accurately for most flow conditions. Four factors that affect the accuracy of the stagnation-pressure reading are: geometry of the pitot tube, misalignment (yaw) of the pitot tube with the flow direction, viscous effects at low Reynolds numbers, and transverse-pressure gradients in flows with high-velocity gradients.

Geometric and misalignment effects are illustrated in Fig. 12.7, where dynamic pressure measurement error $\mathscr{E}$ is plotted as a function of pitot-tube orientation (angle of attack of the pitot tube with respect to the flow direction). In Fig. 12.7,

$$\mathscr{E} = \frac{\Delta p_d}{p_d} = \frac{p_d - p_d'}{p_d}$$

where p_d' is the measured dynamic pressure. It is evident from Fig. 12.7 that a square-end pitot tube with a 15-deg internal bevel angle is capable of providing dynamic pressure data with errors of less than 1 percent, provided that the yaw angle is less than 25 deg. Thus, with reasonable care and with normal flow conditions, small errors in pitot-tube orientation do not produce serious errors in the dynamic pressure measurement.

A pitot-tube coefficient C_P is defined to provide a measure of viscous effects owing to flow around the pitot tube as

$$C_P = \frac{p_s' - p_o'}{p_d} = \frac{2g(p_s - p_s)}{\gamma V_o^2}$$

where p_s' is the measured stagnation pressure in the presence of viscous effects. Experimental results showing the pitot-tube coefficient C_p as a function of Reynolds number Re are presented in Fig. 12.8. From these results it can be seen that for Reynolds numbers greater than 1000, there is no effect of viscosity and C_P is equal to unity. In the range $50 \leq Re \leq 1000$, C_p is slightly less than unity (has a minimum

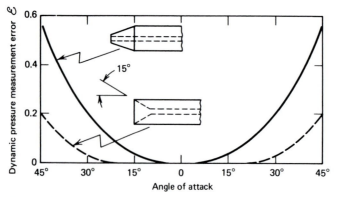

Figure 12.7 Dynamic pressure measurement error as a function of pitot-tube orientation. (See NACA TN 2331, April 1951, for complete details.)

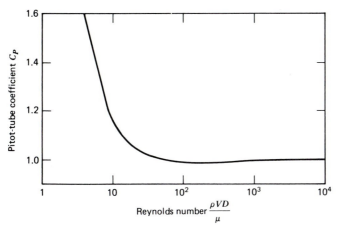

Figure 12.8 Pitot-tube coefficient C_P as a function of Reynolds number.

value of 0.99). For $Re < 50$, C_p is always greater than unity. For $Re < 1$, the coefficient C_p is given by the approximate expression $C_p \approx 5.6/Re$. These results clearly show that viscous effects are important only for Reynolds numbers less than 50.

When a pitot tube is placed in a flow field with a large velocity gradient, the conditions around the tip of the tube can be significantly altered with respect to a uniform flow field. This flow gradient causes the total pressure that is measured to be greater than the true pressure, because the effective center of the pitot tube is shifted from the geometric center toward a region of higher velocity. Since the amount of this shift is limited to approximately $0.2D$ for square-end pitot tubes (where D is the diameter of the pitot tube), very small diameter tubes are required to measure flows with high-velocity gradients.

In Figs. 12.4 through 12.6, the pressures p_s and p_o are illustrated with manometers. In practice, these pressures can be measured with manometers or with pressure gages, as discussed in Section 8.5. Differential pressure transducers are also commercially available for measuring the pressure difference $(p_s - p_o)$ directly. A popular differential pressure transducer utilizes a thin diaphragm as the elastic element. Electrical resistance strain gages used in conjunction with a Wheatstone bridge provide for continuous monitoring of the pressure difference $(p_s - p_o)$. The velocity V_o and mass flow rate $\dot{m}$ are determined by using Eqs. 12.7 and 12.10.

12.2.2 Pitot Tube (Compressible Flow)

In the previous discussion, the fluid was assumed to be incompressible. Measurements of compressible flow are much more complicated and require the use of the equation of motion for a compressible fluid, energy considerations, and a description of the flow process.

The equation of motion (Euler equation) for one-dimensional flow of a compressible ideal (nonviscous or frictionless) fluid can be expressed as

$$\frac{dp}{\rho} + V\,dV = 0 \tag{12.12}$$

where $\rho = \gamma/g$ is the density of the fluid. This equation is based on the assumption that compressible-flow measurements are usually concerned with light gases and with flows where changes of pressure and velocity dominate and changes of elevation are negligible. When there is no heat transfer and no work done by pumps or turbines in the flow of an ideal fluid, the flow is isentropic and the energy equation for steady flow along a streamline reduces to Eq. 12.12. Thus, the energy and Euler equations are identical for isentropic flow of an ideal fluid.

Along a streamline for a perfect gas,

$$\frac{p_1}{\rho_1^k} = \frac{p_2}{\rho_2^k} \tag{12.13}$$

where k is the *adiabatic exponent* (ratio of specific heat at constant pressure to specific heat at constant volume for gas) $k = c_p/c_v$. Integrating Eq. 12.12 with Eq. 12.13 gives the velocity–pressure relations:

$$\frac{V_2^2 - V_1^2}{2} = \frac{p_1}{\rho_1} \frac{k}{k-1} \left[1 - \left(\frac{p_2}{p_1} \right)^{(k-1)/k} \right]$$

$$= \frac{p_2}{\rho_2} \frac{k}{k-1} \left[\left(\frac{p_1}{p_2} \right)^{(k-1)/k} - 1 \right] \tag{12.14}$$

In gas dynamics, the velocity is usually expressed in terms of *Mach number:*

$$M = \frac{V}{c} \tag{12.15a}$$

where the sonic velocity c is given by the expression

$$c = \sqrt{\frac{k\,p}{\rho}} \tag{12.15b}$$

Substituting Eqs. 12.15 into Eq. 12.14 gives

$$\frac{V_2^2}{c_1^2} = M_1^2 + \frac{2}{k-1} \left[1 - \left(\frac{p_2}{p_1} \right)^{(k-1)/k} \right] \tag{12.16}$$

Applying Eq. 12.16 to a stagnation point in a compressible flow ($V_2 = 0$ and $p_2 = p_s$) together with the free-stream conditions ($p_1 = p_o$ and $M_1 = M_o$) yields

$$\frac{p_s}{p_o} = \left[1 + \frac{k-1}{2} M_o^2 \right]^{k/(k-1)} \tag{12.17}$$

This result indicates that measurements of stagnation pressure p_s and static pressure p_o provide sufficient data to determine the free-stream Mach number. Determining the velocity V_o, however, requires measuring a temperature in addition to the static and stagnation pressures. In practice, the stagnation temperature T_s is measured with a thermocouple placed near the stagnation point, as indicated in Fig. 11.36. Once T_s, p_s, and p_o are known, the velocity V_o can be determined by using the expression

$$\frac{V_o^2}{2} = c_p T_s \left[1 - \left(\frac{p_o}{p_s} \right)^{(k-1)/k} \right] \tag{12.18}$$

A comparison of velocities predicted by Eq. 12.10 with those given by Eq. 12.18 shows an agreement within 1 percent for pressure differences ($p_s - p_o$) less than 0.83 psi or Mach numbers less than 0.28. For larger pressure differences or Mach numbers, the agreement becomes less satisfactory and Eq. 12.18 should be used for velocity determinations. This comparison is based on air, where $c = 343$ m/s $= 1226$ ft/s at 68°F.

For supersonic flow ($M_o > 1$), a shock wave forms in front of the pitot tube upstream from the stagnation point. Velocity calculations for this complicated case are beyond the scope of this book.

12.2.3 Hot-Wire and Hot-Film Anemometers

Hot-wire and hot-film anemometers are devices used to measure velocity or velocity fluctuations (at frequencies up to 500 kHz) at a point in the flow field. Typical sensing elements (hot-wire and hot-film) and their supports are shown in Fig. 12.9. Hot-wire sensors are fabricated from platinum, platinum-coated tungsten, or a platinum-iridium alloy. Since the wire sensor is extremely fragile, hot-wire anemometers are usually used only for clean air or gas applications. Hot-film sensors, on the other hand, are extremely rugged; therefore, they can be used in both liquid and contaminated-gas environments. In the hot-film sensor, the high-purity platinum film is bonded to a high-strength, fused-quartz rod. After the platinum film is bonded to the rod, the thin film is protected by a thin coating of alumina if the sensor will be used in a gas or of quartz if the sensor will be used in a liquid. The alumina coatings have a high abrasion resistance and a high thermal conductivity. Quartz coatings are less porous and can be used in heavier layers for electrical insulation. Other hot-film anemometer shapes for special-purpose applications include conical, wedge, and hemispheric probes.

Hot-wire and hot-film anemometers measure velocity indirectly by relating power supplied to the sensor (rate of heat transfer from the sensor to the surrounding cooled fluid) to the velocity of the fluid in a direction normal to the sensor. Heat transfer from a heated wire placed in a flow field was studied by King (Reference 17), who noted that the heat transfer rate q is given by the expression

$$q = (A + B \sqrt{\rho V})(\theta_a - \theta_f) = i_a^2 R_a \qquad \textbf{(12.19)}$$

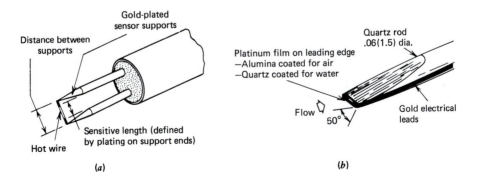

Distance between supports

Gold-plated sensor supports

Sensitive length (defined by plating on support ends)

Hot wire

(a)

Platinum film on leading edge
—Alumina coated for air
—Quartz coated for water

Quartz rod .06(1.5) dia.

Flow

50°

Gold electrical leads

(b)

where

    A and *B* are calibration constants
    θ_a and θ_f are the absolute temperature of the anemometer (hot wire or hot film)
        and the fluid, respectively
        i_a is the current passing through the wire (or film) sensor
        R_a is the resistance of the wire (or foil) sensor

The quantity $(\theta_a - \theta_f)$ is typically maintained at approximately 450°F (232°C) in air and 80°F (27°C) in water.

Materials used for hot-wire and hot-film sensors exhibit a change in resistance with temperature change. The resistance–temperature effect can be represented with sufficient accuracy for thermal-anemometer applications by the linear expression

$$R_a = R_r[1 + \gamma(T_a - T_r)] \tag{12.20}$$

where

    R_r is the resistance of the sensor at reference temperature T_r
    γ is the temperature coefficient of resistance of the wire or foil

From Eq. 12.19 it is evident that the fluid velocity V can be determined by measuring either the current i_a or the resistance R_a. In practice, the velocity is determined by using a hot-wire or hot-film anemometer as the active element in a Wheatstone bridge. With a constant-current bridge, the sensor current i_a is maintained as the sensor resistance R_a changes with the fluid flow to produce an output voltage v_o that is related to the velocity of the fluid. With a constant-temperature bridge, the sensor resistance R_a (and thus the sensor temperature T_a) is held at a constant value by varying the current passing through the sensor as the fluid velocity changes. In this circuit, the current i_a is used to provide a measure of the velocity. A description of each of these systems follows.

A. Constant-Current Bridge–Anemometer System

A constant-current Wheatstone bridge with a hot-wire (or hot-film) anemometer as the sensor is shown schematically in Fig. 12.10. In this circuit, resistances R_2 and R_3 are much larger than sensor resistance R_a; therefore, current i_a is essentially independent

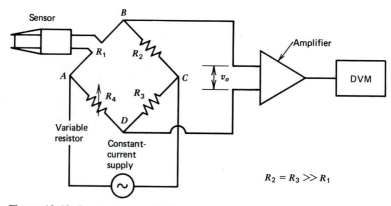

Figure 12.10 Constant-current bridge–anemometer system.

of changes in the sensor resistance R_a. Variable resistor R_4 is used to initially balance the bridge at zero velocity. Any flow past the sensor cools the hot wire (or film), decreases its resistance as indicated by Eq. 12.20, and unbalances the bridge. The unbalanced bridge produces an output voltage v_o, as given by Eq. 6.30, which is related to the fluid velocity V. Because the output voltage v_o from the bridge is small, it must be amplified before it is recorded.

An important application of hot-wire or hot-film anemometry, in addition to steady-flow velocity measurements, is the measurement of turbulence. The voltage output of the bridge (see Fig. 12.10) is a direct indication of velocity fluctuations (intensity of turbulence) in the flow. Thermal lag (the inability of the wire or film to transfer heat fast enough to the surrounding fluid) limits the resolution of these velocity fluctuations to about 300 Hz. Above a cutoff frequency f_c (dependent on sensor material, coating, and diameter), the wire or film acts as an integrator and attenuates the fluctuations, as shown in Fig. 12.11. The frequency range can be extended by passing the bridge output through a low-noise, high-gain amplifier. The high-level signal thus obtained can be differentiated for frequencies above the cutoff frequency to achieve a compensated frequency response that is flat up to approximately 120 kHz. The steady-flow and low-frequency components of the velocity (below 1 Hz) are attenuated as shown in Fig. 12.11. Since the cutoff frequency is different for each hot wire (or film), provision must be made in the frequency-compensation circuit to adjust the cutoff frequency.

Two outstanding features of a constant-current anemometer system are its low noise level and its excellent sensitivity. Turbulence levels less than 0.005 percent of the mean velocity can be resolved in the 10-kHz frequency band. The range of a typical anemometer is from 1 to 120,000 Hz. Also, the constant-current anemometer is very sensitive to small changes in velocity at low velocities, provided that the frequency-compensation circuit is not employed.

The constant-current anemometer system has two disadvantages, which have provided incentive for the development of the constant-temperature anemometer system. First, the frequency response of the constant-current anemometer is separated into two bands: the uncompensated low-frequency band and the compensated high-frequency band. Second, the compensated output is distorted when small high-frequency fluctuations are measured in the presence of large-amplitude, low-frequency oscillations.

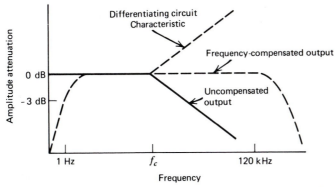

Figure 12.11 Frequency response of a constant-current bridge–anemometer system.

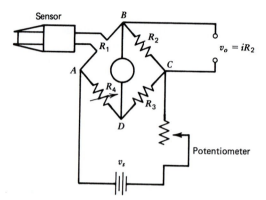

Figure 12.12 Constant-temperature bridge–anemometer system.

B. Constant-Temperature Bridge–Anemometer System

A constant-temperature hot-wire or hot-film anemometer, ideal for steady-flow measurements, is shown schematically in Fig. 12.12. The hot-wire (or hot-film) sensor is used as the active element in a Wheatstone bridge that is initially balanced under no-flow conditions with the variable resistor R_4. As flow past the sensor cools the hot wire (or film), its resistance R_a decreases and the bridge becomes unbalanced. Balance is restored by adjusting the potentiometer to increase the input voltage to the bridge. The increase in the supply voltage increases the current flowing through the sensor and increases both sensor temperature T_a and sensor resistance R_a until they equal their zero-flow values. Under conditions of constant sensor temperature and resistance, Eq. 12.19 reduces to

$$V = C_0 \left[\left(\frac{i}{i_0} \right)^2 - 1 \right]^2 \tag{12.21}$$

where

  C_0 is a calibration constant for the particular hot-wire (or hot-film) probe
  i_0 is the current at zero velocity that gives the desired sensor temperature
  i is the sensor current at velocity V

The operator must determine the constants C_0 and i_0 for the specific application. With precision bridge resistors, a sensitive galvanometer, and constant fluid temperature, accuracy of 1 percent in steady-flow velocity measurements is achieved with constant-temperature bridge–anemometer systems. Significant changes in fluid temperature, however, require rebalance and recalibration of the system, since the quantity $(\theta_a - \theta_f)$ in Eq. 12.19 must not change if Eq. 12.21 is to be valid.

The current i or i_0 passing through the sensor is easily measured by recording the voltage drop across resistance R_2 in the Wheatstone bridge shown in Fig. 12.12. Since the sensor current is also proportional to the input voltage, input voltage is another measure of sensor current. A typical calibration curve showing sensor current i as a function of velocity V is shown in Fig. 12.13.

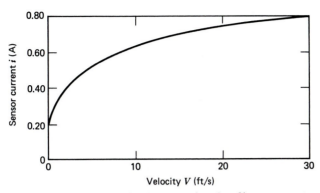

Figure 12.13 Typical calibration curve for a hot-film anemometer in water.

A more advanced constant-temperature bridge–anemometer system is illustrated schematically in Fig. 12.14. Again, the hot-wire (or hot-film) sensor is used as the active element in a Wheatstone bridge that is initially balanced under zero-flow conditions. As flow past the sensor cools the hot wire (or film), its resistance decreases and unbalances the bridge. A high-gain differential amplifier is used to sense the resulting output signal. The amplifier changes the bridge voltage, increasing the current flowing through the sensor. The increased current flow restores sensor temperature T_a and sensor resistance R_a to zero-flow values and rebalances the bridge. This constant-temperature bridge–anemometer system automatically maintains a balance and corrects

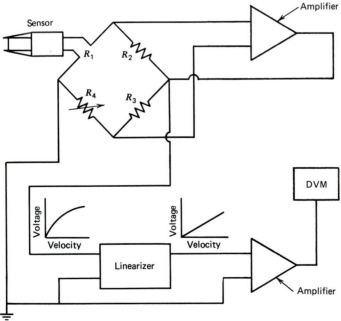

Figure 12.14 Advanced constant-temperature bridge-anemometer system.

for the thermal lag of the wire (or the film). In still more advanced systems, the resistor R_4 is mounted in the probe, where it is used as an RTD to sense and correct for any change in fluid temperature.

A typical constant-temperature anemometer system exhibits a frequency response from dc to approximately 500 kHz and a resolution of approximately 2 percent of the mean velocity in the 10-kHz band. The upper limiting frequency of this constant-temperature system is controlled by sensor size, amplifier gain, and other features of the components of the system.

An important feature of the constant-temperature anemometer system is the provision for linearized output, which permits accurate measurements of large-amplitude, low-frequency velocity fluctuations. In addition, this system is preferred for studies involving steady-flow velocity profiles or flow reversals, where it is impossible to distinguish the direction of flow. However, the constant-current anemometer system is preferred for low-turbulence and high-accuracy measurements. These general distinctions are gradually disappearing as probe design and amplifier technology progress to overcome specific limitations. Eventually the constant-temperature system will be used exclusively because of its more desirable characteristics. The nonlinear problems are accommodated by using specially designed microprocessors for data reduction.

When hot-wire or hot-film sensors are used in liquids, several problems arise. First, liquids often carry dirt particles, lint, or organic matter. These materials quickly coat the hot wire or film and cause significant reductions in heat transfer. Second, the presence of a current-carrying wire in a conducting medium causes electrolysis of metals that shunt the hot wire, thus producing spurious changes in sensitivity. Third, the presence of the hot wire may cause the formation of bubbles, which significantly reduce the heat transfer. Bubbles arise from entrained air or gas in the liquid, from electrolysis, or from boiling of the liquid. Successful use of anemometers in liquids usually requires low wire temperatures, coatings on the hot wires, lower operating voltages, and degasification of the liquid.

12.2.4 Drag-Force Velocity Transducers

Another type of velocity transducer operates on the principle that the drag force F_D on a body in a uniform flow is related to the fluid velocity V. The relationship is quadratic and depends on the fluid density ρ and frontal area A of the body normal to the flow direction.

$$F_D = C_D \frac{\rho V^2 A}{2} \qquad (12.22)$$

where C_D is a nondimensional number known as the drag coefficient.

The drag coefficient depends on Reynolds number Re and the shape of the body. Drag coefficients for a sphere and for a circular disk as a function of Re are shown in Fig. 12.15. Because the drag coefficient for the circular disk is constant over a wide range of velocities ($C_D \approx 1.05$ for $Re \geq 3000$) and because the magnitude of the drag coefficient for the disk is much more than that for the sphere, the circular disk is the preferred body used in a drag-force transducer.

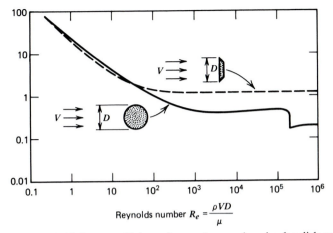

Figure 12.15 Drag coefficients for a sphere and a circular disk as a function of Reynolds number.

A. Rotameter

The *rotameter* is a popular flow-measurement device based on drag principles. The rotameter, shown schematically in Fig. 12.16, consists of a transparent tapered tube and a solid float (bob) that is free to move vertically in the tube. At any flow rate within the range of the meter, fluid entering the bottom of the tube lifts the float (thereby increasing the area between the float and the tube) until the drag and buoyancy forces are balanced by the weight of the float. This balance condition can be expressed by the equation

$$c_D \frac{\rho_f V^2 A_b}{2} + \rho_f \mathcal{V}_b g = \rho_b \mathcal{V}_b g \qquad (12.23)$$

where

       A_b is the frontal area of the float
       $\mathcal{V}_b$ is the volume of the float
       ρ_b is the density of the float
       V is the mean velocity of the fluid in the annular space between the float and the tube

The first term in Eq. 12.23 is the drag, the second term is the buoyancy force, and the third term is the weight of the float. The annular area A for the flow is expressed as

$$A = \frac{\pi}{4}[(D + ay)^2 - d^2]$$

where

    D is the inside diameter at the bottom of the tube
    y is a position coordinate with an origin at the bottom of the tube
    a is a constant that describes the taper of the tube
    d is the diameter of the float

The tubes are constructed so that $D^2 + a^2y^2 \approx d^2$ so that the annular area is

$$A \approx \frac{\pi}{2}Day \qquad (12.24)$$

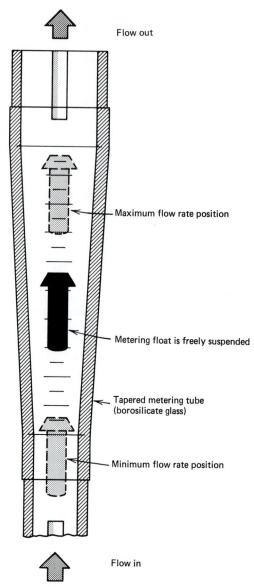

Figure 12.16 Schematic illustration of a glass-tube industrial rotameter-type flow meter. (Courtesy of Fischer & Porter.)

The mass flow rate $\dot{m}$ is obtained from Eqs. 12.23 and 12.24 as

$$
\begin{aligned}
\dot{m} &= \rho_f A V \\
&= \frac{\pi}{2} D a y \sqrt{\frac{2 g \mathcal{V}_b}{C_D A_b}(\rho_b - \rho_f)\rho_f} \\
&= K \sqrt{(\rho_b - \rho_f)\rho_f}\, y
\end{aligned}
\tag{12.25a}
$$

where $K = \sqrt{2\pi D a}\sqrt{g \mathcal{V}_b / C_D A_b}$ is an instrument constant depending on the geometry of the tube and float. The drag coefficient can be made nearly independent of viscosity by using sharp edges on the float, and the influence of fluid density is eliminated by selecting $\rho_b = 2\rho_f$. Equation 12.25a then reduces to

$$\dot{m} = \frac{K\rho_b}{2}y \qquad\qquad (12.25b)$$

The flow rate is indicated by the position of the float, which can be measured on a graduated scale or detected magnetically and transmitted to a remote location for recording.

12.2.5 Current Meters

Current meters are mechanical devices that are widely used to measure water velocities in open rivers, channels, and streams. The rotational speed of the device is proportional to fluid velocity. A direct-reading, cup-type current meter with sensing unit, cable suspension, torpedo-shaped lead weight, and alignment fins is shown in Fig. 12.17. As the cup wheel rotates, a magnetically activated reed switch in the sensing unit produces a train of electrical pulses at a frequency proportional to the speed of the cup wheel. A circuit in the indicating unit, which incorporates a battery power supply and a pulse-rate integrator, converts the train of pulses to a direct display of the fluid velocity. Typically, the range of such instruments is 0 to 25 ft/s. Linearity of the unit shown in Fig. 12.17 is within ±5 percent over the full range of the unit. Error caused by temperature change is approximately 0.05 percent per °F change from 75°F.

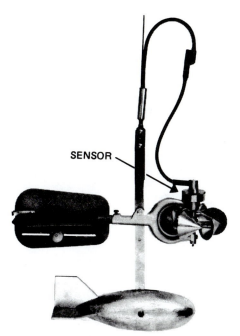

Figure 12.17 Cup-type current meter with sensing unit, cable suspension, torpedo-shaped lead weight, and alignment fins. (Courtesy of Teledyne Gurley.)

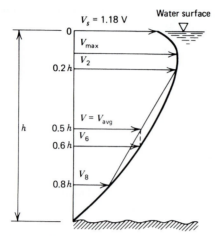

Figure 12.18 The USGS typical velocity profile for a large river.

The U.S. Geological Survey (USGS) has established a typical velocity profile (by using data from thousands of measurements) for use in establishing flow rates (discharge rates) in large rivers, canals, and streams. This profile, shown in Fig. 12.18, indicates that the velocity distribution in a typical large stream is parabolic, with the maximum velocity occurring some distance (from 0.05 to $0.25h$, where h is the depth of the stream) below the free surface. The free-surface velocity is typically 1.18 times the average velocity. The USGS profile indicates that the mean velocity often occurs at $0.6h$. An accurate estimate of the mean velocity can usually be obtained by averaging the velocities V_2 and V_8 measured at points $0.2h$ and $0.8h$, as indicated in Fig. 12.18.

The flow rate in a stream is measured by using the procedure illustrated schematically in Fig. 12.19. First, a segment of the river with a fairly regular cross section is selected as the site for the measurement. The cross section of the river is then divided into vertical strips, as shown in Fig. 12.19. The mean velocity along each vertical

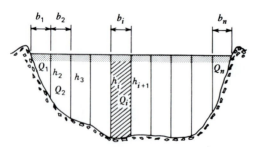

Figure 12.19 Flow-rate measurement procedure for a large river or stream.

line is determined by averaging velocity measurements made at points $0.2h$ and $0.8h$ below the free surface. Thus,

$$v_i = \frac{V_{2i} + V_{8i}}{2} \qquad (12.26)$$

The volume flow rate Q_i in a vertical strip is calculated by using the height of the strip, the width of the strip, and the mean velocity for the strip (averaged from the two vertical lines bounding the strip). Thus,

$$Q_i = b_i \left(\frac{h_i + h_{i+1}}{2}\right)\left(\frac{V_i + V_{i+1}}{2}\right) \qquad (12.27a)$$

The total volume flow rate for the stream is the sum of the flow rates for all of the strips:

$$Q = \sum_{i=1}^{N} Q_i \qquad (12.27b)$$

It is clear that many flow-rate data points are needed in order to establish volume flow rates for rivers and streams with sufficient accuracy.

12.2.6 Turbine Flow meters

A *turbine flow meter* is basically a miniature propeller supported along the center-line of a pipe. The propeller of the axial turbine is freely suspended (see Fig. 12.20)

Figure 12.20 Cutaway view of a turbine flow meter. (Courtesy of Flow Technology, Inc.)

and rotates with the flow of fluid (either liquid or gas) through the flow meter. The rotational speed of the turbine is proportional to the velocity of the fluid. Since the flow passage is fixed, the rotational speed is also an accurate representation of the volume of fluid flowing through the flow meter.

The only physical connection between the turbine and its housing is the turbine bearings. The rotation of the turbine is sensed by a magnetic coil in the flow meter body that responds to the passage of each turbine blade. The output from the coil is a train of voltage pulses with a frequency proportional to the volume flow rate. The pulses are transmitted to an appropriate data-processing system near the flow meter, where they are amplified, counted, and interfaced with a microprocessor to measure or control the fluid flow.

Flowmeters have been developed with outstanding levels of accuracy, linearity, durability, and reliability. They are commercially available to measure fluid flow within the temperature range from $-430°F$ to $750°F$ ($-257°C$ to $400°C$). Accuracy within ± 0.05 percent in liquids and ± 0.5 percent in gases is easily obtained at flow rates from 0.03 to 20,000 gal/min. Turbine flow meters are currently used to monitor and control critical flow rates in a number of industrial processes.

12.2.7 Vortex-Shedding Transducers

When a circular cylinder is placed in a uniform flow with its axis perpendicular to the direction of the flow (as shown in Fig. 12.21), vortices shed regularly (at a frequency f_s) from alternate sides of the cylinder. The pattern is known as a Karman vortex street. The shedding frequency f_s, the diameter of the cylinder D, and the flow velocity V are related by a dimensionless number known as the Strouhal number S_N, defined as

$$S_N = \frac{f_s D}{V} \tag{12.28}$$

Experimental studies indicate that the Strouhal number is relatively constant ($0.20 \leq S_N \leq 0.21$) over a very wide range of Reynolds numbers (300 to 150,000). Since precise frequency measurements are easy to make, Eq. 12.28 provides a means for accurately measuring flow velocities from

$$V = 5f_s D \tag{12.29}$$

This result indicates that small-diameter cylinders should be used in designing the sensor, since they give a higher frequency for a given flow velocity. The natural

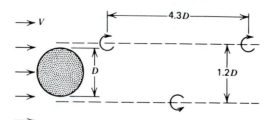

Figure 12.21 Schematic illustration of a Karman vortex street.

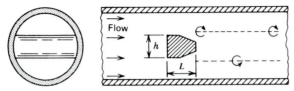

Figure 12.22 Use of a bluff body as a vortex-shedding flow meter.

frequency of the cylinder must be considerably higher than the vortex-shedding frequency $(f_n/f_s > 3)$; otherwise, nonlinear fluid-structure interaction occurs, producing large-amplitude vibrations that can destroy the device.

One design of vortex-shedding flow meter uses a triangular wedge (bluff body) in a pipe, as shown in Fig. 12.22. In order to produce strong and consistent vortex shedding, the wedge must extend across the full width of the pipe. Also, the height h of the bluff body must be approximately one-third the diameter of the pipe, and its length in the direction of flow should be approximately 1.3 times the height. The salient edges at the face of the bluff body fix the location of the vortex shedding, thus giving a consistent Strouhal number of 0.88 ± 0.01 over the range of Reynolds numbers from 10,000 to 1,000,000. The strut that supports the shedding body is fitted with strain gages or with piezoelectric sensors. The sensors are mounted to give two output signals that are out of phase by 180 deg. This arrangement gives a good signal-to-noise ratio, since the vortex-shedding signals add and the common noise components cancel. Calibration studies have shown that air and water have the same calibration constant over a range of pipe Reynolds numbers $(Re)_D$ from 10,000 to 5,000,000. These results support the contention that calibration values for these probes are independent of pressure, temperature, and state of the fluid (liquid or gas). Although Fig. 12.22 implies that such devices can be used only in a closed pipe, they can also be used in a free-flow field by using a short length of pipe as a shroud with the bluff body inside. The system, consisting of the bluff body and short pipe, can be calibrated and suspended in the flow in much the same manner as the current meter.

12.3 FLOW RATES IN CLOSED SYSTEMS BY PRESSURE-VARIATION MEASUREMENTS

Several devices that measure average velocity or flow rate either by insertion of a constriction in the stream tube or by changing the direction of the flow are considered in this section. The devices described are the venturi meter, the flow nozzle, the orifice meter, and the elbow meter. The operation of each of these devices is based on the fact that a change in cross-sectional area of a stream tube causes a corresponding change in velocity and pressure of the fluid within that tube.

The Bernoulli equation applied to an ideal incompressible fluid $(\rho_1 = \rho_2)$ flowing through the tube shown in Fig. 12.23 gives

$$\frac{p_1}{\gamma} + \frac{V_1^2}{2g} + z_1 = \frac{p_2}{\gamma} + \frac{V_2^2}{2g} + z_2 \qquad (12.30)$$

The velocity V_1 can be eliminated from Eq. 12.30 by using the continuity equation, which requires that

$$Q = A_1 V_1 = A_2 V_2 \qquad (12.31)$$

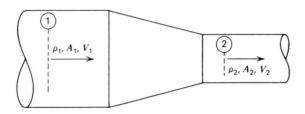

Figure 12.23 Illustration of conservation of mass at two locations in a closed system.

Thus, the ideal volume flow rate Q_1 (for an ideal frictionless fluid) is

$$Q_i = \frac{A_2}{\sqrt{1 - (A_2/A_1)^2}} \sqrt{2g\left(\frac{p_1}{\gamma} + z_1 - \frac{p_2}{\gamma} - z_2\right)} \tag{12.32}$$

For real fluid flow and the same head loss $[(p_1 - p_2)/\gamma + (z_1 - z_2)]$, the flow rate is less than that predicted by Eq. 12.32 because of friction that acts at the walls between the two pressure-measuring points. This energy loss is usually accounted for by introducing an experimentally determined coefficient C_V (coefficient of velocity) into Eq. 12.32. Actual flow rate Q_a is then expressed as

$$Q_a = \frac{C_V A_2}{\sqrt{1 - (A_2/A_1)^2}} \sqrt{2g\left(\frac{p_1}{\gamma} + z_1 - \frac{p_2}{\gamma} - z_2\right)} \tag{12.33}$$

The head-loss difference term $[(p_1 - p_2)/\gamma + (z_1 - z_2)]$ in Eq. 12.33 can be measured with any of the standard pressure gages, such as a differential manometer or a differential pressure transducer. Specific details for different constriction devices are presented in the following subsections.

12.3.1 Venturi Meter

A typical *venturi meter* consists of a cylindrical inlet section, a smooth entrance cone (acceleration cone) with an angle of approximately 21 deg, a short cylindrical throat section, and a diffuser cone (deceleration cone) with an angle between 5 and 15 deg. Recommended proportions and pressure-tap locations for a venturi meter, as specified by the American Society of Mechanical Engineers (ASME) code, are shown in Fig. 12.24. Small diffuser angles tend to minimize the head loss from pipe friction, flow separation, and turbulence. In order for the venturi meter to function properly, the flow must be fully developed as it enters the inlet pressure ring area (1). This developed flow is accomplished by installing the meter downstream from a section of straight and uniform pipe that is about 50 pipe diameters long. Straightening vanes may also be installed upstream of the venturi meter to reduce any rotational motion in the fluid.

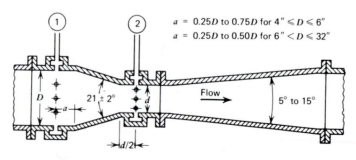

$a = 0.25D$ to $0.75D$ for $4'' \leqslant D \leqslant 6''$
$a = 0.25D$ to $0.50D$ for $6'' < D \leqslant 32''$

Figure 12.24 Recommended proportions and pressure-tap locations for a venturi meter.

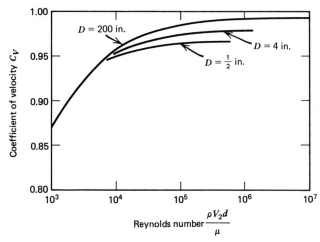

Figure 12.25 Coefficient of velocity C_V as a function of Reynolds number for venturi meters with a diameter ratio $D/d = 2$.

The pressures at the inlet section (1) and at the throat section (2) of the meter are measured with piezometer rings, as indicated in Fig. 12.24.

The coefficient of velocity C_V for different sizes of venturi meters with a pipe-to-throat diameter ratio $D/d = 2$ is shown as a function of Reynolds number at the throat ($Re = \rho V_2 d / \mu$) in Fig. 12.25. These data indicate that C_V ranges from 0.97 to 0.99 over a wide range of sizes for $Re > 10^5$. For $Re < 10^4$, C_V decreases sharply with decreasing Re and becomes a very important correction. Experimental evidence at other diameter ratios indicates that C_V decreases with increasing D/d.

12.3.2 Flow Nozzle

A *flow nozzle* is essentially a venturi meter with the diffuser cone removed. Since the diffuser cone is used to minimize head loss caused by the presence of the meter in the system, larger head losses will occur in flow nozzles than in venturi meters. The flow nozzle is preferred to the venturi meter in many applications because of its lower initial cost and because it can be easily installed between two flanges in any piping system. The geometry recommended by ASME for a long-radius flow nozzle is shown in Fig. 12.26.

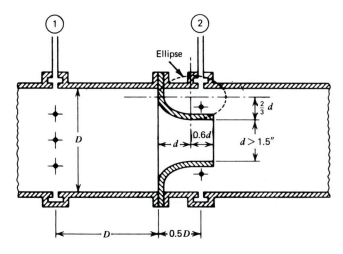

Figure 12.26 Geometry recommended by the ASME for a long-radius flow nozzle.

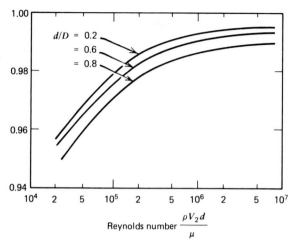

Figure 12.27 Velocity coefficients C_V for ASME long-radius flow nozzles.

The upstream pressure is measured with a piezometer ring located one pipe diameter upstream from the inlet face of the nozzle, and the throat pressure is measured with a piezometer ring located one-half pipe diameter downstream from the inlet face of the nozzle. Errors associated with the measurement of throat pressure at location (2) are corrected for by using either a velocity coefficient C_V for the flow nozzle or a discharge coefficient C_D for the meter, where $C_D = Q_a/Q_i$.

Extensive research on flow nozzles has produced a significant body of reliable data on flow-nozzle installation procedures and velocity coefficients. Flow coefficients for the ASME long-radius flow nozzle are shown in Fig. 12.27.

12.3.3 Orifice Meter

An *orifice meter* is a plate with a sharp-edged circular hole (see Fig. 12.28) that is inserted between two flanges of a piping system. Its purpose is to determine the flow rate from pressure-difference measurements across the orifice. The pattern developed by flow through a sharp-edged orifice plate is also shown in Fig. 12.28. This flow pattern indicates that the streamlines tend to converge a short distance downstream from the plane of the orifice; therefore, the minimum flow area is smaller than the

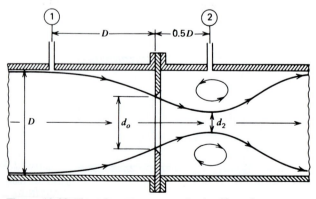

Figure 12.28 Flow through a sharp-edged orifice plate.

area of the opening in the orifice plate. This minimum flow area is known as the *vena contracta*. The area at the vena contracta is accounted for in Eq. 12.33 by defining a contraction coefficient C_C as

$$A_2 = C_C A_0 \tag{12.34}$$

where A_0 is the area of the hole in the orifice plate. When Eq. 12.34 is substituted into Eq. 12.33, the flow rate through the orifice becomes

$$Q_a = C A_0 \sqrt{2g \left(\frac{p_1}{\gamma} + z_1 - \frac{p_2}{\gamma} - z_2 \right)} \tag{12.35a}$$

where the orifice coefficient C is defined as

$$C = \frac{C_V C_C}{\sqrt{1 - (C_C A_0 / A_1)^2}} \tag{12.35b}$$

The value of the orifice coefficient C depends on the velocity coefficient C_V, the contraction coefficient C_C, and the area ratio A_0/A_1 of the installation. The orifice coefficient C is also affected by the location of the pressure taps. Ideally, the pressure p_2 should be measured at the vena contracta; however, this is difficult to implement since for $Re > 10^5$, the location of the vena contracta changes with Reynolds number and area ratio. As a result, pressure taps are usually placed one pipe diameter upstream and one-half pipe diameter downstream from the inlet face of the orifice plate. Variations of orifice coefficient C as a function of the ratio of orifice diameter to pipe diameter are shown in Fig. 12.29 for different Reynolds numbers $(Re)_D$ for these locations of the pressure taps. For $(Re)_D > 100,000$, the value of C remains essentially constant.

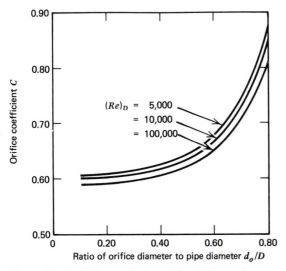

Figure 12.29 Orifice coefficient C for different diameter ratios and Reynolds numbers.

12.3.4 Elbow Meter

The venturi meter, flow nozzle, and orifice meter, which are widely used for measuring flow rates in pipes, all contribute to the energy losses in the system. Elbow meters, however, do not introduce additional energy losses, since they replace an elbow that is already being used in the system to change the direction of flow. The principle of operation of an elbow meter is illustrated in Fig. 12.30, which shows pressure taps located on the inside and outside radii of the elbow. Experimental studies indicate that the flow velocity is related to the pressures and elevations at the taps by

$$C_K \frac{V^2}{2g} = \frac{p_o}{\gamma} + z_o - \frac{p_i}{\gamma} - z_i \tag{12.36}$$

where C_K is a coefficient that depends on the size and shape of the elbow. Nominal values of C_K range from 1.3 to 3.2. The volume flow rate Q is obtained from pressure-difference measurements from the elbow meter by substituting Eq. 12.37 into Eq. 12.31 to give

$$Q = CA \sqrt{2g \left(\frac{p_o}{\gamma} + z_o - \frac{p_i}{\gamma} - z_i \right)} \tag{12.37}$$

where $C = 1/\sqrt{C_k}$ is the elbow-meter coefficient with values ranging from 0.56 to 0.88.

The primary advantage of the elbow meter is energy savings. The primary disadvantage is that each meter must be calibrated in place or in a calibration facility; however, the low operating cost usually justifies the calibration cost. The elbow meter, like the other obstruction meters, requires a minimum of 10 to 30 pipe diameters of unobstructed upstream flow (to reduce large-scale turbulence and swirl) for satisfactory operation and accurate flow measurements. Otherwise, flow straighteners are required to stabilize the flow prior to entry into the flow-metering device.

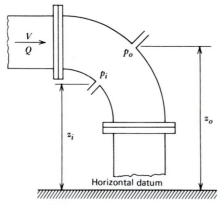

Figure 12.30 Location of pressure taps on an elbow meter.

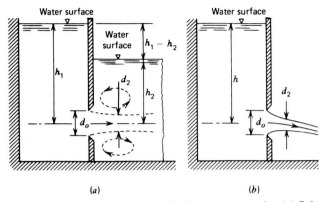

Figure 12.31 Flow through an orifice between reservoirs. (*a*) Submerged orifice. (*b*) Free discharging orifice.

12.4 FLOW RATES IN PARTIALLY CLOSED SYSTEMS

Many variations of the orifice meter are in use. For example, the submerged orifice (Fig. 12.31*a*) and the free discharging orifice (Fig. 12.31*b*) are used to control fluid flow from one large reservoir to another. In both cases, the area of the orifice is so small compared with the face area of the reservoir that ratio $A_0/A_1 \approx 0$; therefore, Eq. 12.35 simplifies to

$$Q_a = C A_0 \sqrt{2g(h_1 - h_2)} \tag{12.38}$$

where $C = C_V C_C$ is the orifice or discharge coefficient, and h_1 and h_2 are the static heads. Orifice coefficient C depends on the design of the orifice and its entrance and exit configuration, as shown in Table 12.1.

The coefficients in Table 12.1 are nominal values for large-diameter ($d > 1$ in. or 25 mm) orifices operating under static heads ($h_1 - h_2$) in excess of 50 in. (1.25 m) of water. Above these limits of diameter and static head, the coefficients are essentially constant. For smaller diameter orifices and lower static heads, both viscous effects

Table 12.1 Orifices and Their Nominal Coefficients

	Sharp edged	Rounded	Short tube[a]	Borda
C	0.61	0.98	0.80	0.51
C_C	0.62	1.00	1.00	0.52
C_V	0.98	0.98	0.80	0.98

[a] $l \approx 2.5d$

and surface-tension effects begin to influence the discharge coefficient and the results in Table 12.1 are not valid.

12.5 FLOW RATES IN OPEN CHANNELS FROM PRESSURE MEASUREMENTS

Accurate measurement of flow rates in open channels is important for navigation, flood control, irrigation, and so on. The two broad classes of devices used for this type of measurement and control are the sluice gate and the weir. Both devices require placement of an obstruction (dam) in the flow channel to alter the flow. Pressures are usually obtained by measuring free-surface elevations to give the heads used in computing flow rates.

12.5.1 Sluice Gate

The *sluice gate* is an open-channel version of the orifice meter. As shown in Fig. 12.32, the flow through the gate exhibits jet contraction on the top surface, which produces a reduced area of flow or vena contracta just downstream from the gate. If it is assumed that there are no energy losses (ideal fluid) and that the pressure in the vena contracta is hydrostatic, the Bernoulli equation with respect to a reference at the floor of the channel gives

$$y_1 + \frac{V_1^2}{2g} = y_2 + \frac{V_2^2}{2g} \tag{12.39}$$

The velocity V_1 is eliminated from Eq. 12.39, friction losses are accounted for by introducing a velocity coefficient C_V. With these substitutions, the actual flow rate Q_a is

$$Q_a = \frac{C_V C_C A}{\sqrt{1 - (y_2/y_1)^2}} \sqrt{2g(y_1 - y_2)} \tag{12.40}$$

where

    A is the area of the sluice-gate opening

    C_C is a contraction coefficient that accounts for the reduced area at the vena contracta

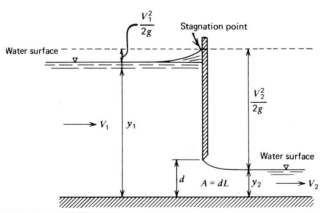

Figure 12.32 Flow through a sluice gate.

Note that the flow rate or discharge through the sluice gate depends on the coefficient of velocity C_V, the contraction coefficient C_C, the depth ratio y_2/y_1, and the difference in depths $(y_1 - y_2)$. Frequently, in practice, all of these effects are combined into a single discharge coefficient C_D, so that Eq. 12.40 is written as

$$Q_a = C_D A \sqrt{2gy_1} \tag{12.41}$$

Values for the discharge coefficient C_D usually range between 0.55 and 0.60, provided that free flow is maintained downstream from the gate. When flow conditions downstream produce submerged flow, the value of the discharge coefficient is reduced significantly.

With a constant upstream head y_1, it is clear from Eq. 12.41 that the flow rate is controlled by the area of the sluice-gate opening. Since the width L of the gate is fixed, the height of the gate opening controls the flow rate. The position of the gate is easily monitored with any displacement transducer, and the flow rate is easily measured and controlled.

12.5.2 Weirs

A *weir* is an obstruction in an open channel over which fluid flows. The flow rate or discharge over a weir is a function of the weir geometry and of the *weir head* H (vertical distance between the weir crest and the liquid surface in the undisturbed region upstream from the weir). The discharge equation for a weir is derived by considering a sharp-crested rectangular weir, illustrated in Fig. 12.33, and applying the Bernoulli equation to a typical streamline. Using the weir crest as the reference,

$$H + \frac{V_1^2}{2g} = (H - h) + \frac{V_2^2}{2g} \tag{12.42}$$

where h is a distance below the free surface where V_2 exists in the weir plane. Solving Eq. 12.42 for V_2 yields

$$V_2 = \sqrt{2g\left(h + \frac{V_1^2}{2g}\right)} \tag{12.43}$$

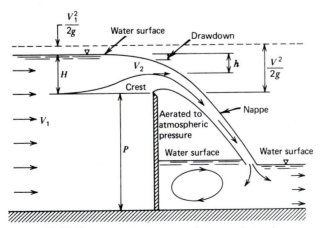

Figure 12.33 Flow over a sharp-crested rectangular weir.

When V_1 is small (which is usually the case), the velocity distribution in the flow plane above the crest of the weir is

$$V_2 = \sqrt{2gh} \tag{12.44}$$

The ideal flow rate Q over the weir is obtained by integrating the flow ($V_2\,dA$) over the area of the flow plane above the weir to obtain

$$Q = \int_a V_2\,dA = \int_0^H \sqrt{2gh}L\,dh$$

$$= \frac{2}{3}\sqrt{2g}LH^{3/2} \tag{12.45}$$

where L is the width of the weir.

The actual flow rate Q_a over a weir is less than the ideal flow rate Q owing to vertical drawdown contraction from the top, friction losses in the flow, and nonhorizontal velocities in the flow plane above the weir. These effects are accounted for by introducing a weir discharge coefficient C_D

$$Q_a = C_D Q = \frac{2}{3}\sqrt{2g}C_D LH^{3/2} \tag{12.46}$$

Values of C_D range from 0.62 to 0.75 as the ratio of weir head H to weir height P ranges from 0.1 to 2.0. The weir must be sharp for these coefficients to be valid. When the rectangular weir does not extend across the full width of the channel, additional contractions occur because of the ends so that the effective width of the weir is $(L - 0.1nH)$, where n is the number of end contractions. Corrosion and algae often cause the weir to appear rough and rounded. This rounding produces an increase in the weir coefficient resulting from a reduction in the edge contraction.

When flow rates are small, a triangular (V-notch) weir, illustrated in Fig. 12.34, is often used. This type of weir exhibits a higher degree of accuracy over a wider range of flow rates than does the rectangular weir. The V-notch weir also has the advantage that the average width of the flow section increases as the head increases. The discharge equation for the triangular weir is derived in the same manner as the equation for the rectangular weir. The results are

$$Q_a = \frac{8}{15}\sqrt{2g}\tan(\theta/2)\,C_D H^{5/2} \tag{12.47}$$

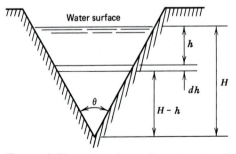

Figure 12.34 A triangular or V-notch weir.

Triangular weirs with included angles θ between 45 and 90 deg have discharge coefficients C_D ranging from 0.58 to 0.60, provided that the head H is in excess of 5 in. of water.

The results indicate that the flow rate depends on the head H when weirs are being used as the measuring device. The head H can be measured manually with a point scale or with a float-activated displacement transducer that serves as the sensor for a data-recording system.

12.6 COMPRESSIBLE FLOW EFFECTS IN CLOSED SYSTEMS

When a gas flows through a gradual contraction in a piping system or through a venturi type of flow meter, compressibility effects occur that must be considered at Mach numbers greater than 0.3 if accurate measurements are to be made. By using the energy equation, the equation of state for a perfect gas, and the continuity relationship for one-dimensional flow, and assuming the process to be isentropic, it can be shown that the mass flow rate $\dot{m}_c$ through a venturi-type contraction is

$$\dot{m}_c = \frac{A_2}{\sqrt{1 - (p_2/p_1)^{2/k}(A_2/A_1)^2}} \sqrt{\frac{2k}{k-1} p_1 \rho_1 \left[\left(\frac{p_2}{p_1} \right)^{2/k} - \left(\frac{p_2}{p_1} \right)^{(k+1)/k} \right]} \quad \textbf{(12.48)}$$

The corresponding relation (Bernoulli's equation) for an incompressible flow is

$$\dot{m}_B = \frac{A_2 \rho_1}{\sqrt{1 - (A_2/A_1)^2}} \sqrt{2g \left(\frac{p_1 - p_2}{\gamma_1} \right)} \quad \textbf{(12.49)}$$

A comparison of Eqs. 12.48 and 12.49 indicates that an expansion factor C_E can be incorporated into Eq. 12.49 to account for the differences between compressible and incompressible flow. Energy losses are accounted for by introducing a velocity coefficient C_V, in the same manner as in Eq. 12.33, to correlate ideal and actual flow rates. With the introduction of these coefficients, the expression for mass flow rate $\dot{m}$ for both compressible and incompressible flow becomes

$$\dot{m} = \frac{C_V C_E A_2 \rho_1}{\sqrt{1 - (A_2/A_1)^2}} \sqrt{2g \left(\frac{p_1 - p_2}{\gamma} \right)} \quad \textbf{(12.50)}$$

Values of the expansion factor C_E for different pressure ratios p_2/p_1 are shown in Fig. 12.35. This relation is limited to subsonic flow ($M < 1$) at the throat. The critical pressure ratio for a gas, above which the flow will be subsonic, is given by

$$\left(\frac{p_2}{p_1} \right)_{critical} = \left(\frac{2}{k+1} \right)^{k/(k-1)} \quad \textbf{(12.51)}$$

For air ($k = 1.4$), the critical pressure ratio is 0.528.

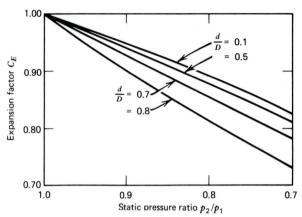

Figure 12.35 Expansion factor C_E as a function of static-pressure ratio for a venturi-type contraction.

12.7   OTHER FLOW-MEASUREMENT METHODS FOR CLOSED SYSTEMS

Several widely used, but markedly different, flow-measurement methods for closed systems are discussed in this section. They include a capillary flow meter, used for very small flow rates; positive-displacement flow meters, used when high accuracy is needed under steady-flow conditions; hot-film mass flow transducers, which are insensitive to temperature and pressure variations; and laser velocimetry systems for noncontacting flow measurements.

12.7.1 Capillary Flow Meter

When very small flow rates must be measured, the capillary flow meter, shown schematically in Fig. 12.36, is useful. The operation of this meter is based on the well-understood and experimentally verified conditions associated with laminar flow in a circular pipe. The Hagen-Poiseuille law, which governs laminar flow, is written as

$$\frac{p_1 - p_2}{\gamma} = \frac{32\mu L}{\gamma D^2} V = \frac{\gamma_m - \gamma}{\gamma} h \tag{12.52}$$

This relation is valid for $(Re)_D < 2000$. The velocity V is expressed in terms of the pressure difference $(p_1 - p_2)$ or a differential manometer head h, as shown in Fig. 12.36, as

$$V = \frac{D^2}{32\mu L}(p_1 - p_2) = \frac{D^2}{32\mu L}(\gamma_m - \gamma)h \tag{12.53}$$

where γ_m is the specific weight of the fluid in the differential manometer.

The volume flow rate, obtained from Eq. 12.53, is

$$Q = \frac{\pi D^4}{128\mu L}(\gamma_m - \gamma)h = Kh \tag{12.54}$$

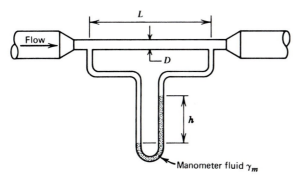

Figure 12.36 Capillary-tube flow meter.

Since the viscosity μ and the specific weight γ of the fluid are temperature dependent, the calibration constant K is a function of temperature. Once the constants have been established for the appropriate temperature, the flow rate Q is determined from a visual inspection or optical measurement of the differential manometer head h or through a measurement of the pressure difference $(p_1 - p_2)$ with a differential pressure transducer and the appropriate recording instrument.

12.7.2 Positive-Displacement Flow meters

Positive-displacement flow meters are normally used where high accuracy is needed (examples are home water meters and gasoline-pump meters). Two common types of positive-displacement flow meters are the nutating-disk meter (wobble meter) and the rotary-vane meter.

The nutating-disk meter, shown schematically in Fig. 12.37, is widely used as the flow sensing unit in home water meters. In this type of meter, an inlet chamber is formed by the housing, a disk, and a partition between the inlet and outlet ports. Water is prevented from leaving the chamber by the disk, which maintains line contact with the upper and lower conical surfaces of the housing. When the pressure is reduced on the outlet side by a demand for water, the pressure difference causes the disk to wobble (but not rotate) about the vertical axis (axis of symmetry of the housing) and thus provide a passage for the flow around the partition. The wobble of the disk causes

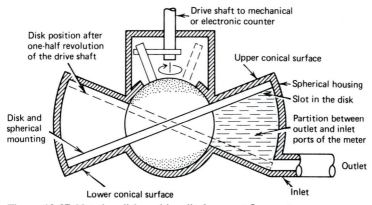

Figure 12.37 Nutating-disk positive-displacement flow meter.

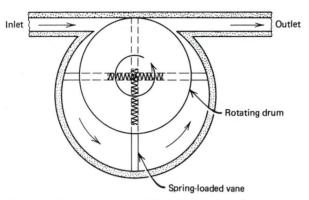

Figure 12.38 Rotary-vane positive-displacement flow meter.

a small pin attached to its spherical mount to trace out a circular path about the vertical axis of the device. This motion of the pin is used to drive the recording mechanism. Since a fixed volume of water moves through the device during each revolution of the drive shaft, a simple mechanical or electronic counter can be used to monitor the flow rate. The nutating-disk flow meter is accurate to within 1 percent when it is in good condition. When it is worn, the accuracy is considerably less, especially for very small flow rates (such as a leaky faucet).

A second type of positive-displacement flow meter is the rotary-vane type illustrated schematically in Fig. 12.38. This type of flow meter consists of a cylindrical housing in which an eccentrically mounted drum with several spring-mounted vane pairs rotates. A fixed volume of fluid is transferred from the inlet port to the outlet port during each rotation of the drum. Thus any type of counter can be used to monitor the flow rate. The rotary-vane flow meter is generally more rugged and more accurate (about $\frac{1}{2}$ percent) than the nutating-disk flow meter.

12.7.3 Hot-Film Mass Flow Transducers

The hot-film sensor, discussed in Section 12.2, provides the basis for a mass flow transducer that is relatively insensitive to variations in gas temperature and pressure. The mass flow rate $\dot{m}$ depends on the cross-sectional area A of the channel, the density ρ of the fluid, and the flow velocity V. With a fixed channel area A, the momentum per unit area ρV provides a measure of the mass flow rate $\dot{m}$. Equation 12.19 applied to mass flow transducer design is written as

$$q = \left[A + B(\rho V)^{1/n} \right](\theta_a - \theta_f) = i_a^2 R_a \tag{12.55}$$

Note that the momentum per unit area (ρV) appears as a product term in Eq. 12.55. A transducer that has been developed to measure ρV directly is shown schematically in Fig. 12.39. The hot-film probe in the center of the venturi throat responds to ρV, and the temperature sensor measures the fluid temperature. The inlet screens align the flow field.

The hot-film sensor is heated by current from the anemometer control circuit to a temperature above that of the fluid. The fluid then transports heat away from the sensor in proportion to the flow rate.

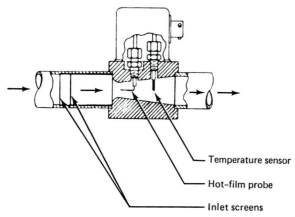

Figure 12.39 Schematic diagram showing the components of a hot-film mass flow transducer. (Courtesy of TSI, Inc.)

The signal is linearized in a microprocessor so that the mass flow rate is related with a calibration constant to the output voltage from the anemometer circuit. The calibration constant is valid for temperatures ranging from 40°F to 100°F, or for pressures ranging from 15 to 30 psia. Flow ranges of 1000 to 1 with an accuracy of 0.5 percent, a repeatability of 0.05 percent, and a response time of 1 ms are possible.

12.7.4 Laser Velocimetry Systems

The coherence of laser light has led to the development of the optical method of velocity measurement known as *laser-Doppler anemometry* or *laser-Doppler velocimetry*. In any form of wave propagation, frequency changes occur as a result of movement of the source, receiver, propagating medium, or intervening reflector or scatterer. Such frequency changes are known as *Doppler shifts* and are named after the Austrian mathematician and physicist Christian Doppler (1803–1853), who first studied the phenomenon.

An example of the *Doppler effect* in the field of acoustics is the increase of the pitch of a train whistle as the train approaches an observer, followed by a decrease in pitch as the train passes and moves away from the observer. The effect, illustrated in Fig. 12.40, is based on the perception that the number of sound waves arriving per unit time at the observer's position represents the frequency of the whistle.

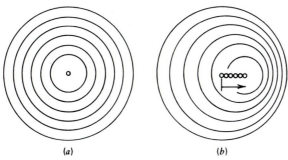

(a)   (b)

Figure 12.40 Schematic illustration of the Doppler effect. (*a*) Wave nodes from a stationary source. (*b*) Wave nodes from a moving source.

The Doppler shift is also well known in the field of astronomy. The optical equivalent of the change in pitch of the train whistle occurs when light from a distant star is observed. If the distance between the star and the earth is decreasing, more light pulses are received in a given time interval; thus, the frequency at the observer increases (wavelength decreases), and the color emitted from the star appears to be shifted toward the violet end of the spectrum. Conversely, when the distance between the star and the earth is increasing, the light is shifted toward the red end of the spectrum. The color shifts of remote galaxies have been accepted as evidence that the universe is expanding.

In laser-Doppler anemometry there is no relative movement between the source and the receiver. Instead, the Doppler shift is produced by the movement of particles (either natural or seeded) in the flow. These particles scatter light from the source and permit it to reach the receiver. This same principle provides the basis for radar; however, in the case of radar, a much lower frequency part of the electromagnetic spectrum is used. The velocities commonly measured by using laser-Doppler anemometry are very small when compared with the velocity of light; therefore, the corresponding Doppler shifts in frequency are very small. With red light from a helium-neon laser, $\lambda = 632.8$ nm, $f = 4.7(10^{14})$ Hz, and a supersonic flow of 500 m/s, the Doppler shift in frequency is approximately 780 MHz. To confirm this result, note that the Doppler frequency shift f_D is

$$\Delta f = f_D = \frac{f V}{c} \tag{12.56}$$

where

f is the frequency of the light source
c is the propagation velocity of light [$2.98(10^8)$ m/s]
V is the particle velocity

Since the resolution of a good-quality optical spectrometer is about 5 MHz, only velocities associated with high-velocity flows can be measured with reasonable accuracy by using direct Doppler-shift measurements as indicated in Eq. 12.56.

An optical beating technique for determining small Doppler shifts, which is equivalent to *heterodyning* in radio (signal mixing to obtain alternating constructive and destructive interference or beating), was first demonstrated by Yeh and Cummins (Reference 39) in 1964. Light scattered from particles seeded in flowing water was mixed (heterodyned) with an unshifted reference beam of light from the laser to produce a beat frequency equal to the Doppler-shift frequency. The result of adding two signals with slightly different frequencies to obtain a beat frequency is illustrated schematically in Fig. 12.41.

A schematic illustration of a simple reference-beam anemometer, see Fig. 12.42 shows the light from the laser divided with a beam splitter into an illuminating beam and a reference beam. Some of the light from the illuminating beam is scattered in the direction of the reference beam by the particles in the flow. The light from the two beams is combined (added) by the photodetector. The output signal contains a beat frequency equal to the Doppler-shift frequency produced by the movement (velocity) of the particles. This frequency is determined using a spectrum analyzer. Optimum results are obtained when the intensity of the reference beam is approximately equal to that of the scattered beam. An attenuator is placed in the path of the reference beam to adjust its intensity. The reference-beam anemometer is simple in principle;

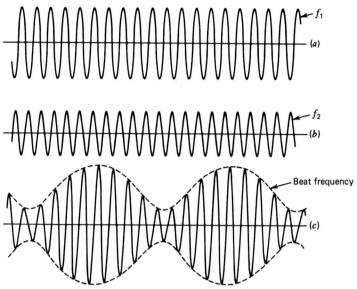

Figure 12.41 Signal addition (heterodyning) yields a combined signal with a beat frequency $f_b = (f_1 - f_2)/2$. (a) Signal No. 1. (b) Signal No. 2. (c) Combined Signal.

however, high signal-to-noise ratios are difficult to obtain in practice. The relationship between Doppler-shift frequency f_D and particle velocity V is

$$f_D = \frac{2V \cos \alpha}{\lambda} \sin \frac{\theta}{2} \qquad (12.57)$$

where

λ is the wavelength of the light

θ is the angle between the illuminating beam and the reference beam

α is the angle between the particle velocity vector and a normal to the bisector of the angle between the illuminating and reference beams

Typical frequency f_D versus velocity V curves for a reference-beam anemometer are shown in Fig. 12.43. Since the wavelength of the helium-neon laser is known

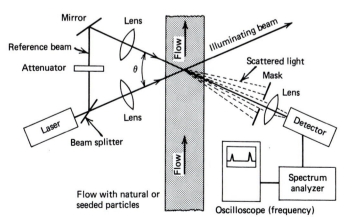

Figure 12.42 Reference-beam anemometer.

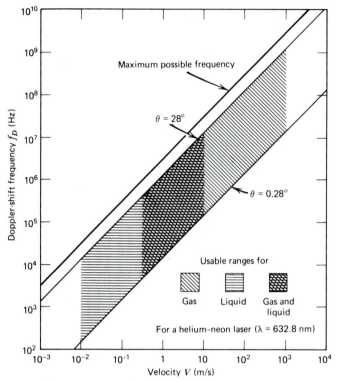

Figure 12.43 Doppler-shift frequency versus velocity for a reference-beam anemometer.

to an accuracy of 0.01 percent and because modern signal-processing electronics provide very accurate determinations of the Doppler-shift frequency f_D, the accuracy of velocity determinations is essentially controlled by the accuracy of the determination of the angle θ. The useful range of θ between 0.28 and 28 deg for gas, liquid, and two-phase flow is shown within the crosshatched area in Fig. 12.43.

A second type of velocity-measuring instrument, known as a *differential-Doppler* anemometer, is shown schematically in Fig. 12.44. In this instrument, scattered light from two equal-intensity beams is combined to produce the beat signal. The frequency

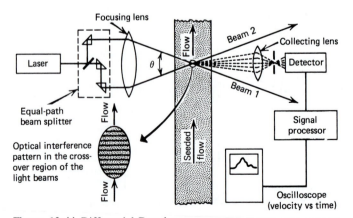

Figure 12.44 Differential-Doppler anemometer.

of this beat signal is equal to the difference between the Doppler shifts for the two angles of scattering. The primary advantage of this mode of operation is that the beat frequency is independent of the receiving direction. The light is collected with a large-aperture lens that is focused on the photodetector. All of this light contributes to the output signal, enhancing the signal-to-noise ratio. When the particle concentration is low, the differential-Doppler anemometer is preferred to the reference-beam anemometer because of its sensitivity to low-intensity signals.

Operation of the differential-Doppler anemometer is based on the optical interference pattern (fringe pattern) formed in the crossover region of the two beams, as shown in Fig. 12.45. The spacing of the interference fringes is given by the expression

$$s = \frac{\lambda}{2\sin(\theta/2)} \tag{12.58}$$

where

    s is the spacing between fringes
    λ is the wavelength of the light producing the fringes
    θ is the angle between the two intersecting light beams

A particle in the flow moving with a velocity V in a direction that makes an angle α, as defined in Fig. 12.44, produces a modulation of light intensity as it moves through the fringes. Since the light scattered from the particle depends on the intensity associated with the fringes, it is also modulated at the frequency f_D and will be independent of the direction of observation. The frequency f_D obtained from the particle velocity and fringe spacing is identical to that given by Eq. 12.57.

Output signals from the photodetectors can be processed in many ways to obtain the Doppler frequency f_D required for velocity determinations. Included are spectrum analysis, frequency tracking, counter processing, filter bank processing, and photon correlation. Although details of these procedures are beyond the scope of this book, additional information is found in Reference 6.

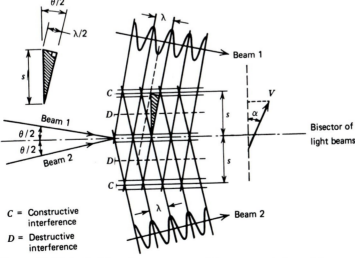

Figure 12.45 Optical interference pattern (fringe pattern) formed by constructive and destructive interference in the crossover region of the two light beams.

The advantages of laser-Doppler measurements include the following:

1. The method is nonobtrusive, and the flow is not disturbed by the presence of a probe.
2. Velocity is measured directly, and calibration is not required.
3. A component of velocity in a specified direction can be measured.
4. System output is a linear function of the velocity component being measured.
5. Velocities can be measured in flows exhibiting high turbulence.
6. The method is suitable for a very wide range of velocities.

12.8 SUMMARY

The methods employed to measure flow depend primarily on five factors:

1. Is the fluid incompressible, or is it necessary to account for compressibility?
2. Is the system closed, partially closed, or is a body moving through a boundless medium?
3. To what degree is it permissible to disturb the flow field in making the measurement?
4. How important are the energy losses owing to the insertion of flow meters?
5. What are the requirements for frequency response and accuracy?

It is relatively easy to measure steady-state flow of an incompressible fluid transported in a closed system by using a calibrated flow meter (Section 12.3) that has been standardized and measures differential pressure.

When the fluid is compressible, it is necessary to measure temperature in addition to pressure differences across the obstruction to predict mass flow rates. In this situation, the flow meter calibration combines the effects of energy losses and expansion with coefficients C_V and C_E.

Often, disturbances to the flow are not important, and it is a simple matter, for example, to build a sluice gate to measure flow in an irrigation ditch. However, in other instances, even a minor disturbance can alter the flow field. Probes and obstructions cannot be employed, and noncontact Doppler velocimetry is required.

For high-frequency phenomena, where small probes are possible, hot-wire or hot-film anemometers provide high accuracy to frequencies of 500 kHz. These probes are also widely employed in the study of turbulence to determine either the mean velocity or the spectrum of the turbulence.

For low-velocity airflow (probably the most common application), the pitot tube, with simple manometers to measure the differential pressures, remains the preferred approach.

Accuracy in the measurement of flow and mass rates depends heavily on calibration. The calibration process includes the flow sensing device, usually one or two pressure sensors, and a voltage record. When possible, system calibration techniques should be employed. However, when system calibration is not feasible, serious attention must be given to establish each of the individual calibration constants.

REFERENCES

1. Bean, H. S. (ed): *Fluid Meters: Their Theory and Application,* 6th ed., American Society of Mechanical Engineers, New York, 1971.
2. Benedict, R. P.: "Most Probable Discharge Coefficients for ASME Flow Nozzles," *Trans. ASME, Journal of Basic Engineering,* December 1966, p. 734.

3. Benedict, R. P.: *Fundamentals of Temperature, Pressure, and Flow Measurement,* 3rd ed., Wiley, New York, 1984.

4. Daugherty, R. L., and J. R. Franzini: *Fluid Mechanics with Engineering Applications,* 8th ed., McGraw–Hill, New York, 1985.

5. Dean, R. C.: "On the Necessity of Unsteady Flow in Fluid Mechanics," *ASME Journal of Basic Engineering* 81D, March 1959, pp. 24–28.

6. Drain, L. E.: *The Laser-Doppler Technique,* Wiley, New York, 1980.

7. Durrani, T. S., and C. A. Greated: *Laser Systems in Flow Measurement,* Plenum, New York, 1977.

8. Durst, F., A. Melling, and J. H. Whitelaw: *Principles and Practice of Laser-Doppler Anemometry,* 2nd ed., Academic Press, New York, 1981.

9. French, R. H.: *Open Channel Hydraulics,* McGraw–Hill, New York, 1985.

10. Freymuth, P.: "A Bibliography of Thermal Anemometry," *TSI Quarterly,* vol. 4, May/June 1978, available from Thermo Systems Incorporated, St. Paul, Minn.

11. Goldstein, R. J. (ed.): *Fluid Mechanics Measurements,* Hemisphere, New York, 1983.

12. Gracey, W., W. Letko, and W. R. Russel: "Wind-Tunnel Investigation of a Number of Total-Pressure Tubes at High Angles of Attack," NACA TN2331, National Advisory Committee for Aeronautics, Washington, D.C., April 1951.

13. Hall, I. M.: "The Displacement Effect of a Sphere in a Two-dimensional Shear Flow," *J. Fluid Mech.,* vol. 1, part 2, 1956, p. 142.

14. Holman, J. P.: *Experimental Methods for Engineers,* 4th ed., McGraw–Hill, New York, 1984.

15. Hurd, C. W., K. P. Chesky, and A. H. Shapiro: "Influence of Viscous Effects on Impact Tubes," *Trans. ASME, Journal of Applied Mechanics,* June 1953, p. 253.

16. John, J. E., and W. L. Haberman: *Introduction to Fluid Mechanics,* 3rd ed., Prentice–Hall, Englewood Cliffs, N.J., 1988.

17. King, L. V.: "On the Convection of Heat from Small Cylinders in a Stream of Fluid, with Applications to Hot-Wire Anemometry," *Philos. Trans. R. Soc. London,* vol. 214, no. 14, 1914, pp. 373–432.

18. Lansford, W. M.: "The Use of an Elbow in a Pipe Line for Determining the Flow in a Pipe," Bulletin 289, Engineering Experiment Station, University of Illinois, Urbana, Ill. 1936.

19. Lenz, A. T.: "Viscosity and Surface-Tension Effects on V-Notch Weir Coefficients," *Transactions of the American Society of Civil Engineers,* vol. 108, 1943, pp. 759–802.

20. Lighthill, M. J.: "Contributions to the Theory of Pitot-tube Displacement Effects," *J. Fluid Mech.,* vol. 2, part 2, 1956, p. 142.

21. Marris, A. W., and O. G. Brown: "Hydrodynamically Excited Vibrations of Cantilever-Supported Rods," ASME Paper 62-HYD-7, American Society of Mechanical Engineers, New York, 1962.

22. Miller, R. W.: *Flow Measurement Engineering Handbook,* 2nd ed., McGraw–Hill, New York, 1989.

23. Moody, L. F., "Friction Factors for Pipe Flow," *Trans. ASME,* vol. 66, 1944.

24. Munson, B. R., D. F. Young, and T. H. Okiishi,: *Fundamentals of Fluid Mechanics,* Wiley, New York, 1990.

25. Olson, R. M.: *Essentials of Engineering Fluid Mechanics,* 4th ed., Harper & Row, New York, 1980.

26. Panton, R. L.: *Incompressible Flow,* Wiley, New York, 1984.

27. Perry, A. E.: *Hot-Wire Anemometry,* Oxford Univ. Press (Clarendon Press), London/New York, 1982.

28. Rayle, R. E.: "Influence of Orifice Geometry on Static Pressure Measurements," ASME Paper 59-A-234, American Society of Mechanical Engineers, New York, December 1959.

29. Robertson, J. A., and C. T. Crowe: *Engineering Fluid Mechanics,* 2nd ed., Houghton Mifflin, Boston, 1980.

30. Schlicting, H.: *Boundary Layer Theory,* 7th ed., McGraw–Hill, New York, 1979.

31. Spitzer, David W.: *Industrial Flow Measurement,* Instrument Society of America, Research Triangle Park, N.C., 1990.
32. Streeter, V. L., and E. B. Wylie: *Fluid Mechanics,* 8th ed., McGraw–Hill, New York, 1985.
33. Tuve, G. L., and L. C. Domholdt: *Engineering Experimentation,* McGraw–Hill, New York, 1966.
34. Tuve, G. L., and R. E. Sprenkle: "Orifice Discharge Coefficients for Viscous Liquids," *Instruments,* vol. 6, 1933, p. 201; also vol. 8, 1935, pp. 202, 225, 232.
35. *The U.S. Standard Atmosphere,* U.S. Government Printing Office, Washington, D.C., 1976.
36. Vennard, J. K., and R. L. Street: *Elementary Fluid Mechanics,* 6th ed., Wiley, New York, 1982.
37. White, D. F., A. E. Rodley, and C. L. McMurtrie: "The Vortex Shedding Flowmeter," *Flow, Its Measurement and Control in Science and Industry,* vol. 1, part 2, Instrument Society of America, Pittsburgh, 1974, p. 967.
38. White, F. M.: *Viscous Fluid Flow,* 2nd ed., McGraw–Hill, New York, 1991.
39. Yeh, Y., and H. Z. Cummins: "Localized Flow Measurements with a He-Ne Laser Spectrometer," *Appl. Phys. Lett.,* vol. 4, 1964, pp. 176–178.
40. Yothers, M. T. (ed.): *Standards and Practices for Instrumentation,* 5th ed., Instrument Society of America, Pittsburgh, 1977.

EXERCISES

12.1 For laminar flow in a circular pipe, show that the average velocity V_{av} is one-half of the centerline velocity V_0.

12.2 The velocity profile for fully developed flow in circular pipes is given by Eq. 12.5. Show that the average velocity V_{av} is related to the centerline velocity V_0 by Eq. 12.6.

12.3 Water at 60°F flows through a 10-in.-diameter pipe with an average velocity of 20 ft/s. Determine the weight flow rate, the mass flow rate, the energy per second being transmitted through the pipe in the form of kinetic energy $(\gamma Q V^2/2g)$, the velocity profile exponent n, and the centerline velocity.

12.4 For the rectangular duct shown in Fig. E12.4, the velocity profile can be approximated by the expression

$$V = V_C[1 - (x/a)^2][1 - (y/b)^2]$$

Determine V_{av} in terms of V_C for this case.

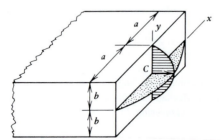

Figure E12.4

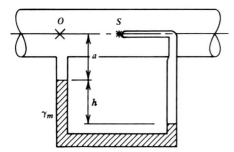

Figure E12.5

12.5   A pitot tube is connected to a manometer filled with fluid of specific weight γ_m, as shown in Fig. E12.5. When the manometer connection pipes are filled with the flowing fluid of specific weight γ, the pressure differential is independent of dimension a. Show that the free-stream velocity is given by the expression

$$V_o = C_1 \sqrt{2g \frac{\gamma_m - \gamma}{\gamma} h} = K \sqrt{h}$$

Determine the constant K when the flowing fluid is glycerin, the manometer fluid is mercury, $C_1 = 0.98$, V_o is expressed in meters per second, and h is measured in centimeters. How would these results be altered if the pipe were inclined at an angle of 60 deg from the horizontal?

12.6   Compare the velocity-measurement error associated with a 40-deg misalignment for the two types of pitot tubes shown in Fig. 12.7 when the true dynamic pressure is 4 in. of water for carbon dioxide flowing at 68°F.

12.7   A 0.125-in.-diameter pitot tube is to be used to measure the velocity of two liquids (water and glycerin). What is the minimum velocity for each liquid below which viscous effects must be considered? Assume that both liquids are at room temperature.

12.8   A pitot-static tube is used to measure the speed of an aircraft. The air temperature and pressure are 30°F and 12.3 psia. What is the aircraft speed in miles per hour if the differential pressure is 35 in. of water? Solve by using both Eq. 12.10 and Eq. 12.18. Compare the results.

12.9   Show that the fundamental form of Eq. 12.21 follows directly from King's law as expressed by Eq. 12.19 and that

$$C_0 = A^2/\rho B^2 \quad \text{and} \quad i_0 = \sqrt{A(\theta_a - \theta_f)/R_a}$$

12.10   From the typical calibration curve for a hot-film anemometer operating in water, as shown in Fig. 12.13, estimate the reference current i_0 and the calibration constant C_0. Assuming a nominal temperature difference of 75°F during calibration, how would you correct for a fluid temperature drop of 8°F occurring after start-up of the experiment?

12.11   A circular plate having a diameter of 25 mm is to be used as a drag-force velocity transducer to measure the velocity of water at 68°F. Determine the velocity sensitivity of the plate and the minimum velocity for which this sensitivity is valid.

12.12   Current-meter data were collected at 13 vertical locations similar to those shown in Fig. 12.19 on a river that is 144 ft wide. The calibration constant for the current meter can be expressed as $V = 2.45N$, where V is the flow velocity in feet per second and N is the speed of the rotating element in revolutions per second. Determine the river flow rate if the data collected at the 13 locations are as follows:

Station	River Depth (ft)	Rotating Element (rpm) 0.2h	0.8h
1	0.0	—	—
2	3.0	40.1	31.2
3	3.5	51.1	41.5
4	4.2	59.0	43.0
5	3.7	62.1	48.2
6	5.1	68.3	50.2
7	4.6	65.6	48.2
8	3.8	60.2	45.8
9	4.0	56.5	48.0
10	3.2	57.3	39.8
11	3.1	48.8	38.0
12	2.0	41.2	29.8
13	0.0	—	—

12.13   A vortex-shedding transducer is being designed to measure the velocity of water flowing in an open channel. The velocities to be measured range from 0.2 to 10.0 ft/s. The relationship between shedding frequency f_s and velocity V (subject to calibration) should be approximately $f_s = 10V$. Show that a cylinder having a diameter of 0.25 in. approximates these design requirements. What other factors must be considered?

12.14   Design a bluff-body vortex-shedding flow meter similar to the one shown in Fig. 12.22 for use in a 12-in.-diameter pipe that is carrying water. Estimate the meter sensitivity (feet per second per hertz) and the range of velocities over which this sensitivity should be nearly constant. What is the equivalent volume flow-rate sensitivity?

12.15   Show that the ideal and actual volume flow rates for the closed system shown in Fig. E12.15 are

$$Q_a = C_V Q_i$$

$$= \frac{C_V A_2}{\sqrt{1 - (A_2/A_1)^2}} \sqrt{\frac{2g(\gamma_m - \gamma)h}{\gamma}}$$

where γ_m is the specific weight of the manometer fluid ($\gamma_m > \gamma$).
(a)   Why aren't dimensions a and b in this equation?
(b)   How would you change the manometer connection if $\gamma_m < \gamma$?
(c)   What effect would $\gamma_m < \gamma$ have on the preceding equation?

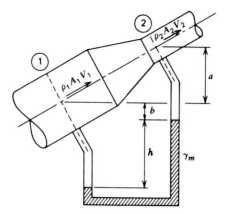

Figure E12.15

12.16 A 3.0-in. by 1.5-in.-diameter venturi meter is used to measure the volume flow rate of turpentine in a chemical processing plant. Pressure differences are measured with a manometer having water as the manometer fluid.
  (a) Determine the ideal flow-rate sensitivity when the pressure drop is measured in inches of water.
  (b) How can the actual flow-rate sensitivity be estimated by using Fig. 12.25?
  (c) For a manometer reading of 18 in. of water and a fluid temperature of 68°F, determine the actual flow rate.

12.17 The flow of water at 150°F in an existing 10-in.-diameter pipeline is to be measured by using a 6-in.-diameter ASME long-radius flow nozzle. The flow rate will vary from 0.05 to 3.0 ft³/s.
  (a) What range of pressure drops should the differential pressure transducer be able to measure?
  (b) If a manometer is to be used to measure these pressure drops, select a reasonable manometer fluid for use in this application if it is assumed that a manometer can be easily read to ±0.05 in.

12.18 A mercury manometer is connected to a standard orifice meter with a 30-mm-diameter hole that has been placed in a 100-mm-diameter pipe. What is the flow rate of crude oil at 20°C if the manometer reading across the orifice plate is 240 mm of Hg.

12.19 A 6-in.-diameter elbow meter has a coefficient $C = 0.75$ when installed in a water line. The meter is connected to a mercury manometer having a 24-in. scale that is graduated in units of 0.05 in.
  (a) Determine the sensitivity of the instrument in terms of cubic feet per second and inches of mercury.
  (b) When $h = 18$ in. of mercury, what is the change in volume flow rate ΔQ corresponding to a scale reading error of 0.05 in.?
  (c) How significant is the 0.05-in. reading error when $h = 9$ in.?

12.20 Compare flow rates through 50-mm-diameter openings under 1.5-m static heads if the openings have been constructed as sharp-edged, rounded, short-tube, or Borda orifices. Ancient Rome's famous water system used sharp-edged orifices to meter water to Roman citizens. The clever citizens found that they could obtain 30 percent more water by inserting a short tube into the orifice and thus cheat Caesar out of significant water revenues.

12.21   An 8-in.-diameter opening is to be located in the side of a large tank. Water flows from the tank into a large reservoir. Estimate the flow between the tank and the reservoir when the difference in free-surface elevations is 20 ft. Assume that:

(a)   the most efficient orifice construction from Table 12.1 is used

(b)   the most inefficient orifice construction from Table 12.1 is used

What assumptions are made when the terms *large tank* and *large reservoir* are used?

12.22   A 3.0-m-wide by 0.4-m-deep sluice gate is used to control the overflow of water from a small reservoir with a surface area of 100,000 m². When the water surface in the reservoir is 3 m above the bottom of the spillway (see Fig. 12.32), estimate the flow rate through the sluice gate and the rate at which the reservoir surface is falling. Assume that $C_C = 0.61$ and $C_V = 0.96$.

12.23   Derive Eq. 12.47 for the flow rate over a rectangular weir and carefully list any assumptions required for the derivation. If C_D is assumed to be equal to 0.623 and if end-contraction effects are neglected, show that

(a) $Q_a = 3.33 \, LH^{3/2}$ for the English System of units

(b) $Q_a = 1.84 \, LH^{3/2}$ for the SI system of units

12.24   T. Rehbock of the Karlsruhe Laboratory in Germany developed the following empirical expression for the weir discharge coefficient C_D, which yields good results for rectangular weirs with good ventilation, sharp edges, smooth weir faces, and adequate water stilling.

$$C_D = 0.605 + 0.08H/P + 1/305H$$

Plot values of C_D as a function of H (0.08 ft $< H <$ 2.0 ft) for P equal to 0.2 ft, 0.5 ft, 1.0 ft, 2.0 ft, and 5.0 ft. Note H/P must be less than 2 for the Rehbock equation to retain an accuracy of 1 percent.

12.25   A rectangular weir is to be placed in a 5-m-wide channel to measure a nominal flow rate of 6 m³/s while maintaining a minimal channel depth of 4 m. Determine a suitable rectangular weir (width L and height P) if H/P must be less than 0.4 to ensure that $V_1^2/2g$ is negligible. What would the height P be for a 90-deg V-notch weir? If the flow rate doubles, which weir would experience the smaller change in weir head H?

12.26   The flow rate in a rectangular open channel of width L must be measured while a nearly constant fluid depth y is maintained. A floating sluice gate and a weir have been proposed as methods to achieve these goals. The two methods are shown in Fig. E12.26.

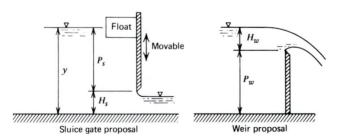

Sluice gate proposal   Weir proposal   **Figure E12.26**

(a) Show that the flow rate under the sluice gate as given by the following linear equation between Q and H is accurate within 5 percent, provided that C_{Ds} is constant and $H_s/P_s < 0.1$.

$$Q = C_{Ds} \sqrt{2gP_sL}H_s$$

(b) Compare the sluice-gate and weir-discharge equations and show that the sluice gate and weir readings are related by the expression

$$H_s = \frac{2C_{Dw}}{3C_{Ds}} \frac{H_w^{3/2}}{\sqrt{P_s}}$$

(c) Based on the information of (b), which method will give the least variation of H with flow rate?

(d) Which unit is least expensive to install and maintain?

(e) Which proposal would you select, and why?

12.27 Air stored in a tank at 300 psia and 150°F flows into a second tank at 250 psia through a 1-in.-diameter flow nozzle. Determine
(a) the mass flow rate of the air moving from one tank to the other
(b) the minimum pressure in the second tank for subsonic flow.

12.28 Oxygen at 70°F and 150 psia is flowing in a 3-in.-diameter line at a rate of 6 lb/s. Estimate the pressure drop across a 1.5-in.-diameter venturi meter that would be available for measuring the flow rate.

12.29 A capillary-tube flow meter is being constructed to measure the flow rate of water. The glass tubing has an inside diameter of 1.0 mm, and the pressure taps are located 0.1 m apart.
(a) Estimate the flow rate if the manometer reading is 200 mm of mercury and the water temperature is 20°C.
(b) How much error results if the water temperature increases to 25°C? Neglect any effects resulting from expansion of the glass with temperature change.

12.30 Calibration of flow meters is often performed with an experimental facility consisting of a pump, a valve to regulate the flow from the pump, a test meter and manometer, a weight tank and scale, a stopwatch, and a reservoir, as shown in Fig. E12.30.
(a) What minimum length L of pipe should be used on the inlet side of the test meter?

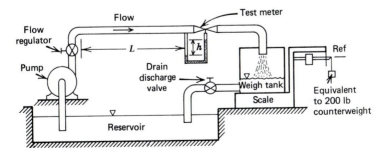

Figure E12.30

(b)   What is the maximum flow rate that can be measured if errors are to be limited to ±1 percent, if water caught in the weigh tank is limited to 200 lb, and if the stopwatch can be started and stopped within 0.05 s of the correct time?

(c)   What additional information must be collected in order to properly calibrate the flow meter?

(d)   How should flow rate Q versus manometer reading h be plotted in order to obtain the meter calibration relationship?

(e)   From sources available in the library, find a standard relating to calibration of flow meters and study the calibration procedures recommended.

12.31   Write an engineering brief for a laboratory technician describing the operation of a hot-film mass flow transducer.

12.32   Prepare a graph showing particle velocity in a flow field as a function of the Doppler shift in frequency. Assume that a helium-neon laser is used as the light source.

Chapter **13**

Statistical Methods

13.1 INTRODUCTION

Experimental measurements of quantities such as pressure, temperature, length, force, stress, or strain will always exhibit some variation if the measurements are repeated a number of times with precise instruments. This variability, which is fundamental to all measuring systems, is due to two causes. First, the quantity being measured may exhibit significant variation. For example, in a materials study to determine fatigue life at a specified stress level, large differences in the number of cycles to failure are noted when a number of specimens are tested. This variation is inherent in the fatigue process and is observed in all fatigue life measurements. Second, the measuring system, which includes the transducer, signal conditioning equipment, analog-to-digital converter, recording instrument, and the operator may introduce error into the measurement. This error may be systematic or random, depending on its source. An instrument operated out of calibration produces a systematic error, whereas reading errors owing to interpolation on a chart are random. The accumulation of random errors in a measuring system produces a variation that must be examined in relation to the magnitude of the quantity being measured.

The data obtained from repeated measurements represent an array of readings, not an exact result. Maximum information can be extracted from such an array of readings by employing statistical methods. The first step in the statistical treatment of data is to establish the distribution. A graphic representation of the distribution is usually the most useful form for initial evaluation. Next, the statistical distribution is characterized with a measure of its central value, such as the mean, the median, or the mode. Finally, the spread or dispersion of the distribution is determined in terms of the variance or the standard deviation.

With elementary statistical methods, the experimentalist can reduce a large amount of data to a compact and useful form by defining the type of distribution, establishing the single value that best represents the central value of the distribution (mean), and determining the variation from the mean value (standard deviation). Summarizing data in this manner is the most meaningful form of presentation for application to design problems or for communication to others who need the results of the experiments.

The treatment of statistical methods presented in this chapter is relatively brief; therefore, only the most commonly employed techniques for representing and interpreting data are presented. A formal course in statistics, which covers these techniques in much greater detail as well as many other useful techniques, should be included in the program of study of all engineering students.

13.2 CHARACTERIZING STATISTICAL DISTRIBUTIONS

For purposes of this discussion, consider that an experiment has been conducted n times to determine the yield strength of a particular type of cold-drawn mild steel. The data obtained represent a sample of size n from an infinite population of all possible measurements that could have been made. The simplest way to present these data is to list the strength measurements in order of increasing magnitude, as shown in Table 13.1.

These data can be arranged in five groups to give a frequency distribution as shown in Table 13.2. The advantage of representing data in a frequency distribution is that the central tendency is more clearly illustrated.

13.2.1 Graphic Representations of the Distribution

The shape of the distribution function representing the yield strength of the cold-drawn mild steel is indicated by the data groupings of Table 13.2. A graphic presentation of this group data, known as a histogram, is shown in Fig. 13.1. The histogram method of presentation shows the central tendency and variability of the distribution much more clearly than the tabular method of presentation of Table 13.2. Superimposed on the histogram is a curve showing the relative frequency of the occurrence of a group of measurements. Note that the points for the relative frequency are plotted at the midpoint of the group interval.

A cumulative-frequency diagram, shown in Fig. 13.2, is another way of representing the yield-strength data from the experiments. The cumulative frequency is the number

Table 13.1 Listing of Data in Order of Increasing Magnitude. Yield Strength of Cold-Drawn Mild Steel

Sample Number	Strength (ksi:MPa)	Sample Number	Strength (ksi:MPa)
1	65.0 : 448	11	79.0 : 545
2	68.3 : 471	12	79.2 : 546
3	72.2 : 498	13	79.9 : 551
4	73.5 : 507	14	80.3 : 554
5	74.0 : 510	15	81.1 : 559
6	75.2 : 519	16	82.6 : 570
7	76.8 : 530	17	84.0 : 579
8	77.7 : 536	18	85.5 : 590
9	78.1 : 539	19	87.0 : 600
10	78.8 : 543	20	89.8 : 619

Table 13.2 Frequency Distribution of Yield Strength

Group Intervals (ksi : MPa)	Observations in the Group	Relative Frequency	Cumulative Frequency
65.0–69.9 : 448–482	2	0.10	0.10
70.0–74.9 : 483–516	3	0.15	0.25
75.0–79.9 : 517–551	8	0.40	0.65
80.0–84.9 : 552–585	4	0.20	0.85
85.0–89.9 : 586–620	3	0.15	1.00
Total	20		

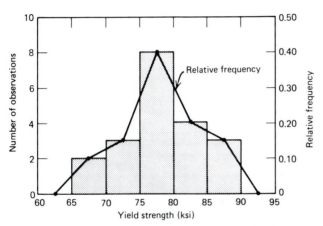

Figure 13.1 Histogram with a superimposed relative-frequency diagram.

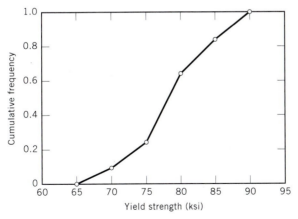

Figure 13.2 Cumulative-frequency diagram.

of readings having a value less than a specified value of the quantity being measured (yield strength) divided by the total number of measurements. As indicated in Table 13.2, the cumulative frequency is the running sum of the relative frequencies. When the graph of cumulative frequency versus the quantity being measured is prepared, the end value for the group intervals is used to position the point along the abscissa.

13.2.2   Measures of Central Tendency

Whereas histograms or frequency distributions are used to provide a visual representation of a distribution, numerical measures are used to define the characteristics of the distribution. One basic characteristic is the central tendency of the data. The most commonly employed measure of the central tendency of a distribution of data is the sample mean $\overline{x}$, which is defined as

$$\overline{x} = \sum_{i=1}^{n} \frac{x_i}{n} \tag{13.1}$$

where

x_i is the ith value of the quantity being measured
n is the total number of measurements

Because of time and costs involved in conducting tests, the number of measurements is usually limited; therefore, the sample mean $\overline{x}$ is only an estimate of the true arithmetic mean μ of the population. It is shown later that $\overline{x}$ approaches μ as the number of measurements increases. The mean value of the yield-strength data presented in Table 13.1 is $\overline{x} = 78.4$ ksi (541 MPa).

The median and mode are also measures of central tendency. The median is the central value in a group of ordered data. For example, in an ordered set of 21 readings, the eleventh reading represents the median value with 10 readings lower than the median and 10 readings higher than the median. When an even number of readings is taken, the median is obtained by averaging the two middle values. For example, in an ordered set of 20 readings, the median is the average of the tenth and eleventh readings. Thus, for the yield-strength data presented in Table 13.1, the median is $\frac{1}{2}(78.8 + 79.0) = 78.9$ ksi (544 MPa).

The mode is the most frequent value of the data; therefore, it is the peak value on the relative-frequency curve. In Fig. 13.1, the peak of the relative-probability curve occurs at a yield strength $S_y = 77.5$ ksi (535 MPa); therefore, this value is the mode of the data set presented in Table 13.1.

A typical set of data gives different values for the three measures of central tendency. There are two reasons for this difference. First, the population from which the samples are drawn may not be Gaussian where the three measures are expected to coincide. Second, even if the population is Gaussian, the number of measurements n is usually small and deviations are to be expected from a small sample size.

13.2.3   Measures of Dispersion

It is possible for two different distributions of data to have the same mean but different dispersions, as shown in the relative-frequency diagrams of Fig. 13.3. Different mea-

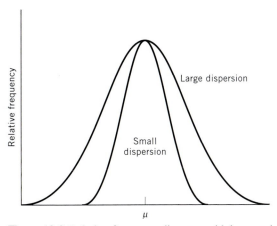

Figure 13.3 Relative-frequency diagrams with large and small dispersions.

sures of dispersion are the range, the mean deviation, the variance, and the standard deviation. The standard deviation S_x is the most popular and is defined as

$$S_x = \left[\sum_{i=1}^{n} \frac{(x_i - \overline{x})^2}{n-1} \right]^{1/2} \tag{13.2}$$

Since the sample size n is small, the standard deviation S_x of the sample represents an estimate of the true standard deviation σ of the population. Computation of S_x and $\overline{x}$ from a data sample is easily performed with most scientific calculators.

Expressions for the other measures of dispersion, namely, range R, mean deviation d_x, and variance S_x^2 are as follows:

$$R = x_L - x_S \tag{13.3}$$

where

  x_L is the largest value of the quantity in the distribution
  x_S is the smallest value of the quantity in the distribution

$$d_x = \frac{\sum\limits_{i=1}^{n} x_i - \overline{x}}{n} \tag{13.4}$$

Equation 13.4 indicates that the deviation of each reading from the mean is determined and summed. The average of the n deviations is the mean deviation. The absolute value of the difference $(x_i - \overline{x})$ must be used in the summing process to avoid cancellation of positive and negative deviations.

$$S_x^2 = \frac{\sum\limits_{i=1}^{n} (x_i - \overline{x})^2}{n-1} \tag{13.5}$$

The variance of the population σ^2 is estimated by S_x^2 where the denominator $(n-1)$ in Eqs. 13.2 and 13.5 serves to reduce error introduced by approximating the true mean μ with the estimate $\overline{x}$. As the sample size n is increased, the estimates of $\overline{x}$, S_x, and S_x^2 improve, as shown in Section 13.4. Variance is an important measure of dispersion because it is used in defining the normal distribution function.

Finally, a measure known as the coefficient of variation C_v is used to express the standard deviation S_x as a percentage of the mean $\overline{x}$. Thus,

$$C_v = \frac{S_x}{\overline{x}}(100) \tag{13.6}$$

The coefficient of variation represents a normalized parameter that indicates the variability of the data in relation to its mean.

13.3   STATISTICAL DISTRIBUTION FUNCTIONS

As the sample size is increased, it is possible in tabulating the data to increase the number of group intervals and to decrease their width. The corresponding relative-frequency diagram, similar to the one illustrated in Fig. 13.1, will approach a smooth curve (a theoretical distribution curve) known as a distribution function.

Several different distribution functions are used in statistical analyses. The best-known and most widely used distribution in engineering is the Gaussian, or normal, distribution. This distribution is important because it describes random errors in measurements and variations observed in strength determinations. Other useful distributions include binomial, exponential, hypergeometric, chi-square (χ^2), F, Gumbel, Poisson, Student's t, and Weibull distributions. References 1 through 5 provide a complete description of these distributions. Emphasis here will be on Gaussian and Weibull distribution functions because of their wide range of applications in engineering.

13.3.1   Gaussian Distribution

The Gaussian, or normal, distribution function, as represented by a normalized relative-frequency diagram, is shown in Fig. 13.4. The Gaussian distribution is com-

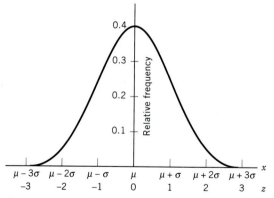

Figure 13.4 The Gaussian, or normal, distribution function.

pletely defined by two parameters: the mean μ and the standard deviation σ. The equation for the relative frequency f in terms of these two parameters is

$$f(z) = \frac{1}{\sqrt{2\pi}} e^{-(z^2/2)} \tag{13.7}$$

where

$$z = \frac{x - \mu}{\sigma} \qquad \frac{x - \bar{x}}{S_x} \tag{13.8}$$

Experimental data (with finite sample sizes) can be analyzed to obtain $\bar{x}$ as an estimate of μ and S_x as an estimate of σ. This procedure permits the experimentalist to use data drawn from small samples to represent the entire population.

The method for predicting population properties from a Gaussian distribution function utilizes the normalized relative-frequency diagram shown in Fig. 13.4. The area A under the entire curve is given by Eq. 13.7 as

$$A = \frac{1}{\sqrt{2\pi}} \int_{-\infty}^{+\infty} e^{-(z^2/2)} dz = 1 \tag{13.9}$$

Equation 13.9 implies that the population has a value z between $-\infty$ and $+\infty$ and that the probability of making a single observation from the population with a value $-\infty \leq z \leq +\infty$ is 100 percent. Although this statement may appear trivial, it serves to illustrate the concept of using the area under the normalized relative-frequency curve to determine the probability P of observing a measurement within a specific interval. Figure 13.5 shows graphically (shaded area under the curve) the probability that a measurement will occur within the interval between z_1 and z_2. Thus, from Eq. 13.7,

$$P(z_1, z_2) = \int_{z_1}^{z_2} f(z) \, dz = \frac{1}{\sqrt{2\pi}} \int_{z_1}^{z_2} e^{-(z^2/2)} dz \tag{13.10}$$

Evaluation of Eq. 13.10 is most easily made by using tables that list the areas under the normalized relative-frequency curve as a function of z. Table 13.3 lists one-side areas between limits of $z_1 = 0$ and z_2 for the normal distribution function.

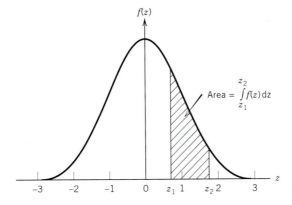

Figure 13.5 Probability of a measurement of x between limits of z_1 and z_2. The total area under the curve $f(z)$ is 1.

Table 13.3 Areas under the Normal Distribution Curve from $z_1 = 0$ to z_2 (One Side)

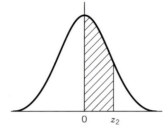

$z_2 = \frac{x-\bar{x}}{S_x}$	0.00	0.01	0.02	0.03	0.04	0.05	0.06	0.07	0.08	0.09
0.0	.0000	.0040	.0080	.0120	.0160	.0199	.0239	.0279	.0319	.0359
0.1	.0398	.0438	.0478	.0517	.0557	.0596	.0636	.0675	.0714	.0753
0.2	.0793	.0832	.0871	.0910	.0948	.0987	.1026	.1064	.1103	.1141
0.3	.1179	.1217	.1255	.1293	.1331	.1368	.1406	.1443	.1480	.1517
0.4	.1554	.1591	.1628	.1664	.1700	.1736	.1772	.1808	.1844	.1879
0.5	.1915	.1950	.1985	.2019	.2054	.2088	.2123	.2157	.2190	.2224
0.6	.2257	.2291	.2324	.2357	.2389	.2422	.2454	.2486	.2517	.2549
0.7	.2580	.2611	.2642	.2673	.2704	.2734	.2764	.2794	.2823	.2852
0.8	.2881	.2910	.2939	.2967	.2995	.3023	.3051	.3078	.3106	.3233
0.9	.3159	.3186	.3212	.3238	.3264	.3289	.3315	.3340	.3365	.3389
1.0	.3413	.3438	.3461	.3485	.3508	.3531	.3554	.3577	.3599	.3621
1.1	.3643	.3655	.3686	.3708	.3729	.3749	.3770	.3790	.3810	.3830
1.2	.3849	.3869	.3888	.3907	.3925	.3944	.3962	.3980	.3997	.4015
1.3	.4032	.4049	.4066	.4082	.4099	.4115	.4131	.4147	.4162	.4177
1.4	.4192	.4207	.4222	.4236	.4251	.4265	.4279	.4292	.4306	.4319
1.5	.4332	.4345	.4357	.4370	.4382	.4394	.4406	.4418	.4429	.4441
1.6	.4452	.4463	.4474	.4484	.4495	.4505	.4515	.4525	.4535	.4545
1.7	.4554	.4564	.4573	.4582	.4591	.4599	.4608	.4616	.4625	.4633
1.8	.4641	.4649	.4656	.4664	.4671	.4678	.4686	.4693	.4699	.4706
1.9	.4713	.4719	.4726	.4732	.4738	.4744	.4750	.4758	.4761	.4767
2.0	.4772	.4778	.4783	.4788	.4793	.4799	.4803	.4808	.4812	.4817
2.1	.4821	.4826	.4830	.4834	.4838	.4842	.4846	.4850	.4854	.4857
2.2	.4861	.4864	.4868	.4871	.4875	.4878	.4881	.4884	.4887	.4890
2.3	.4893	.4896	.4898	.4901	.4904	.4906	.4909	.4911	.4913	.4916
2.4	.4918	.4920	.4922	.4925	.4927	.4929	.4931	.4932	.4934	.4936
2.5	.4938	.4940	.4941	.4943	.4945	.4946	.4948	.4949	.4951	.4952
2.6	.4953	.4955	.4956	.4957	.4959	.4960	.4961	.4962	.4963	.4964
2.7	.4965	.4966	.4967	.4968	.4969	.4970	.4971	.4972	.4973	.4974
2.8	.4974	.4975	.4976	.4977	.4977	.4978	.4979	.4979	.4980	.4981
2.9	.4981	.4982	.4982	.4983	.4984	.4984	.4985	.4985	.4986	.4986
3.0	.49865	.4987	.4987	.4988	.4988	.4988	.4989	.4989	.4989	.4990

Since the distribution function is symmetric about $z = 0$, this one-side table is sufficient for all evaluations. For example,

$$A(-1, 0) = A(0, +1)$$

Therefore

$$A(-1, +1) = p(-1, +1) = 0.3413 + 0.3413 = 0.6826$$
$$A(-2, +2) = p(-2, +2) = 0.4772 + 0.4772 = 0.9544$$
$$A(-3, +3) = p(-3, +3) = 0.49865 + 0.49865 = 0.9973$$
$$A(-1, +2) = p(-1, +2) = 0.3413 + 0.4772 = 0.8185$$

Since the normal distribution function has been well characterized, predictions can be made regarding the probability of a specific strength value or measurement error. For example, one may anticipate that 68.3 percent of the data will fall between limits of $\bar{x} \pm S_x$, 95.4 percent between limits of $\bar{x} \pm 2s_x$, and 99.7 percent between limits of $\bar{x} \pm 3S_x$. Also, 81.9 percent of the data should fall between limits of $\bar{x} - S_x$ and $\bar{x} + 2S_x$.

In many problems, the probability of a single sample exceeding a specified value z_2 must be determined. It is possible to determine this probability by using Table 13.3 together with the fact that the area under the entire curve is unity ($A = 1$); however, Table 13.4, which lists one-side areas between limits of $z_1 = z$ and $z_2 \to \infty$, yields the results more directly.

Use of Tables 13.3 and 13.4 can be illustrated by considering the yield-strength data presented in Table 13.1. By using Eqs. 13.1 and 13.2, it is easy to establish estimates for the mean and standard deviation as $\bar{x} = 78.4$ ksi (541 MPa) and $S_x = 6.04$ ksi (41.7 MPa). These values of $\bar{x}$ and S_x characterize the population from which the data of Table 13.1 were drawn. It is possible to establish the probability that the yield strength of a single specimen drawn randomly from the population will be between specified limits (by using Table 13.3) or that the yield strength of a single sample will not be above or below a specified value (by using Table 13.4). For example, the probability that a single sample will exhibit a yield strength between 66 and 84 ksi is determined by computing z_1 and z_2 and using Table 13.3. Thus,

$$z_1 = \frac{66 - 78.4}{6.04} = -2.05 \qquad z_2 = \frac{84 - 78.4}{6.04} = 0.93$$
$$p(-2.05, 0.93) = A(-2.05, 0) + A(0, 0.93)$$
$$= 0.4798 + 0.3238 = 0.8036$$

This simple calculation shows that the probability of obtaining a yield strength between 66 and 84 ksi from a single specimen is 80.4 percent. The probability of the yield strength of a single specimen being less than 65 ksi is determined by computing z_1 and using Table 13.4. Thus,

$$z_1 = \frac{65 - 78.4}{6.04} = -2.22$$
$$p(-\infty, -2.22) = A(-\infty, -2.22) = A(2.22, \infty) = 0.0132$$

Table 13.4 Areas under the Normal-Distribution Curve from z_1 to $z_2 \to \infty$ (One Side)

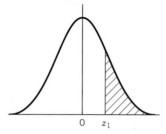

$z_1 = \frac{x-\bar{x}}{S_x}$	0.00	0.01	0.02	0.03	0.04	0.05	0.06	0.07	0.08	0.09
0.0	.5000	.4960	.4920	.4880	.4840	.4801	.4761	.4721	.4681	.4641
0.1	.4602	.4562	.4522	.4483	.4443	.4404	.4364	.4325	.4286	.4247
0.2	.4207	.4168	.4129	.4090	.4052	.4013	.3974	.3936	.3897	.3859
0.3	.3821	.3783	.3745	.3707	.3669	.3632	.3594	.3557	.3520	.3483
0.4	.3446	.3409	.3372	.3336	.3300	.3264	.3228	.3192	.3156	.3121
0.5	.3085	.3050	.3015	.2981	.2946	.2912	.2877	.2843	.2810	.2776
0.6	.2743	.2709	.2676	.2643	.2611	.2578	.2546	.2514	.2483	.2451
0.7	.2420	.2389	.2358	.2327	.2296	.2266	.2236	.2206	.2177	.2148
0.8	.2119	.2090	.2061	.2033	.2005	.1977	.1949	.1922	.1984	.1867
0.9	.1841	.1814	.1788	.1762	.1736	.1711	.1685	.1660	.1635	.1611
1.0	.1587	.1562	.1539	.1515	.1492	.1469	.1446	.1423	.1401	.1379
1.1	.1357	.1335	.1314	.1292	.1271	.1251	.1230	.1210	.1190	.1170
1.2	.1151	.1131	.1112	.1093	.1075	.1056	.1038	.1020	.1003	.0985
1.3	.0968	.0951	.0934	.0918	.0901	.0885	.0869	.0853	.0838	.0823
1.4	.0808	.0793	.0778	.0764	.0749	.0735	.0721	.0708	.0694	.0681
1.5	.0668	.0655	.0643	.0630	.0618	.0606	.0594	.0582	.0571	.0559
1.6	.0548	.0537	.0526	.0516	.0505	.0495	.0485	.0475	.0465	.0455
1.7	.0446	.0436	.0427	.0418	.0409	.0401	.0392	.0384	.0375	.0367
1.8	.0359	.0351	.0344	.0336	.0329	.0322	.0314	.0307	.0301	.0294
1.9	.0287	.0281	.0274	.0268	.0262	.0256	.0250	.0244	.0239	.0233
2.0	.0228	.0222	.0217	.0212	.0207	.0202	.0197	.0192	.0188	.0183
2.1	.0179	.0174	.0170	.0166	.0162	.0158	.0154	.0150	.0146	.0143
2.2	.0139	.0136	.0132	.0129	.0125	.0122	.0119	.0116	.0113	.0110
2.3	.0107	.0104	.0102	.00990	.00964	.00939	.00914	.00889	.00866	.0084
2.4	.00820	.00798	.00776	.00755	.00734	.00714	.00695	.00676	.00657	.00639
2.5	.00621	.00604	.00587	.00570	.00554	.00539	.00523	.00508	.00494	.00480
2.6	.00466	.00453	.00440	.00427	.00415	.00402	.00391	.00379	.00368	.00357
2.7	.00347	.00336	.00326	.00317	.00307	.00298	.00288	.00280	.00272	.00264
2.8	.00256	.00248	.00240	.00233	.00226	.00219	.00212	.00205	.00199	.00193
2.9	.00187	.00181	.00175	.00169	.00164	.00159	.00154	.00149	.00144	.00139

Thus, the probability of drawing a single sample with a yield strength less than 65 ksi is 1.3 percent.

13.3.2 Weibull Distribution

In investigations of the strength of materials as a result of brittle fracture, crack-initiation toughness, or fatigue life, researchers often find that the Weibull distribution provides a more suitable approach to the statistical analysis of the available data. The Weibull distribution function $P(x)$ is defined as

$$P(x) = 1 - e^{-[(x-x_0)/b]^m} \qquad \text{for } x > x_0 \qquad (13.11)$$

$$P(x) = 0 \qquad \text{for } x < x_0$$

where x_0, b, and m are the three parameters that define this distribution function. In studies of strength, $P(x)$ is taken as the probability of failure when a stress x is placed on the specimen. The parameter x_0 is the zero strength, since $P(x) = 0$ for $x < x_0$. The constants b and m are known as the scale parameter and the Weibull slope parameter (modulus), respectively.

Four Weibull distribution curves are presented in Fig. 13.6 for the case where $x_0 = 3$, $b = 10$, and $m = 2, 5, 10$, and 20. These curves illustrate two important features of the Weibull distribution. First, there is a threshold strength x_0, and if the applied stress is less than x_0, the probability of failure is zero. Second, the Weibull distribution curves are not symmetric, and the distortion in the S-shaped curves is controlled by the Weibull slope parameter m. Application of the Weibull distribution to predict failure rates of 1 percent or less of the population is particularly important in engineering projects that require reliabilities of 99 percent or greater.

Use of the Weibull distribution requires knowledge of the Weibull parameters. It is necessary to conduct experiments and obtain a relatively large data set for the determination of x_0, b, and m. Consider as an illustration Weibull's own work in statistically characterizing the fiber strength of Indian cotton. An unusually large sample ($n = 3000$) was studied by measuring the load to fracture (in grams) for each fiber. The strength data obtained were placed in sequential order with the lowest value corresponding to $k = 1$ first and the largest value corresponding to $k = 3000$ last. Using this example, the probability of failure $P(x)$ at a load x can be determined from

$$P = \frac{k}{n + 1} \qquad (13.12)$$

where

k is the order number of the sequenced data
n is the total sample size

At this stage it is possible to prepare a graph of probability of failure $P(x)$ as a function of strength x to obtain a curve similar to that shown in Fig. 13.6. However, to determine the Weibull parameters x_0, b, and m requires additional conditioning of the data. From Eq. 13.11, it is evident that

$$e^{[(x-x_0)/b]^m} = [1 - P(x)]^{-1} \qquad (13.13)$$

Taking the natural log of both sides of Eq. 13.13 yields

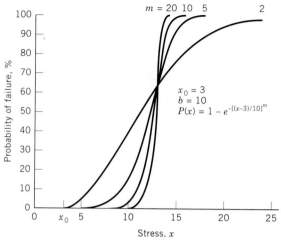

Figure 13.6 The Weibull distribution function.

$$[(x - x_0)/b]^m = \ln[1 - P(x)]^{-1} \tag{13.14}$$

Taking $\log_{10}$ of both sides of Eq. 13.14 gives a relation for the slope parameter m. Thus,

$$m = \frac{\log_{10} \ln[1 - P(x)]^{-1}}{\log_{10}(x - x_0) - \log_{10} b} \tag{13.15}$$

The numerator of Eq. 13.15 is the reduced variate $y = \log_{10} \ln[1 - P(x)]^{-1}$ used for the ordinate in preparing a graph of the conditioned data as indicated in Fig. 13.7. Note that y is a function of P alone and for this reason both the P and y scales can be displayed on the ordinates (see Fig. 13.7). The lead term in the denominator of Eq. 13.15 is the reduced variate $x = \log_{10}(x - x_0)$ used for the abscissa in Fig. 13.7.

In the Weibull example, x_0 was adjusted to 0.46 gr so that the data would fall on a straight line when plotted against the reduced x and y variates. The constant b is determined from the condition that

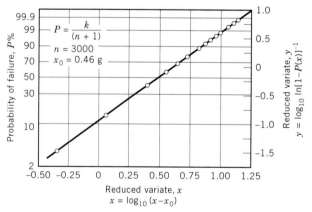

Figure 13.7 Fiber strength of Indian cotton shown as a graph with Weibull's reduced variate (from data by Weibull).

$$\log_{10} b = \log_{10}(x - x_0) \qquad \text{when } y = 0 \qquad \textbf{(13.16)}$$

Note from Fig. 13.7 that $y = 0$ when $\log_{10}(x - x_0) = 0.54$, which gives $b = 0.54$. Finally, m is given by the slope of the straight line when the data are plotted in terms of the reduced variates x and y. In this example problem, $m = 1.48$.

13.4   CONFIDENCE INTERVALS FOR PREDICTIONS

Once experimental data are represented with a normal distribution by using estimates of the mean $\bar{x}$ and standard deviation S_x and predictions are made about the occurrence of measurements, questions arise concerning the confidence that can be placed on either the estimates or the predictions. One cannot be totally confident in the predictions or estimates because of the effects of sampling error.

Sampling error can be illustrated by drawing a series of samples (each containing n measurements) from the same population and determining several estimates of the mean $\bar{x}_1, \bar{x}_2, \bar{x}_3, \cdots$. A variation in $\bar{x}$ will occur, but, fortunately, this variation can also be characterized by a normal distribution function, as shown in Fig. 13.8. The mean of the x and $\bar{x}$ distributions is the same; however, the standard deviation of the $\bar{x}$ distribution $S_{\bar{x}}$ (sometimes referred to as the standard error) is less than S_x, since

$$S_{\bar{x}} = \frac{S_x}{\sqrt{n}} \qquad \textbf{(13.17)}$$

Once the standard deviation of the population of $\bar{x}$'s is known, it is possible to place confidence limits on the determination of the true population mean μ from a sample of size n, provided that n is large ($n > 25$). The confidence interval within which the true population mean μ is located is given by the expression

$$(\bar{x} - z S_{\bar{x}}) < \mu < (\bar{x} + z S_{\bar{x}}) \qquad \textbf{(13.18)}$$

where

$\bar{x} - z S_{\bar{x}}$ is the lower confidence limit
$\bar{x} + z S_{\bar{x}}$ is the upper confidence limit

The width of the confidence interval depends on the confidence level required. For instance, if $z = 3$ in Eq. 13.18, a relatively wide confidence interval exists; therefore, the probability that the population mean μ will be located within the confidence interval is high (99.7 percent). As the width of the confidence interval

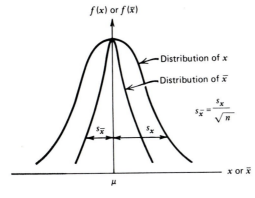

Figure 13.8 Normal distribution of individual measurements of the quantity x and of measurements of the mean $\bar{x}$ from samples of size n.

Table 13.5 Confidence Interval Variation
with Confidence Level Interval $= \bar{x} + zS_{\bar{x}}$

Confidence Level, %	z	Confidence Level, %	z
99.9	3.30	90.0	1.65
99.7	3.00	80.0	1.28
99.0	2.57	68.3	1.00
95.0	1.96	60.0	0.84

decreases, the probability that the population mean μ will fall within the interval decreases. Commonly used confidence levels and their associated intervals are shown in Table 13.5.

When the sample size is very small ($n < 20$), the standard deviation S_x does not provide a reliable estimate of the standard deviation σ of the population and Eq. 13.18 should not be employed. The bias introduced by small sample size can be removed by modifying Eq. 13.18 to read

$$[\bar{x} - t(\alpha)S_{\bar{x}}] < \mu < [\bar{x} + t(\alpha)S_{\bar{x}}] \tag{13.19}$$

where

$t(\alpha)$ is the statistic known as *Student's t*
α is the level of significance (the probability of exceeding a given value of t)

The Student's t distribution is defined by a relative-frequency equation $f(t)$, which can be expressed as

$$f(t) = F_0\left(1 + \frac{t^2}{\nu}\right)^{(\nu+1)/2} \tag{13.20}$$

where

F_0 is the relative frequency at $t = 0$ required to make the total area under the $f(t)$ curve equal to unity
ν is the number of degrees of freedom

The distribution function $f(t)$ is shown in Fig. 13.9 for several different degrees of freedom ν. The degrees of freedom equal the number of independent measurements

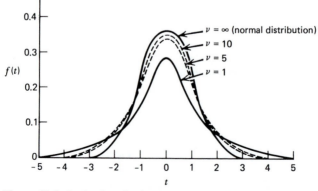

Figure 13.9 Student's t distribution for several degrees of freedom ν.

employed in the determination. It is evident that as ν becomes large, Student distribution approaches the normal distribution. One-side areas for the t distribution are listed in Table 13.6 and illustrated in Fig. 13.10.

The term $t(\alpha)S_{\bar{x}}$ in Eq. 13.19 represents the measure from the estimated mean $\bar{x}$ to one or the other of the confidence limits. This term may be used to estimate the sample size required to produce an estimate of the mean $\bar{x}$ with a specified reliability.

Table 13.6 Student's t Distribution for ν Degrees of Freedom Showing $t(\alpha)$ as a Function of Area A (One Side)

ν	0.995	0.99	0.975	0.95	0.90	0.80	0.75	0.70	0.60	0.55
					Confidence level α					
1	63.66	31.82	12.71	6.31	3.08	1.376	1.000	.727	.325	.158
2	9.92	6.96	4.30	2.92	1.89	1.061	.816	.617	.289	.142
3	5.84	4.54	3.18	2.35	1.64	.978	.765	.584	.277	.137
4	4.60	3.75	2.78	2.13	1.53	.941	.741	.569	.271	.134
5	4.03	3.36	2.57	2.02	1.48	.920	.727	.559	.267	.132
6	3.71	3.14	2.45	1.94	1.44	.906	.718	.553	.265	.131
7	3.50	3.00	2.36	1.90	1.42	.896	.711	.549	.263	.130
8	3.36	2.90	2.31	1.86	1.40	.889	.706	.546	.262	.130
9	3.25	2.82	2.26	1.83	1.38	.883	.703	.543	.261	.129
10	3.17	2.76	2.23	1.81	1.37	.879	.700	.542	.260	.129
11	3.11	2.72	2.20	1.80	1.36	.876	.697	.540	.260	.129
12	3.06	2.68	2.18	1.78	1.36	.873	.695	.539	.259	.128
13	3.01	2.65	2.16	1.77	1.35	.870	.694	.538	.259	.128
14	2.98	2.62	2.14	1.76	1.34	.868	.692	.537	.258	.128
15	2.95	2.60	2.13	1.75	1.34	.866	.691	.536	.258	.128
16	2.92	2.58	2.12	1.75	1.34	.865	.690	.535	.258	.128
17	2.90	2.57	2.11	1.74	1.33	.863	.689	.534	.257	.128
18	2.88	2.55	2.10	1.73	1.33	.862	.688	.534	.257	.127
19	2.86	2.54	2.09	1.73	1.33	.861	.688	.533	.257	.127
20	2.84	2.53	2.09	1.72	1.32	.860	.687	.533	.257	.127
21	2.83	2.52	2.08	1.72	1.32	.859	.686	.532	.257	.127
22	2.82	2.51	2.07	1.72	1.32	.858	.686	.532	.256	.127
23	2.81	2.50	2.07	1.71	1.32	.858	.685	.532	.256	.127
24	2.80	2.49	2.06	1.71	1.32	.857	.685	.531	.256	.127
25	2.79	2.48	2.06	1.71	1.32	.856	.684	.531	.256	.127
26	2.78	2.48	2.06	1.71	1.32	.856	.684	.531	.256	.127
27	2.77	2.47	2.05	1.70	1.31	.855	.684	.531	.256	.127
28	2.76	2.47	2.05	1.70	1.31	.855	.683	.530	.256	.127
29	2.76	2.46	2.04	1.70	1.31	.854	.683	.530	.256	.127
30	2.75	2.46	2.04	1.70	1.31	.854	.683	.530	.256	.127
40	2.70	2.42	2.02	1.68	1.30	.851	.681	.529	.255	.126
60	2.66	2.39	2.00	1.67	1.30	.848	.679	.527	.254	.126
120	2.62	2.36	1.98	1.66	1.29	.845	.677	.526	.254	.126
∞	2.58	2.33	1.96	1.65	1.28	.842	.674	.524	.253	.126

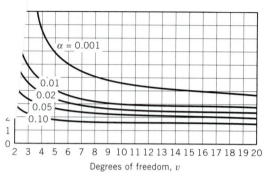

Degrees of freedom, v

Figure 13.10 Student's t statistic as a function of degrees of freedom v with α, the probability of exceeding t, as a parameter.

Noting that one-half the bandwidth of the confidence interval is $\delta = t(\alpha)S_{\bar{x}}$ and using Eq. 13.17, it is apparent that the sample size is given by

$$n = \left[\frac{t(\alpha)S_x}{\delta}\right]^2 \tag{13.21}$$

Use of Eq. 13.21 can be illustrated by considering the data in Table 13.1, where $S_x = 6.04$ ksi and $\bar{x} = 78.4$ ksi. If this estimate of μ is to be accurate to ± 5 percent with a reliability of 95 percent, then

$$\delta = (0.05)(78.4) = 3.92$$

Since $t(\alpha)$ depends on n, a trial-and-error solution is needed to establish the sample size n needed to satisfy the specifications. For the data of Table 13.1, $n = 20$; therefore, $v = 19$ and $t(\alpha) = t(0.975) = 2.09$, from Table 13.6. The value $t(\alpha) = t(0.975)$ is used, since 2.5 percent of the distribution must be excluded on each end of the curve to give a two-sided area corresponding to a reliability of 95 percent. Substituting into Eq. 13.21 yields

$$n = [(2.09)(6.04)/(3.92)^2] = 10.4$$

With $n = 11$, $v = 10$, and $t(\alpha) = 2.23$,

$$n = [(2.23)(6.04)/(3.92)^2] = 11.8$$

Finally, with $n = 12$, $v = 11$, and $t(\alpha) = 2.20$,

$$n = [(2.20)(6.04)/(3.92)^2] = 11.5$$

Thus, a sample size of 12 would be sufficient to ensure an accuracy of ± 5 percent with a confidence level of 95 percent. The sample size of 20 listed in Table 13.1 is too large for the degree of accuracy and confidence level specified. This simple example illustrates how sample size can be reduced and cost savings effected by using statistical methods to determine sample size.

13.5   COMPARISON OF MEANS

Since Student's t distribution compensates for the effect of small-sample bias and converges to the normal distribution in large samples, it is a useful statistic in engineering applications. A second important application utilizes the t distribution as the basis for a test to determine if the difference between two means is significant or due to random variation. For example, consider the yield-strength data of Table 13.1, where $n_1 = 20$, $\overline{x}_1 = 78.4$ ksi, and $S_{x1} = 6.04$ ksi. Suppose that a second sample from another supplier is tested to determine the yield strength and the results are $n_2 = 25$, $\overline{x}_2 = 81.6$ ksi, and $S_{x2} = 5.56$ ksi. Is the steel from the second supplier superior in terms of yield strength? The standard deviation of the difference in means $S_{(\overline{x}_2 - \overline{x}_1)}$ can be expressed as

$$S^2_{(\overline{x}_2 - \overline{x}_1)} = S^2_p \left(\frac{1}{n_1} + \frac{1}{n_2} \right) = S^2_p \frac{n_1 + n_2}{n_1 n_2} \tag{13.22}$$

where S^2_p is the pooled variance that can be expressed as

$$S^2_p = \frac{(n_1 - 1)S^2_{x1} + (n_2 - 1)S^2_{x2}}{n_1 + n_2 - 2} \tag{13.23}$$

The statistic t can be computed from the expression

$$t = \frac{|\overline{x}_2 - \overline{x}_1|}{S_{(\overline{x}_2 - \overline{x}_1)}} \tag{13.24}$$

A comparison of the value of t determined from Eq. 13.24 with a value of $t(\alpha)$ obtained from Table 10.6 provides a statistical basis for deciding whether the difference in means is real or due to random variations. The value of $t(\alpha)$ to be used depends on the degrees of freedom $\nu = n_1 + n_2 - 2$ and the level of significance required. Levels of significance commonly employed are 5 percent and 1 percent. The 5 percent level of significance means that the probability of a random variation being taken for a real difference is only 5 percent. Comparisons at the 1 percent level of significance are 99 percent certain; however, in such a strong test, real differences can often be attributed to random error.

In the example being considered Eq. 13.23 yields $S^2_p = 33.37$ ksi, Eq. 13.22 yields $S^2_{(\overline{x}_2 - \overline{x}_1)} = 3.00$ ksi, and Eq. 13.24 yields $t = 1.848$. For a 5 percent level of significance test with $\nu = 43$ and $\alpha = 0.05$ (the comparison is one-sided, since the t test is for superiority), Table 13.6 indicates that $t(\alpha) = 1.68$. Since $t > t(\alpha)$, it can be concluded with a 95 percent level of confidence that the yield strength of the steel from the second supplier is higher than the yield strength of steel from the first supplier.

13.6   STATISTICAL CONDITIONING OF DATA

As previously indicated, measurement error can be characterized by a normal distribution function and the standard deviation of the estimated mean $S_{\overline{x}}$ can be reduced by increasing the number of measurements. In most situations, cost places an upper limit on the number of measurements made. Also, because systematic error is not a random variable, statistical procedures cannot serve as a substitute for precise, accurately calibrated, and properly zeroed measuring instruments.

One area in which statistical procedures can be used very effectively to condition experimental data is with the erroneous data point that results from a measuring or recording mistake. Often, this data point appears questionable when it is compared with the other data collected, and the experimentalist must decide whether the deviation of the data point is due to a mistake (and, hence, to be rejected) or due to some unusual but real condition (and, hence, to be retained). A statistical procedure known as Chauvenet's criterion provides a consistent basis for making the decision to reject or retain such a point from a sample that contains several readings.

Application of Chauvenet's criterion requires computation of a deviation ratio DR for each data point, followed by comparison with a standard deviation ratio DR_0. The standard deviation ratio DR_0 is a statistic that depends on the number of measurements, whereas the deviation ratio DR for a point is defined as

$$DR = \frac{x_i - \overline{x}}{S_x} \tag{13.25}$$

The data point is rejected when $DR > DR_0$ and retained when $DR \leq DR_0$. Values for the standard deviation ratio DR_0 are listed in Table 13.7.

If the statistical test of Eq. 13.25 indicates that a single data point in a sequence of n data points should be rejected, then the data point should be removed from the sequence and the mean $\overline{x}$ and the standard deviation S_x should be recalculated. Chauvenet's method can be applied only once to reject a data point that is questionable from a sequence of points. If several data points indicate that $DR > DR_0$, then it is likely that the instrumentation system is inadequate or that the process being investigated is extremely variable.

13.7   REGRESSION ANALYSIS

Many experiments involve the measurement of one dependent variable, for example, y, which may depend on one or more independent variables, $x_1, x_2, \cdots, x_k$. Regression analysis provides a statistical approach for conditioning the data obtained from experiments in which two or more related quantities are measured.

Table 13.7 Deviation Ratio DR_0 Used for Statistical Conditioning of Data

Number of Measurements n	Deviation Ratio DR_0	Number of Measurements n	Deviation Ratio DR_0
4	1.54	25	2.33
5	1.65	50	2.57
7	1.80	100	2.81
10	1.96	300	3.14
15	2.13	500	3.29

13.7.1 Linear Regression Analysis

Suppose measurements are made of two quantities that describe the behavior of a process exhibiting variation. Let y be the dependent variable and x the independent variable. Since the process exhibits variation, there is not a unique relationship between x and y, and the data, when plotted, exhibit scatter, as illustrated in Fig. 13.11. Frequently, the relation between x and y that most closely represents the data, even with the scatter, is a linear function. Thus,

$$Y_i = m x_i + b \qquad (13.26)$$

where Y_i is the predicted value of the dependent variable y_i for a given value of the independent variable x_i.

A statistical procedure used to fit a straight line through scattered data points is called the *least-squares method*. With the least-squares method, the slope m and the intercept b in Eq. 13.26 are selected to minimize the sum of the squared deviations of the data points from the straight line, as shown in Fig. 13.11. In utilizing the least-squares method, it is assumed that the independent variable x is free of measurement error and the quantity

$$\Delta^2 = \sum (y_i - Y_i)^2 \qquad (13.27)$$

is minimized at fixed values of x. After substituting Eq. 13.26 into Eq. 13.27 this implies that

$$\frac{\partial \Delta^2}{\partial b} = \frac{\partial}{\partial b} \sum (y_i - m x - b)^2 = 0$$

$$\frac{\partial \Delta^2}{\partial m} = \frac{\partial}{\partial m} \sum (y_i - m x - b)^2 = 0$$

$$\text{(a)}$$

Differentiating yields

$$2 \sum (y_i - m x - b)(-x) = 0$$

$$2 \sum (y_i - m x - b)(-1) = 0 \qquad \text{(b)}$$

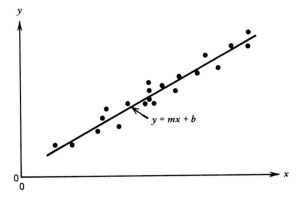

Figure **13.11** Linear regression analysis is used to fit a least-squares line through scattered data points.

Solving Eqs. b for m and b yields

$$m = \frac{\sum x \sum y - n \sum xy}{(\sum x)^2 - n \sum x^2}$$

$$b = \frac{\sum y - m \sum x}{n}$$

(13.28)

where n is the number of data points. The slope m and intercept b define a straight line through the scattered data points such that Eq. 13.27 is minimized.

In any regression analysis it is important to establish the correlation between x and y. Equation 13.26 does not exactly predict the values that were measured because of the variation in the process. To illustrate, assume that the independent quantity x is fixed at a value x_1 and that a sequence of measurements is made of the dependent quantity y. The data obtained would give a distribution of y as illustrated in Fig. 13.12. The dispersion of the distribution of y is a measure of the correlation. When the dispersion is small, the correlation is good and the regression analysis is effective in describing the variation in y. When the dispersion is large, the correlation is poor and the regression analysis may not be adequate to describe the variation in y.

The adequacy of a regression analysis can be evaluated by determining a correlation coefficient R^2 that is given by the following expression:

$$R^2 = 1 - \frac{n-1}{n-2}\left[\frac{\{y^2\} - m\{xy\}}{\{y^2\}}\right]$$

(13.29)

where

$$\{y^2\} = \sum y^2 - (\sum y)^2/n$$

$$\{xy\} = \sum xy - (\sum x)(\sum y)/n$$

When the value of the correlation coefficient $R^2 = 1$, perfect correlation exists between y and x. If $R^2 = 0$, no correlation exists and the variations observed in y are

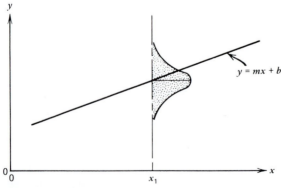

Figure 13.12 Distribution of y at a fixed value of x superimposed on the linear regression display.

Table 13.8 Probability of Obtaining a
Correlation Coefficient R^2 Due to Random
Variations in y

	Probability			
n	0.10	0.05	0.02	0.01
5	0.805	0.878	0.934	0.959
6	0.729	0.811	0.882	0.917
7	0.669	0.754	0.833	0.874
8	0.621	0.707	0.789	0.834
10	0.549	0.632	0.716	0.765
15	0.441	0.514	0.592	0.641
20	0.378	0.444	0.516	0.561
30	0.307	0.362	0.423	0.464
40	0.264	0.312	0.367	0.403
60	0.219	0.259	0.306	0.337
80	0.188	0.223	0.263	0.291
100	0.168	0.199	0.235	0.259

due to random fluctuations and not changes in x. Because random variations in y exist, a value of $R^2 = 1$ is not obtained even if $y(x)$ is linear. To interpret correlation coefficients $0 < R^2 < 1$, the data in Table 13.8 are used to establish the probability of obtaining a given R^2 owing to random variations in y.

As an example, consider a regression analysis with $n = 15$, which gives $R^2 = 0.65$ as determined by Eq. 13.29. Reference to Table 13.8 indicates that the probability of obtaining $R^2 = 0.65$ as a result of random variations is slightly less than 1 percent. Thus one can be 99 percent certain that the regression analysis represents a true correlation between y and x.

13.7.2 Multivariate Regression

Many experiments involve measurements of a dependent variable y, which depends on several independent variables x_1, x_2, x_3, and so on. It is possible to represent y as a function of x_1, x_2, x_3, and so on by employing the multivariate regression equation

$$Y_i = a + b_1 x_1 + b_2 x_2 + \cdots + b_k x_k \tag{13.30}$$

where a, b_1, b_2, $\cdots$, b_k are regression coefficients.

The regression coefficients a, b_1, b_2, $\cdots$, b_k are determined by using the method of least squares in a manner similar to that employed for linear regression analysis where the quantity $\Delta^2 = \sum (y_i - Y_i)^2$ is minimized. Substituting Eq. 13.27 into Eq. 13.30 yields

$$\Delta^2 = \sum (y_i - a - b_1 x_1 - b_2 x_2 - \cdots - b_k x_k)^2 \tag{13.31}$$

Differentiating yields

$$\frac{\partial \Delta^2}{\partial a} = 2\left[\sum(y_i - a - b_1 x_1 - b_2 x_2 - \cdots - b_k x_k)(-1)\right] = 0$$

$$\frac{\partial \Delta^2}{\partial b_1} = 2\left[\sum(y_i - a - b_1 x_1 - b_2 x_2 - \cdots - b_k x_k)(-x_1)\right] = 0$$

$$\frac{\partial \Delta^2}{\partial b_2} = 2\left[\sum(y_i - a - b_1 x_1 - b_2 x_2 - \cdots - b_k x_k)(-x_2)\right] = 0 \quad \textbf{(13.32)}$$

$$\cdots = \qquad\qquad\qquad\qquad\qquad\qquad \cdots = 0$$

$$\frac{\partial \Delta^2}{\partial b_k} = 2\left[\sum(y_i - a - b_1 x_1 - b_2 x_2 - \cdots - b_k x_k)(-x_k)\right] = 0$$

Equations 13.32 lead to the following set of $k + 1$ equations, which can be solved for the unknown regression coefficients $a, b_1, b_2, \cdots, b_k$.

$$an + b_1 \sum x_1 + b_2 \sum x_2 + \cdots + b_k \sum x_k = \sum y_i$$

$$a \sum x_1 + b_1 \sum x_1^2 + b_2 \sum x_1 x_2 + \cdots + b_k \sum x_1 x_k = \sum y_i x_i$$

$$a \sum x_2 + b_1 \sum x_1 x_2 + b_2 \sum x_2^2 + \cdots + b_k \sum x_1 x_k = \sum y_i x_2 \quad \textbf{(13.33)}$$

$$\cdots = \cdots$$

$$a \sum x_k + b_1 \sum x_1 x_k + b_2 \sum x_2 x_k + \cdots + b_k \sum x_k^2 = \sum y_i x_k$$

The correlation coefficient R^2 is again used to determine the degree of association between the dependent and independent variables. For multiple regression equations, the correlation coefficient R^2 is given as

$$R^2 = 1 - \frac{n - 1}{n - k}\left[\frac{\{y^2\} - b_1\{y x_1\} - b_2\{y x_2\} - \cdots - b_k\{y x_k\}}{\{y^2\}}\right] \quad \textbf{(13.34)}$$

where

$$\{y x_k\} = \sum y x_k - \frac{(\sum y)(\sum x_k)}{n}$$

$$\{y^2\} = \sum y^2 - \frac{(\sum y)^2}{n}$$

This analysis is for linear, noninteracting, independent variables; however, the analysis can be extended to include cases in which the regression equations have higher order and cross-product terms. The nonlinear terms can enter the regression equation in an additive manner and are treated as extra variables. With well-established computer routines for regression analysis, the set of $(k+1)$ simultaneous equations given by Eqs. 13.33 can be solved quickly and inexpensively and no difficulties are encountered in adding extra terms to account for nonlinearities and interactions.

13.8 CHI-SQUARE TESTING

The chi-square (χ^2) test is used in statistics to verify the use of a specific distribution function to represent the population from which a set of data has been obtained. The chi-square statistic is defined as

$$\chi^2 = \sum_{i=1}^{k} \left[\frac{(n_o - n_e)^2}{n_e} \right]_i \qquad (13.35)$$

where

n_o is the actual number of observations in the ith group interval
n_e is the expected number of observations in the ith group interval based on the specified distribution
k is the total number of group intervals

The value of χ^2 is computed to determine how well the data fit the assumed statistical distribution. If $\chi^2 = 0$, the match is perfect. Values of $\chi^2 > 0$ indicate the possibility that the data are not represented by the specified distribution. The probability P that the value of χ^2 is due to random variation is listed in Table 13.9 and illustrated in Fig. 13.13. The degree of freedom is defined as

$$\nu = n - k \qquad (13.36)$$

where

n is the number of observations
k is the number of conditions imposed on the distribution

As an example of the χ^2 test, consider the yield-strength data presented in Table 13.1 and judge the adequacy of representing the yield strength with a normal probability distribution described with $\bar{x} = 78.4$ ksi and $S_x = 6.04$ ksi. By using the properties of a normal distribution function, the number of specimens expected to fall in any strength group can be computed. The observed number of specimens in Table 13.1 exhibiting yield strengths within each of three group intervals, together with the computed number of specimens in a normal distribution in the same group intervals, are listed in Table 13.10.

The number of groups is 3 $(n = 3)$, and since the two distribution parameters $\bar{x}$ and S_x were determined by using these data $(k = 2)$, the number of degrees of freedom is $\nu = n - k = 3 - 2 = 1$.

Plotting these results $(\nu = 1$ and $\chi^2 = 0.67)$ in Fig. 13.13 shows that the point falls in the region where there is no reason to expect that the hypothesis is not correct. The hypothesis is to represent the yield strength with a normal probability function. The χ^2 test does not prove the validity of this hypothesis, but fails to disprove it.

The lines dividing the $\chi^2 - \nu$ graph of Fig. 13.13 into five different regions are based on probabilities of obtaining χ^2 values greater than the values shown by the

Table 13.9 Chi-Squared (χ^2) Values with Different Degrees
of Freedom for Different Probability Levels

Degrees of Freedom	Probability (%)					
	1.0	2.5	5.0	10.0	20.0	30.0
1	.00016	.00098	.00393	.0158	.0642	.148
2	.0201	.0506	.103	.211	.446	.713
3	.115	.216	.352	.584	1.00	1.42
4	.297	.484	.711	1.06	1.65	2.19
5	.554	.831	1.15	1.61	2.34	3.00
6	.872	1.24	1.64	2.20	3.07	3.83
7	1.24	1.69	2.17	2.83	3.82	4.67
8	1.65	2.18	2.73	3.49	4.59	5.53
9	2.09	2.70	3.33	4.17	5.38	6.39
10	2.56	3.25	3.94	4.87	6.18	7.27
11	3.05	3.82	4.57	5.58	6.99	8.15
12	3.57	4.40	5.23	6.30	7.81	9.03
13	4.11	5.01	5.89	7.04	8.63	9.93
14	4.66	5.63	6.57	7.79	9.47	10.8
15	5.23	6.26	7.26	8.55	10.3	11.7
20	8.26	9.59	10.9	12.4	14.6	16.3
25	11.5	13.1	14.6	16.5	18.9	20.9
30	15.0	16.8	18.5	20.6	23.4	25.5
40	22.2	24.4	26.5	29.1	32.3	34.9
50	29.7	32.4	34.8	37.7	41.4	44.3
100	70.1	74.2	77.9	82.4	87.9	92.1

curves. For example, the line dividing the regions "No reason to reject hypothesis" and "Hypothesis probably not correct" has been selected at a probability level of 10 percent. Thus, there is only one chance in 10 that data drawn from a population correctly represented by the hypothesis would give a χ^2 value exceeding that specified by the $P > 0.10$ curve. The "Hypothesis not correct" region is defined with the $P > 0.01$ curve, indicating only one chance in 100 of obtaining a χ^2 value exceeding those shown by this curve.

The χ^2 function can also be used to question whether the data have been adjusted. Probability levels of 0.90 and 0.99 have been used to define the regions "Data suspect as too good" and "Data may be falsified." For the latter classification there are 99 chances out of 100 that the χ^2 value will exceed that determined by a χ^2 analysis of the data.

The χ^2 statistic can also be used in contingency testing where the sample is classified under one of two categories, such as pass or fail. Consider, for example, an inspection procedure with a particular type of strain gage where 10 percent of the gages are rejected because of etching imperfections in the grid. In an effort to reduce this rejection rate, the manufacturer has introduced new clean-room techniques that are expected to improve the quality of the grids. On the first lot of 2000 gages, the failure

Table 13.9 (*continued*)

Degrees of Freedom	Probability (%)							
	40.0	50.0	60.0	70.0	80.0	90.0	95.0	99.0
1	.275	.455	.708	1.07	1.64	2.71	3.84	6.63
2	1.02	1.39	1.83	2.41	3.22	4.61	5.99	9.21
3	1.87	2.37	2.95	3.67	4.64	6.25	7.81	11.3
4	2.75	3.36	4.04	4.88	5.99	7.78	9.49	13.3
5	3.66	4.35	5.13	6.06	7.29	9.24	11.1	15.1
6	4.57	5.35	6.21	7.23	8.56	10.6	12.6	16.8
7	5.49	6.35	7.28	8.38	9.80	12.0	14.1	18.5
8	6.42	7.34	8.35	9.52	11.0	13.4	15.5	20.1
9	7.36	8.34	9.41	10.7	12.2	14.7	16.9	21.7
10	8.30	9.34	10.5	11.8	13.4	16.0	18.3	23.2
11	9.24	10.3	11.5	12.9	14.6	17.3	19.7	24.7
12	10.2	11.3	12.6	14.0	15.8	18.5	21.0	26.2
13	11.1	12.3	13.6	15.1	17.0	19.8	22.4	27.7
14	12.1	13.3	14.7	16.2	18.2	21.1	23.7	29.1
15	13.0	14.3	15.7	17.3	19.3	22.3	25.0	30.6
20	17.8	19.3	21.0	22.8	25.0	28.4	31.4	37.6
25	22.6	24.3	26.1	28.2	30.7	34.4	37.7	44.3
30	27.4	29.3	31.3	33.5	36.3	40.3	43.8	50.9
40	37.1	39.3	41.6	44.2	47.3	51.8	55.8	63.7
50	46.9	49.3	51.9	54.7	58.2	63.2	67.5	76.2
100	95.8	99.3	102.9	106.9	111.7	118.5	124.3	135.8

rate was reduced to 8 percent. Is this reduced failure rate the result of chance variation, or have the new clean-room techniques improved the manufacturing process? A χ^2 test can establish the probability of the improvement being the result of random variation. The computation of χ^2 for this example is illustrated in Table 13.11.

Plotting the results from Table 13.11 on Fig. 13.13 after noting that $\nu = 1$ shows that $\chi^2 = 8.89$ falls into the region "Hypothesis not correct." In this case the hypothesis is that there has been no improvement. The χ^2 test has shown that there is less than one chance in 100 of the improvement in rejection rate (8 percent instead of 10 percent) being the result of random variables. Thus, it can be concluded with confidence that the new clean-room techniques were effective in improving yield.

13.9 ERROR ACCUMULATION AND PROPAGATION

Previous discussions of error have been limited to error arising in the measurement of a single quantity: however, in many engineering applications, several quantities are measured (each with its associated error) and another quantity is predicted on the basis of these measurements. For example, the volume V of a cylinder could be predicted on the basis of measurements of two quantities (diameter D and length L). Thus, errors in the measurements of diameter and length will propagate through

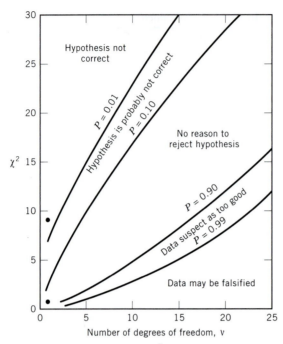

Figure 13.13 Probability of χ^2 values exceeding those shown as a function of the number of degrees of freedom.

Table 13.10 Chi-Square $\left(\chi^2\right)$ Computation for Grouped Yield-Strength Data

Group Interval	Number Observed	Number Expected	$(n_0 - n_e)^2 / n_e$
$0 - 74.99$	5	5.754	0.0988
$75.0 - 79.99$	8	6.298	0.4600
$80.0 - \infty$	7	7.948	0.1131
		Thus, from Eq. 13.35, $\chi^2 =$	0.6719

Table 13.11 Observed and Expected Inspection Results

Group Interval	Number Observed	Number Expected	$(n_0 - n_e)^2 / n_e$
Passed	1840	1800	0.89
Failed	160	200	8.00
		Thus, from Eq. 13.35, $\chi^2 =$	8.89

the governing mathematical formula $V = \pi D^2 L/4$ to the quantity (volume, i
case) being predicted. Since the propagation of error depends on the form of
mathematical expression used to predict the reported quantity, standard deviations for
several different mathematical operations are listed here.

For addition of quantities or subtraction of quantities ($y = x_1 \pm x_2 \pm \cdots \pm x_n$), the
standard deviation $S_{\bar{y}}$ of the mean $\bar{y}$ of the projected quantity y is

$$S_{\bar{y}} = \sqrt{S_{\bar{x}1}^2 + S_{\bar{x}2}^2 + \cdots + S_{\bar{x}n}^2} \qquad (13.37)$$

For multiplication of quantities ($y = x_1 x_2 \cdots x_n$), the standard deviation $S_{\bar{y}}$ is

$$S_{\bar{y}} = (\bar{x}_1 \bar{x}_2 \cdots \bar{x}_n)\sqrt{\frac{S_{\bar{x}1}^2}{\bar{x}_1^2} + \frac{S_{\bar{x}2}^2}{\bar{x}_2^2} + \cdots + \frac{S_{\bar{x}n}^2}{\bar{x}_n^2}} \qquad (13.38)$$

For division of quantities ($y = x_1/x_2$), the standard deviation $S_{\bar{y}}$ is

$$S_{\bar{y}} = \frac{\bar{x}_1}{\bar{x}_2}\sqrt{\frac{S_{\bar{x}1}^2}{\bar{x}_1^2} + \frac{S_{\bar{x}2}^2}{\bar{x}_2^2}} \qquad (13.39)$$

For calculations of the form ($y = x_1^k$), the standard deviation $S_{\bar{y}}$ is

$$S_{\bar{y}} = k\bar{x}_1^{k-1}S_{\bar{x}1} \qquad (13.40)$$

For calculations of the form ($y = x_1^{1/k}$), the standard deviation $S_{\bar{y}}$ is

$$S_{\bar{y}} = \frac{\bar{x}_1^{1/k}}{k\bar{x}_1}S_{\bar{x}1} \qquad (13.41)$$

Consider, for example, a rectangular rod where independent measurements of its
width, thickness, and length have yielded $\bar{x}_1 = 2.0$ with $S_{\bar{x}1} = 0.005$; $\bar{x}_2 = 0.5$
with $S_{\bar{x}2} = 0.002$; and $\bar{x}_3 = 16.5$ with $S_{\bar{x}3} = 0.040$, with all dimensions in inches.
Since the volume of the bar is

$$V = \bar{x}_1 \bar{x}_2 \bar{x}_3$$

the standard error of the volume can be determined by using Eq. 13.38. Thus,

$$S_{\bar{y}} = (\bar{x}_1 \bar{x}_2 \bar{x}_3)\sqrt{\frac{S_{\bar{x}1}^2}{\bar{x}_1^2} + \frac{S_{\bar{x}2}^2}{\bar{x}_2^2} + \frac{S_{\bar{x}3}^2}{\bar{x}_3^2}}$$

$$= (2.0)(0.5)(16.5)\sqrt{\frac{(0.005)^2}{(2.0)^2} + \frac{(0.002)^2}{(0.5)^2} + \frac{(0.040)^2}{(16.5)^2}}$$

$$= 0.0875 \text{ in.}^3$$

This determination of $S_{\bar{y}}$ for the volume of the bar can be used together with the
properties of a normal probability distribution to predict the number of bars with
volumes within specific limits.

od of computing the standard error of a quantity $S_{\bar{y}}$ as given by Eqs.
h 13.41, which are based on the properties of the normal probability
unction, should be used where possible. However, in many engineering
the number of measurements that can be made is small; therefore, the
$\cdots$, $\bar{x}_n$ and $S_{\bar{x}1}$, $S_{\bar{x}2}$, $\cdots$, $S_{\bar{x}n}$ needed for statistical estimates of the error
ble. In these instances, error estimates can still be made, but the results
... ress reliable.

A second method of estimating error, used when data are limited, is based on the
chain rule of differential calculus. For example, consider a quantity y that is a function
of several variables

$$y = f(x_1, x_2, \cdots, x_n) \tag{13.42}$$

Differentiating yields

$$dy = \frac{\partial y}{\partial x_1} dx_1 + \frac{\partial y}{\partial x_2} dx_2 + \cdots + \frac{\partial y}{\partial x_n} dx_n \tag{13.43}$$

In Eq. 13.43, dy is the error in y and dx_1, dx_2, $\cdots$, dx_n are errors involved in
the measurements of x_1, x_2, $\cdots$, x_n. The partial derivatives $\partial y/\partial x_1$, $\partial y/\partial x_2$, $\cdots$,
$\partial y/\partial x_n$ can be determined exactly from Eq. 13.42. Frequently, the errors dx_1, dx_2,
$\cdots$, dx_n are estimates based on the experience and judgment of the experimentalist.
An estimate of the maximum possible error can be obtained by summing the individual
error terms in Eq. 13.43. Thus,

$$dy]_{max} = |\frac{\partial y}{\partial x_1} dx_1| + |\frac{\partial y}{\partial x_2} dx_2| + \cdots + |\frac{\partial y}{\partial x_n} dx_n| \tag{13.44}$$

Use of Eq. 13.44 gives a worst-case estimate of error, since the maximum errors dx_1,
dx_2, $\cdots$, dx_n are assumed to occur simultaneously and with the same sign.

A more realistic equation for estimating error is obtained by squaring both sides of
Eq. 13.43 to give

$$(dy)^2 = \sum_{i=1}^{n} \left(\frac{\partial y}{\partial x_i}\right)^2 (dx_i)^2 + \sum_{i=1, j=1}^{n} \left(\frac{\partial y}{\partial x_i}\right)\left(\frac{\partial y}{\partial x_j}\right) dx_i dx_j \tag{13.45}$$

where $i \neq j$.

If the errors dx_i are independent and symmetrical with regard to positive and
negative values, then the cross-product terms will tend to cancel and Eq. 13.45 reduces
to

$$dy = \sqrt{\left(\frac{\partial y}{\partial x_1} dx_1\right)^2 + \left(\frac{\partial y}{\partial x_2} dx_2\right)^2 + \cdots + \left(\frac{\partial y}{\partial x_n} dx_n\right)^2} \tag{13.46}$$

13.10 SUMMARY

Statistical methods are extremely important in engineering, since they provide a means
for representing large amounts of data in a concise form that is easily interpreted and
understood. Usually, the data are represented with a statistical distribution function that

can be characterixed by a measure of central tendency (the mean $\bar{x}$) and a measure of dispersion (the standard deviation S_x). A normal, or Gaussian, probability distribution is by far the most commonly employed; however, in some cases, other distribution functions may have to be employed to adequately represent the data.

The most significant advantage resulting from use of a probability distribution function in engineering applications is the ability to predict the occurrence of an event based on a relatively small sample. The effects of sampling error are accounted for by placing confidence limits on the predictions and establishing the associated confidence levels. Sampling error can be controlled if the sample size is adequate. Use of the Student's t distribution function, which characterizes sampling error, provides a basis for determining sample size consistent with specified levels of confidence. The Student's t distribution also permits a comparison to be made of two means to determine whether the observed difference is significant or whether it is due to random variation.

Statistical methods can also be employed to condition data and to eliminate an erroneous data point (one) from a series of measurements. This is a useful technique that improves the data base by providing strong evidence when something unanticipated is affecting an experiment.

Regression analysis can be used effectively to interpret data when the behavior of one quantity (for example, y) depends on variations in one or more independent quantities (for example, $x_1, x_2, \cdots, x_n$). Even though the functional relationship between quantities exhibiting variation remains unknown, it can be characterized statistically. Regression analysis provides a method to fit a straight line or a curve through a series of scattered data points on a graph. The adequacy of the regression analysis can be evaluated by determining a correlation coefficient. Methods for extending regression analysis to multivariate functions exist. In principle, these methods are identical to linear regression analysis; however, the analysis becomes much more complex. The increase in complexity is not a concern, however, since computer subroutines are available that solve the tedious equations and provide the results in a convenient format.

Many probability functions are used in statistical analyses to represent data and predict population properties. Once a probability function has been selected to represent a population, any series of measurements can be subjected to a χ^2 test to check the validity of the assumed function. Accurate predictions can be made only if the proper probability function has been selected.

Finally, statistical methods for assessing error propagation provide a means for determining error in a quantity of interest y based on measurements of related quantities $x_1, x_2, \cdots, x_n$ and the functional relationship $y = f(x_1, x_2, \cdots, x_n)$ between quantities.

REFERENCES

1. Bethea, R. M., B. S. Duran, and T. L. Boullion: *Statistical Methods for Engineers and Scientists*, 2nd ed., Dekker, New York, 1985.
2. Bethea, R. M., and R. R. Rhinehart: *Applied Engineering Statistics*, Dekker, New York, 1991.
3. Blackwell, D.: *Basic Statistics*, McGraw–Hill, New York, 1969.
4. Bragg, G. M.: *Principles of Experimentation and Measurement*, Prentice–Hall, Englewood Cliffs, N.J., 1974.

5. Chou, Y.: *Probability and Statistics for Decision Making,* Holt, Rinehart & Winston, New York, 1972.

6. Davies, O. L., and P. L. Goldsmith: *Statistical Methods in Research and Production,* Hafner, New York, 1972.

7. McCall, C. H., Jr.: *Sampling and Statistics Handbook for Research,* Iowa State University Press, Ames, Iowa, 1982.

8. Snedecor, G. W., and W. G. Cochran: *Statistical Methods,* 8th ed., Iowa State University Press, Ames, Iowa, 1989.

9. Weibull, W.: *Fatigue Testing and Analysis of Results,* Pergamon Press, Elmsford, N.Y., 1961.

10. Young, H. D.: *Statistical Treatment of Experimental Data,* McGraw–Hill, New York, 1962.

11. Zehna, P. W.: *Probability Distributions and Statistics,* Allyn & Bacon, Boston, 1970.

12. Zehna, P. W.: *Introductory Statistics,* Prindle, Weber & Schmidt, Boston, 1974.

EXERCISES

13.1   The air pressure (in psi) at a point near the end of an air-supply line is monitored at 15-min intervals over an 8-h period. The readings are listed in the four columns that follow:

Column 1	Column 2	Column 3	Column 4
113	105	100	97
107	101	96	100
109	98	93	103
97	123	87	105
95	118	81	110
92	112	85	113
103	103	91	120
117	101	94	128

List the pressure readings in order of increasing magnitude. Rearrange the data in five-group intervals to obtain a frequency distribution. Show both the relative frequency and the cumulative frequency in this tabular rearrangement. Select the median pressure reading from the data.

13.2   Construct a histogram for the data listed in Exercise 13.1. Superimpose a plot of the relative frequency on the histogram.

13.3   Prepare a cumulative-frequency curve for the data of Exercise 13.1.

13.4   For the air-pressure data listed in Exercise 13.1,

(a)   compute the sample mean $\bar{x}$ of the individual columns

(b)   compute the sample means of columns 1 + 2 and columns 3 + 4

(c)   compute the sample mean of the complete set of data

(d)   comment on the results of (a), (b), and (c)

13.5   Determine the mode for the air-pressure data of Exercise 13.1 and compare it with the median and the mean of the data.

13.6   A quality-control laboratory monitors the tensile strength of paper by testing a small sample every 10 min. The data reported over an 8-h shift are listed in the following table.

Column 1	Column 2	Column 3	Column 4	Column 5	Column 6
980	887	1043	968	920	933
992	913	1031	963	897	944
969	972	999	950	871	981
929	987	990	942	847	1013
892	1021	983	934	843	1043
860	1066	982	936	868	1072
820	1079	977	939	884	1091
862	1080	971	922	907	1115

List the strength readings in order of increasing magnitude. Rearrange the data in seven-group intervals to obtain a frequency distribution. Show both the relative frequency and the cumulative frequency in this tabular rearrangement. Select the median strength reading from the data.

13.7 Construct a histogram for the data listed in Exercise 13.6. Superimpose a plot of relative frequency on the histogram.

13.8 Prepare a cumulative-frequency curve for the data of Exercise 13.6.

13.9 For the tensile-strength data listed in Exercise 13.6,
 (a) compute the sample mean $\bar{x}$ of the individual columns
 (b) compute the sample means of columns $1 + 2$, $3 + 4$, and $5 + 6$
 (c) compute the sample mean of the complete set of data
 (d) comment on the results of (a), (b), and (c)

13.10 Determine the mode for the tensile-strength data of Exercise 13.6 and compare it with the median and the mean of the data.

13.11 The accuracy of a new flow meter for diesel fuel is being checked by pumping 100 gal of the fuel into a tank and measuring the true volume with a calibrated sight glass. The readings from the sight glass are as follows:

Column 1	Column 2	Column 3	Column 4	Column 5	Column 6
100.89	100.07	100.47	99.99	100.37	99.92
100.67	100.02	99.75	100.12	100.16	99.68
99.93	100.28	99.86	99.68	99.99	99.97
99.69	100.21	99.68	99.72	99.87	100.09
99.77	99.68	100.18	99.84	100.02	99.85
99.86	99.42	100.37	99.98	99.91	100.19
100.04	99.85	100.29	100.04	100.06	99.71
100.05	99.81	99.79	99.87	99.85	99.95
100.68	99.35	99.97	99.83	100.17	99.90
100.27	99.68	100.21	99.57	100.28	100.08

List the volume readings in order of increasing magnitude. Rearrange the data in eight-group intervals to obtain a frequency distribution. Show both the relative frequency and the cumulative frequency in this tabular rearrangement. Select the median volume delivered. Comment on the merits of the new flow meter.

13.12   Construct a histogram for the data listed in Exercise 13.11. Superimpose a plot of relative frequency on the histogram.

13.13   Prepare a cumulative-frequency curve for the data of Exercise 13.11.

13.14   For the volume-delivered data listed in Exercise 13.11,
    (a)   compute the sample mean $\bar{x}$ of the individual columns
    (b)   compute the sample means of columns $1 + 2$, $3 + 4$, and $5 + 6$
    (c)   compute the sample means of columns $1 + 2 + 3$, and $4 + 5 + 6$
    (d)   compute the sample mean for the complete set of data
    (e)   comment on the results of (a), (b), (c), and (d)

13.15   Determine the mode for the volume-delivered data of Exercise 13.11 and compare it with the median and the mean of the data.

13.16   For the data presented in Exercise 13.1, determine the
    (a)   standard deviation S_x
    (b)   range R
    (c)   mean deviation d_x
    (d)   variance S_x^2
    (e)   coefficient of variation C_v

13.17   For the data presented in Exercise 13.6, determine the
    (a)   standard deviation S_x
    (b)   range R
    (c)   mean deviation d_x
    (d)   variance S_x^2
    (e)   coefficient of variation C_v

13.18   For the data presented in Exercise 13.11, determine the
    (a)   standard deviation S_x
    (b)   range R
    (c)   mean deviation d_x
    (d)   variance S_x^2
    (e)   coefficient of variation C_v

13.19   A quality-control laboratory associated with a manufacturing operation periodically makes a measurement that has been characterized as normal or Gaussian with a mean of 80 and a standard deviation of 3. Determine the probability of a single measurement
    (a)   falling between 73 and 83
    (b)   falling between 70 and 86
    (c)   falling between 67 and 89
    (d)   falling between 80 and 82
    (e)   falling between 79 and 80
    (f)   being less than 77
    (g)   being greater than 86

13.20   Careful measurement of diameter was made after a large shipment of electric-motor shafts was received. Measurements of the bearing journal indicate that the mean is 7 mm, with a standard deviation of 0.015 mm. Determine the probability of a journal measurement being
    (a)   greater than 7.04 mm
    (b)   greater than 7.03 mm
    (c)   less than 7.00 mm
    (d)   less than 6.94 mm

     (e)   between 7.01 and 7.03 mm

     (f)   between 6.96 and 6.99 mm

     (g)   between 6.97 and 7.03 mm

13.21   Use the data for the flowmeter in Exercise 13.11 to determine the probability of the pump delivering a volume that is

     (a)   in excess of the measured volume by 1.0 percent

     (b)   in excess of the measured volume by 0.5 percent

     (c)   in excess of the measured volume by 0.2 percent

     (d)   in excess of the measured volume by 0.1 percent

     (e)   less than the measured volume by 1.0 percent

     (f)   less than the measured volume by 0.5 percent

     (g)   less than the measured volume by 0.2 percent

     (h)   less than the measured volume by 0.1 percent

13.22   Prepare graphs showing the Weibull distribution function $P(x)$ if $x_0 = 2$, $b = 5$, and $m = 1, 2, 5$, and 10.

13.23   Use the graph of Exercise 13.22 for $m = 5$ to indicate the probability of failure at a stress of 2.5.

13.24   Use the graph of Exercise 13.22 for $m = 10$ to indicate the probability of failure at a stress of 3.0.

13.25   Use the graph of Exercise 13.22 for $m = 1$ to indicate the probability of failure at a stress of 4.0.

13.26   Determine the standard deviation of the mean $S_{\bar{x}}$ (standard error) for the data of

     (a)   Exercise 13.1   (b) Exercise 13.6   (c) Exercise 13.11

13.27   The data presented in Exercise 13.1 were drawn from a large population and provided an estimate $\bar{x}$ of the true mean μ of the population. Determine the confidence interval if the mean is to be stated with a confidence level of

     (a)   99 percent   (c) 90 percent

     (b)   95 percent   (d) 80 percent

13.28   Repeat Exercise 13.27 with the data presented in Exercise 13.6.

13.29   Repeat Exercise 13.27 with the data presented in Exercise 13.11.

13.30   A small sample ($n < 20$) of measurements from a drag-coefficient C_D determination for an airfoil are as follows:

     0.045   0.050

     0.052   0.047

     0.055   0.053

     0.049   0.044

     0.051   0.049

     (a)   Estimate the mean drag coefficient.

     (b)   Estimate the standard deviation.

     (c)   Determine the confidence level for the statement that $C_D = 0.0495$.

13.31   Determine the sample size needed in Exercise 13.30 to permit specification of the mean drag coefficient C_D with a confidence level of

     (a)   90 percent   (c) 99 percent

     (b)   95 percent   (d) 99.9 percent

13.32 Measurements of flue-gas temperature from a power-plant stack are as follows:

275°F   279°F
277°F   289°F
282°F   284°F
289°F   289°F
284°F   274°F

(a) Estimate the mean flue-gas temperature.
(b) Estimate the standard deviation.
(c) Determine the confidence level for the statement that the flue-gas temperature is 282.2°F.

13.33 Determine the sample size required in Exercise 13.32 to permit specification of the flue-gas temperature with a confidence level of

(a) 90 percent   (c) 99 percent
(b) 95 percent   (d) 99.9 percent

13.34 A manufacturing process yields aluminum rods with a mean yield strength of 36,000 psi and a standard deviation of 1250 psi. A customer places a large order for rods with a minimum yield strength of 32,000 psi. Prepare a letter for submission to the customer that describes the yield strength to be expected, and outline your firm's procedures for ensuring that this quality level will be achieved and maintained.

13.35 An inspection laboratory samples two large shipments of dowel pins by measuring both length and diameter. For shipment A, the sample size is 40; the mean diameter, 6.13 mm; the mean length, 25.4 mm; the estimated standard deviation on diameter, 0.022 mm; and the estimated standard deviation on length, 0.140 mm. For shipment B, the sample size is 60; the mean diameter, 6.06 mm; the mean length, 25.0 mm; the estimated standard deviation on diameter, 0.034 mm; and the estimated standard deviation on length, 0.203 mm.

(a) Are the two shipments of dowel pins the same?
(b) What is the level of confidence in your prediction?
(c) Would it be safe to mix the two shipments of pins? Explain.

13.36 Repeat Exercise 13.31 for the following two shipments of dowel pins.

	Shipment A	Shipment B
Number	25	15
Diameter	$\bar{x} = 6.05$ mm, $S_x = 0.02$ mm	$\bar{x} = 5.98$ mm, $S_x = 0.05$ mm
Length	$\bar{x} = 24.9$ mm, $S_x = 0.20$ mm	$\bar{x} = 25.4$ mm, $S_x = 0.22$ mm

13.37 Employ Chauvenet's criterion to statistically condition the following sequence of measurements of atmospheric pressure (millimeters of mercury) obtained with a mercury barometer.

763.2   763.5   763.3
764.1   763.4   763.4
764.6   764.3   763.7
764.2   764.1   763.8
763.5   764.0   763.5

13.38   After conditioning the data in Exercise 13.37, determine the mean and the standard deviation.

13.39   The weight of a shipment of gold has been measured n times to obtain $\bar{x}$ and S_x. If the precision of the stated weight must be increased by a factor of 1.2, how many additional measurements of the weight should be made?

13.40   Repeat Exercise 13.39 if the required factor of improvement is

    (a)   1.25   (c)   1.75   (e)   5.0
    (b)   1.5   (d)   2.0   (f)   10.0

13.41   Prepare a graph showing relative precision as a function of the number of measurements for a population with $\mu = 100$ and $\sigma = 1$.

13.42   Determine the slope m and the intercept b for a linear regression equation $y = mx + b$ representing the following set of data.

x	y	x	y
0.5	1.4	2.8	8.4
0.9	2.9	3.2	9.8
1.4	4.4	3.5	10.7
2.0	6.2	3.9	11.4
2.3	7.1	4.2	12.4

13.43   Determine the slope m and the intercept b for a linear regression equation $y = mx + b$ representing the following set of data.

x	y	x	y	x	y
11.7	23.0	17.2	34.8	21.8	43.7
13.0	26.4	18.0	36.2	23.2	46.0
13.9	28.1	18.9	38.0	24.5	49.6
14.7	29.1	19.7	39.3	25.1	52.5
16.1	32.4	20.9	42.1	27.4	54.4

13.44   Determine the slope m and the intercept b for a linear regression equation $y = mx + b$ representing the following set of data.

x	y	x	y	x	y
11.7	58.2	16.8	83.7	21.5	108.1
13.1	65.6	17.5	87.0	23.0	114.0
14.0	71.1	18.9	95.1	24.1	121.2
14.8	74.3	19.6	97.3	25.6	127.4
15.6	77.7	20.5	102.2	27.2	135.3

13.45   Determine the correlation coefficient R^2 for the regression analysis of Exercise 13.42.

13.46   Determine the correlation coefficient R^2 for the regression analysis of Exercise 13.43.

13.47   Determine the correlation coefficient R^2 for the regression analysis of Exercise 13.44.

13.48 The solvent concentration in a coating as a function of time is governed by an equation of the form

$$C = ae^{-mt}$$

where

    C is the concentration
    t is the time
    a and m are constants that depend on the diffusion process

t (s)	C (%)	t (s)	C (%)
10	2.05	50	1.89
20	2.00	100	1.79
30	1.96	200	1.68
40	1.92	500	1.45

For the given data, determine the constants a and m that provide the best fit.

13.49 Determine the regression coefficients a, b_1, b_2, and b_3 for the following data set.

y	x_1	x_2	x_3
6.9	1.0	2.0	1.0
7.9	1.5	2.0	1.5
8.0	2.0	2.0	2.0
8.1	2.5	3.0	1.0
8.4	3.0	3.0	1.5
8.5	3.5	3.0	2.0
8.6	4.0	4.0	1.0
8.7	4.5	4.0	1.5
9.0	5.0	4.0	2.0
9.2	5.5	5.0	1.0
9.4	6.0	5.0	1.5
9.7	6.5	5.0	2.0

13.50 Determine the correlation coefficient R^2 for the solution of Exercise 13.49.

13.51 Batteries from a production process are weighed prior to packing as a routine quality check. The data from a typical 8-h run are as follows:

Weight (lb)	Number
Less than 30.60	22
30.6 to 30.79	480
30.8 to 30.99	5,106
31.0 to 31.19	10,461
31.2 to 31.39	4,618
31.4 to 31.59	619
Greater than 31.60	15

(a) Find the mean and the standard deviation of the weights.

(b) Determine whether the weights of the batteries can be expressed as a Gaussian distribution function.

(c) Would you expect quality-control problems with batteries weighing less than 30.6 lb or more than 31.6 lb?

13.52 A die-casting operation produces bearing housings with a rejection rate of 4 percent when the machine is operated over an 8-h shift to produce a total output of 3200 housings. The method of die cooling was changed in an attempt to reduce the rejection rate. After 2 h of operation under the new cooling conditions, 778 acceptable castings and 22 rejects had been produced.

(a) Did the change in the process improve the output?

(b) How certain are you of your answer?

13.53 A gear-shaft assembly consists of a shaft with a shoulder, a bearing, a sleeve, a gear, a second sleeve, and a nut. Dimensional tolerances for each of these components are as follows:

Components	Tolerance (mm)
Shoulder	0.040
Bearing	0.030
First sleeve	0.080
Gear	0.050
Second sleeve	0.700
Nut	0.080

(a) Determine the anticipated tolerance of the series assembly.

(b) What will be the frequency of occurrence of the tolerance of part (a)?

13.54 Estimate the error in a determination of the volume of a sphere having a diameter of 100 mm if the diameter is measured with a standard deviation of the mean (standard error) of $S_{\bar{x}} = 0.05$ mm.

13.55 The stress σ_x at a point on the free surface of a structure or machine component can be expressed in terms of the normal strains ε_x and ε_y measured with electrical resistance strain gages as

$$\sigma_x = \frac{E}{1 - \nu^2}(\varepsilon_x + \nu\varepsilon_y)$$

If ε_x and ε_y are measured within ± 3 percent and E and ν are measured within ± 5 percent, estimate the error in σ_x.

Temperature-Resistance Data for Thermistors and Thermoelectric Voltages for Thermocouples

Table A.1 Temperature-Resistance Data for a Thermistor

Temperature °C	Resistance	Temperature °C	Resistance	Temperature °C	Resistance
− 80	2,210,400	− 41	107,910	− 2	10,857
− 79	2,022,100	− 40	100,950	− 1	10,311
− 78	1,851,100	− 39	94,470	0	9,795.0
− 77	1,695,800	− 38	88,440	1	9,309.0
− 76	1,554,500	− 37	82,860	2	8,850.0
− 75	1,425,900	− 36	77,640	3	8,415.0
− 74	1,308,900	− 35	72,810	4	8,007.0
− 73	1,202,200	− 34	68,280	5	7,617.0
− 72	1,105,000	− 33	64,080	6	7,251.0
− 71	1,016,300	− 32	60,150	7	6,903.0
− 70	935,250	− 31	56,490	8	6,576.0
− 69	861,240	− 30	53,100	9	6,264.0
− 68	793,590	− 29	49,890	10	5,970.0
− 67	731,700	− 28	46,920	11	5,691.0
− 66	675,060	− 27	44,160	12	5,427.0
− 65	623,160	− 26	41,550	13	5,175.0
− 64	575,610	− 25	39,120	14	4,938.0
− 63	531,990	− 24	36,840	15	4,713.0
− 62	491,970	− 23	34,710	16	4,500.0
− 61	455,220	− 22	32,730	17	4,296.0
− 60	421,470	− 21	30,870	18	4,104.0
− 59	390,420	− 20	29,121	19	3,921.0
− 58	361,890	− 19	27,483	20	3,747.0
− 57	335,610	− 18	25,947	21	3,582.0
− 56	311,400	− 17	24,507	22	3,426.0
− 55	289,110	− 16	23,154	23	3,276.0
− 54	268,560	− 15	21,885	24	3,135.0
− 53	249,600	− 14	20,694	25	3,000.0
− 52	232,110	− 13	19,572	26	2,871.9
− 51	215,970	− 12	18,519	27	2,750.1
− 50	201,030	− 11	17,529	28	2,633.1
− 49	187,230	− 10	16,599	29	2,522.1
− 48	174,450	− 9	15,720	30	2,417.1
− 47	162,660	− 8	14,895	31	2,316.9
− 46	151,710	− 7	14,118	32	2,220.9
− 45	141,570	− 6	13,386	33	2,129.1
− 44	132,180	− 5	12,699	34	2,042.1
− 43	123,480	− 4	12,048	35	1,959.0
− 42	115,410	− 3	11,433	36	1,880.1

Table A.1 (*Continued*)

Temperature °C	Resistance	Temperature °C	Resistance	Temperature °C	Resistance
37	1,805.1	75	443.70	113	141.00
38	1,733.1	76	429.30	114	137.19
39	1,664.1	77	415.20	115	133.50
40	1,598.1	78	402.00	116	129.99
41	1,535.1	79	389.10	117	126.51
42	1,475.1	80	376.50	118	123.21
43	1,418.1	81	364.50	119	120.00
44	1,362.9	82	353.10	120	116.79
45	1,311.0	83	342.00	121	113.79
46	1,260.0	84	331.20	122	110.91
47	1,212.0	85	321.00	123	108.00
48	1,167.0	86	310.80	124	105.18
49	1,122.9	87	301.20	125	102.51
50	1,080.9	88	292.11	126	99.930
51	1,040.1	89	283.20	127	97.410
52	1,002.0	90	274.59	128	94.950
53	965.10	91	266.31	129	92.580
54	929.70	92	258.30	130	90.279
55	895.80	93	250.59	131	88.041
56	863.40	94	243.09	132	85.869
57	832.20	95	236.01	133	83.751
58	802.50	96	228.99	134	81.609
59	773.70	97	222.30	135	79.710
60	746.40	98	215.79	136	77.790
61	720.00	99	209.61	137	75.900
62	694.80	100	203.49	138	74.079
63	670.50	101	197.70	139	72.309
64	647.10	102	192.09	140	70.581
65	624.90	103	186.60	141	68.910
66	603.30	104	181.29	142	67.290
67	582.60	105	176.19	143	65.700
68	562.80	106	171.30	144	64.170
69	543.90	107	166.50	145	62.661
70	525.60	108	161.91	146	61.209
71	507.90	109	157.50	147	59.799
72	490.80	110	153.09	148	58.431
73	474.60	111	149.01	149	57.099
74	459.00	112	144.90	150	55.791

Table A.2 Thermoelectric Voltages for Chromel-Alumel Thermocouples
with the Reference Junction at 0°C (32°F)

°C	0	1	2	3	4	5	6	7	8	9	10	°C
					Thermoelectric Voltage (absolute mV)							
−270	−6.458											−270
−260	−6.441	−6.444	−6.446	−6.448	−6.450	−6.452	−6.453	−6.455	−6.456	−6.457	−6.458	−260
−250	−6.404	−6.408	−6.413	−6.417	−6.421	−6.425	−6.429	−6.432	−6.435	−6.438	−6.441	−250
−240	−6.344	−6.351	−6.358	−6.364	−6.371	−6.377	−6.382	−6.388	−6.394	−6.399	−6.404	−240
−230	−6.262	−6.271	−6.280	−6.289	−6.297	−6.306	−6.314	−6.322	−6.329	−6.337	−6.344	−230
−220	−6.158	−6.170	−6.181	−6.192	−6.202	−6.213	−6.223	−6.233	−6.243	−6.253	−6.262	−220
−210	−6.035	−6.048	−6.061	−6.074	−6.087	−6.099	−6.111	−6.123	−6.135	−6.147	−6.158	−210
−200	−5.891	−5.907	−5.922	−5.936	−5.951	−5.965	−5.980	−5.994	−6.007	−6.021	−6.035	−200
−190	−5.730	−5.747	−5.763	−5.780	−5.796	−5.813	−5.829	−5.845	−5.860	−5.876	−5.891	−190
−180	−5.550	−5.569	−5.587	−5.606	−5.624	−5.642	−5.660	−5.678	−5.695	−5.712	−5.730	−180
−170	−5.354	−5.374	−5.394	−5.414	−5.434	−5.454	−5.474	−5.493	−5.512	−5.531	−5.550	−170
−160	−5.141	−5.163	−5.185	−5.207	−5.228	−5.249	−5.271	−5.292	−5.313	−5.333	−5.354	−160
−150	−4.912	−4.936	−4.959	−4.983	−5.006	−5.029	−5.051	−5.074	−5.097	−5.119	−5.141	−150
−140	−4.669	−4.694	−4.719	−4.743	−4.768	−4.792	−4.817	−4.841	−4.865	−4.889	−4.912	−140
−130	−4.410	−4.437	−4.463	−4.489	−4.515	−4.541	−4.567	−4.593	−4.618	−4.644	−4.669	−130
−120	−4.138	−4.166	−4.193	−4.221	−4.248	−4.276	−4.303	−4.330	−4.357	−4.384	−4.410	−120
−110	−3.852	−3.881	−3.910	−3.939	−3.968	−3.997	−4.025	−4.053	−4.082	−4.110	−4.138	−110
−100	−3.553	−3.584	−3.614	−3.644	−3.674	−3.704	−3.734	−3.764	−3.793	−3.823	−3.852	−100
−90	−3.242	−3.274	−3.305	−3.337	−3.368	−3.399	−3.430	−3.461	−3.492	−3.523	−3.553	−90
−80	−2.920	−2.953	−2.985	−3.018	−3.050	−3.082	−3.115	−3.147	−3.179	−3.211	−3.242	−80
−70	−2.586	−2.620	−2.654	−2.687	−2.721	−2.754	−2.788	−2.821	−2.854	−2.887	−2.920	−70
−60	−2.243	−2.277	−2.312	−2.347	−2.381	−2.416	−2.450	−2.484	−2.518	−2.552	−2.586	−60
−50	−1.889	−1.925	−1.961	−1.996	−2.032	−2.067	−2.102	−2.137	−2.173	−2.208	−2.243	−50
−40	−1.527	−1.563	−1.600	−1.636	−1.673	−1.709	−1.745	−1.781	−1.817	−1.853	−1.889	−40
−30	−1.156	−1.193	−1.231	−1.268	−1.305	−1.342	−1.379	−1.416	−1.453	−1.490	−1.527	−30
−20	−0.777	−0.816	−0.854	−0.892	−0.930	−0.968	−1.005	−1.043	−1.081	−1.118	−1.156	−20
−10	−0.392	−0.431	−0.469	−0.508	−0.547	−0.585	−0.624	−0.662	−0.701	−0.739	−0.777	−10
0	0.000	−0.039	−0.079	−0.118	−0.157	−0.197	−0.236	−0.275	−0.314	−0.353	−0.392	0
0	0.000	0.039	0.079	0.119	0.158	0.198	0.238	0.277	0.317	0.357	0.397	0
10	0.397	0.437	0.477	0.517	0.557	0.597	0.637	0.677	0.718	0.758	0.798	10
20	0.798	0.838	0.879	0.919	0.960	1.000	1.041	1.081	1.122	1.162	1.203	20
30	1.203	1.244	1.285	1.325	1.366	1.407	1.448	1.489	1.529	1.570	1.611	30
40	1.611	1.652	1.693	1.734	1.776	1.817	1.858	1.899	1.940	1.981	2.022	40
50	2.022	2.064	2.105	2.146	2.188	2.229	2.270	2.312	2.353	2.394	2.436	50
60	2.436	2.477	2.519	2.560	2.601	2.643	2.684	2.726	2.767	2.809	2.850	60
70	2.850	2.892	2.933	2.975	3.016	3.058	3.100	3.141	3.183	3.224	3.266	70
80	3.266	3.307	3.349	3.390	3.432	3.473	3.515	3.556	3.598	3.639	3.681	80
90	3.681	3.722	3.764	3.805	3.847	3.888	3.930	3.971	4.012	4.054	4.095	90
100	4.095	4.137	4.178	4.219	4.261	4.302	4.343	4.384	4.426	4.467	4.508	100
110	4.508	4.549	4.590	4.632	4.673	4.714	4.755	4.796	4.837	4.878	4.919	110
120	4.919	4.960	5.001	5.042	5.083	5.124	5.164	5.205	5.246	5.287	5.327	120
130	5.327	5.368	5.409	5.450	5.490	5.531	5.571	5.612	5.652	5.693	5.733	130
140	5.733	5.774	5.814	5.855	5.895	5.936	5.976	6.016	6.057	6.097	6.137	140
150	6.137	6.177	6.218	6.258	6.298	6.338	6.378	6.419	6.459	6.499	6.539	150
160	6.539	6.579	6.619	6.659	6.699	6.739	6.779	6.819	6.859	6.899	6.939	160
170	6.939	6.979	7.019	7.059	7.099	7.139	7.179	7.219	7.259	7.299	7.338	170
180	7.338	7.378	7.418	7.458	7.498	7.538	7.578	7.618	7.658	7.697	7.737	180
190	7.737	7.777	7.817	7.857	7.897	7.937	7.977	8.017	8.057	8.097	8.137	190
200	8.137	8.177	8.216	8.256	8.296	8.336	8.376	8.416	8.456	8.497	8.537	200
210	8.537	8.577	8.617	8.657	8.697	8.737	8.777	8.817	8.857	8.898	8.938	210
220	8.938	8.978	9.018	9.058	9.099	9.139	9.179	9.220	9.260	9.300	9.341	220
230	9.341	9.381	9.421	9.462	9.502	9.543	9.583	9.624	9.664	9.705	9.745	230
240	9.745	9.786	9.826	9.867	9.907	9.948	9.989	10.029	10.070	10.111	10.151	240
250	10.151	10.192	10.233	10.274	10.315	10.355	10.396	10.437	10.478	10.519	10.560	250
260	10.560	10.600	10.641	10.682	10.723	10.764	10.805	10.846	10.887	10.928	10.969	260
270	10.969	11.010	11.051	11.093	11.134	11.175	11.216	11.257	11.298	11.339	11.381	270
280	11.381	11.422	11.463	11.504	11.546	11.587	11.628	11.669	11.711	11.752	11.793	280
290	11.793	11.835	11.876	11.918	11.959	12.000	12.042	12.083	12.125	12.166	12.207	290

Table A.2 (*Continued*)

°C	0	1	2	3	4	5	6	7	8	9	10	°C
					Thermoelectric Voltage (absolute mV)							
300	12.207	12.249	12.290	12.332	12.373	12.415	12.456	12.498	12.539	12.581	12.623	300
310	12.623	12.664	12.706	12.747	12.789	12.831	12.872	12.914	12.955	12.997	13.039	310
320	13.039	13.080	13.122	13.164	13.205	13.247	13.289	13.331	13.372	13.414	13.456	320
330	13.456	13.497	13.539	13.581	13.623	13.665	13.706	13.748	13.790	13.832	13.874	330
340	13.874	13.915	13.957	13.999	14.041	14.083	14.125	14.167	14.208	14.250	14.292	340
350	14.292	14.334	14.376	14.418	14.460	14.502	14.544	14.586	14.628	14.670	14.712	350
360	14.712	14.754	14.796	14.838	14.880	14.922	14.964	15.006	15.048	15.090	15.132	360
370	15.132	15.174	15.216	15.258	15.300	15.342	15.384	15.426	15.468	15.510	15.552	370
380	15.552	15.594	15.636	15.679	15.721	15.763	15.805	15.847	15.889	15.931	15.974	380
390	15.974	16.016	16.058	16.100	16.142	16.184	16.227	16.269	16.311	16.353	16.395	390
400	16.395	16.438	16.480	16.522	16.564	16.607	16.649	16.691	16.733	16.776	16.818	400
410	16.818	16.860	16.902	16.945	16.987	17.029	17.072	17.114	17.156	17.199	17.241	410
420	17.241	17.283	17.326	17.368	17.410	17.453	17.495	17.537	17.580	17.622	17.664	420
430	17.664	17.707	17.749	17.792	17.834	17.876	17.919	17.961	18.004	18.046	18.088	430
440	18.088	18.131	18.173	18.216	18.258	18.301	18.343	18.385	18.428	18.470	18.513	440
450	18.513	18.555	18.598	18.640	18.683	18.725	18.768	18.810	18.853	18.895	18.938	450
460	18.938	18.980	19.023	19.065	19.108	19.150	19.193	19.235	19.278	19.320	19.363	460
470	19.363	19.405	19.448	19.490	19.533	19.576	19.618	19.661	19.703	19.746	19.788	470
480	19.788	19.831	19.873	19.916	19.959	20.001	20.044	20.086	20.129	20.172	20.214	480
490	20.214	20.257	20.299	20.342	20.385	20.427	20.470	20.512	20.555	20.598	20.640	490
500	20.640	20.683	20.725	20.768	20.811	20.853	20.896	20.938	20.981	21.024	21.066	500
510	21.066	21.109	21.152	21.194	21.237	21.280	21.322	21.365	21.407	21.450	21.493	510
520	21.493	21.535	21.578	21.621	21.663	21.706	21.749	21.791	21.834	21.876	21.919	520
530	21.919	21.962	22.004	22.047	22.090	22.132	22.175	22.218	22.260	22.303	22.346	530
540	22.346	22.388	22.431	22.473	22.516	22.559	22.601	22.644	22.687	22.729	22.772	540
550	22.772	22.815	22.857	22.900	22.942	22.985	23.028	23.070	23.113	23.156	23.198	550
560	23.198	23.241	23.284	23.326	23.369	23.411	23.454	23.497	23.539	23.582	23.624	560
570	23.624	23.667	23.710	23.752	23.795	23.837	23.880	23.923	23.965	24.008	24.050	570
580	24.050	24.093	24.136	24.178	24.221	24.263	24.306	24.348	24.391	24.434	24.476	580
590	24.476	24.519	24.561	24.604	24.646	24.689	24.731	24.774	24.817	24.859	24.902	590
600	24.902	24.944	24.987	25.029	25.072	25.114	25.157	25.199	25.242	25.284	25.327	600
610	25.327	25.369	25.412	25.454	25.497	25.539	25.582	25.624	25.666	25.709	25.751	610
620	25.751	25.794	25.836	25.879	25.921	25.964	26.006	26.048	26.091	26.133	26.176	620
630	26.176	26.218	26.260	26.303	26.345	26.387	26.430	26.472	26.515	26.557	26.599	630
640	26.599	26.642	26.684	26.726	26.769	26.811	26.853	26.896	26.938	26.980	27.022	640
650	27.022	27.065	27.107	27.149	27.192	27.234	27.276	27.318	27.361	27.403	27.445	650
660	27.445	27.487	27.529	27.572	27.614	27.656	27.698	27.740	27.783	27.825	27.867	660
670	27.867	27.909	27.951	27.993	28.035	28.078	28.120	28.162	28.204	28.246	28.288	670
680	28.288	28.330	28.372	28.414	28.456	28.498	28.540	28.583	28.625	28.667	28.709	680
690	28.709	28.751	28.793	28.835	28.877	28.919	28.961	29.002	29.044	29.086	29.128	690
700	29.128	29.170	29.212	29.254	29.296	29.338	29.380	29.422	29.464	29.505	29.547	700
710	29.547	29.589	29.631	29.673	29.715	29.756	29.798	29.840	29.882	29.924	29.965	710
720	29.965	30.007	30.049	30.091	30.132	30.174	30.216	30.257	30.299	30.341	30.383	720
730	30.383	30.424	30.466	30.508	30.549	30.591	30.632	30.674	30.716	30.757	30.799	730
740	30.799	30.840	30.882	30.924	30.965	31.007	31.048	31.090	31.131	31.173	31.214	740
750	31.214	31.256	31.297	31.339	31.380	31.422	31.463	31.504	31.546	31.587	31.629	750
760	31.629	31.670	31.712	31.753	31.794	31.836	31.877	31.918	31.960	32.001	32.042	760
770	32.042	32.084	32.125	32.166	32.207	32.249	32.290	32.331	32.372	32.414	32.455	770
780	32.455	32.496	32.537	32.578	32.619	32.661	32.702	32.743	32.784	32.825	32.866	780
790	32.866	32.907	32.948	32.990	33.031	33.072	33.113	33.154	33.195	33.236	33.277	790
800	33.277	33.318	33.359	33.400	33.441	33.482	33.523	33.564	33.604	33.645	33.686	800
810	33.686	33.727	33.768	33.809	33.850	33.891	33.931	33.972	34.013	34.054	34.095	810
820	34.095	34.136	34.176	34.217	34.258	34.299	34.339	34.380	34.421	34.461	34.502	820
830	34.502	34.543	34.583	34.624	34.665	34.705	34.746	34.787	34.827	34.868	34.909	830
840	34.909	34.949	34.990	35.030	35.071	35.111	35.152	35.192	35.233	35.273	35.314	840
850	35.314	35.354	35.395	35.435	35.476	35.516	35.557	35.597	35.637	35.678	35.718	850
860	35.718	35.758	35.799	35.839	35.880	35.920	35.960	36.000	36.041	36.081	36.121	860
870	36.121	36.162	36.202	36.242	36.282	36.323	36.363	36.403	36.443	36.483	36.524	870
880	36.524	36.564	36.604	36.644	36.684	36.724	36.764	36.804	36.844	36.885	36.925	880
890	36.925	36.965	37.005	37.045	37.085	37.125	37.165	37.205	37.245	37.285	37.325	890

Table A.2 (*Continued*)

°C	0	1	2	3	4	5	6	7	8	9	10	°C
					Thermoelectric Voltage (absolute mV)							
900	37.325	37.365	37.405	37.445	37.484	37.524	37.564	37.604	37.644	37.684	37.724	900
910	37.724	37.764	37.803	37.843	37.883	37.923	37.963	38.002	38.042	38.082	38.122	910
920	38.122	38.162	38.201	38.241	38.281	38.320	38.360	38.400	38.439	38.479	38.519	920
930	38.519	38.558	38.598	38.638	38.677	38.717	38.756	38.796	38.836	38.875	38.915	930
940	38.915	38.954	38.994	39.033	39.073	39.112	39.152	39.191	39.231	39.270	39.310	940
950	39.310	39.349	39.388	39.428	39.467	39.507	39.546	39.585	39.625	39.664	39.703	950
960	39.703	39.743	39.782	39.821	39.861	39.900	39.939	39.979	40.018	40.057	40.096	960
970	40.096	40.136	40.175	40.214	40.253	40.292	40.332	40.371	40.410	40.449	40.488	970
980	40.488	40.527	40.566	40.605	40.645	40.684	40.723	40.762	40.801	40.840	40.879	980
990	40.879	40.918	40.957	40.996	41.035	41.074	41.113	41.152	41.191	41.230	41.269	990
1,000	41.269	41.308	41.347	41.385	41.424	41.463	41.502	41.541	41.580	41.619	41.657	1,000
1,010	41.657	41.696	41.735	41.774	41.813	41.851	41.890	41.929	41.968	42.006	42.045	1,010
1,020	42.045	42.084	42.123	42.161	42.200	42.239	42.277	42.316	42.355	42.393	42.432	1,020
1,030	42.432	42.470	42.509	42.548	42.586	42.625	42.663	42.702	42.740	42.779	42.817	1,030
1,040	42.817	42.856	42.894	42.933	42.971	43.010	43.048	43.087	43.125	43.164	43.202	1,040
1,050	43.202	43.240	43.279	43.317	43.356	43.394	43.432	43.471	43.509	43.547	43.585	1,050
1,060	43.585	43.624	43.662	43.700	43.739	43.777	43.815	43.853	43.891	43.930	43.968	1,060
1,070	43.968	44.006	44.044	44.082	44.121	44.159	44.197	44.235	44.273	44.311	44.349	1,070
1,080	44.349	44.387	44.425	44.463	44.501	44.539	44.577	44.615	44.653	44.691	44.729	1,080
1,090	44.729	44.767	44.805	44.843	44.881	44.919	44.957	44.995	45.033	45.070	45.108	1,090
1,100	45.108	45.146	45.184	45.222	45.260	45.297	45.335	45.373	45.411	45.448	45.486	1,100
1,110	45.486	45.524	45.561	45.599	45.637	45.675	45.712	45.750	45.787	45.825	45.863	1,110
1,120	45.863	45.900	45.938	45.975	46.013	46.051	46.088	46.126	46.163	46.201	46.238	1,120
1,130	46.238	46.275	46.313	46.350	46.388	46.425	46.463	46.500	46.537	46.575	46.612	1,130
1,140	46.612	46.649	46.687	46.724	46.761	46.799	46.836	46.873	46.910	46.948	46.985	1,140
1,150	46.985	47.022	47.059	47.096	47.134	47.171	47.208	47.245	47.282	47.319	47.356	1,150
1,160	47.356	47.393	47.430	47.468	47.505	47.542	47.579	47.616	47.653	47.689	47.726	1,160
1,170	47.726	47.763	47.800	47.837	47.874	47.911	47.948	47.985	48.021	48.058	48.095	1,170
1,180	48.095	48.132	48.169	48.205	48.242	48.279	48.316	48.352	48.389	48.426	48.462	1,180
1,190	48.462	48.499	48.536	48.572	48.609	48.645	48.682	48.718	48.755	48.792	48.828	1,190
1,200	48.828	48.865	48.901	48.937	48.974	49.010	49.047	49.083	49.120	49.156	49.192	1,200
1,210	49.192	49.229	49.265	49.301	49.338	49.374	49.410	49.446	49.483	49.519	49.555	1,210
1,220	49.555	49.591	49.627	49.663	49.700	49.736	49.772	49.808	49.844	49.880	49.916	1,220
1,230	49.916	49.952	49.988	50.024	50.060	50.096	50.132	50.168	50.204	50.240	50.276	1,230
1,240	50.276	50.311	50.347	50.383	50.419	50.455	50.491	50.526	50.562	50.598	50.633	1,240
1,250	50.633	50.669	50.705	50.741	50.776	50.812	50.847	50.883	50.919	50.954	50.990	1,250
1,260	50.990	51.025	51.061	51.096	51.132	51.167	51.203	51.238	51.274	51.309	51.344	1,260
1,270	51.344	51.380	51.415	51.450	51.486	51.521	51.556	51.592	51.627	51.662	51.697	1,270
1,280	51.697	51.733	51.768	51.803	51.838	51.873	51.908	51.943	51.979	52.014	52.049	1,280
1,290	52.049	52.084	52.119	52.154	52.189	52.224	52.259	52.294	52.329	52.364	52.398	1,290
1,300	52.398	52.433	52.468	52.503	52.538	52.573	52.608	52.642	52.677	52.712	52.747	1,300
1,310	52.747	52.781	52.816	52.851	52.886	52.920	52.955	52.989	53.024	53.059	53.093	1,310
1,320	53.093	53.128	53.162	53.197	53.232	53.266	53.301	53.335	53.370	53.404	53.439	1,320
1,330	53.439	53.473	53.507	53.542	53.576	53.611	53.645	53.679	53.714	53.748	53.782	1,330
1,340	53.782	53.817	53.851	53.885	53.920	53.954	53.988	54.022	54.057	54.091	54.125	1,340
1,350	54.125	54.159	54.193	54.228	54.262	54.296	54.330	54.364	54.398	54.432	54.466	1,350
1,360	54.466	54.501	54.535	54.569	54.603	54.637	54.671	54.705	54.739	54.773	54.807	1,360
1,370	54.807	54.841	54.875									1,370

Table A.3 Thermoelectric Voltages for Chromel-Constantan Thermocouples with the Reference Junction at 0°C (32°F)

°C	0	1	2	3	4	5	6	7	8	9	10	°C
					Thermoelectric Voltage (absolute mV)							
−270	−9.835											−270
−260	−9.797	−9.802	−9.808	−9.813	−9.817	−9.821	−9.825	−9.828	−9.831	−9.833	−9.835	−260
−250	−9.719	−9.728	−9.737	−9.746	−9.754	−9.762	−9.770	−9.777	−9.784	−9.791	−9.797	−250
−240	−9.604	−9.617	−9.630	−9.642	−9.654	−9.666	−9.677	−9.688	−9.699	−9.709	−9.719	−240
−230	−9.455	−9.472	−9.488	−9.503	−9.519	−9.534	−9.549	−9.563	−9.577	−9.591	−9.604	−230
−220	−9.274	−9.293	−9.313	−9.332	−9.350	−9.368	−9.386	−9.404	−9.421	−9.438	−9.455	−220
−210	−9.063	−9.085	−9.107	−9.129	−9.151	−9.172	−9.193	−9.214	−9.234	−9.254	−9.274	−210
−200	−8.824	−8.850	−8.874	−8.899	−8.923	−8.947	−8.971	−8.994	−9.017	−9.040	−9.063	−200
−190	−8.561	−8.588	−8.615	−8.642	−8.669	−8.696	−8.722	−8.748	−8.774	−8.799	−8.824	−190
−180	−8.273	−8.303	−8.333	−8.362	−8.391	−8.420	−8.449	−8.477	−8.505	−8.533	−8.561	−180
−170	−7.963	−7.995	−8.027	−8.058	−8.090	−8.121	−8.152	−8.183	−8.213	−8.243	−8.273	−170
−160	−7.631	−7.665	−7.699	−7.733	−7.767	−7.800	−7.833	−7.866	−7.898	−7.931	−7.963	−160
−150	−7.279	−7.315	−7.351	−7.387	−7.422	−7.458	−7.493	−7.528	−7.562	−7.597	−7.631	−150
−140	−6.907	−6.945	−6.983	−7.020	−7.058	−7.095	−7.132	−7.169	−7.206	−7.243	−7.279	−140
−130	−6.516	−6.556	−6.596	−6.635	−6.675	−6.714	−6.753	−6.792	−6.830	−6.869	−6.907	−130
−120	−6.107	−6.149	−6.190	−6.231	−6.273	−6.314	−6.354	−6.395	−6.436	−6.476	−6.516	−120
−110	−5.680	−5.724	−5.767	−5.810	−5.853	−5.896	−5.938	−5.981	−6.023	−6.065	−6.107	−110
−100	−5.237	−5.282	−5.327	−5.371	−5.416	−5.460	−5.505	−5.549	−5.593	−5.637	−5.680	−100
−90	−4.777	−4.824	−4.870	−4.916	−4.963	−5.009	−5.055	−5.100	−5.146	−5.191	−5.237	−90
−80	−4.301	−4.350	−4.398	−4.446	−4.493	−4.541	−4.588	−4.636	−4.683	−4.730	−4.777	−80
−70	−3.811	−3.860	−3.910	−3.959	−4.009	−4.058	−4.107	−4.156	−4.204	−4.253	−4.301	−70
−60	−3.306	−3.357	−3.408	−3.459	−3.509	−3.560	−3.610	−3.661	−3.711	−3.761	−3.811	−60
−50	−2.787	−2.839	−2.892	−2.944	−2.996	−3.048	−3.100	−3.152	−3.203	−3.254	−3.306	−50
−40	−2.254	−2.308	−2.362	−2.416	−2.469	−2.522	−2.575	−2.628	−2.681	−2.734	−2.787	−40
−30	−1.709	−1.764	−1.819	−1.874	−1.929	−1.983	−2.038	−2.092	−2.146	−2.200	−2.254	−30
−20	−1.151	−1.208	−1.264	−1.320	−1.376	−1.432	−1.487	−1.543	−1.599	−1.654	−1.709	−20
−10	−0.581	−0.639	−0.696	−0.754	−0.811	−0.868	−0.925	−0.982	−1.038	−1.095	−1.151	−10
0	0.000	−0.059	−0.117	−0.176	−0.234	−0.292	−0.350	−0.408	−0.466	−0.524	−0.581	0
0	0.000	0.059	0.118	0.176	0.235	0.295	0.354	0.413	0.472	0.532	0.591	0
10	0.591	0.651	0.711	0.770	0.830	0.890	0.950	1.011	1.071	1.131	1.192	10
20	1.192	1.252	1.313	1.373	1.434	1.495	1.556	1.617	1.678	1.739	1.801	20
30	1.801	1.862	1.924	1.985	2.047	2.109	2.171	2.233	2.295	2.357	2.419	30
40	2.419	2.482	2.544	2.607	2.669	2.732	2.795	2.858	2.921	2.984	3.047	40
50	3.047	3.110	3.173	3.237	3.300	3.364	3.428	3.491	3.555	3.619	3.683	50
60	3.683	3.748	3.812	3.876	3.941	4.005	4.070	4.134	4.199	4.264	4.329	60
70	4.329	4.394	4.459	4.524	4.590	4.655	4.720	4.786	4.852	4.917	4.983	70
80	4.983	5.049	5.115	5.181	5.247	5.314	5.380	5.446	5.513	5.579	5.646	80
90	5.646	5.713	5.780	5.846	5.913	5.981	6.048	6.115	6.182	6.250	6.317	90
100	6.317	6.385	6.452	6.520	6.588	6.656	6.724	6.792	6.860	6.928	6.996	100
110	6.996	7.064	7.133	7.201	7.270	7.339	7.407	7.476	7.545	7.614	7.683	110
120	7.683	7.752	7.821	7.890	7.960	8.029	8.099	8.168	8.238	8.307	8.377	120
130	8.377	8.447	8.517	8.587	8.657	8.727	8.797	8.867	8.938	9.008	9.078	130
140	9.078	9.149	9.220	9.290	9.361	9.432	9.503	9.573	9.644	9.715	9.787	140
150	9.787	9.858	9.929	10.000	10.072	10.143	10.215	10.286	10.358	10.429	10.501	150
160	10.501	10.573	10.645	10.717	10.789	10.861	10.933	11.005	11.077	11.150	11.222	160
170	11.222	11.294	11.367	11.439	11.512	11.585	11.657	11.730	11.803	11.876	11.949	170
180	11.949	12.022	12.095	12.168	12.241	12.314	12.387	12.461	12.534	12.608	12.681	180
190	12.681	12.755	12.828	12.902	12.975	13.049	13.123	13.197	13.271	13.345	13.419	190
200	13.419	13.493	13.567	13.641	13.715	13.789	13.864	13.938	14.012	14.087	14.161	200
210	14.161	14.236	14.310	14.385	14.460	14.534	14.609	14.684	14.759	14.834	14.909	210
220	14.909	14.984	15.059	15.134	15.209	15.284	15.359	15.435	15.510	15.585	15.661	220
230	15.661	15.736	15.812	15.887	15.963	16.038	16.114	16.190	16.266	16.341	16.417	230
240	16.417	16.493	16.569	16.645	16.721	16.797	16.873	16.949	17.025	17.101	17.178	240
250	17.178	17.254	17.330	17.406	17.483	17.559	17.636	17.712	17.789	17.865	17.942	250
260	17.942	18.018	18.095	18.172	18.248	18.325	18.402	18.479	18.556	18.633	18.710	260
270	18.710	18.787	18.864	18.941	19.018	19.095	19.172	19.249	19.326	19.404	19.481	270
280	19.481	19.558	19.636	19.713	19.790	19.868	19.945	20.023	20.100	20.178	20.256	280
290	20.256	20.333	20.411	20.488	20.566	20.644	20.722	20.800	20.877	20.955	21.033	290

°C	0	1	2	3	4	5	6	7	8	9	10	°C
					Thermoelectric Voltage (absolute mV)							
300	21.033	21.111	21.189	21.267	21.345	21.423	21.501	21.579	21.657	21.735	21.814	300
310	21.814	21.892	21.970	22.048	22.127	22.205	22.283	22.362	22.440	22.518	22.597	310
320	22.597	22.675	22.754	22.832	22.911	22.989	23.068	23.147	23.225	23.304	23.383	320
330	23.383	23.461	23.540	23.619	23.698	23.777	23.855	23.934	24.013	24.092	24.171	330
340	24.171	24.250	24.329	24.408	24.487	24.566	24.645	24.724	24.803	24.882	24.961	340
350	24.961	25.041	25.120	25.199	25.278	25.357	25.437	25.516	25.595	25.675	25.754	350
360	25.754	25.833	25.913	25.992	26.072	26.151	26.230	26.310	26.389	26.469	26.549	360
370	26.549	26.628	26.708	26.787	26.867	26.947	27.026	27.106	27.186	27.265	27.345	370
380	27.345	27.425	27.504	27.584	27.664	27.744	27.824	27.903	27.983	28.063	28.143	380
390	28.143	28.223	28.303	28.383	28.463	28.543	28.623	28.703	28.783	28.863	28.943	390
400	28.943	29.023	29.103	29.183	29.263	29.343	29.423	29.503	29.584	29.664	29.744	400
410	29.744	29.824	29.904	29.984	30.065	30.145	30.225	30.305	30.386	30.466	30.546	410
420	30.546	30.627	30.707	30.787	30.868	30.948	31.028	31.109	31.189	31.270	31.350	420
430	31.350	31.430	31.511	31.591	31.672	31.752	31.833	31.913	31.994	32.074	32.155	430
440	32.155	32.235	32.316	32.396	32.477	32.557	32.638	32.719	32.799	32.880	32.960	440
450	32.960	33.041	33.122	33.202	33.283	33.364	33.444	33.525	33.605	33.686	33.767	450
460	33.767	33.848	33.928	34.009	34.090	34.170	34.251	34.332	34.413	34.493	34.574	460
470	34.574	34.655	34.736	34.816	34.897	34.978	35.059	35.140	35.220	35.301	35.382	470
480	35.382	35.463	35.544	35.624	35.705	35.786	35.867	35.948	36.029	36.109	36.190	480
490	36.190	36.271	36.352	36.433	36.514	36.595	36.675	36.756	36.837	36.918	36.999	490
500	36.999	37.080	37.161	37.242	37.323	37.403	37.484	37.565	37.646	37.727	37.808	500
510	37.808	37.889	37.970	38.051	38.132	38.213	38.293	38.374	38.455	38.536	38.617	510
520	38.617	38.698	38.779	38.860	38.941	39.022	39.103	39.184	39.264	39.345	39.426	520
530	39.426	39.507	39.588	39.669	39.750	39.831	39.912	39.993	40.074	40.155	40.236	530
540	40.236	40.316	40.397	40.478	40.559	40.640	40.721	40.802	40.883	40.964	41.045	540
550	41.045	41.125	41.206	41.287	41.368	41.449	41.530	41.611	41.692	41.773	41.853	550
560	41.853	41.934	42.015	42.096	42.177	42.258	42.339	42.419	42.500	42.581	42.662	560
570	42.662	42.743	42.824	42.904	42.985	43.066	43.147	43.228	43.308	43.389	43.470	570
580	43.470	43.551	43.632	43.712	43.793	43.874	43.955	44.035	44.116	44.197	44.278	580
590	44.278	44.358	44.439	44.520	44.601	44.681	44.762	44.843	44.923	45.004	45.085	590
600	45.085	45.165	45.246	45.327	45.407	45.488	45.569	45.649	45.730	45.811	45.891	600
610	45.891	45.972	46.052	46.133	46.213	46.294	46.375	46.455	46.536	46.616	46.697	610
620	46.697	46.777	46.858	46.938	47.019	47.099	47.180	47.260	47.341	47.421	47.502	620
630	47.502	47.582	47.663	47.743	47.824	47.904	47.984	48.065	48.145	48.226	48.306	630
640	48.306	48.386	48.467	48.547	48.627	48.708	48.788	48.868	48.949	49.029	49.109	640
650	49.109	49.189	49.270	49.350	49.430	49.510	49.591	49.671	49.751	49.831	49.911	650
660	49.911	49.992	50.072	50.152	50.232	50.312	50.392	50.472	50.553	50.633	50.713	660
670	50.713	50.793	50.873	50.953	51.033	51.113	51.193	51.273	51.353	51.433	51.513	670
680	51.513	51.593	51.673	51.753	51.833	51.913	51.993	52.073	52.152	52.232	52.312	680
690	52.312	52.392	52.472	52.552	52.632	52.711	52.791	52.871	52.951	53.031	53.110	690
700	53.110	53.190	53.270	53.350	53.429	53.509	53.589	53.668	53.748	53.828	53.907	700
710	53.907	53.987	54.066	54.146	54.226	54.305	54.385	54.464	54.544	54.623	54.703	710
720	54.703	54.782	54.862	54.941	55.021	55.100	55.180	55.259	55.339	55.418	55.498	720
730	55.498	55.577	55.656	55.736	55.815	55.894	55.974	56.053	56.132	56.212	56.291	730
740	56.291	56.370	56.449	56.529	56.608	56.687	56.766	56.845	56.924	57.004	57.083	740
750	57.083	57.162	57.241	57.320	57.399	57.478	57.557	57.636	57.715	57.794	57.873	750
760	57.873	57.952	58.031	58.110	58.189	58.268	58.347	58.426	58.505	58.584	58.663	760
770	58.663	58.742	58.820	58.899	58.978	59.057	59.136	59.214	59.293	59.372	59.451	770
780	59.451	59.529	59.608	59.687	59.765	59.844	59.923	60.001	60.080	60.159	60.237	780
790	60.237	60.316	60.394	60.473	60.551	60.630	60.708	60.787	60.865	60.944	61.022	790
800	61.022	61.101	61.179	61.258	61.336	61.414	61.493	61.571	61.649	61.728	61.806	800
810	61.806	61.884	61.962	62.041	62.119	62.197	62.275	62.353	62.432	62.510	62.588	810
820	62.588	62.666	62.744	62.822	62.900	62.978	63.056	63.134	63.212	63.290	63.368	820
830	63.368	63.446	63.524	63.602	63.680	63.758	63.836	63.914	63.992	64.069	64.147	830
840	64.147	64.225	64.303	64.380	64.458	64.536	64.614	64.691	64.769	64.847	64.924	840
850	64.924	65.002	65.080	65.157	65.235	65.312	65.390	65.467	65.545	65.622	65.700	850
860	65.700	65.777	65.855	65.932	66.009	66.087	66.164	66.241	66.319	66.396	66.473	860
870	66.473	66.551	66.628	66.705	66.782	66.859	66.937	67.014	67.091	67.168	67.245	870
880	67.245	67.322	67.399	67.476	67.553	67.630	67.707	67.784	67.861	67.938	68.015	880
890	68.015	68.092	68.169	68.246	68.323	68.399	68.476	68.553	68.630	68.706	68.783	890
900	68.783	68.860	68.936	69.013	69.090	69.166	69.243	69.320	69.396	69.473	69.549	900
910	69.549	69.626	69.702	69.779	69.855	69.931	70.008	70.084	70.161	70.237	70.313	910
920	70.313	70.390	70.466	70.542	70.618	70.694	70.771	70.847	70.923	70.999	71.075	920
930	71.075	71.151	71.227	71.304	71.380	71.456	71.532	71.608	71.683	71.759	71.835	930
940	71.835	71.911	71.987	72.063	72.139	72.215	72.290	72.366	72.442	72.518	72.593	940
950	72.593	72.669	72.745	72.820	72.896	72.972	73.047	73.123	73.199	73.274	73.350	950
960	73.350	73.425	73.501	73.576	73.652	73.727	73.802	73.878	73.953	74.029	74.104	960
970	74.104	74.179	74.255	74.330	74.405	74.480	74.556	74.631	74.706	74.781	74.857	970
980	74.857	74.932	75.007	75.082	75.157	75.232	75.307	75.382	75.458	75.533	75.608	980
990	75.608	75.683	75.758	75.833	75.908	75.983	76.058	76.133	76.208	76.283	76.358	990
1,000	76.358											1,000

Table A.4 Thermoelectric Voltages for Copper-Constantan Thermocouples with the Reference Junction at 0°C (32°F)

°C	0	1	2	3	4	5	6	7	8	9	10	°C
					Thermoelectric Voltage (absolute mV)							
−270	−6.258											−270
−260	−6.232	−6.236	−6.239	−6.242	−6.245	−6.248	−6.251	−6.253	−6.255	−6.256	−6.258	−260
−250	−6.181	−6.187	−6.193	−6.198	−6.204	−6.209	−6.214	−6.219	−6.224	−6.228	−6.232	−250
−240	−6.105	−6.114	−6.122	−6.130	−6.138	−6.146	−6.153	−6.160	−6.167	−6.174	−6.181	−240
−230	−6.007	−6.018	−6.028	−6.039	−6.049	−6.059	−6.068	−6.078	−6.087	−6.096	−6.105	−230
−220	−5.889	−5.901	−5.914	−5.926	−5.938	−5.950	−5.962	−5.973	−5.985	−5.996	−6.007	−220
−210	−5.753	−5.767	−5.782	−5.795	−5.809	−5.823	−5.836	−5.850	−5.863	−5.876	−5.889	−210
−200	−5.603	−5.619	−5.634	−5.650	−5.665	−5.680	−5.695	−5.710	−5.724	−5.739	−5.753	−200
−190	−5.439	−5.456	−5.473	−5.489	−5.506	−5.522	−5.539	−5.555	−5.571	−5.587	−5.603	−190
−180	−5.261	−5.279	−5.297	−5.315	−5.333	−5.351	−5.369	−5.387	−5.404	−5.421	−5.439	−180
−170	−5.069	−5.089	−5.109	−5.128	−5.147	−5.167	−5.186	−5.205	−5.223	−5.242	−5.261	−170
−160	−4.865	−4.886	−4.907	−4.928	−4.948	−4.969	−4.989	−5.010	−5.030	−5.050	−5.069	−160
−150	−4.648	−4.670	−4.693	−4.715	−4.737	−4.758	−4.780	−4.801	−4.823	−4.844	−4.865	−150
−140	−4.419	−4.442	−4.466	−4.489	−4.512	−4.535	−4.558	−4.581	−4.603	−4.626	−4.648	−140
−130	−4.177	−4.202	−4.226	−4.251	−4.275	−4.299	−4.323	−4.347	−4.371	−4.395	−4.419	−130
−120	−3.923	−3.949	−3.974	−4.000	−4.026	−4.051	−4.077	−4.102	−4.127	−4.152	−4.177	−120
−110	−3.656	−3.684	−3.711	−3.737	−3.764	−3.791	−3.818	−3.844	−3.870	−3.897	−3.923	−110
−100	−3.378	−3.407	−3.435	−3.463	−3.491	−3.519	−3.547	−3.574	−3.602	−3.629	−3.656	−100
−90	−3.089	−3.118	−3.147	−3.177	−3.206	−3.235	−3.264	−3.293	−3.321	−3.350	−3.378	−90
−80	−2.788	−2.818	−2.849	−2.879	−2.909	−2.939	−2.970	−2.999	−3.029	−3.059	−3.089	−80
−70	−2.475	−2.507	−2.539	−2.570	−2.602	−2.633	−2.664	−2.695	−2.726	−2.757	−2.788	−70
−60	−2.152	−2.185	−2.218	−2.250	−2.283	−2.315	−2.348	−2.380	−2.412	−2.444	−2.475	−60
−50	−1.819	−1.853	−1.886	−1.920	−1.953	−1.987	−2.020	−2.053	−2.087	−2.120	−2.152	−50
−40	−1.475	−1.510	−1.544	−1.579	−1.614	−1.648	−1.682	−1.717	−1.751	−1.785	−1.819	−40
−30	−1.121	−1.157	−1.192	−1.228	−1.263	−1.299	−1.334	−1.370	−1.405	−1.440	−1.475	−30
−20	−0.757	−0.794	−0.830	−0.867	−0.903	−0.940	−0.976	−1.013	−1.049	−1.085	−1.121	−20
−10	−0.383	−0.421	−0.458	−0.496	−0.534	−0.571	−0.608	−0.646	−0.683	−0.720	−0.757	−10
0	0.000	−0.039	−0.077	−0.116	−0.154	−0.193	−0.231	−0.269	−0.307	−0.345	−0.383	0
0	0.000	0.039	0.078	0.117	0.156	0.195	0.234	0.273	0.312	0.351	0.391	0
10	0.391	0.430	0.470	0.510	0.549	0.589	0.629	0.669	0.709	0.749	0.789	10
20	0.789	0.830	0.870	0.911	0.951	0.992	1.032	1.073	1.114	1.155	1.196	20
30	1.196	1.237	1.279	1.320	1.361	1.403	1.444	1.486	1.528	1.569	1.611	30
40	1.611	1.653	1.695	1.738	1.780	1.822	1.865	1.907	1.950	1.992	2.035	40
50	2.035	2.078	2.121	2.164	2.207	2.250	2.294	2.337	2.380	2.424	2.467	50
60	2.467	2.511	2.555	2.599	2.643	2.687	2.731	2.775	2.819	2.864	2.908	60
70	2.908	2.953	2.997	3.042	3.087	3.131	3.176	3.221	3.266	3.312	3.357	70
80	3.357	3.402	3.447	3.493	3.538	3.584	3.630	3.676	3.721	3.767	3.813	80
90	3.813	3.859	3.906	3.952	3.998	4.044	4.091	4.137	4.184	4.231	4.277	90

Table A.4 (*Continued*)

°C	0	1	2	3	4	5	6	7	8	9	10	°C
					Thermoelectric Voltage (absolute mV)							
100	4.277	4.324	4.371	4.418	4.465	4.512	4.559	4.607	4.654	4.701	4.749	100
110	4.749	4.796	4.844	4.891	4.939	4.987	5.035	5.083	5.131	5.179	5.227	110
120	5.227	5.275	5.324	5.372	5.420	5.469	5.517	5.566	5.615	5.663	5.712	120
130	5.712	5.761	5.810	5.859	5.908	5.957	6.007	6.056	6.105	6.155	6.204	130
140	6.204	6.254	6.303	6.353	6.403	6.452	6.502	6.552	6.602	6.652	6.702	140
150	6.702	6.753	6.803	6.853	6.903	6.954	7.004	7.055	7.106	7.156	7.207	150
160	7.207	7.258	7.309	7.360	7.411	7.462	7.513	7.564	7.615	7.666	7.718	160
170	7.718	7.769	7.821	7.872	7.924	7.975	8.027	8.079	8.131	8.183	8.235	170
180	8.235	8.287	8.339	8.391	8.443	8.495	8.548	8.600	8.652	8.705	8.757	180
190	8.757	8.810	8.863	8.915	8.968	9.021	9.074	9.127	9.180	9.233	9.286	190
200	9.286	9.339	9.392	9.446	9.499	9.553	9.606	9.659	9.713	9.767	9.820	200
210	9.820	9.874	9.928	9.982	10.036	10.090	10.144	10.198	10.252	10.306	10.360	210
220	10.360	10.414	10.469	10.523	10.578	10.632	10.687	10.741	10.796	10.851	10.905	220
230	10.905	10.960	11.015	11.070	11.125	11.180	11.235	11.290	11.345	11.401	11.456	230
240	11.456	11.511	11.566	11.622	11.677	11.733	11.788	11.844	11.900	11.956	12.011	240
250	12.011	12.067	12.123	12.179	12.235	12.291	12.347	12.403	12.459	12.515	12.572	250
260	12.572	12.628	12.684	12.741	12.797	12.854	12.910	12.967	13.024	13.080	13.137	260
270	13.137	13.194	13.251	13.307	13.364	13.421	13.478	13.535	13.592	13.650	13.707	270
280	13.707	13.764	13.821	13.879	13.936	13.993	14.051	14.108	14.166	14.223	14.281	280
290	14.281	14.339	14.396	14.454	14.512	14.570	14.628	14.686	14.744	14.802	14.860	290
300	14.860	14.918	14.976	15.034	15.092	15.151	15.209	15.267	15.326	15.384	15.443	300
310	15.443	15.501	15.560	15.619	15.677	15.736	15.795	15.853	15.912	15.971	16.030	310
320	16.030	16.089	16.148	16.207	16.266	16.325	16.384	16.444	16.503	16.562	16.621	320
330	16.621	16.681	16.740	16.800	16.859	16.919	16.978	17.038	17.097	17.157	17.217	330
340	17.217	17.277	17.336	17.396	17.456	17.516	17.576	17.636	17.696	17.756	17.816	340
350	17.816	17.877	17.937	17.997	18.057	18.118	18.178	18.238	18.299	18.359	18.420	350
360	18.420	18.480	18.541	18.602	18.662	18.723	18.784	18.845	18.905	18.966	19.027	360
370	19.027	19.088	19.149	19.210	19.271	19.332	19.393	19.455	19.516	19.577	19.638	370
380	19.638	19.699	19.761	19.822	19.883	19.945	20.006	20.068	20.129	20.191	20.252	380
390	20.252	20.314	20.376	20.437	20.499	20.560	20.622	20.684	20.746	20.807	20.869	390
400	20.869											400

Table A.5 Thermoelectric Voltages for Iron-Constantan Thermocouples
with the Reference Junction at 0°C (32°F)

°C	0	1	2	3	4	5	6	7	8	9	10	°C
					Thermoelectric Voltage (absolute mV)							
−210	−8.096											−210
−200	−7.890	−7.912	−7.934	−7.955	−7.976	−7.996	−8.017	−8.037	−8.057	−8.076	−8.096	−200
−190	−7.659	−7.683	−7.707	−7.731	−7.755	−7.778	−7.801	−7.824	−7.846	−7.868	−7.890	−190
−180	−7.402	−7.429	−7.455	−7.482	−7.508	−7.533	−7.559	−7.584	−7.609	−7.634	−7.659	−180
−170	−7.122	−7.151	−7.180	−7.209	−7.237	−7.265	−7.293	−7.321	−7.348	−7.375	−7.402	−170
−160	−6.821	−6.852	−6.883	−6.914	−6.944	−6.974	−7.004	−7.034	−7.064	−7.093	−7.122	−160
−150	−6.499	−6.532	−6.565	−6.598	−6.630	−6.663	−6.695	−6.727	−6.758	−6.790	−6.821	−150
−140	−6.159	−6.194	−6.228	−6.263	−6.297	−6.331	−6.365	−6.399	−6.433	−6.466	−6.499	−140
−130	−5.801	−5.837	−5.874	−5.910	−5.946	−5.982	−6.018	−6.053	−6.089	−6.124	−6.159	−130
−120	−5.426	−5.464	−5.502	−5.540	−5.578	−5.615	−5.653	−5.690	−5.727	−5.764	−5.801	−120
−110	−5.036	−5.076	−5.115	−5.155	−5.194	−5.233	−5.272	−5.311	−5.349	−5.388	−5.426	−110
−100	−4.632	−4.673	−4.714	−4.755	−4.795	−4.836	−4.876	−4.916	−4.956	−4.996	−5.036	−100
−90	−4.215	−4.257	−4.299	−4.341	−4.383	−4.425	−4.467	−4.508	−4.550	−4.591	−4.632	−90
−80	−3.785	−3.829	−3.872	−3.915	−3.958	−4.001	−4.044	−4.087	−4.130	−4.172	−4.215	−80
−70	−3.344	−3.389	−3.433	−3.478	−3.522	−3.566	−3.610	−3.654	−3.698	−3.742	−3.785	−70
−60	−2.892	−2.938	−2.984	−3.029	−3.074	−3.120	−3.165	−3.210	−3.255	−3.299	−3.344	−60
−50	−2.431	−2.478	−2.524	−2.570	−2.617	−2.663	−2.709	−2.755	−2.801	−2.847	−2.892	−50
−40	−1.960	−2.008	−2.055	−2.102	−2.150	−2.197	−2.244	−2.291	−2.338	−2.384	−2.431	−40
−30	−1.481	−1.530	−1.578	−1.626	−1.674	−1.722	−1.770	−1.818	−1.865	−1.913	−1.960	−30
−20	−0.995	−1.044	−1.093	−1.141	−1.190	−1.239	−1.288	−1.336	−1.385	−1.433	−1.481	−20
−10	−0.501	−0.550	−0.600	−0.650	−0.699	−0.748	−0.798	−0.847	−0.896	−0.945	−0.995	−10
0	0.000	−0.050	−0.101	−0.151	−0.201	−0.251	−0.301	−0.351	−0.401	−0.451	−0.501	0
0	0.000	0.050	0.101	0.151	0.202	0.253	0.303	0.354	0.405	0.456	0.507	0
10	0.507	0.558	0.609	0.660	0.711	0.762	0.813	0.865	0.916	0.967	1.019	10
20	1.019	1.070	1.122	1.174	1.225	1.277	1.329	1.381	1.432	1.484	1.536	20
30	1.536	1.588	1.640	1.693	1.745	1.797	1.849	1.901	1.954	2.006	2.058	30
40	2.058	2.111	2.163	2.216	2.268	2.321	2.374	2.426	2.479	2.532	2.585	40
50	2.585	2.638	2.691	2.743	2.796	2.849	2.902	2.956	3.009	3.062	3.115	50
60	3.115	3.168	3.221	3.275	3.328	3.381	3.435	3.488	3.542	3.595	3.649	60
70	3.649	3.702	3.756	3.809	3.863	3.917	3.971	4.024	4.078	4.132	4.186	70
80	4.186	4.239	4.293	4.347	4.401	4.455	4.509	4.563	4.617	4.671	4.725	80
90	4.725	4.780	4.834	4.888	4.942	4.996	5.050	5.105	5.159	5.213	5.268	90
100	5.268	5.322	5.376	5.431	5.485	5.540	5.594	5.649	5.703	5.758	5.812	100
110	5.812	5.867	5.921	5.976	6.031	6.085	6.140	6.195	6.249	6.304	6.359	110
120	6.359	6.414	6.468	6.523	6.578	6.633	6.688	6.742	6.797	6.852	6.907	120
130	6.907	6.962	7.017	7.072	7.127	7.182	7.237	7.292	7.347	7.402	7.457	130
140	7.457	7.512	7.567	7.622	7.677	7.732	7.787	7.843	7.898	7.953	8.008	140
150	8.008	8.063	8.118	8.174	8.229	8.284	8.339	8.394	8.450	8.505	8.560	150
160	8.560	8.616	8.671	8.726	8.781	8.837	8.892	8.947	9.003	9.058	9.113	160
170	9.113	9.169	9.224	9.279	9.335	9.390	9.446	9.501	9.556	9.612	9.667	170
180	9.667	9.723	9.778	9.834	9.889	9.944	10.000	10.055	10.111	10.166	10.222	180
190	10.222	10.277	10.333	10.388	10.444	10.499	10.555	10.610	10.666	10.721	10.777	190
200	10.777	10.832	10.888	10.943	10.999	11.054	11.110	11.165	11.221	11.276	11.332	200
210	11.332	11.387	11.443	11.498	11.554	11.609	11.665	11.720	11.776	11.831	11.887	210
220	11.887	11.943	11.998	12.054	12.109	12.165	12.220	12.276	12.331	12.387	12.442	220
230	12.442	12.498	12.553	12.609	12.664	12.720	12.776	12.831	12.887	12.942	12.998	230
240	12.998	13.053	13.109	13.164	13.220	13.275	13.331	13.386	13.442	13.497	13.553	240
250	13.553	13.608	13.664	13.719	13.775	13.830	13.886	13.941	13.997	14.052	14.108	250
260	14.108	14.163	14.219	14.274	14.330	14.385	14.441	14.496	14.552	14.607	14.663	260
270	14.663	14.718	14.774	14.829	14.885	14.940	14.995	15.051	15.106	15.162	15.217	270

Table A.5 *(Continued)*

°C	0	1	2	3	4	5	6	7	8	9	10	°C
					Thermoelectric Voltage (absolute mV)							
280	15.217	15.273	15.328	15.383	15.439	15.494	15.550	15.605	15.661	15.716	15.771	280
290	15.771	15.827	15.882	15.938	15.993	16.048	16.104	16.159	16.214	16.270	16.325	290
300	16.325	16.380	16.436	16.491	16.547	16.602	16.657	16.713	16.768	16.823	16.879	300
310	16.879	16.934	16.989	17.044	17.100	17.155	17.210	17.266	17.321	17.376	17.432	310
320	17.432	17.487	17.542	17.597	17.653	17.708	17.763	17.818	17.874	17.929	17.984	320
330	17.984	18.039	18.095	18.150	18.205	18.260	18.316	18.371	18.426	18.481	18.537	330
340	18.537	18.592	18.647	18.702	18.757	18.813	18.868	18.923	18.978	19.033	19.089	340
350	19.089	19.144	19.199	19.254	19.309	19.364	19.420	19.475	19.530	19.585	19.640	350
360	19.640	19.695	19.751	19.806	19.861	19.916	19.971	20.026	20.081	20.137	20.192	360
370	20.192	20.247	20.302	20.357	20.412	20.467	20.523	20.578	20.633	20.688	20.743	370
380	20.743	20.798	20.853	20.909	20.964	21.019	21.074	21.129	21.184	21.239	21.295	380
390	21.295	21.350	21.405	21.460	21.515	21.570	21.625	21.680	21.736	21.791	21.846	390
400	21.846	21.901	21.956	22.011	22.066	22.122	22.177	22.232	22.287	22.342	22.397	400
410	22.397	22.453	22.508	22.563	22.618	22.673	22.728	22.784	22.839	22.894	22.949	410
420	22.949	23.004	23.060	23.115	23.170	23.225	23.280	23.336	23.391	23.446	23.501	420
430	23.501	23.556	23.612	23.667	23.722	23.777	23.833	23.888	23.943	23.999	24.054	430
440	24.054	24.109	24.164	24.220	24.275	24.330	24.386	24.441	24.496	24.552	24.607	440
450	24.607	24.662	24.718	24.773	24.829	24.884	24.939	24.995	25.050	25.106	25.161	450
460	25.161	25.217	25.272	25.327	25.383	25.438	25.494	25.549	25.605	25.661	25.716	460
470	25.716	25.772	25.827	25.883	25.938	25.994	26.050	26.105	26.161	26.216	26.272	470
480	26.272	26.328	26.383	26.439	26.495	26.551	26.606	26.662	26.718	26.774	26.829	480
490	26.829	26.885	26.941	26.997	27.053	27.109	27.165	27.220	27.276	27.332	27.388	490
500	27.388	27.444	27.500	27.556	27.612	27.668	27.724	27.780	27.836	27.893	27.949	500
510	27.949	28.005	28.061	28.117	28.173	28.230	28.286	28.342	28.398	28.455	28.511	510
520	28.511	28.567	28.624	28.680	28.736	28.793	28.849	28.906	28.962	29.019	29.075	520
530	29.075	29.132	29.188	29.245	29.301	29.358	29.415	29.471	29.528	29.585	29.642	530
540	29.642	29.698	29.755	29.812	29.869	29.926	29.983	30.039	30.096	30.153	30.210	540
550	30.210	30.267	30.324	30.381	30.439	30.496	30.553	30.610	30.667	30.724	30.782	550
560	30.782	30.839	30.896	30.954	31.011	31.068	31.126	31.183	31.241	31.298	31.356	560
570	31.356	31.413	31.471	31.528	31.586	31.644	31.702	31.759	31.817	31.875	31.933	570
580	31.933	31.991	32.048	32.106	32.164	32.222	32.280	32.338	32.396	32.455	32.513	580
590	32.513	32.571	32.629	32.687	32.746	32.804	32.862	32.921	32.979	33.038	33.096	590
600	33.096	33.155	33.213	33.272	33.330	33.389	33.448	33.506	33.565	33.624	33.683	600
610	33.683	33.742	33.800	33.859	33.918	33.977	34.036	34.095	34.155	34.214	34.273	610
620	34.273	34.332	34.391	34.451	34.510	34.569	34.629	34.688	34.748	34.807	34.867	620
630	34.867	34.926	34.986	35.046	35.105	35.165	35.225	35.285	35.344	35.404	35.464	630
640	35.464	35.524	35.584	35.644	35.704	35.764	35.825	35.885	35.945	36.005	36.066	640
650	36.066	36.126	36.186	36.247	36.307	36.368	36.428	36.489	36.549	36.610	36.671	650
660	36.671	36.732	36.792	36.853	36.914	36.975	37.036	37.097	37.158	37.219	37.280	660
670	37.280	37.341	37.402	37.463	37.525	37.586	37.647	37.709	37.770	37.831	37.893	670
680	37.893	37.954	38.016	38.078	38.139	38.201	38.262	38.324	38.386	38.448	38.510	680
690	38.510	38.572	38.633	38.695	38.757	38.819	38.882	38.944	39.006	39.068	39.130	690
700	39.130	39.192	39.255	39.317	39.379	39.442	39.504	39.567	39.629	39.692	39.754	700
710	39.754	39.817	39.880	39.942	40.005	40.068	40.131	40.193	40.256	40.319	40.382	710
720	40.382	40.445	40.508	40.571	40.634	40.697	40.760	40.823	40.886	40.950	41.013	720
730	41.013	41.076	41.139	41.203	41.266	41.329	41.393	41.456	41.520	41.583	41.647	730
740	41.647	41.710	41.774	41.837	41.901	41.965	42.028	42.092	42.156	42.219	42.283	740
750	42.283	42.347	42.411	42.475	42.538	42.602	42.666	42.730	42.794	42.858	42.922	750
760	42.922											760

Table A.6 Polynomial Coefficients for Six Different Types of Thermocouples

	Type of Thermocouple					
	Temperature Range					
	E	J	K	R	S	T
Coefficient	-100° to 1000°C	0° to 760°C	0° to 1370°C	0° to 1000°C	0° to 1750°C	-160° to 400°C
a_0	0.104967248	-0.048868252	0.226584602	0.263632917	0.927763167	0.100860910
a_1	17189.45282	19873.14503	24152.10900	179075.491	169526.5150	25727.94369
a_2	-282639.0850	-218614.5353	67233.4248	-48840341.37	-31568363.94	-767345.8295
a_3	12695339.5	11569199.78	2210340.682	$1.90002E+10$	8990730663	78025595.81
a_4	-448703084.6	-264917531.4	-860963914.9	$-4.82704E+12$	$-1.63565E+12$	-9247486589
a_5	$1.10866E+10$	2018441314	$4.83506E+10$	$7.62091E+14$	$1.88027E+14$	$6.97688E+11$
a_6	$-1.76807E+11$		$-1.18452E+12$	$-7.20026E+16$	$-1.37241E+16$	$-2.66192E+13$
a_7	$1.71842E+12$		$1.38690E+13$	$3.71496E+18$	$6.17501E+17$	$3.94078E+14$
a_8	$-9.19278E+12$		$-6.33708E+13$	$-8.03104E+19$	$-1.56105E+19$	
a_9	$2.06132E+13$				$1.69535E+20$	

Table A.7 Physical Properties of Several Thermocouple Alloys

Property	Thermocouple Alloy							
	Iron	Constantan	Copper	Chromel	Alumel	Platinum 13% Rhodium	Platinum 10% Rhodium	Platinum
Melting Point (°C)	1490	1220	1083	1427	1399	1860	1850	1769
Resistivity ($\mu\Omega\cdot$m) (at 20°C)	0.0967	0.489	0.01724	0.706	0.294	0.196	0.189	0.104
Temperature Coefficient of Resistance ($10^{-4}/$°C) (0 to 100°C)	65	−0.1	43	4.1	23.9	15.6	16.6	39.2
Coefficient of Thermal Expansion ($10^{-6}/$°C) (20 to 100°C)	11.7	14.9	16.6	13.1	12.0	9.0	9.0	9.0
Thermal Conductivity at 100°C (W/m·°C)	67.8	21.2	377.0	19.2	29.7	36.8	37.7	71.5
Specific Heat at 20°C (J/kg·°C)	448	394	385	448	523	—	—	134
Density (kg/m^3)	7860	8920	8920	8730	8600	19610	19970	21450
Tensile Strength (MPa)	345	550	245	655	585	315	315	135
Magnetic Attraction	strong	none	none	none	moderate	none	none	none

Tables of Properties of Some Common Liquids and Gases

Table B.1 Physical Properties of Water

Temperature		Specific Weight[a] lb/ft^3 γ	Density[a] slug/ft^3 ρ	Viscosity lb-s/ft^2 $\mu \times 10^5$	Kinematic Viscosity ft^2/s $\nu \times 10^5$
English units:	°F				
	32	62.42	1.940	3.746	1.931
	40	62.43	1.940	3.229	1.664
	50	62.41	1.940	2.735	1.410
	60	62.37	1.938	2.359	1.217
	70	62.30	1.936	2.050	1.059
	80	62.22	1.934	1.799	0.930
	90	62.11	1.931	1.595	0.826
	100	62.00	1.927	1.424	0.739
	110	61.86	1.923	1.284	0.667
	120	61.71	1.918	1.168	0.609
	130	61.55	1.913	1.069	0.558
	140	61.38	1.908	0.981	0.514
	150	61.20	1.902	0.905	0.476
	160	61.00	1.896	0.838	0.442
	170	60.80	1.890	0.780	0.413
	180	60.58	1.883	0.726	0.385
	190	60.36	1.876	0.678	0.362
	200	60.12	1.868	0.637	0.341
	212	59.83	1.860	0.593	0.319
SI units:	°C	kN/m^3 γ	kg/m^3 ρ	N · s/m^2 $\mu \times 10^3$	m^2/s $\nu \times 10^6$
	0	9.805	999.8	1.781	1.785
	5	9.807	000.0	1.518	1.519
	10	9.804	999.7	1.307	1.306
	15	9.798	999.1	1.139	1.139
	20	9.789	998.2	1.002	1.003
	25	9.777	997.0	0.890	0.893
	30	9.764	995.7	0.798	0.800
	40	9.730	992.2	0.653	0.658
	50	9.689	988.0	0.547	0.553
	60	9.642	983.2	0.466	0.474
	70	9.589	977.8	0.404	0.413
	80	9.530	971.8	0.354	0.364
	90	9.466	965.3	0.315	0.326
	100	9.399	958.4	0.282	0.294

[a] At 14.7 psia (standard atmospheric pressure).

Table B.2 Approximate Properties of Some Common Liquids at Standard Atmospheric Pressure

Liquid	Temper- ature	Specific Weight	Density	Viscosity	Kinematic Viscosity
English Units:	°F	lb/ft^3 γ	slug/ft^3 ρ	lb-s/ft^2 $\mu \times 10^5$	ft^2/s $\nu \times 10^5$
Benzene	68	54.8	1.702	1.37	0.80
Castor oil	68	59.8	1.858	2060.0	1109.0
Crude oil	68	53.6	1.665	15.0	9.0
Ethyl alcohol	68	49.2	1.528	2.51	1.64
Gasoline	68	42.4	1.317	0.61	0.46
Glycerin	68	78.5	2.439	3120.0	1280.0
Kerosene	68	50.5	1.569	4.00	2.55
Linseed oil	68	58.6	1.820	92.0	50.0
Mercury	68	844.0	26.2	3.24	0.124
Olive oil	68	56.7	1.761	176.0	100.0
Turpentine	68	53.6	1.665	3.11	1.87
Water	68	62.32	1.936	2.10	1.085
SI units:	°C	kN/m^3 γ	kg/m^3 ρ	N · s/m^2 $\mu \times 10^3$	m^2/s $\nu \times 10^6$
Benzene	20	8.77	895	0.65	0.73
Castor oil	20	9.39	958	979.0	1025.0
Crude oil	20	8.42	858	7.13	8.32
Ethyl alcohol	20	7.73	787	1.19	1.52
Gasoline	20	6.66	679	0.29	0.43
Glycerin	20	12.33	1257	1495.0	1189.0
Kerosene	20	7.93	809	1.90	2.36
Linseed oil	20	9.21	938	43.7	46.2
Mercury	20	132.6	13500	1.54	0.12
Olive oil	20	8.91	908	83.7	92.4
Turpentine	20	8.42	858	1.48	1.73
Water	20	9.79	998	1.00	1.00

Table B.3 Approximate Properties of Air at Standard Atmospheric Pressure

	Temper-ature	Specific Weight[a]	Density[a]	Viscosity	Kinematic Viscosity
English Units:	°F	lb/ft^3 γ	slug/ft^3 ρ	lb-s/ft^2 $\mu \times 10^7$	ft^2/s $\nu \times 10^5$
	0	0.0866	0.00269	3.39	1.26
	10	0.0847	0.00263	3.45	1.31
	20	0.0828	0.00257	3.51	1.37
	30	0.0811	0.00252	3.57	1.42
	40	0.0794	0.00247	3.63	1.47
	50	0.0779	0.00242	3.68	1.52
	60	0.0764	0.00237	3.74	1.58
	70	0.0750	0.00233	3.79	1.63
	80	0.0735	0.00228	3.85	1.69
	90	0.0722	0.00224	3.90	1.74
	100	0.0709	0.00220	3.96	1.80
	150	0.0651	0.00202	4.23	2.09
	200	0.0601	0.00187	4.48	2.40
	300	0.0522	0.00162	4.96	3.05
	400	0.0462	0.00143	5.40	3.77
SI units:	°C	CN/m^3 γ	kg/m^3 ρ	N · s/m^2 $\mu \times 10^5$	m^2/s $\nu \times 10^5$
	−20	13.7	1.40	1.61	1.16
	−10	13.2	1.34	1.67	1.24
	0	12.7	1.29	1.72	1.33
	10	12.2	1.25	1.76	1.41
	20	11.8	1.20	1.81	1.51
	30	11.4	1.17	1.86	1.60
	40	11.1	1.13	1.91	1.69
	50	10.7	1.09	1.95	1.79
	60	10.4	1.06	2.00	1.89
	70	10.1	1.03	2.04	1.99
	80	9.81	1.00	2.09	2.09
	90	9.54	0.97	2.13	2.19
	100	9.28	0.95	2.17	2.29
	120	8.82	0.90	2.26	2.51
	140	8.38	0.85	2.34	2.74
	160	7.99	0.81	2.42	2.97
	180	7.65	0.78	2.50	3.20
	200	7.32	0.75	2.57	3.44

[a] At 14.7 psia (standard atmospheric pressure).

Table B.4 Approximate Properties of Some Common Gases at Standard Atmospheric Pressure

Gas	Density[a]	Engineering Gas Constant	Specific Heat		Adiabatic Exponent	Viscosity
	slug/ft^3	ft · lb/slug °R	ft-lb/slug° R		ft^2/s	lb = s/ft^2
English units:	ρ	R	c_p	c_v	k	$\mu \times 10^5$
Air	0.00234	1715	6000	4285	1.40	0.0376
Carbon dioxide	0.00354	1123	5132	4009	1.28	0.0310
Carbon monoxide	0.00225	1778	6218	4440	1.40	0.0380
Helium	0.00032	12420	31230	18810	1.66	0.0411
Hydrogen	0.00016	24680	86390	61710	1.40	0.0189
Methane	0.00129	3100	13400	10300	1.30	0.0280
Nitrogen	0.00226	1773	6210	4437	1.40	0.0368
Oxygen	0.00258	1554	5437	3883	1.40	0.0418
Water vapor	0.00145	2760	11110	8350	1.33	0.0212
	kg/m^3	N · m/kg K	N · m/kg K			N · s/m^2
SI units:	ρ	R	c_p	c_v	k	$\mu \times 10^5$
Air	1.205	287	1003	716	1.40	1.80
Carbon dioxide	1.84	188	858	670	1.28	1.48
Carbon monoxide	1.16	297	1040	743	1.40	1.82
Helium	0.166	2077	5220	3143	1.66	1.97
Hydrogen	0.0839	4120	14450	10330	1.40	0.90
Methane	0.668	520	2250	1730	1.30	1.34
Nitrogen	1.16	297	1040	743	1.40	1.76
Oxygen	1.33	260	909	649	1.40	2.00
Water vapor	0.747	462	1862	1400	1.33	1.01

[a] At 14.7 psia (standard atmospheric pressure).

Author Index

Subject Index